CHICAGO PUBLIC LIBRARY
HAROLD WASHINGTON LIBRARY CENTER

Atlas of
Stereochemistry

Atlas of Stereochemistry

*Absolute Configurations
of Organic Molecules*

W. Klyne and J. Buckingham
Westfield College, University of London

1974
Oxford University Press
New York

First published in 1974
by Chapman and Hall Ltd
© *1974 W. Klyne and J. Buckingham*
Published in the U.S.A. by
Oxford University Press
New York
ISBN 0 19 519 747
Library of Congress Catalog Card Number 73 90484

Printed in Great Britain

All rights reserved. No part of this book may be reprinted, or reproduced or utilized in any form or by any electronic, mechanical or other means, now known or hereafter invented, including photocopying and recording, or in any information storage and retrieval system, without permission in writing from the Publisher.

Contents

Introduction	*page*	vii
Key		x
A.	Fundamental Chiral Compounds	1
C.	Carbohydrates	69
T.	Terpenes (including Steroids)	75
K.	Alkaloids	137
Y.	Miscellaneous Natural Products	177
D.	Compounds with Chirality due to Isotopic Substitution	211
X.	Compounds containing Chiral Axes, Planes, etc.	215
Z.	Compounds containing Chiral Atoms other than Carbon	229
Formulae Index		241
Author Index		285
Subject Index		301

Introduction

A striking feature in the development of Organic Chemistry over the last twenty years has been a steady growth in the study of steric factors—that is, of three-dimensional relationships within and between molecules. One such aspect of this work lies in stereochemical correlations, the determination of relationships between the absolute and relative configurations of optically active compounds. Until recently, the absolute configuration was considered all too often as a sideline, to be determined when the detailed structure of the compound in all other respects was known, but as a result of the growing interest in biological chemistry, knowledge of the absolute configuration is now considered an essential part of any structure determination.

The study of absolute configurations has its greatest importance in connection with biochemical problems, because all living organisms are composed largely of chiral substances. (The term 'chiral' is preferred to 'optically active', since 'chiral' reflects the fundamental nature of the compounds with which we are concerned, whilst 'optically active' refers to only one physical property, albeit an important one.) No biochemical study of any kind can be complete without three-dimensional knowledge of the relationships between the molecules concerned, and this cannot be achieved until the absolute configurations of all the compounds are known or at any rate all have been referred to one arbitrary standard.

We may refer here to Sir Frederick Gowland Hopkins's classical definition of biochemistry as 'an adequate and acceptable description of molecular dynamics in cells and tissues' and repeat that this definition must include three-dimensional knowledge. As a striking example of an area where three-dimensional knowledge was lacking until recently the following may be cited. Many discussions are found in biochemical literature on the role of biotin in intermediary metabolism, and formulae are drawn showing the way in which biotin is linked with various other biologically active molecules, but until the absolute configuration of biotin was determined in 1966, these formulae were inadequate in that they could not represent the three-dimensional relationships of the substances involved.

A full description of the mode of action of an enzyme must entail detailed knowledge of the relative stereochemistry of the enzyme and its substrate. Each year more precise knowledge of the fit between enzyme and substrate appears in the literature.

Knowledge of absolute configurations is also needed in the pharmaceutical industry, both for compounds which are the organic chemist's attempts to improve on nature, for example, analogues of oestrone, morphine or oxytocin, and for totally artificial structures. The suggestion might be seriously considered that no chiral pharmaceutical product should be used without (a) resolution of a racemate into its components, and (b) determination of the absolute configuration of the active enantiomer.

To stress this importance of chirality in biological problems is not to minimize its significance and usefulness in many problems of pure chemistry, such as the study of reaction mechanism, and of the relationship of optical activity to structure.

The purpose of this book is to bring together in a readily accessible form a proportion of the vast mass of data which exists in the literature concerning the absolute configurations of chiral molecules.

Previous surveys of absolute configurations of organic compounds are few in number, including two reviews from this Department (Mills & Klyne, Klyne & Scopes in *Progress in Stereochemistry*); a chapter in Eliel's book *Stereochemistry of Carbon Compounds*; and parts of the excellent two-volume work entitled *Molecular Asymmetry in Biology* by Bentley.

The connection of one of us with Heilbron's *Dictionary of Organic Compounds*, which provides a bird's-eye view of the whole range of organic compounds at a price accessible to the small library, suggested the use of the configurational data in that book as the basis for this 'Atlas'. We have been encouraged in this work by the comments of colleagues on the editorial board of Heilbron's *Dictionary*, and by other stereochemical friends, notably Professors D. H. R. Barton and V. Prelog. The approximately 3000 compounds listed in the 'Atlas' still represent only an outline of the field, since to cover all chiral compounds would be to re-write about one tenth of Beilstein's *Handbuch*. We have concentrated first on compounds containing one or two centres of chirality. Secondly, we give a selection of the main groups of natural products; among the latter we list key compounds only for each group, since the links between individual members within a group are readily available in well-known reference books and reviews. Finally, special sections of the 'Atlas' deal with configuration around chiral axes and chiral planes, chiral centres other than carbon atoms, and chirality due to isotopic substitution.

Methods of correlation

These have been dealt with in some detail in the review articles mentioned above, and here we discuss only their relative importance.

The Bijvoet method for determination of absolute configuration by X-ray crystallography is the only sound and widely applicable direct method of determination. (For the past four years, Professor D. Rogers (Imperial College) and his colleagues have produced lists of compounds to which the Bijvoet method has been applied; we are grateful to Professor Rogers for supplying us with copies of these lists before publication.) The compounds to which the Bijvoet method has been applied are all marked specially in the 'Atlas'; they are so to speak the fundamental 'triangulation points' of the survey on which all others must in one way or another depend.

The linking of other compounds with these fundamental ones, the configurations of which have been established by the Bijvoet method, is most reliably carried out by chemical interconversion *not* involving the centre of chirality concerned. With certain simple precautions, this procedure is entirely safe. Much classical structure determination indirectly provides links of this kind.

The remaining methods are all of varying reliability, and all depend on argument by analogy. These methods include chemical interconversions involving the chiral centre(s) accompanied by mechanistic interpretation of the reaction pathway, asymmetric synthesis, chiroptical methods (ORD and CD) and the method of quasi-racemates, etc. In all of these the reliability depends directly on care in the choice of analogies. For example, in the case of asymmetric syntheses, the following quotation by Mislow and his co-workers (*J. Amer. Chem. Soc.*, 1965, 87, 1958) is relevant: 'Thus, in the assignment of configurations, a heavy burden of proof rests on the credibility of the transition state proposed in asymmetric syntheses, or in the initial analysis which underlies the weighting of conformational populations in mobile equilibria, whichever applies.' Similar cautions could be made to cover correlations involving the quasi-racemate method and the chiroptical methods.

We indicate by means of broken arrows that a somewhat lesser reliability is considered to be attached to *all* correlations involving asymmetric syntheses or comparison of monochromatic rotations; the same symbolism is used in a few other cases where the correlation is clearly not fully reliable, either as a result of comments made by the original authors, or where subsequent work has cast doubt on the original correlation.

The correlations given in the 'Atlas' are in general the most recent and reliable available. Where a chemical correlation between substances has been carried out, this is given preference over correlation by chiroptical methods, asymmetric synthesis, quasi-racemate formation, etc., and these in turn are given precedence over any older work involving comparison of monochromatic rotations. The latter is included only in a few cases where no more recent information is available and where the correlation is of considerable interest.

Introduction

Absolute configurations of metal complexes

In general, the 'Atlas' covers the field of organic and organometallic (e.g. metallocene) stereochemistry. Metal complexes are not included; they have been comprehensively covered by a recent monograph (C. J. Hawkins, *Absolute Configuration of Metal Complexes* (Interscience Monographs on Chemistry), Wiley-Interscience, 1971.)

Key

Scope

The 'Atlas' is intended to cover the most important simple chiral compounds and the main structural types of natural products.

Correlations within a 'stereochemically homogeneous' series of compounds, such as monosaccharides and steroids, which are well documented elsewhere in the literature are considered in outline only. Series of compounds which are essentially dimeric or polymeric types built up from chiral monomeric units are not covered in detail, since their configurations follow readily from those of the monomeric unit. Examples of such polymeric-type compounds are di-, oligo- and polysaccharides; bis (benzylisoquinoline) and other dimeric alkaloids; biflavonoids and polypeptides. Series of natural products having essentially the same carbon skeleton but known in a variety of stereochemical types due to epimerism at one or more chiral centres are frequently covered by general notes rather than by large numbers of examples. This treatment is given, for example, to the labdane-type diterpenes and the yohimbane-type alkaloids.

If a chiral compound, which does not fall into one of these categories cannot be found in the 'Atlas', the most likely explanations are (a) its absolute configuration has not been determined, or rests only on inconclusive data such as biogenetic analogy or specific rotation comparisons, or (b) it has been related by a fairly simple sequence to a compound which does appear. (We would, of course, be grateful to hear of any compounds of sufficient importance which we have not included.)

Literature coverage

The literature has been scanned to the end of 1971, and a few important references from 1972 have been incorporated.

We acknowledge the invaluable help rendered in the compilation of the 'Atlas' by the following standard works:

Dictionary of Organic Compounds, 2nd edition and supplements, Eyre & Spottiswoode, 1965-71.
Rodd's Chemistry of Carbon Compounds (2nd edition, S. Coffey, Ed.), Elsevier (1967-)
Molecular Asymmetry in Biology, by R. Bentley, 2 vols., Academic Press, 1970.

Other more specialized books and reviews pertaining to individual series of compounds are noted at the beginning of the appropriate sections.

Arrangement of material

The 'Atlas' is divided into the following chapters:

- A – fundamental chiral compounds mostly containing one or two chiral carbon atoms (but also including the cyclitols, for convenience of presentation).
- C – carbohydrates.
- T – terpenoids, including steroids.
- K – alkaloids.
- Y – other classes of natural products.
- D – chirality due to deuterium or other isotopes.
- X – compounds containing chiral axes, planes, etc.
- Z – chirality at atoms other than carbon (especially sulphur, phosphorus and silicon).

Key

Some compounds may appear out of place because they have been correlated with compounds in another chapter; while others, for example alkaloids which contain only one chiral centre, belong strictly speaking in two different sections (in this case A and K). No hard-and-fast rules have been used to cover these situations and such compounds have been placed where it seemed most convenient, usually with notes at the other possible point of placement to assist the reader. Notes on the arrangement of material within each chapter appear at the head of the chapter.

References

References are given for each correlation between substances against the arrow joining their formulae and for each determination of structure by X-ray analysis, etc. Extra references to the clarification of some point of stereochemistry are given in some cases beneath the structural formula of a compound and immediately after its name.

In general, however, *references are not given to work establishing the relative configuration of individual centres within a molecule*. This point is sufficiently important to justify further explanation by means of an example. On page **K1**, the cyclization of (*R*)–(+) 2-amino-1-(3-hydroxyphenyl)-ethanol **K1.1**, studied by Kametani and his co-workers, is shown to produce two tetrahydroisoquinoline derivatives epimeric at the newly-created C_1 chiral centre. One of these, assigned the configuration (1*R*,4*R*) (**K1.2**) is then converted by removal of the original chiral centre, into (*R*)–(+) 1,2,3,4-tetrahydro-6-methoxy-2-methyl-1-phenylisoquinoline (**K1.3**)

K1.1   **K1.2**   **K1.3**

Clearly the absolute configuration assigned to the new substance **K1.3** depends as much on correct establishment of the *relative* configuration between the two chiral centres in **K1.2** (in this case, principally by n.m.r. spectroscopy) as on a correct assignment of absolute configuration to the starting material **K1.1**. The details of how this relative configuration is determined lie outside the scope of the 'Atlas'.

Numbering of formulae and cross-references

At the first appearance of a given compound in the 'Atlas' it is given a unique reference number consisting of the page number followed by a number showing its position on the page, e.g. **A46.11**, the 11th formula on Page A46. On subsequent appearances the original reference number is given. Immediately after the compound name at the point of first mention, all subsequent appearances of the compound are listed, indexed by page number. Thus to trace all entries for a given compound it is necessary to find the original entry, and thence turn forward to all subsequent entries.

Combination of formulae

Derivatives such as esters of acids, acetates of alcohols, ring-substituted derivatives of aromatic compounds, etc. are usually listed under the parent compound. Similarly the formulae of closely related natural products are often combined, e.g. where one is a dihydro-derivative of the other. In such cases, a direct unequivocal correlation between the substances thus combined may be assumed to exist in the literature.

Nomenclature

No attempt has been made to systematize nomenclature throughout. The 'Atlas' covers a very wide range of simple and complex skeletal systems; for many of the latter the nomenclature has not been formalized. In general the nomenclature of natural products follows that in the Heilbron *Dictionary* and the original publications.

In naming compounds containing two or more chiral centres, attempts have been made to avoid partially systematic forms, frequently encountered in the literature, such as '(2S,3S)-(-) hydroxycitric acid'. In the absence of further information, this can be misleading since it implies a numbering system for a non-systematically named compound (citric acid). We have preferred forms such as either (a) the systematic (2S,3S)-(-) 3-carboxy-2,3-dihydroxyadipic acid or (b) the trivial (-)-hydroxycitric acid.

Designation of configurations; The Sequence Rule

Absolute configurations are generally described by the Sequence Rule of R. S. Cahn, C. K. Ingold and V. Prelog, commonly called the (R,S)-system, which is the only completely general and unambiguous form of designation. (R. S. Cahn, C. K. Ingold & V. Prelog, *J. Chem. Soc.*, 1951, 612; *Experientia*, 1956, **12**, 81; *Angew. Chem. Internat. Edn.*, 1966, **5**, 385.) No attempt will be made to describe the system in detail here, since to attempt explanation without reproducing a large part of the original papers would tend to give rise to over-simplification.

For convenience, however, the following list of common substituents in order of fiducial precedence is presented:

$-I > -Br > -Cl > -SO_2R > -SOR > -SR > -SH > -F > -OCOR > -OR > -OH > -NO_2 >$
$-NHCOR > -NR_2 > -NHR > -NH_2 > -CH_2Br > -CCl_3 > -CHCl_2 > -COCl > -CH_2Cl >$
$-COOR > -COOH > -CONH_2 > -COR > -CH_2OH > -CN > -CH_2NH_2 > -CH_2COOH >$
α–naphthyl $>$ β-naphthyl $>$ Ph $> -CMe_3 > -CH=CH_2 >$ cyclohexyl $> -CHMe_2 >$
$-CH_2-CH=CH_2 > -CH_2CH_2CH_3 > -C_2H_5 > -CH_3 > -D > -H >$ lone-pair.

In cases where (R,S) nomenclature is applied to complex chiral structures (especially in Section X), brief notes on its application in particular cases are appended where appropriate, but for a fuller description the user is again referred to the Cahn-Ingold-Prelog papers.

D,L *nomenclature*

The configuration of a few key compounds in Section A are also given according to the older D,L system, following the glyceraldehyde or serine conventions as appropriate. (This juxtaposition of the (R,S) and (D,L) conventions is contrary to normal practice and would not be accepted by journal editors; it is justified here purely on grounds of convenience to the user.)

The D,L system is now redundant for all compounds except carbohydrates, amino acids and closely related substances, but its limited inclusion here may assist those who have to consult the older literature. For a helpful discussion of the present state of the D,L system and its ambiguities, see D. W. Slocum, D. Sugarmann and S. P. Tucker, *J. Chem. Educ.*, 1971, **48**, 597.

Formulae

(a) General

The purpose of the 'Atlas' is to depict *absolute configurations* at chiral carbon atoms or other features of dissymmetry. It is not concerned with *conformations, bond-lengths* or *bond angles*. Even where 'perspective' type drawings have been included to improve the representation of a particular compound, distortions such as alteration of bond angles and flattening of rings may have been incorporated in order to improve general clarity and the depiction of configurations at individual centres of chirality.

(b) Fischer-type projections

These are used throughout Chapters A, C, D and Z.

The projections used in the 'Atlas' do not in general obey the Fischer convention that the principal carbon chain of the molecule should be vertical with the lowest-numbered carbon atom at the top. To follow this convention in the 'Atlas' would have meant that a complex and confusing series of substituent interchanges would have been necessary on most pages.

Interconversion of Fischer-type formulae Interchanging any one pair of substituents in a Fischer-type drawing of compound (+)X produces the enantiomer, (−)X. Interchanging two pairs of substituents reverts to the original enantiomer (+)X. Special cases of this are (i) interchange of a with e then e with b, which has the effect of a cyclic interchange of a, b and e (Fig. 1), and (ii) interchange of a with d and of b with e, which has the effect of a 180° rotation in the *xy* plane. (Fig. 2).

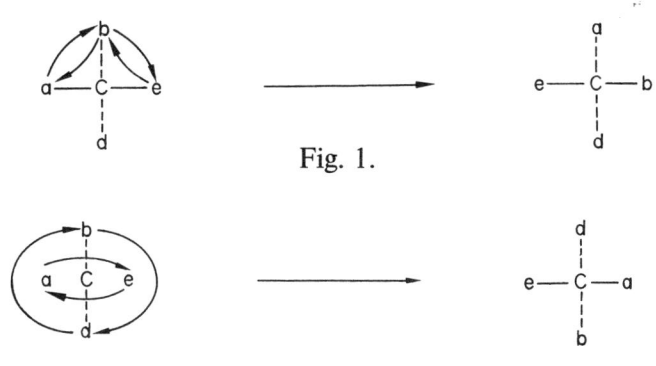

Fig. 1.

Fig. 2.

Assignment of (R,S) symbols to Fischer-type formulae If d is the group of lowest precedence, then examination of Fig. 3 shows that (R) or (S) chirality may be readily assigned by inspection.

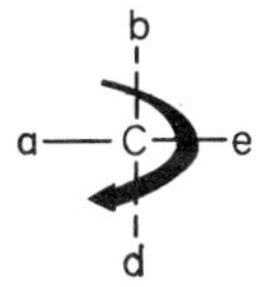

(R)-chirality with precedence a > b > e > d.

Fig. 3.

If d is not the group of lowest precedence, with a little practice it is again possible to assign chirality directly by inspection; the following table (Fig. 4) demonstrates the twelve ways of writing the formula having (R)-chirality.

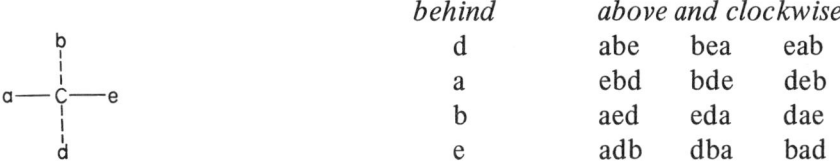

	behind	above and clockwise		
	d	abe	bea	eab
	a	ebd	bde	deb
	b	aed	eda	dae
	e	adb	dba	bad

Fig. 4.

(c) Cyclic structures

Substituents lying above and below the *xy* plane in more complex structures are indicated by the use of wedges and dashed bonds respectively, in accordance with usual practice.

Interconversion of Fischer and cyclic formulae (Fig. 5) Consider the tetrahedral molecule (A). Rotation produces the Fischer formula (B). Substituents a and b can clearly be portions of a cyclic residue as well as cyclic groups; for example when a and b become incorporated into a six-membered ring, (C) results.

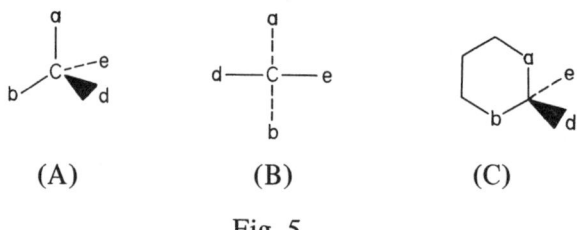

Fig. 5.

(d) *More complex structures*

In the depiction of complex structures, as much care as possible has been taken to give unambiguous representation of the correct configurations at all chiral centres. Deviations from the true bond angles are inevitable in any attempt to represent three-dimensional structures in two dimensions, and further distortions result from various conventions in common use, for example the representation of cyclohexane rings as planar hexagons with 120° valency angles. These distortions cannot be eliminated fully, but the aim has been to present drawings in which the amount of deviation, from the drawing required to produce the 'correct' configuration at a given chiral centre, is always considerably smaller than that required to produce the opposite, 'incorrect' configuration. It should be reiterated that in complex structures containing several chiral centres, the aim has been to show the correct absolute configuration at each chiral centre considered individually, and in cases where there is any possibility of ambiguity, *each chiral centre should be inspected independently of other chiral centres in the molecule.*

Symbols and abbreviations

Abs. X-ray. Substance for which the Bijvoet method for determination of absolute configuration by the anomalous dispersion method has been applied.

A recent publication has pointed out a possible source of error in one method of assigning absolute configurations by the anomalous dispersion method, and has indicated that *some* of the previous determinations are conceivably in error.

Rel. X-ray. X-ray determination of structure without application of anomalous dispersion. This determines the configurations of all chiral centres in the molecule relative to each other but does not place them on an absolute basis.

Conversion of compounds to derivatives, often those containing a heavy atom for use in X-ray structure determination, is a very common procedure. No special mention has been made of such procedure unless it causes a significant change in configuration.

$$A \xrightarrow{C(n)} B.$$ Chemical conversion (C) of A to B in approximately *n* steps, i.e. A → K → L → M → N → B

$$A \xleftrightarrow{C(n)} B.$$ Chemical conversion of both A and B to a common intermediate, or of a common intermediate to both A and B, in a total of approximately *n* steps, i.e. A → K → L → M ← N ← P ← B or A ← K ← L ← M → N → P → B

ORD – optical rotatory dispersion

CD – circular dichroism

AS Asymmetric synthesis. The two most frequently encountered types of asymmetric synthesis are further distinguished as follows.

AS(H) Asymmetric synthesis by the method of Horeau (*Tetrahedron Letters*, 1961, 506). This involves reaction of the chiral alcohol with racemic 2-phenylbutyric anhydride. The absolute configuration at the hydroxyl-bearing chiral centre is determined by measurement of the specific rotation of the recovered unreacted 2-phenylbutyric acid.

AS(P) Asymmetric synthesis by the methods of Prelog (*Helv. Chim. Acta*, 1953, **36**, 308), Cram (D. J. Cram and F. A. A. Elhafez, *J. Amer. Chem. Soc.*, 1952, **74**, 5828), and related methods. A

Key

chiral alcohol is converted to its α-ketoester and this is allowed to react with a Grignard reagent. Hydrolysis of the product produces a chiral α-hydroxy acid. The method may be used to compare the absolute configuration of an unknown alcohol with that of a known α-hydroxy acid (usually atrolactic), or that of a known alcohol (e.g. menthol) with that of an unknown α-hydroxy acid.

Where an asymmetric synthesis is of neither of these two types, the original reference should be consulted for details of the precise method employed and the reasoning used to interpret the observed stereospecificity.

- AC — absolute configuration
- —O→— chemical conversion with inversion at a chiral centre
- ----→— dashed arrow indicating a correlation which is considered to be of lower reliability (see Introduction)
- —8— 'Enantiomer of' sign. For example A —8—$^{C(n)}$→ B means 'conversion of A to the *enantiomer* of B'
- * — denotes natural products both enantiomers of which are known by us to occur naturally. This symbol is used only in the sections dealing predominantly with natural products, namely T (terpenoids), K (alkaloids) and Y (other natural products).
- § — indicates that the absolute configuration shown in the 'Atlas' differs from that appearing in one or more of the original references quoted. This may be due to one of several factors.
    - (a) a misprint or error in the paper.
    - (b) a relatively recent change in the generally accepted configuration due to further work by other or the same authors.
    - (c) personal correspondence between the authors and ourselves.
    - (d) in a few cases in Chapter X, changes in the accepted (R,S) convention since the date of the original publications have led to a change in (R,S) assignment although there has been no change in the accepted absolute configuration.

    In all cases the absolute configuration given here is correct according to the best available information (January 1972); other sources showing a different configuration are believed to be in error.
- ΔM_D — comparison by the Freudenberg method of molecular rotation shifts.
- QR — comparison by the quasi-racemate method.
- > — has precedence over, according to the Cahn, Ingold, Prelog system.

Other abbreviations have their usual chemical significance.

Choice of enantiomer depicted.

(a) As a general rule, *the enantiomer shown in the 'Atlas' may not be the one on which the correlations shown were originally carried out*. No special indication is given where this is the case. This is for convenience in fitting large numbers of formulae together, and to minimize use of the '—8—' sign.

(b) In the *chapters which deal with natural products*, namely T, K and Y, the enantiomer shown is the naturally occurring one unless otherwise stated.

When both enantiomers are known to occur as natural products, this is shown by the sign * after the name of the compound. Refer to the original publications to ascertain which enantiomer is obtained from which natural source.

(c) In the other chapters (A, C, X, D and Z) no distinction is made between natural and synthetic compounds. Check with the literature to determine which enantiomer, if either, is a natural product.

Rotations

The sign of rotation given for each compound is that shown in the Heilbron 'Dictionary' or in the original publication. Rotations are understood to be at the sodium D line unless stated otherwise.

An empty bracket ( ), means that the specific rotation has not been determined, or cannot be found.

In most cases, changes in sign of rotation with solvent or pH are noted where known, but the absence of such indication cannot of course be taken to mean that no such anomalous behaviour occurs under any circumstances. It is necessary to consult the original publication for details of solvent and temperature.

-A-

Fundamental Chiral Compounds

Introductory Notes to Chapter A

Chapter A contains a wide selection of the most important fundamental chiral molecules. The great majority contain one or two chiral centres, but a number (notably cyclitols, decalins and norbornanes) contain three or more.

Sub-classification of compounds in Chapter A

To facilitate the finding of a particular compound, the contents of Chapter A are subdivided according to the nature of the chiral carbon atom(s).
Classes 0, 1, 2, 3, 4 are designated according to the number of other *carbon* atoms directly attached to the chiral centre C* under consideration.
R^1, R^2, R^3, R^4 are groups of all kinds bonded by *carbon* atoms, including oxidised groups such as –COOH; W, X, Y, Z are groups bonded to the chiral centre by *non-carbon* atoms. (H, O, N, S, Halogen, etc.)

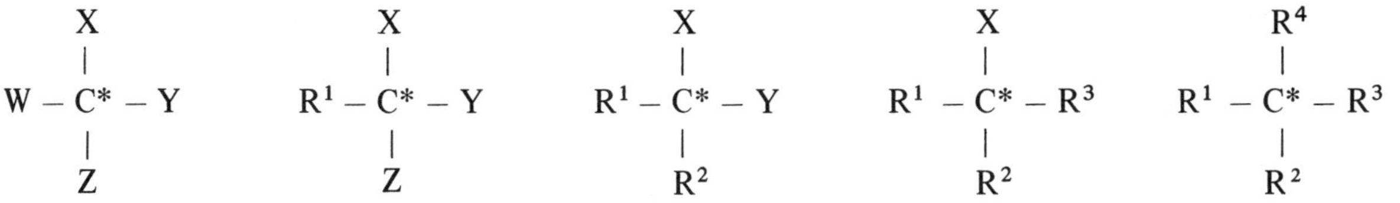

$$
\begin{array}{ccccc}
\text{X} & \text{X} & \text{X} & \text{X} & \text{R}^4 \\
| & | & | & | & | \\
\text{W} - \text{C*} - \text{Y} & \text{R}^1 - \text{C*} - \text{Y} & \text{R}^1 - \text{C*} - \text{Y} & \text{R}^1 - \text{C*} - \text{R}^3 & \text{R}^1 - \text{C*} - \text{R}^3 \\
| & | & | & | & | \\
\text{Z} & \text{Z} & \text{R}^2 & \text{R}^2 & \text{R}^2 \\
\end{array}
$$

Class 0   Class 1   Class 2   Class 3   Class 4

The classes are further divided as follows: 1a, 2a, 3a, 4a; all R groups aliphatic. 1b, 2b, 3b, 4b; at least *one* R group aromatic or heteroaromatic.

Compounds containing two or more chiral centres, are indexed under the highest applicable number, with subgroup b taking precedence over subgroup a. For example:

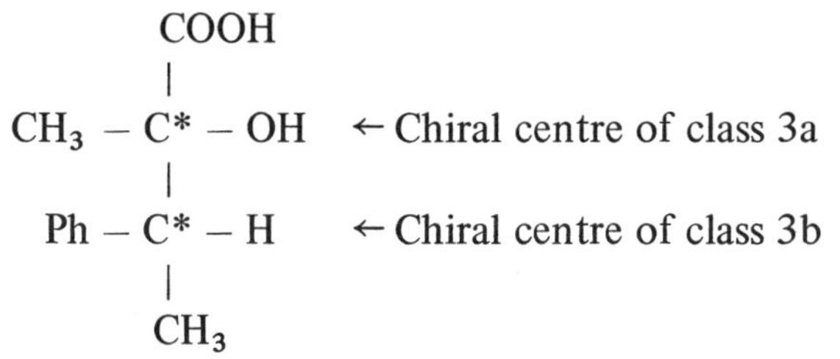

$$
\begin{array}{l}
\quad\quad\text{COOH} \\
\quad\quad\quad| \\
\text{CH}_3 - \text{C*} - \text{OH} \quad \leftarrow \text{Chiral centre of class 3a}\\
\quad\quad\quad| \\
\text{Ph} - \text{C*} - \text{H} \quad\quad \leftarrow \text{Chiral centre of class 3b}\\
\quad\quad\quad| \\
\quad\quad\text{CH}_3
\end{array}
$$

is indexed under 3b.

Location of compounds by class

Classes 0, 1a, 1b;
No optically active compounds belonging to these classes (with the exception of deuterated com-

pounds, which appear in Chapter D) appear to be of known absolute configuration.
Class 2a; Pages **A1-A26** and **A58**
Class 2b; Pages **A19-A25** and **A58**
Class 3a; Pages **A26-A47** and **A59-A61**
Class 3b; Pages **A40-A52** and **A61**
Class 4a; Pages **A53-A56**
Class 4b; Pages **A55-A56**

Simple compounds found in other chapters

The following is a list of simple compounds containing one or two chiral centres which would normally be considered to belong in Chapter A, but which because of difficulties of arrangement, appear in other chapters.

Class 2a

2,5-dioxo-5-methylhexanoic acid **T17.5**
2-hydroxymethyl-1-methylcyclohex-1-en-4-ol **T40.6**
cyclodopa **K17.9**
2-(dimethylamino)heptane **K19.3**
nonan-2-ol **K19.5**
2,6-dimethylpiperidine **K19.6**
3,4-dihydroxy-4-methylpentanoic acid lactone **Y1.6**
5-hydroxyhexanoic acid lactone **Y14.9**
3-aminoproline **Y22.7**
2,2-dimethyl-thiazolidine-4-carboxylic acid **Y29.3**
cyclooct-1-en-3-ol **X2.5**
3-acetoxycylooctyne **X2.6**
hexane-3,4-diol **X9.8**

Class 2b
2-amino-1-(3-hydroxyphenyl)ethanol **K1.1**
1,2,3,4-tetrahydro-4,6-dihydroxy-1-phenylisoquinoline **K1.2, K1.5**
1,2,3,4-tetrahydro-6-methoxy-2-methyl-1-phenylisoquinoline **K1.3**
3-methylphthalide **Y14.5**
3-butylphthalide **Y14.6**
2-mercapto-4-phenylimidazoline **Y30.6**
3-bromo-4-phenylcrotonolactones **X3.4**

Class 3a
2,6-dimethylheptanoic acid **T1.7**
4-hydroxy-4-methylhexanoic acid lactone **T3.9**
cinenic acid **T3.10**
3-carboxy-2,2-dimethylglutaric acid **T8.9**
pinonic acid **T8.14**
3-isopropenylcyclopentanone **T9.1**
3-isopropenylcyclopentanols **T9.2, T9.3**
paraconic acid **T15.14**
4-methyl-3-oxocyclohexanecarboxylic acid **T17.10**
dihydrohaematinic acid **T22.6**
trans-norcaryophyllenic acid **T28.8**
trans-caryophyllenic acid **T28.10**
4,8-dimethylnonanal **T46.1**
5-ethyl-6-methyl-heptan-2-one **T48.3**

Introductory Notes to Chapter A

4-methylcyclohexane-1,2-dione **T58.4**
3-ethylglutaric acid mononitrile **T58.7**
4-methyl-2-pyrrolidone **K35.8**
2-hydroxy-3-methylsuccinic acid **Y5.4**
4,8-dimethyl-4-hydroxynonanoic acid lactone **Y9.6**
1-hydroxymethylcyclohexane-1,2-diol **Y19.11**
5-hydroxy-3-methylhexanoic acid lactone **Y21.5, Y21.10**
2,4-dimethylcyclohexanone **Y21.8, Y21.11**
3-methylheptanedioic acid **Y27.4**

Class 3b

2-(*p*-methoxyphenyl)-propionic acid **Y5.7**
(*p*-methoxyphenyl)-succinic acid **Y6.8**
2,4-di(*p*-methoxyphenyl)-butyric acid **Y9.4**
2-(3,5-dimethoxy-2-methylphenyl)-butane **Y13.10**
3-bromo-4-alkyl-4-phenylcrotonolactone **X3.4**

Class 4a

3,4-di(hydroxymethyl)-4-methylcyclohexanone **T40.2**
4-carbethoxy-2-pyrrolidone-4-acetic acid **X4.5**

Chapters T, K and Y contain many natural products with one or two chiral centres, and the user scanning the 'Atlas' for a compound of a particular stereochemical type should search these chapters as well as Chapter A. In order to find any individual compound, the Index should be used.

Amino acids

All the common amino acids have been chemically correlated with other chiral substances, and are included in the 'Atlas', as are many rarer amino acids. A large number of less common amino acids have been isolated from natural sources (for a review, see L. Fowden, *Progr. Phytochem.*, 1970, 2, 203); the absolute configurations of many of these have been assigned indirectly, by one of the following methods.

(1) Enzymatic methods

(2) Molecular rotation shift methods. The most frequently used is the Clough-Lutz-Jirgenson rule which states that acids of the L series (which corresponds to (*S*) in most but not all cases) show a positive shift in molecular rotation on passing from neutral to acid solution. Studies of chiroptical properties at various pHs (P. M. Scopes and co-workers, *J. Chem. Soc., (C)*, 1971, 833) are a development of this method. These procedures rely on the availability as reference substances of amino acids which have been chemically correlated and which appear in Chapter A.

See J. P. Greenstein and M. Winitz, *Chemistry of the Amino Acids*, Wiley, 1961, for descriptions of the enzymatic and molecular rotation shift methods.

Class 2a — Important one-centre compounds including glyceraldehyde, lactic and malic acids, alanine and butan-2-ol

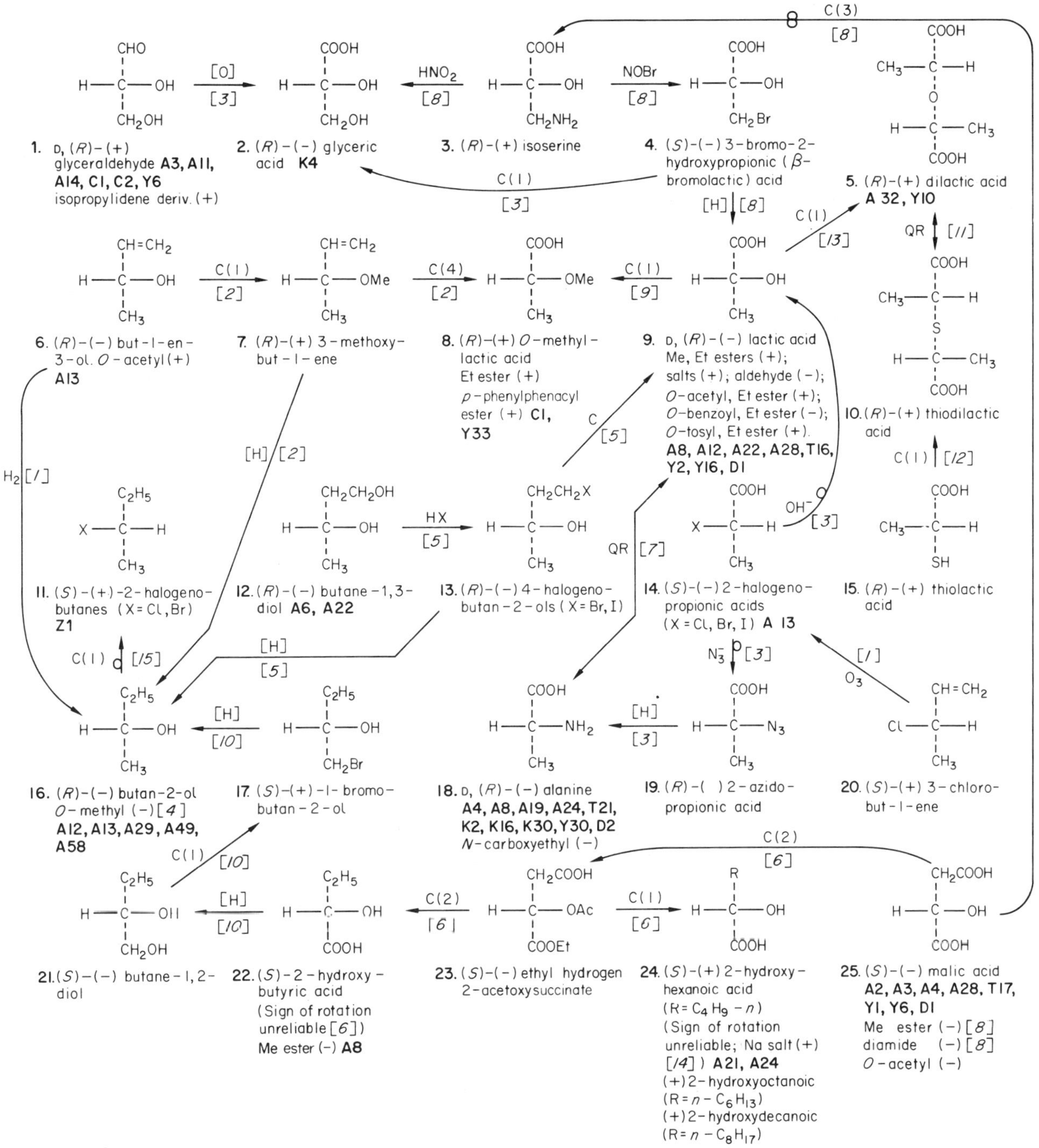

1. D, (R)-(+) glyceraldehyde **A3, A11, A14, C1, C2, Y6** isopropylidene deriv. (+)
2. (R)-(−) glyceric acid **K4**
3. (R)-(+) isoserine
4. (S)-(−) 3-bromo-2-hydroxypropionic (β-bromolactic) acid
5. (R)-(+) dilactic acid **A 32, Y10**
6. (R)-(−) but-1-en-3-ol. O-acetyl (+) **A13**
7. (R)-(+) 3-methoxy-but-1-ene
8. (R)-(+) O-methyl-lactic acid Et ester (+); p-phenylphenacyl ester (+) **C1, Y33**
9. D, (R)-(−) lactic acid Me, Et esters (+); salts (+); aldehyde (−); O-acetyl, Et ester (+); O-benzoyl, Et ester (−); O-tosyl, Et ester (+). **A8, A12, A22, A28, T16, Y2, Y16, D1**
10. (R)-(+) thiodilactic acid
11. (S)-(+)-2-halogeno-butanes (X = Cl, Br) **Z1**
12. (R)-(−) butane-1,3-diol **A6, A22**
13. (R)-(−) 4-halogeno-butan-2-ols (X = Br, I)
14. (S)-(−) 2-halogeno-propionic acids (X = Cl, Br, I) **A 13**
15. (R)-(+) thiolactic acid
16. (R)-(−) butan-2-ol O-methyl (−) [4] **A12, A13, A29, A49, A58**
17. (S)-(+)-1-bromo-butan-2-ol
18. D, (R)-(−) alanine **A4, A8, A19, A24, T21, K2, K16, K30, Y30, D2** N-carboxyethyl (−)
19. (R)-( ) 2-azido-propionic acid
20. (S)-(+) 3-chloro-but-1-ene
21. (S)-(−) butane-1,2-diol
22. (S)-2-hydroxy-butyric acid (Sign of rotation unreliable [6]) Me ester (−) **A8**
23. (S)-(−) ethyl hydrogen 2-acetoxysuccinate
24. (S)-(+) 2-hydroxyhexanoic acid (R = C_4H_9-n) (Sign of rotation unreliable; Na salt (+) [14]) **A21, A24** (+) 2-hydroxyoctanoic (R = n-C_6H_{13}) (+) 2-hydroxydecanoic (R = n-C_8H_{17})
25. (S)-(−) malic acid **A2, A3, A4, A28, T17, Y1, Y6, D1** Me ester (−) [8] diamide (−) [8] O-acetyl (−)

1. W. G. Young and F. F. Caserio, *J. Org. Chem.*, 1961, **26**, 245.
2. K. B. Wiberg, *J. Amer. Chem. Soc.*, 1952, **74**, 3891.
3. P. Brewster, E. D. Hughes, C. K. Ingold and P. A. D. S. Rao, *Nature*, 1950, **166**, 178.
4. W. von E. Doering and R. W. Young, *J. Amer. Chem. Soc.*, 1952, **74**, 2997.
5. P. A. Levene, A. Walti and H. L. Haller, *J. Biol. Chem.*, 1927, **71**, 465.
6. D. H. S. Horn and Y. Y. Pretorius, *J. Chem. Soc.*, 1954, 1460.
7. A. Fredga, *Svensk kem. Tidskr.*, 1942, **54**, 26.
8. K. Freudenberg, *Ber.*, 1914, **47**, 2027.
9. M. L. Wolfrom, R. U. Lemieux, S. M. Olin and D. I. Weisblat, *J. Amer. Chem. Soc.*, 1949, **71**, 4057.
10. P. A. Levene and H. L. Haller, *J. Biol. Chem.*, 1927, **74**, 343.
11. A. Fredga, *Arkiv Kemi*, 1940, **14B**, no. 12.
12. J. M. Lovén, *J. prakt. Chem.*, 1908, **78**, 63.
13. P. Vièles, *Ann. Chim. (France)*, 1935, **3**, 143.
14. S. Gronowitz, *Arkiv Kemi*, 1958, **13**, 87, 231.
15. W. A. Cowdrey, E. D. Hughes, C. K. Ingold, S. Masterman and A. D. Scott, *J. Chem. Soc.*, 1937, 1252.

A[2] Class 2a
Mainly two-centre compounds related to tartaric acid

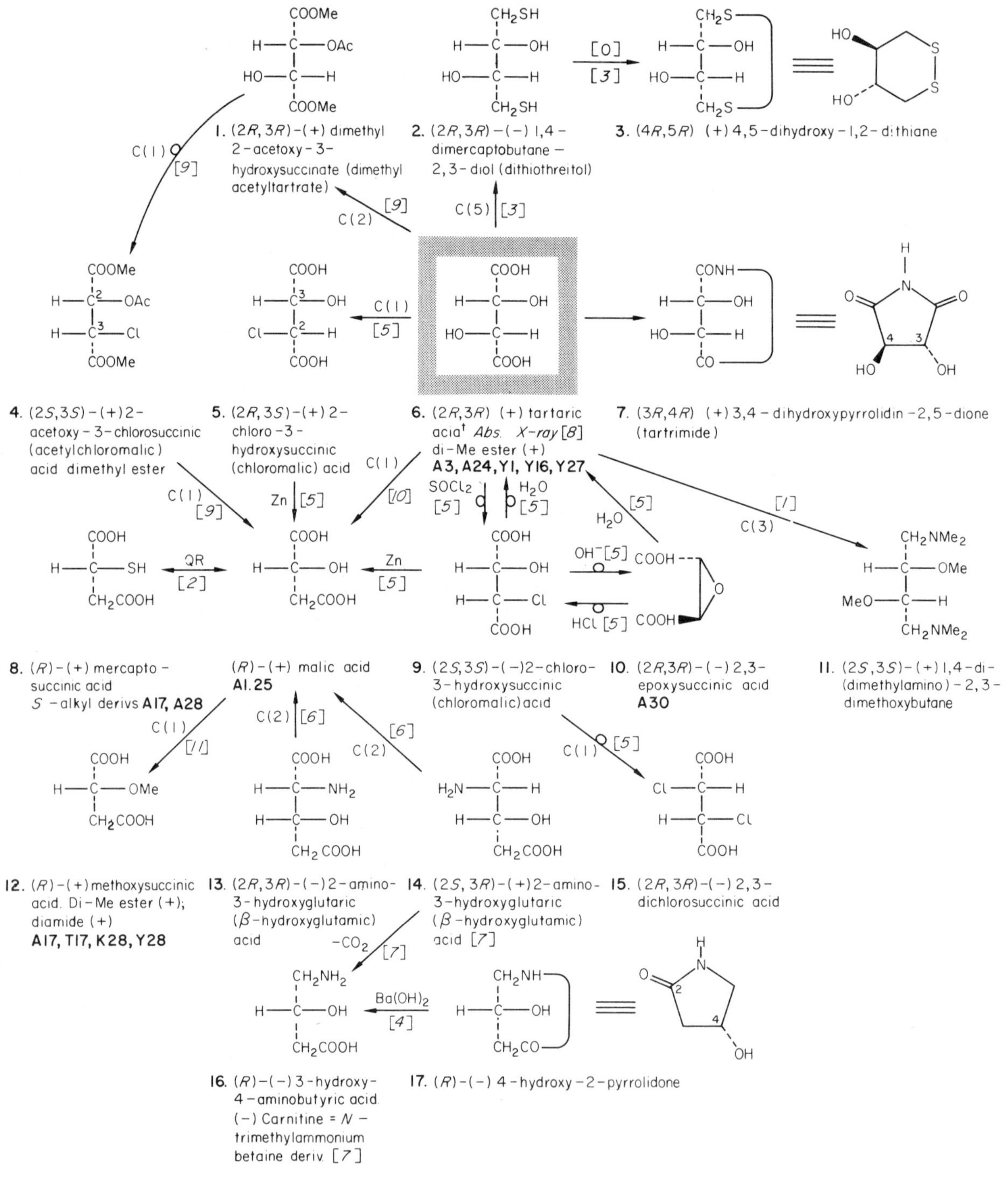

1. (2R,3R)-(+) dimethyl 2-acetoxy-3-hydroxysuccinate (dimethyl acetyltartrate)
2. (2R,3R)-(−) 1,4-dimercaptobutane-2,3-diol (dithiothreitol)
3. (4R,5R) (+) 4,5-dihydroxy-1,2-dithiane
4. (2S,3S)-(+) 2-acetoxy-3-chlorosuccinic (acetylchloromalic) acid dimethyl ester
5. (2R,3S)-(+) 2-chloro-3-hydroxysuccinic (chloromalic) acid
6. (2R,3R) (+) tartaric acid† Abs. X-ray [8] di-Me ester (+) **A3, A24, Y1, Y16, Y27**
7. (3R,4R) (+) 3,4-dihydroxypyrrolidin-2,5-dione (tartrimide)
8. (R)-(+) mercaptosuccinic acid S-alkyl derivs **A17, A28**
9. (2S,3S)-(−)-2-chloro-3-hydroxysuccinic (chloromalic) acid
10. (2R,3R)-(−) 2,3-epoxysuccinic acid **A30**
11. (2S,3S)-(+) 1,4-di-(dimethylamino)-2,3-dimethoxybutane
12. (R)-(+) methoxysuccinic acid. Di-Me ester (+); diamide (+) **A17, T17, K28, Y28**
13. (2R,3R)-(−) 2-amino-3-hydroxyglutaric (β-hydroxyglutamic) acid
14. (2S,3R)-(+) 2-amino-3-hydroxyglutaric (β-hydroxyglutamic) acid [7]
15. (2R,3R)-(−) 2,3-dichlorosuccinic acid
16. (R)-(−) 3-hydroxy-4-aminobutyric acid. (−) Carnitine = N-trimethylammonium betaine deriv. [7]
17. (R)-(−) 4-hydroxy-2-pyrrolidone

† For a discussion of the ambiguity of D,L-nomenclature as applied to tartaric acid, see [12]

1. D. Seebach, H. Dörr, B. Bastani and V. Ehrig, *Angew. Chem., Internat. Edn.*, 1969, **8**, 982.
2. A. Fredga, *Arkiv Kemi*, 1941, **B14**, no. 27.
3. M. Carmack & C. J. Kelley, *J. Org. Chem.*, 1968, **33**, 2171.
4. T. Matsumoto, W. Trueb, R. Gwinner and C. H. Eugster, *Helv. Chim. Acta*, 1969, **52**, 716.
5. R. Kuhn & T. Wagner-Jauregg, *Ber.*, 1928, **61**, 504.
6. T. Kaneko, R. Yoshida and H. Katsura, *Nippon Kagaku Zasshi*, 1959, **80**, 316 (*Chem. Abs.*, 1960, **54**, 24423b)
7. T. Kaneko and R. Yoshida, *Bull. Chem. Soc. Japan*, 1962, **35**, 1153.
8. A. F. Peerdeman, A. J. van Bommel and J. M. Bijvoet, *Proc. k. ned. Akad Wetenschap.*, 1951, **B54**, 16; H. Hope and V. de la Camp, *Nature*, 1969, **222**, 54.
9. K. Freudenberg and F. Brauns, *Ber.*, 1922, **55**, 1339.
10. K. Freudenberg, *Ber.*, 1914, **47**, 2027.
11. G. Fodor and S. Sóti, *J. Chem. Soc.*, 1965, 6830.
12. C. Buchanan, *Nature*, 1951, **167**, 689.

Class 2a Further one- and two-centre compounds related to tartaric acid

1. (−) chorismic acid
2. (4R,5R)-(−)4,5-dihydroxycyclohexene 1-carboxylic acid
3. (3S,4S)-(−)3,4-dihydroxyadipic acid
4. (2S,3S)-(−)2,3-methoxybutane-1,4-diamine
5. (R)-(+)2-hydroxy-3-phenylpropionic acid A19, K28
6. (R)-(+)3-hydroxypyrrolidin-2,5-dione (malimide)
   (R)-(+) malic acid A1.25
7. (2S,3S)-(+) butane-2,3-diol A13 di-OAc (−)
   (2R,3R)-(+) tartaric acid A2.6
8. (2R,3R)-( ) dimethoxysuccinic acid Di-Me ester (+); bis-methylamide (+) A16
9. (2S,3S)-(+)2,3-dimethoxybutane-1,4-diol
10. (R)-(+)2,4-dihydroxybutyric acid Y15
11. (R)-(−)2,3-dihydroxy-3-methylbutyric acid Me ester (−); p-phenylphenacyl ester (−) T57
12. (2R,3S)-( )2,3,4-trihydroxybutyric acid
    (S) (−) glyceraldehyde A1.1
14. (S)-(−)3-methyl butane 1,2,3-triol
15. (1S,2S)-(+)2-amino cyclohexanol N-methyl (+); N,N-dimethyl (+)
16. (1S,2S)-(+) cyclooctane-1,2-diol X2
13. (4S,5S)-( )4,5-dimethoxycycloheptanone

Various carbohydrate derivatives cf. C1

CD (complexes)

1. A. Wohl and F. Momber, *Ber.*, 1917, **50**, 455.
2. S. T. K. Bukhari, R. D. Guthrie, A. I. Scott and A. D. Wrixon, *Tetrahedron*, 1970, **26**, 3653.
3. A. C. Cope and A. S. Mehta, *J. Amer. Chem. Soc.*, 1964, **86**, 1268.
4. B. E. Neilsen and J. Lemmich, *Acta Chem. Scand.*, 1969, **23**, 967.
5. T. Posternak and J-P. Susz, *Helv. Chim. Acta*, 1956, **39**, 2032.
6. J. J. Plattner and H. Rapoport, *J. Amer. Chem. Soc.*, 1971, **93**, 1758.
7. W. N. Haworth and D. I. Jones, *J. Chem. Soc.*, 1927, 2349.
8. H. Arakawa, *Naturwiss.*, 1963, **50**, 441.
9. J. W. E. Glattfield and F. V. Sander, *J. Amer. Chem. Soc.*, 1921, **43**, 2675.

A⁴ Amino acids; alanine, serine and aspartic acid — Class 2a

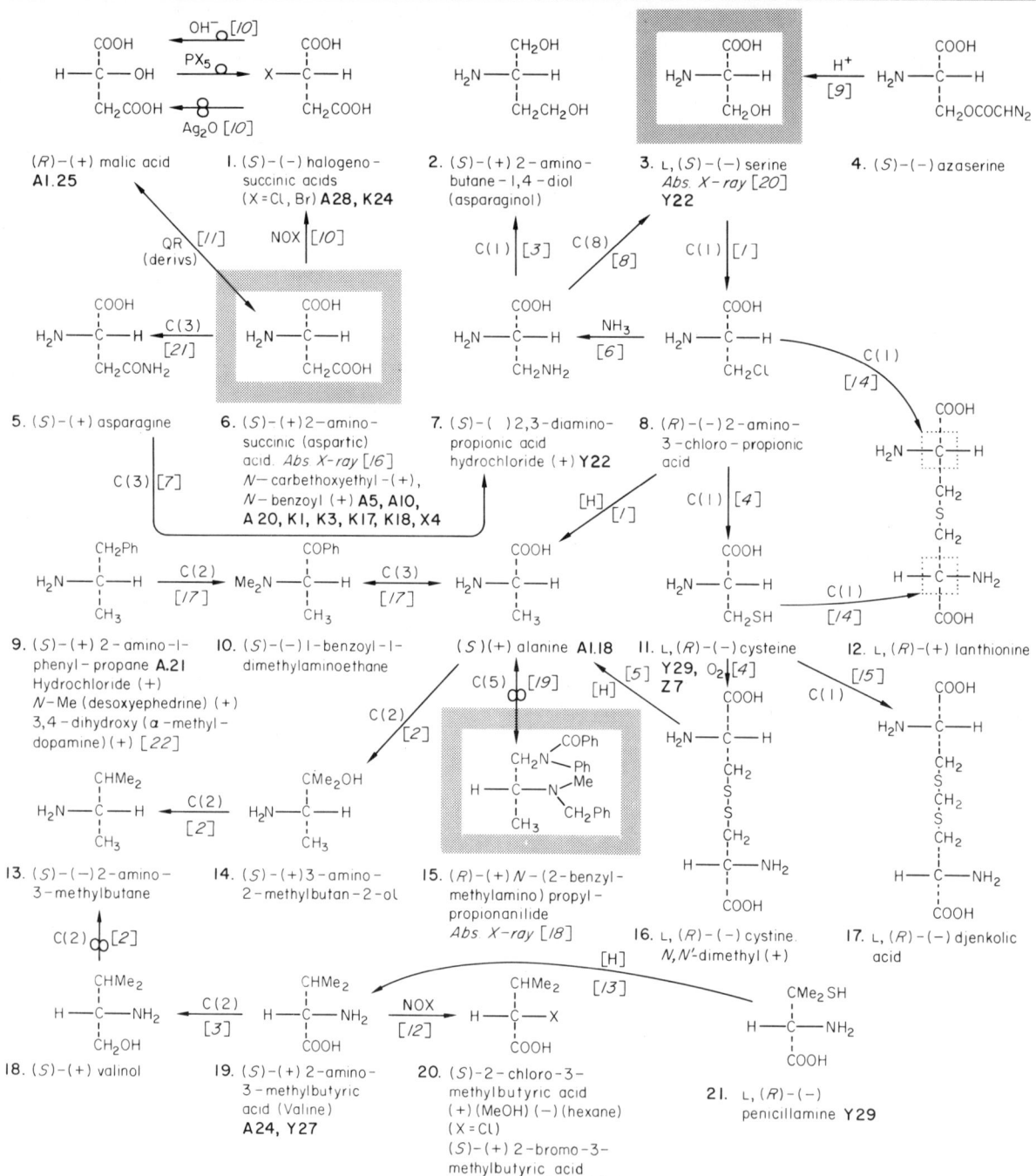

1. E. Fischer and K. Raske, *Ber.*, 1907, **40**, 3717.
2. F. Barrow and G. W. Ferguson, *J. Chem. Soc.*, 1935, 410.
3. P. Karrer, P. Portmann and M. Suter, *Helv. Chim. Acta*, 1949, **32**, 1156; 1948, **31**, 1617.
4. E. Fischer and K. Raske, *Ber.*, 1908, **41**, 893.
5. R. Mozingo, D. E. Wolf, S. A. Harris and K. Folkers, *J. Amer. Chem. Soc.*, 1943, **65**, 1013.
6. P. Karrer, *Helv. Chim. Acta*, 1923, **6**, 957.
7. P. Karrer and A. Schlosser, *Helv. Chim. Acta*, 1923, **6**, 411.
8. F. Schneider, *Annalen*, 1937, **529**, 1.
9. S. A. Fusari, T. H. Haskell, R. P. Frohardt and Q. R. Bartz, *J. Amer. Chem. Soc.*, 1954, **76**, 2881.
10. W. A. Cowdrey, E. D. Hughes and C. K. Ingold, *J. Chem. Soc.*, 1937, 1208.
11. A. Fredga, *Svensk kem. Tidskr.*, 1941, **53**, 221.
12. W. Gaffield and W. G. Galetto, *Tetrahedron*, 1971, **27**, 915.
13. Merck report, October 1943; H. M. Crooks in 'The Chemistry of Penicillin', Princeton Univ. Press, 1943.
14. G. B. Brown and V. du Vigneaud, *J. Biol. Chem.*, 1941, **140**, 767.
15. V. du Vigneaud and W. I. Patterson, *J. Biol. Chem.*, 1936, **114**, 533.
16. T. Doyne, R. Pepinsky and T. Watanabe, *Acta Cryst.*, 1957, **10**, 438.
17. K. Freudenberg and F. Nikolai, *Annalen*, 1934, **510**, 223.
18. P. Singh and F. R. Ahmed, *Acta Cryst.*, 1969, **B25**, 1901.

References continued on page 9

Class 2a — Phenylalanine, tyrosine and related one-centre nitrogen compounds — A⁵

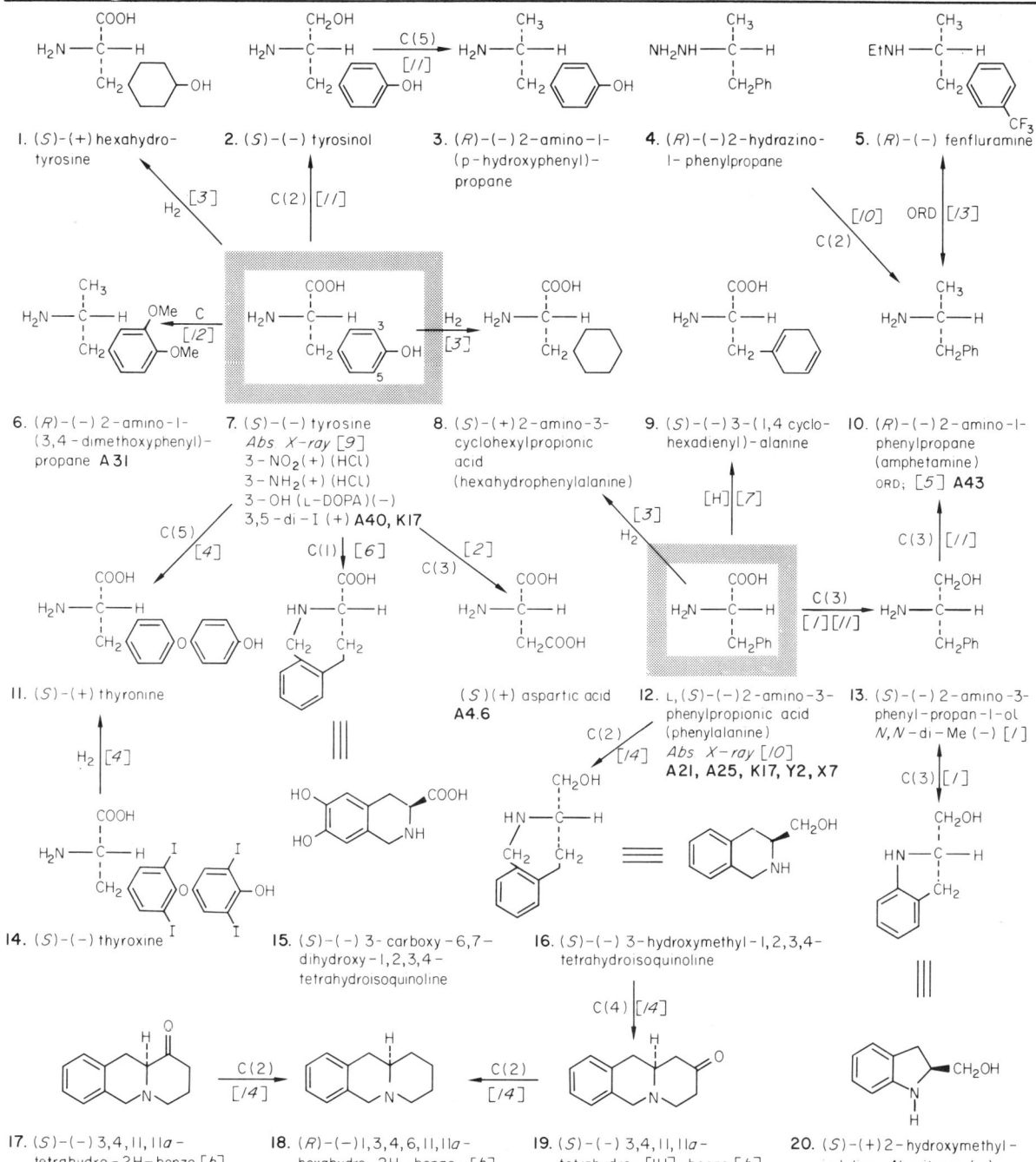

1. E. J. Corey, R. J. McCaully and H. S. Sachdev, *J. Amer. Chem. Soc.*, 1970, **92**, 2476.
2. S. Goldschmidt and G. Freyss, *Ber.*, 1933, **66**, 784.
3. E. Waser and E. Brauchli, *Helv. Chim. Acta*, 1924, **7**, 740.
4. A. Canzanelli, C. R. Harington and S. S. Randall, *Biochemical J.*, 1934, **28**, 68.
5. J. C. Craig, R. P. K. Chan and S. K. Roy, *Tetrahedron*, 1967, **23**, 3573.
6. E. A. Bell, J. R. Nulu and C. Cone, *Phytochemistry*, 1971, **10**, 2191.
7. M. L. Snow, C. Lauinger and C. Ressler, *J. Org. Chem.*, 1968, **33**, 1774.
8. W. Gulewitsch, *Z. physiol. Chem.*, 1907, **50**, 535.
9. R. Parthasarathy, *Acta Cryst.*, 1962, **15**, 41.
10. M. Mallikarjunan, S. T. Rao, K. Venkatesan and V. R. Sarma, *Acta Cryst.*, 1969, **B25**, 220.
11. P. Karrer and K. Ehrhardt, *Helv. Chim. Acta*, 1951, **34**, 2202.
12. A. W. Schrecker and J. L. Hartwell, *J. Amer. Chem. Soc.*, 1957, **79**, 3827.
13. A. H. Beckett and L. G. Brookes, *Tetrahedron*, 1968, **24**, 1283.
14. S. Yamada and T. Kuneida, *Chem. Pharm. Bull. (Japan)*, 1967, **15**, 491.

◀ 19. P. S. Portoghese and D. L. Larson, *J. Pharm. Sci.*, 1964, **53**, 302.
20. A. Zalkin, J. D. Forrester and D. H. Templeton, *Science*, 1964, **146**, 261.
21. F. E. King and D. A. A. Kidd, *J. Chem. Soc.*, 1951, 2976.
22. A. H. Beckett, G. Kirk and A. J. Sharpen, *Tetrahedron*, 1965, **21**, 1489.

A⁶ Pentan-2-ol and related alcohols — Class 2a

(S)-(+) butane-1,3-diol **A1.12**

1. *(S)-(+) 1-aminobutan-3-ol*
2. *(S)-(+) 4-hydroxypentanoic acid* **A12, T25, Y19**
3. *(S)-(+) pentane-1,4-diol* **Y19**
4. *(S)-(+) 3-hydroxybutyric acid Nitrile (+)* **A22, A34, Y2**
5. *(R)-(−) 2-hydroxypentanoic acid* **A24**
6. *(R)-(+) pentane-1,2-diol*
7. *(S)-(+) pentan-2-ol O-methyl (+)* **A12, A34, A58**
8. *(S)-(+) pent-4-en-2-ol O-methyl (−)* **A58**
9. *(2S,4S)-(+) pentane-2,4-diol*
10. *(R)-1-acetamido-2-acetoxypentane (+) (CHCl₃), (−) (EtOH)*
11. *(S)-(+) hex-1-en-3-ol* **A58**
12. *(R)-(−) hexan-3-ol* **A58**
13. *(R)-(−) 3-hydroxy hexanoic acid*
14. *(R)-(+) methyl hydrogen 3-acetoxy glutarate* **A7**
15. *(S)-(−) hex-1-yn-3-ol*
16. *(S)-(+) 2-methyl-tetrahydrofuran*
17. *(2R,4R)-(−) 2,4-dibromopentane*
18. *(R)-(−) methyl hydrogen 3-hydroxy-glutarate*
19. *(R)-(−) 3-hydroxy-pentanoic acid*
20. *(3S,5S)-(−) 3,5-dimethylpyrazoline*

---

1. S. G. Cohen and E. Khedouri, *J. Amer. Chem. Soc.*, 1961, **83**, 4228.
2. P. A. Levene and H. L. Haller, *J. Biol. Chem.*, 1926, **69**, 165, 569; 1928, **77**, 555.
3. R. U. Lemieux and J. Giguere, *Canad. J. Chem.*, 1951, **29**, 678.
4. K. Serck-Hanssen, *Arkiv Kemi*, 1956, **10**, 135.
5. E. R. Novak and D. S. Tarbell, *J. Amer. Chem. Soc.*, 1967, **89**, 73.
6. P. A. Levene and H. L. Haller, *J. Biol. Chem.*, 1929, **81**, 425.
7. A. Mishra and R. J. Crawford, *Canad. J. Chem.*, 1969, **47**, 1515.
8. S. R. Landor, B. J. Miller and A. R. Tatchell, *J. Chem. Soc. (C)*, 1971, 2339.
9. P. A. Levene and H. L. Haller, *J. Biol. Chem.*, 1928, **76**, 415; 1929, **83**, 579, 591.

Class 2a Long-chain hydroxyacids A⁷

1. (R)-9-hydroxy-octadecanoic acid. $[\alpha]_D \simeq 0$ Me ester (−)[5]
Structure: (CH₂)₇COOH / H—C—OH / (CH₂)₈CH₃

2. (R)-(+) heptadecane-1,3-diol
Structure: CH₂CH₂OH / H—C—OH / (CH₂)₁₃CH₃

3. (S)-(+) heptadecan-3-ol
Structure: C₂H₅ / H—C—OH / (CH₂)₁₃CH₃

4. (R)-(−) falcarinol (heptadeca-1,9-dien-4,6-diyn-3-ol)
Structure: CH=CH₂ / HO—C—H / (C≡C)₂CH₂CH=CH(CH₂)₆CH₃

Arrows: 2 → C(2)[3] → 3 ← H₂[3] ← 4
1 → C(1)[2] → 5

5. (R)-(−)3-hydroxydodecanoic acid
Structure: CH₂COOH / H—C—OH / (CH₂)₈CH₃

(R)-(+) methyl hydrogen 3-acetoxyglutarate **A6.14**
Structure: CH₂COOMe / H—C—OAc / CH₂COOH

5 ← C(2)[2] ← A6.14 → C(1)[1] → 6

6. (R)-(+) 3-hydroxynonanoic acid
Structure: CH₂COOH / H—C—OH / (CH₂)₅CH₃

7. (R)-(+)12-hydroxyoctadeca-9-enoic (ricinoleic) acid
Structure: CH₂CH=CH(CH₂)₇COOH / H—C—OH / (CH₂)₅CH₃

7 → C(1)[1] and H₂[1] → 10

8. (R)-( ) dimorphecolic (10-hydroxyoctadeca-6,8-diendioic) acid Me ester (−)
Structure: (CH=CH)₂(CH₂)₄COOH / H—C—OH / (CH₂)₇COOH

9. (S)-( )9-hydroxyoctadecanedioic acid
Structure: (CH₂)₈COOH / H—C—OH / (CH₂)₇COOH

10. (R)-(−)12-hydroxyoctadecanoic acid
Structure: (CH₂)₁₀COOH / H—C—OH / (CH₂)₅CH₃

11. (R)-( )13-hydroxyoctadecanoic acid
Structure: (CH₂)₁₁COOH / H—C—OH / (CH₂)₄CH₃

12. (R)-(−) coriolic (13-hydroxyoctadeca-9,11-dienoic) acid
Structure: COOH / (CH=CH)₂(CH₂)₇ / H—C—OH / (CH₂)₄CH₃

8 → H₂[4] → 9 ←ORD[4]→ 10 ←ORD[6]→ 11 ← H₂[6] ← 12

13. (R)-densipolic (12-hydroxyoctadeca-9,15-dienoic) acid $[\alpha]_D \simeq 0$.[4]
Structure: (CH₂)₂CH=CHC₂H₅ / H—C—OH / CH₂CH=CH(CH₂)₇COOH

14. (S)-( )12-hydroxyoctadecanoic acid
Structure: (CH₂)₅CH₃ / H—C—OH / (CH₂)₁₀COOH

15. (R)-( )14-hydroxyeicosanoic acid
Structure: (CH₂)₁₂COOH / H—C—OH / (CH₂)₅CH₃

16. (R)-(+) lesquerolic (14-hydroxyeicos-11-enoic) acid
Structure: (CH₂)₉COOH / CH₂CH=CH / H—C—OH / (CH₂)₅CH₃

13 → H₂[4] → 14 ←ORD[4]→ 10 ; 15 ← H₂[4] ← 16 ; 11 ←ORD[4]→ 15

---

1. K. Serck-Hanssen, *Chem. and Ind.*, 1958, 1554.
2. C. D. Baker and F. D. Gunstone, *J. Chem. Soc.*, 1963, 759.
3. P. K. Larsen, B. E. Nielsen and J. Lemmich, *Acta Chem. Scand.*, 1969, **23**, 2552.
4. T. H. Applewhite, R. G. Binder and W. Gaffield, *Chem. Comm.*, 1965, 255.
5. G. J. Schroepfer and K. Bloch, *J. Biol. Chem.*, 1965, **240**, 54.
6. W. H. Tallent, J. Harris, I. A. Wolff and R. E. Lundin, *Tetrahedron Letters*, 1966, 4329.

A⁸ One- and two-centre compounds related to threonine and 2-hydroxybutyric acid — Class 2a

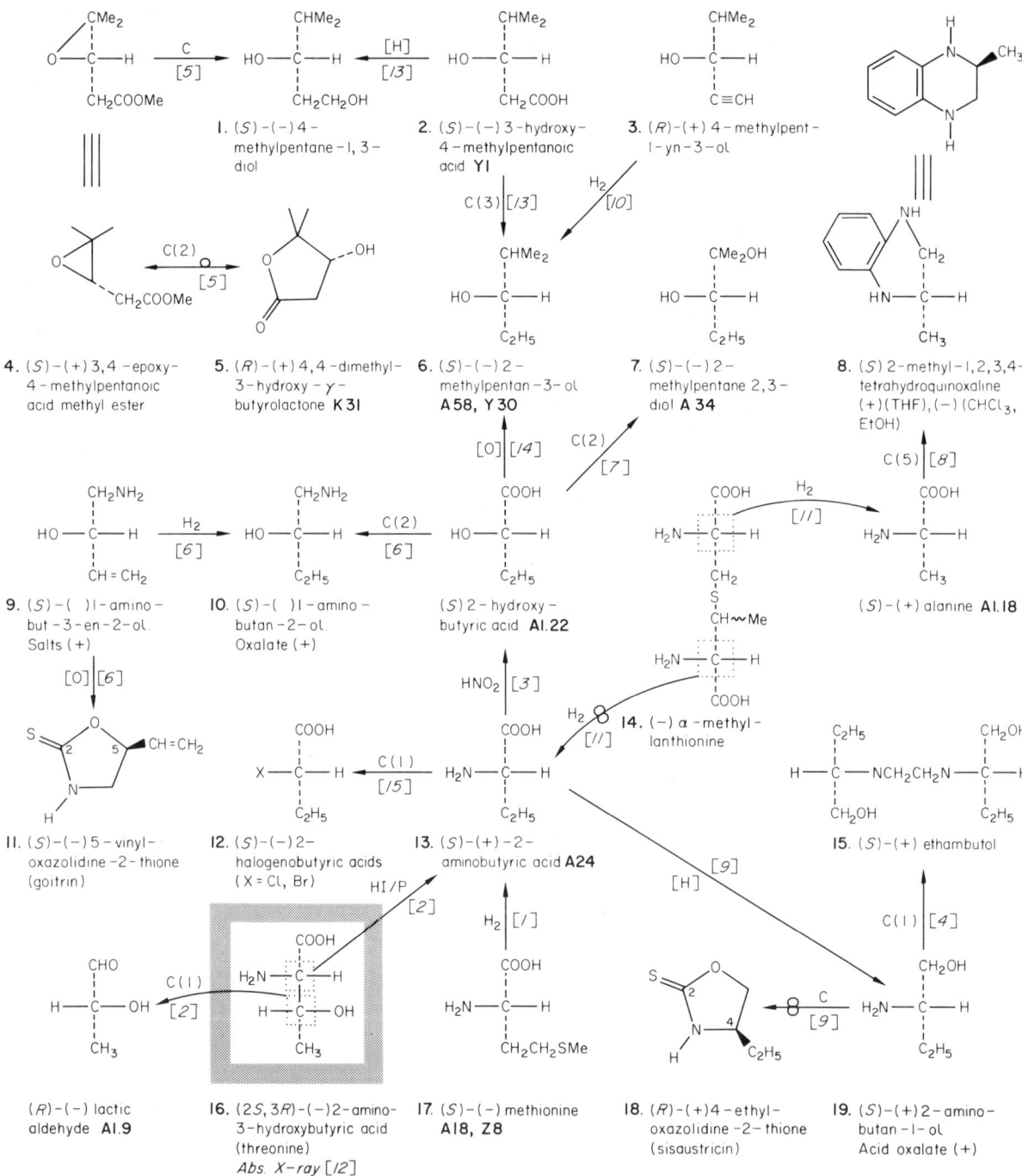

1. G. S. Fonken and R. Mozingo, *J. Amer. Chem. Soc.*, 1947, **69**, 1212.
2. C. E. Meyer and W. C. Rose, *J. Biol. Chem.*, 1936, **115**, 721.
3. C. G. Baker and A. Meister, *J. Amer. Chem. Soc.*, 1951, **73**, 1336.
4. R. G. Wilkinson, R. G. Shepherd, J. P. Thomas and C. Baughn, *J. Amer. Chem. Soc.*, 1961, **83**, 2213.
5. J. F. Collins and M.F. Grundon, *Chem. Comm.*, 1969, 1078.
6. A. Kjaer, B. W. Christensen and S. E. Hansen, *Acta Chem. Scand.*, 1959, **13**, 144.
7. D. G. Manwaring, R. W. Rickards and R. M. Smith, *Tetrahedron Letters*, 1970, 1029.
8. G. H. Fisher, P. J. Whitman and H. P. Schultz, *J. Org. Chem.*, 1970, **35**, 2240.
9. A. Kjaer and B. W. Christensen, *Acta Chem. Scand.*, 1962, **16**, 71.
10. S. R. Landor, B. J. Miller and A. R. Tatchell, *J. Chem. Soc. (C).*, 1971, 2339.
11. G. Alderton, *J. Amer. Chem. Soc.*, 1953, **75**, 2391.
12. M. Mallikarjunan, S. T. Rao, K. Venkatesan and V. R. Sarma, *Acta Cryst.*, 1969, **B25**, 220.
13. G. Büchi, L. Crombie, P. J. Godin, J. S. Kaltenbronn, K. S. Siddalingaiah and D. A. Whiting, *J. Chem. Soc.*, 1961, 2843.
14. P. A. Levene and R. E. Marker, *J. Biol. Chem.*, 1933, **101**, 413.
15. W. Gaffield and W. G. Galetto, *Tetrahedron*, 1971, **27**, 915.

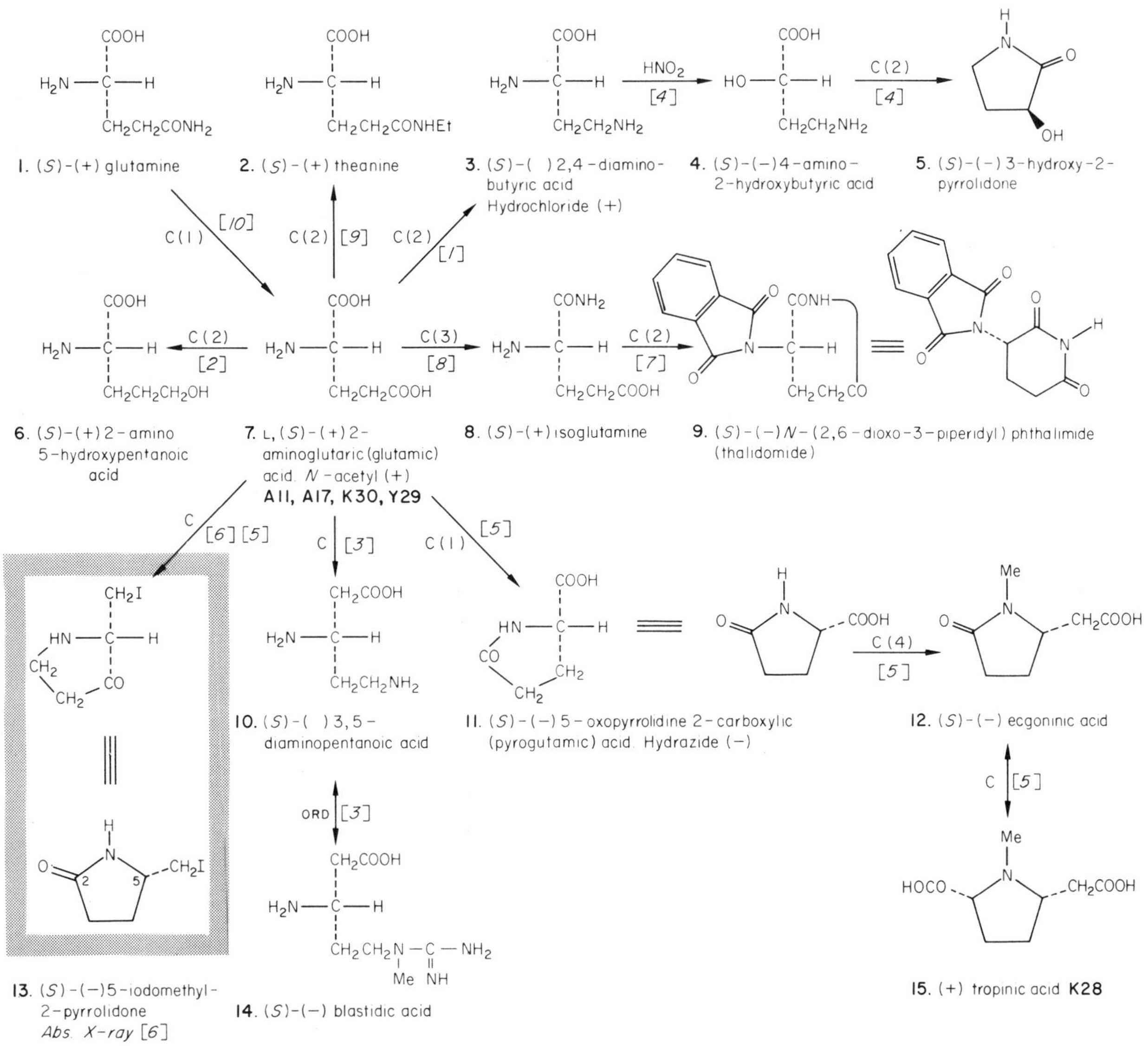

Class 2a — Glutamic acid and related compounds (including pyrrolidines) — **A⁹**

1. (S)-(+) glutamine
2. (S)-(+) theanine
3. (S)-( ) 2,4-diamino-butyric acid Hydrochloride (+)
4. (S)-(−) 4-amino-2-hydroxybutyric acid
5. (S)-(−) 3-hydroxy-2-pyrrolidone
6. (S)-(+) 2-amino 5-hydroxypentanoic acid
7. L,(S)-(+) 2-aminoglutaric (glutamic) acid. N-acetyl (+) **A11, A17, K30, Y29**
8. (S)-(+) isoglutamine
9. (S)-(−) N-(2,6-dioxo-3-piperidyl) phthalimide (thalidomide)
10. (S)-( ) 3,5-diaminopentanoic acid
11. (S)-(−) 5-oxopyrrolidine 2-carboxylic (pyroglutamic) acid. Hydrazide (−)
12. (S)-(−) ecgoninic acid
13. (S)-(−) 5-iodomethyl-2-pyrrolidone Abs. X-ray [6]
14. (S)-(−) blastidic acid
15. (+) tropinic acid **K28**

1. H. Paulus and E. Gray, *J. Biol. Chem.*, 1964, **239**, 865.
2. J. F. Thompson, C. J. Morris and G. E. Hunt, *J. Biol. Chem.*, 1964, **239**, 1122.
3. H. Yonehara and N. Otake, *Tetrahedron Letters*, 1966, 3785.
4. P. W. K. Woo, H. W. Dion and Q. R. Bartz, *Tetrahedron Letters*, 1971, 2617.
5. E. Hardegger and H. Ott, *Helv. Chim. Acta*, 1955, **38**, 312.
6. J. A. Molin-Case, E. Fleischer and D. W. Urry, *J. Amer. Chem. Soc.*, 1971, **92**, 4728.
7. Y. F. Shealy, C. E. Opliger and J. A. Montgomery, *Chem. and Ind.*, 1965, 1030.
8. C. Ressler, *J. Amer. Chem. Soc.*, 1960, **82**, 1641.
9. N. Lichtenstein, *J. Amer. Chem. Soc.*, 1942, **64**, 1021.
10. E. Schulze and G. Trier, *Ber.*, 1912, **45**, 257.

A10 Aspartic acid, lysine and pipecolic acid and derivatives — Class 2a

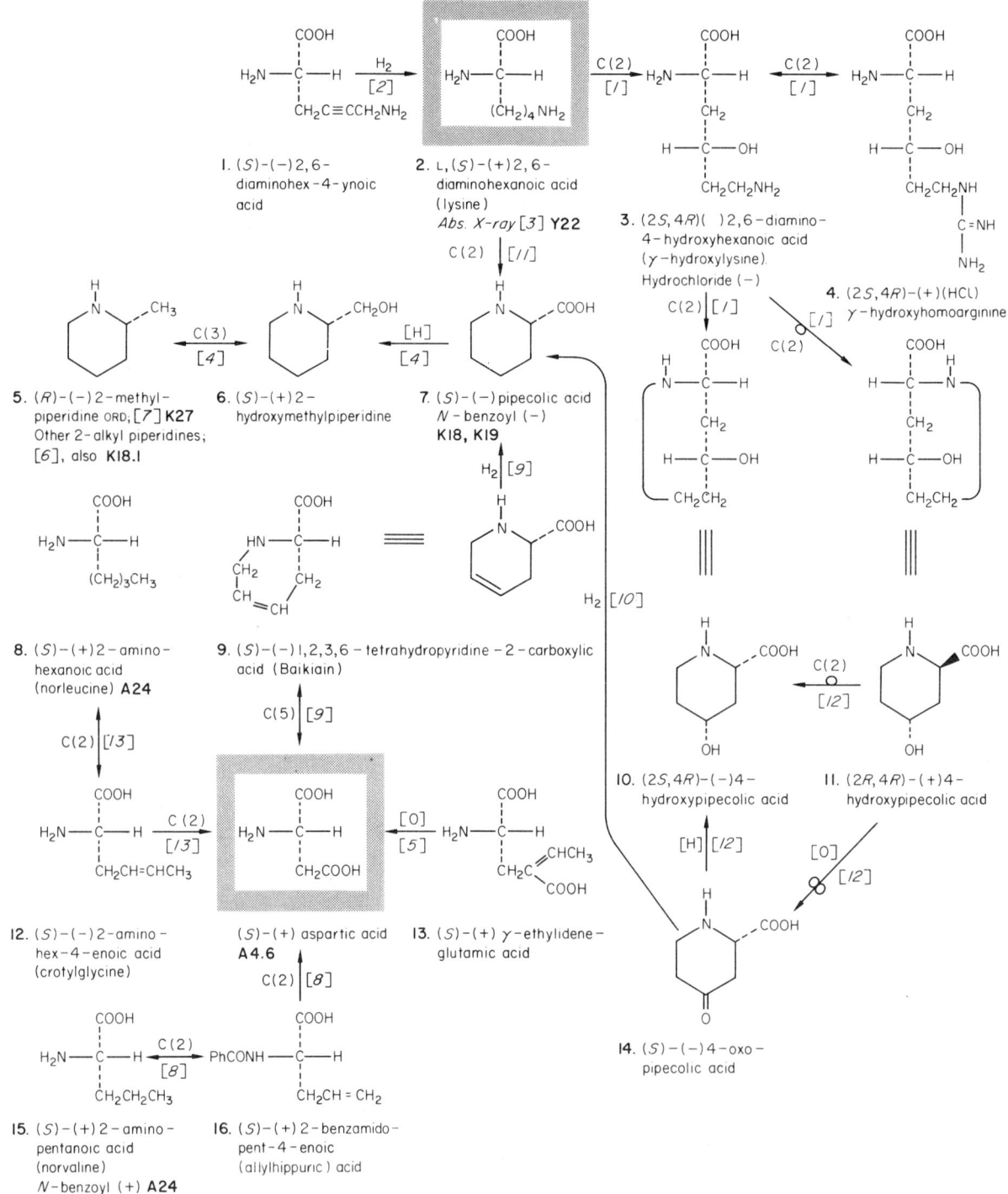

1. Y. Fujita, J. Kollonitsch and B. Witkop, *J. Amer. Chem. Soc.*, 1965, 87, 2030.
2. A. C. A. Jansen, K. E. T. Kerling and E. Havinga, *Rec. Trav. chim.*, 1970, 89, 861.
3. S. Raman, *Z. Kristallogr. Kristallgeom.*, 1959, 111, 301.
4. H. Ripperger and K. Schreiber, *Tetrahedron*, 1965, 21, 1485.
5. L. Fowden, *Biochem. J.*, 1966, 98, 57.
6. J. F. Archer, D. R. Boyd, W. R. Jackson, M. F. Grundon and W. A. Khan, *J. Chem. Soc. (C)*, 1971, 2560; H. C. Beyerman, S. van den Bosch, J. H. Breuker and L. Maat, *Rec. Trav. chim.*, 1971, 90, 755.
7. J. C. Craig and S. K. Roy, *Tetrahedron*, 1965, 21, 401.
8. P. Karrer and H. Schneider, *Helv. Chim. Acta*, 1930, 13, 1281.
9. F. E. King, T. J. King and A. J. Warwick, *J. Chem. Soc.*, 1950, 3590.
10. F. W. Eastwood, B. K. Snell and A. Todd, *J. Chem. Soc.*, 1960, 2286.
11. U. Schiedt and H. G. Höss, *Z. physiol. Chem.*, 1957, 308, 179.
12. J. W. Clark-Lewis and P. I. Mortimer, *J. Chem. Soc.*, 1961, 189.
13. P. Karrer and V. Itschner, *Helv. Chim. Acta*, 1935, 18, 782.

Class 2a
Compounds related to arginine and glutamic acid

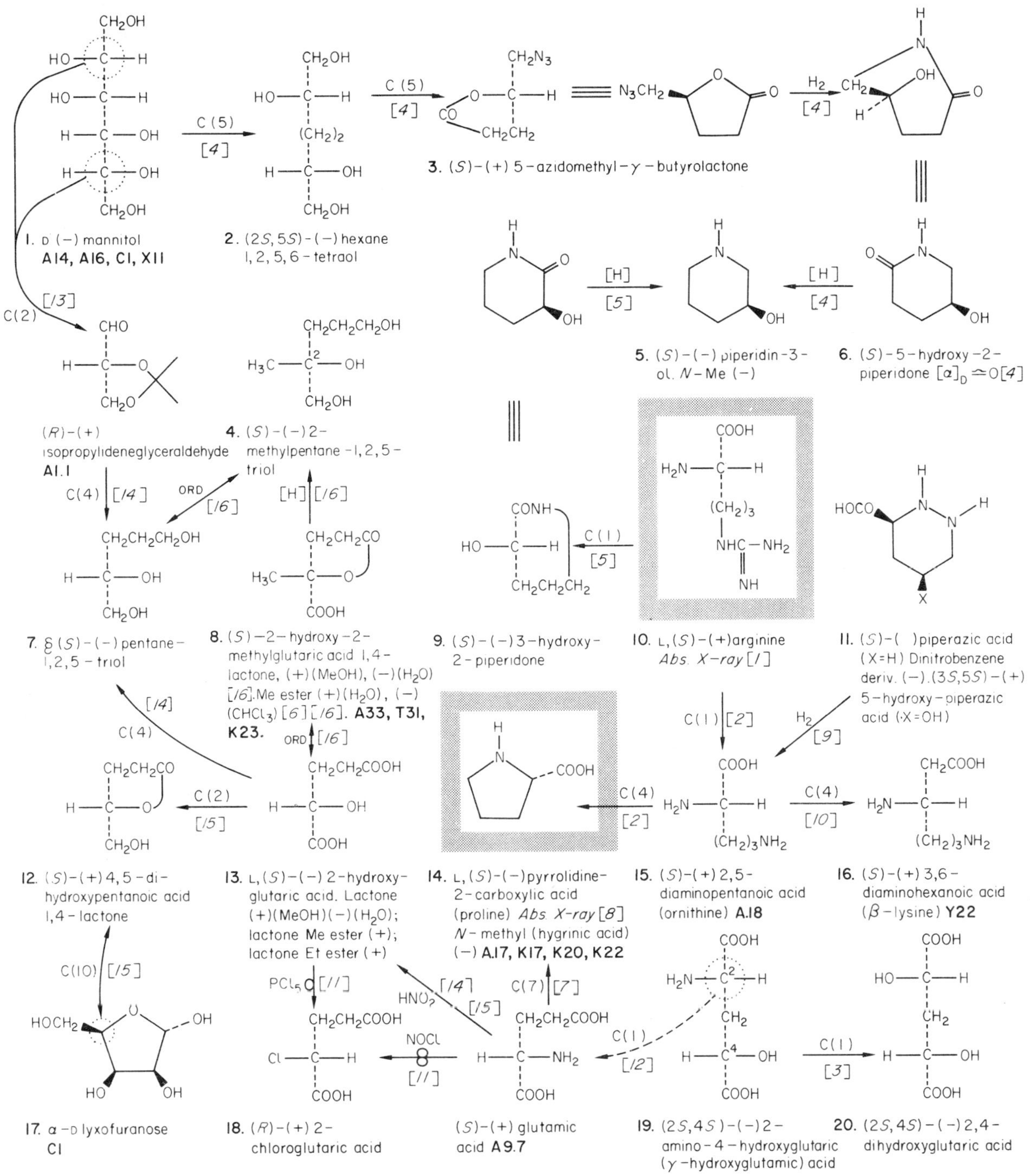

A11

1. D(−) mannitol A14, A16, C1, XII
2. (2S,5S)-(−) hexane 1,2,5,6-tetraol
3. (S)-(+) 5-azidomethyl-γ-butyrolactone
4. (S)-(−) 2-methylpentane-1,2,5-triol
5. (S)-(−) piperidin-3-ol, N-Me (−)
6. (S)-5-hydroxy-2-piperidone $[\alpha]_D \simeq 0$ [4]
7. (S)-(−) pentane-1,2,5-triol
8. (S)-2-hydroxy-2-methylglutaric acid 1,4-lactone, (+)(MeOH), (−)(H_2O) [6].Me ester (+)(H_2O), (−)($CHCl_3$) [6] [/6]. A33, T31, K23.   ORD [/6]
9. (S)-(−) 3-hydroxy-2-piperidone
10. L,(S)-(+) arginine Abs. X-ray [/]
11. (S)-(−) piperazic acid (X=H) Dinitrobenzene deriv. (−).(3S,5S)-(+) 5-hydroxy-piperazic acid (·X=OH)
12. (S)-(+) 4,5-dihydroxypentanoic acid 1,4-lactone
13. L,(S)-(−) 2-hydroxyglutaric acid. Lactone (+)(MeOH)(−)(H_2O); lactone Me ester (+); lactone Et ester (+)
14. L,(S)-(−) pyrrolidine-2-carboxylic acid (proline) Abs. X-ray [8] N-methyl (hygrinic acid) (−) A.17, K17, K20, K22
15. (S)-(+) 2,5-diaminopentanoic acid (ornithine) A.18
16. (S)-(+) 3,6-diaminohexanoic acid (β-lysine) Y22
17. α-D lyxofuranose C1
18. (R)-(+) 2-chloroglutaric acid
    (S)-(+) glutamic acid A9.7
19. (2S,4S)-(−) 2-amino-4-hydroxyglutaric acid (γ-hydroxyglutamic) acid
20. (2S,4S)-(−) 2,4-dihydroxyglutaric acid

(R)-(+) isopropylideneglyceraldehyde A1.1

1. A. Zalkin, J. D. Forrester and D. H. Templeton, *Science*, 1964, 146, 261.
2. P. Karrer and M. Ehrenstein, *Helv. Chim. Acta*, 1926, 9, 323.
3. L. Benoiton, M. Winitz, S. M. Birnbaum and J. P. Greenstein, *J. Amer. Chem. Soc.*, 1957, 79, 6192.
4. C. C. Deane and T. D. Inch, *Chem. Comm.*, 1969, 813.
5. R. E. Lyle and C. K. Spicer, *Tetrahedron Letters*, 1970, 1133.
6. S. Lindstedt and G. Linstedt, *J. Org. Chem.*, 1963, 28, 251.
7. Z. Pravda and J. Rudinger, *Coll. Czech. Chem. Comm.*, 1955, 20, 1.
8. D. A. Buckingham, L. G. Marzilli, I. E. Maxwell, A. M. Sargeson and H. C. Freeman, *Chem. Comm.*, 1969, 583.
9. K. Bevan, J. S. Davies, C. H. Hassall, R. B. Morton and D. A. S. Phillips, *J. Chem. Soc. (C)*, 1971, 514.
10. H. E. Carter, W. R. Hearn, E. M. Lansford, A. C. Page, N. P. Salzman, D. Shapiro, and W. R. Taylor, *J. Amer. Chem. Soc.*, 1952, 74, 3704; E. E. Van Tamelen and E. E. Smissman, *ibid.*, 3713.
11. W. A. Cowdrey, E. D. Hughes, C. K. Ingold, S. Masterman and A. D. Scott, *J. Chem. Soc.*, 1937, 1252.
12. A. I. Virtanen and P. K. Hietala, *Acta Chem. Scand.*, 1955, 9, 175.
13. E. Baer and H. O. L. Fischer, *J. Amer. Chem. Soc.*, 1948, 70, 609.
14. R. Kuhn and R. Brossmer, *Angew. Chem.*, 1962, 74, 252.
15. K. Koga, M. Taniguchi and S. Yamada, *Tetrahedron Letters*, 1971, 263.
16. O. Cervinka and L. Hub, *Coll. Czech. Chem. Comm.* 1968, 33, 2927.

A[12] Class 2a
Lactic acid, octan-2-ol and related compounds

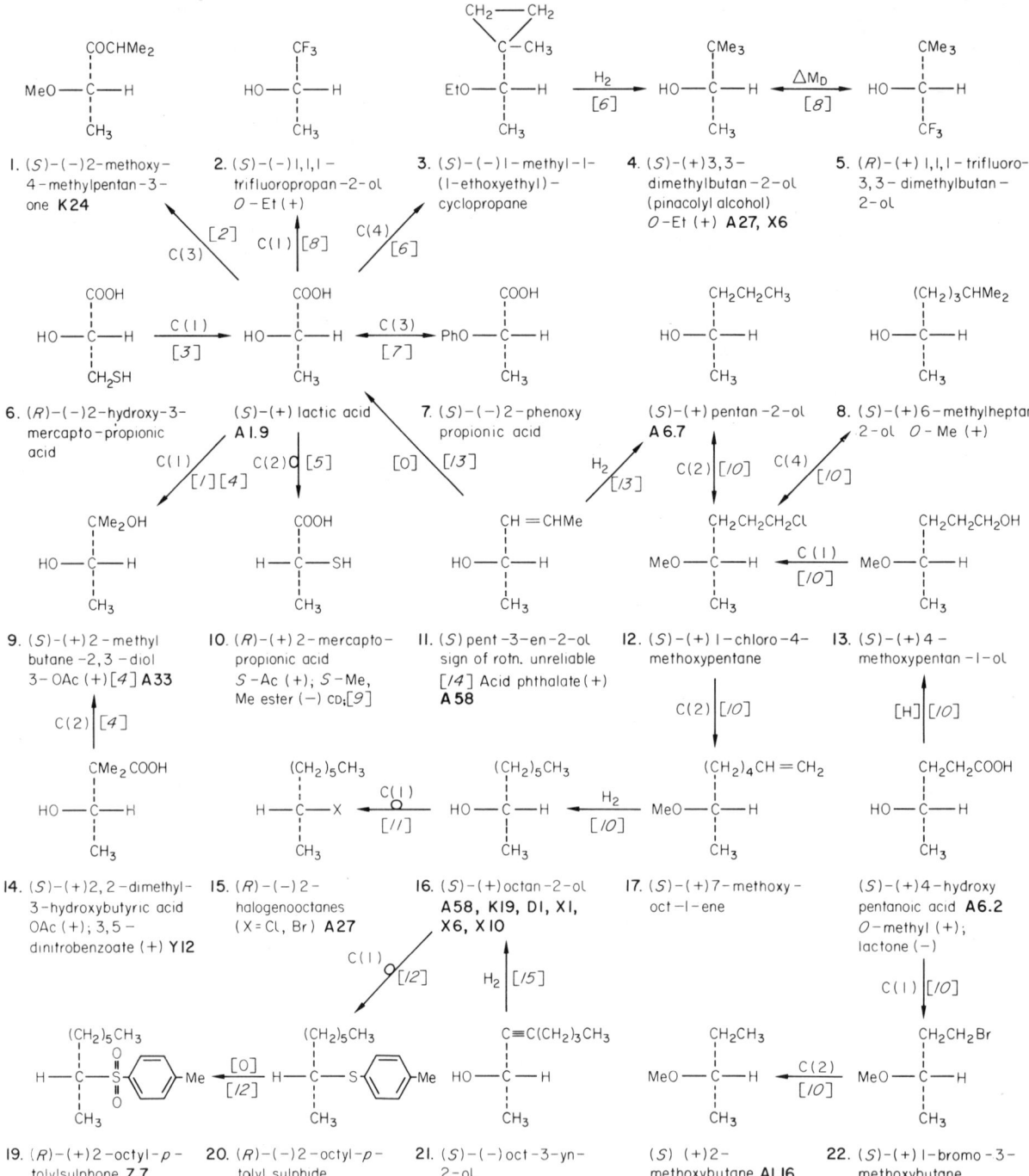

1. B. W. Christensen and A. Kjaer, *Proc. Chem. Soc.*, 1962, 307.
2. N. K. Kochetkov, A. M. Likhosherstov and V. N. Kulakov, *Tetrahedron*, 1969, 25, 2313.
3. D. B. Hope and M. Wälti, *J. Chem. Soc. (C)*, 1970, 2475.
4. J. S. Brooks and G. A. Morrison, *Chem. Comm.*, 1971, 1359.
5. L. N. Owen and M. B. Rahman, *J. Chem. Soc. (C)*, 1971, 2432.
6. J. Jacobus, Z. Majerski, K. Mislow and P. von R. Schleyer, *J. Amer. Chem. Soc.*, 1969, 91, 1998.
7. M. Matell, *Arkiv Kemi*, 1953, 6, 355.
8. H. M. Peters, D. M. Feigl and H. S. Mosher, *J. Org. Chem.*, 1968, 33, 4245.
9. P. M. Scopes, R. N. Thomas and M. B. Rahman, *J. Chem. Soc. (C)*, 1971, 1671.
10. W. von E. Doering and R. W. Young, *J. Amer. Chem. Soc.*, 1952, 74, 2997; P. A. Levene and H. L. Haller, *J. Biol. Chem.*, 1927, 72, 591.
11. W. A. Cowdrey, E. D. Hughes, C. K. Ingold, S. Masterman and A. D. Scott, *J. Chem. Soc.*, 1937, 1252.
12. K. Mislow, M. M. Green, P. Laur, J. T. Melillo, T. Simmons and A. L. Ternay, *J. Amer. Chem. Soc.*, 1965, 87, 1958.
13. P. A. Levene and H. L. Haller, *J. Biol. Chem.*, 1929, 81, 703.
14. H. L. Goering and W. I. Kimoto, *J. Amer. Chem. Soc.*, 1965, 87, 1748.
15. S. R. Landor, B. J. Miller and A. R. Tatchell, *J. Chem. Soc. (C)*, 1971, 2339.

Class 2a — One- and two-centre compounds related to butan-2-ol and butane-2,3-diol

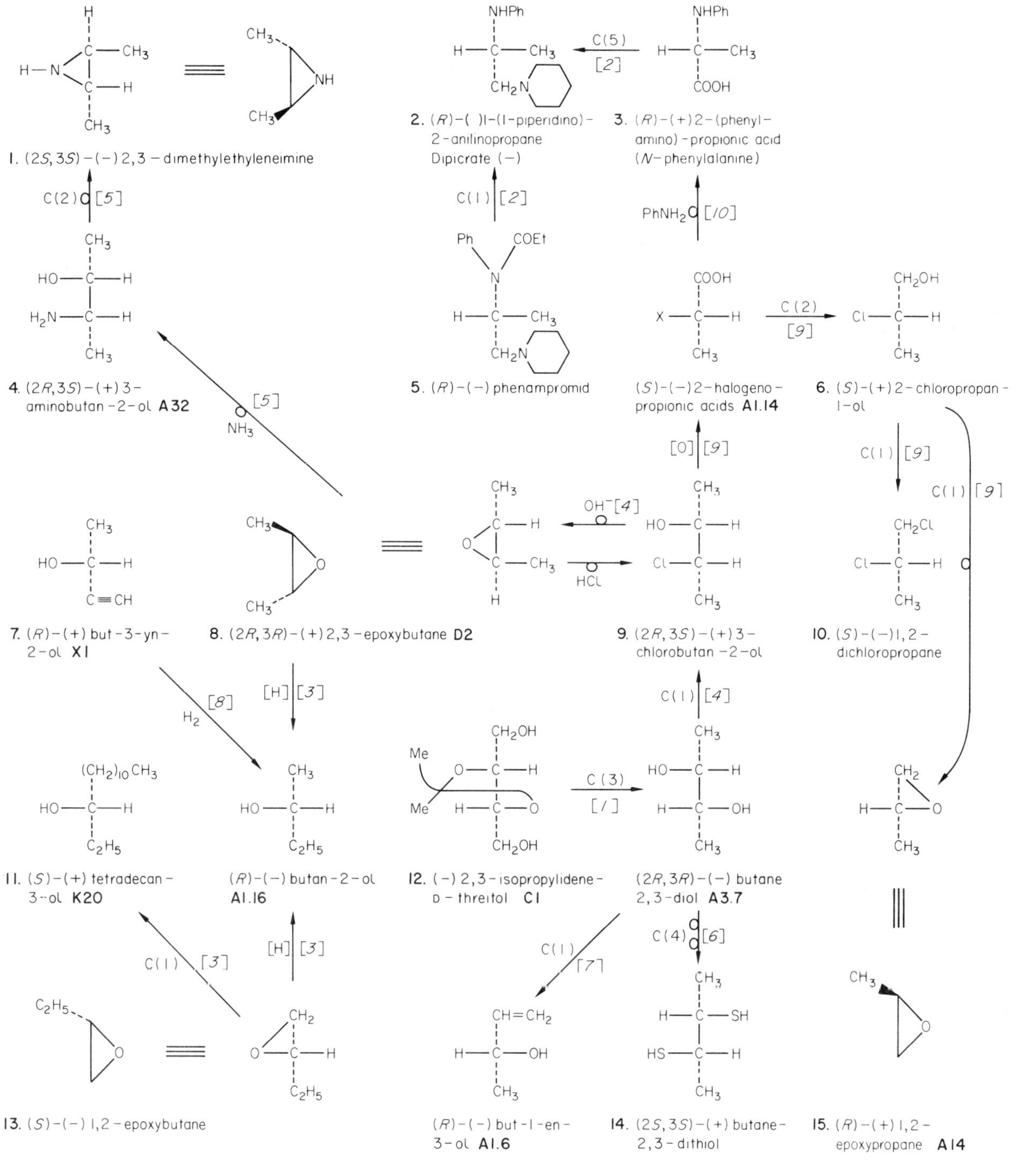

1. (2S,3S)-(−) 2,3-dimethylethyleneimine
2. (R)-(−)1-(1-piperidino)-2-anilinopropane Dipicrate (−)
3. (R)-(+)2-(phenyl-amino)-propionic acid (N-phenylalanine)
4. (2R,3S)-(+)3-aminobutan-2-ol **A32**
5. (R)-(−) phenampromid
6. (S)-(+)2-chloropropan-1-ol
7. (R)-(+) but-3-yn-2-ol **X1**
8. (2R,3R)-(+)2,3-epoxybutane **D2**
9. (2R,3S)-(+)3-chlorobutan-2-ol
10. (S)-(−)1,2-dichloropropane
11. (S)-(+) tetradecan-3-ol **K20**
12. (−) 2,3-isopropylidene-D-threitol **C1**
13. (S)-(−) 1,2-epoxybutane
14. (2S,3S)-(+) butane-2,3-dithiol
15. (R)-(+) 1,2-epoxypropane **A14**

(S)-(−)2-halogeno-propionic acids **A1.14**

(2R,3R)-(−) butane 2,3-diol **A3.7**

(R)-(−) but-1-en-3-ol **A1.6**

---

1. L. J. Rubin, H. A. Lardy and H. O. L. Fischer, *J. Amer. Chem. Soc.*, 1952, 74, 425.
2. P. S. Portoghese, *Chem. and Ind.*, 1964, 582.
3. J. L. Coke and W. Y. Rice, *J. Org. Chem.*, 1965, 30, 3420.
4. P. J. Leroux and H. J. Lucas, *J. Amer. Chem. Soc.*, 1951, 73, 41.
5. F. H. Dickey, W. Fickett and H. J. Lucas, *J. Amer. Chem. Soc.*, 1952, 74, 944.
6. E. J. Corey and R. B. Mitra, *J. Amer. Chem. Soc.*, 1962, 84, 2939.
7. S. A. Morell and A. H. Auernheimer, *J. Amer. Chem. Soc.*, 1944, 66, 792.
8. S. R. Landor, B. J. Miller and A. R. Tatchell, *J. Chem. Soc. (C)*, 1971, 2339.
9. W. Fickett, H. K. Garner and H. J. Lucas, *J. Amer. Chem. Soc.*, 1951, 73, 5063.
10. P. S. Portoghese, *J. Pharm. Sci.*, 1964, 53, 229.

A[14] Glycerides and other one-centre compounds derived from propane-1,2-diol — Class 2a

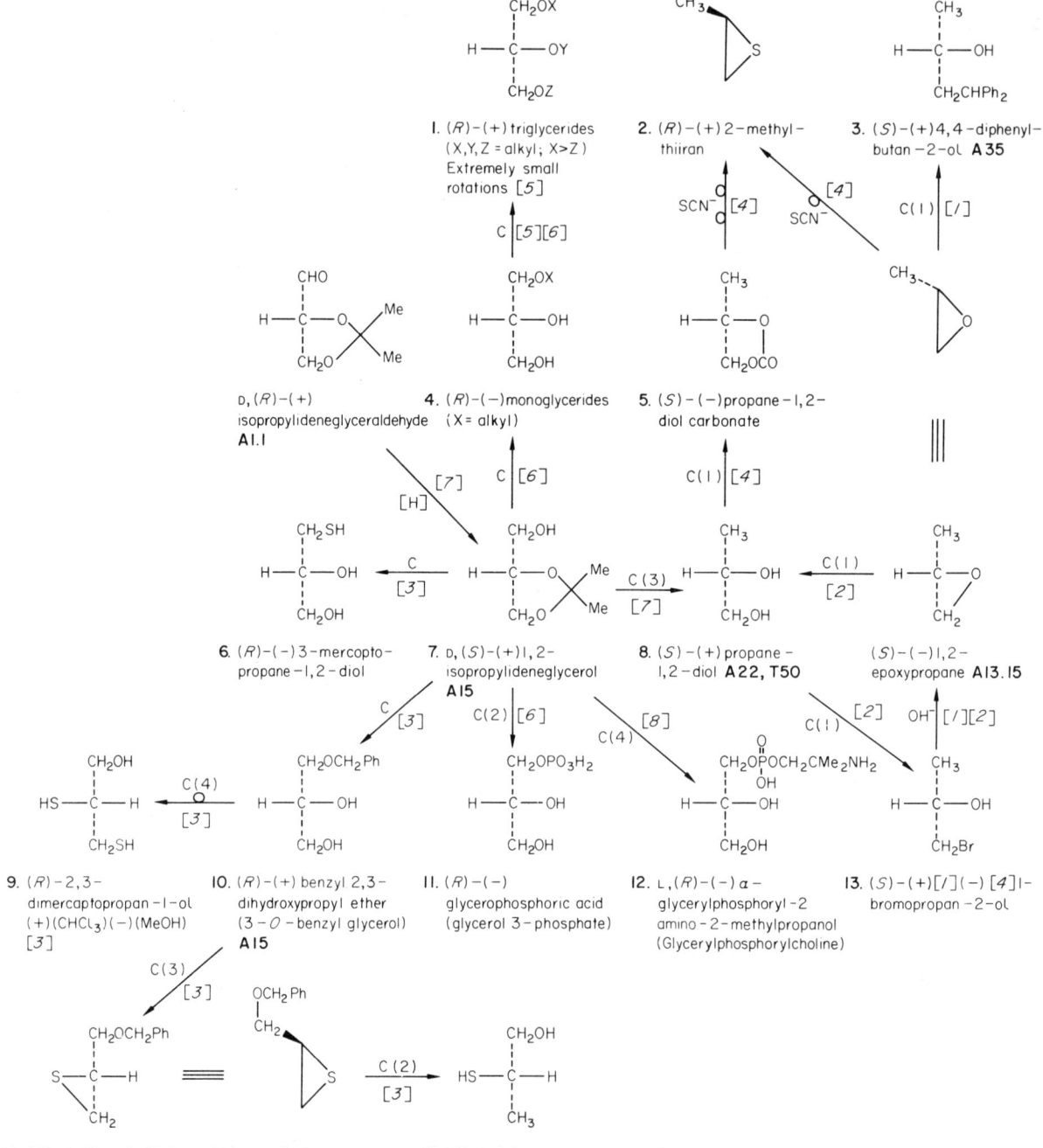

1. (R)-(+) triglycerides (X,Y,Z = alkyl; X>Z) Extremely small rotations [5]
2. (R)-(+) 2-methyl-thiiran
3. (S)-(+) 4,4-diphenyl-butan-2-ol **A 35**
4. (R)-(–) monoglycerides (X = alkyl)
5. (S)-(–) propane-1,2-diol carbonate
6. (R)-(–) 3-mercapto-propane-1,2-diol
7. D,(S)-(+)1,2-isopropylideneglycerol **A15**
8. (S)-(+) propane-1,2-diol **A22, T50**
   (S)-(–)1,2-epoxypropane **A13.15**
9. (R)-2,3-dimercaptopropan-1-ol (+)(CHCl₃)(–)(MeOH) [3]
10. (R)-(+) benzyl 2,3-dihydroxypropyl ether (3-O-benzyl glycerol) **A15**
11. (R)-(–) glycerophosphoric acid (glycerol 3-phosphate)
12. L,(R)-(–) α-glycerylphosphoryl-2-amino-2-methylpropanol (Glycerylphosphorylcholine)
13. (S)-(+)[/](–)[4]1-bromopropan-2-ol
14. (R)-(+) benzyl (2,3-epithiopropyl) ether
15. (S)-(+) 2-mercaptopropan-1-ol

D,(R)-(+) isopropylideneglyceraldehyde **A1.1**

---

1. H. M. Walborsky and C. G. Pitt, *J. Amer. Chem. Soc.*, 1962, **84**, 4831.
2. P. A. Levene and A. Walti, *J. Biol. Chem.*, 1926, **68**, 415.
3. A. K. M. Anisuzzaman and L. N. Owen, *J. Chem. Soc. (C)*, 1967, 1021.
4. N. Spassky and P. Sigwalt, *Tetrahedron Letters*, 1968, 3541.
5. W. Schlenk, *J. Amer. Oil Chemists' Soc.*, 1965, **42**, 945.
6. H. O. L. Fischer and E. Baer, *Chem. Rev.*, 1941, **29**, 287.
7. E. Baer and H. O. L. Fischer, *J. Amer. Chem. Soc.*, 1948, **70**, 609.
8. E. Baer and G. V. Rao, *Canad. J. Biochem. Physiol.*, 1964, **42**, 1547.

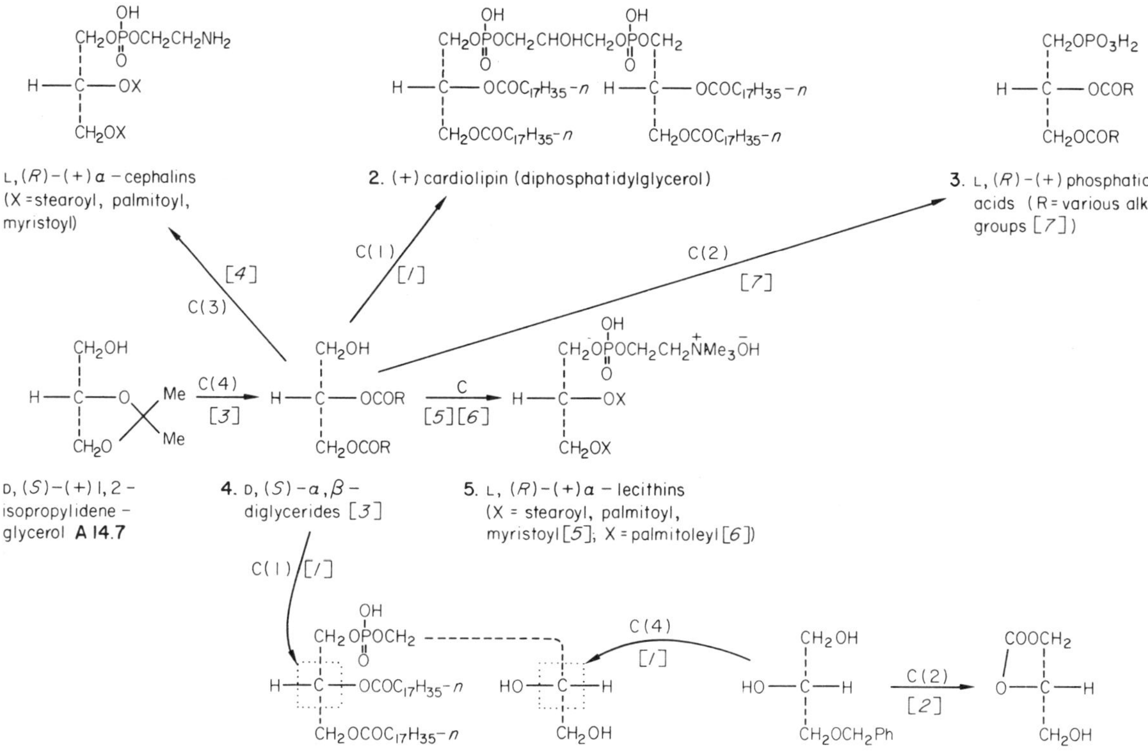

Class 2a
(i) Further glycerides (ii) Quinolizidines

A[15]

1. L,(R)-(+)α-cephalins (X = stearoyl, palmitoyl, myristoyl)

2. (+) cardiolipin (diphosphatidylglycerol)

3. L,(R)-(+) phosphatidic acids (R = various alkyl groups [7])

D,(S)-(+)1,2-isopropylidene-glycerol A14.7

4. D,(S)-α,β-diglycerides [3]

5. L,(R)-(+)α-lecithins (X = stearoyl, palmitoyl, myristoyl[5]; X = palmitoleyl[6])

6. (+) phosphatidylglycerol

(R)-(+) 3-O benzyl-glycerol A14.10

7. (S)-(-)glycerol-1,2-carbonate

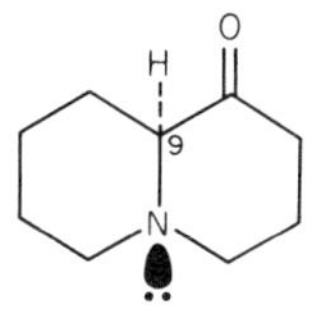

8. (9S)-(-)1-oxo-quinolizidine. By CD [8] (octant rule)

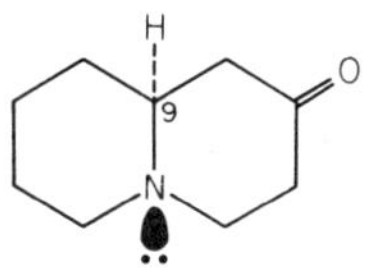

9. (9S)-(+)2-oxo-quinolizidine. By CD [8] (octant rule)

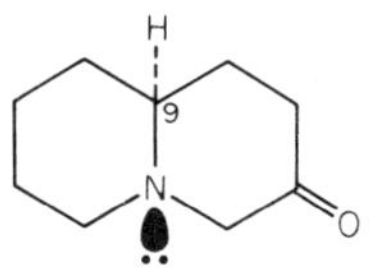

10. (9S)-(+)3-oxo-quinolizidine. By CD [8] (octant rule)

1. R. M. Saunders and H. P. Schwarz, *J. Amer. Chem. Soc.*, 1966, **88**, 3844.
2. J. Gigg and R. Gigg, *J. Chem. Soc.*, 1967, 1865.
3. J. C. Sowden and H. O. L. Fischer, *J. Amer. Chem. Soc.*, 1941, **63**, 3244.
4. E. Baer, J. Marukas and M. Russell, *Science*, 1951, **113**, 12.
5. E. Baer and M. Kates, *J. Amer. Chem. Soc.*, 1950, **72**, 942.
6. D. J. Hanahan and M. E. Jayko, *J. Amer. Chem. Soc.*, 1952, **74**, 5070.
7. E. Baer, *J. Biol. Chem.*, 1951, **189**, 235.
8. S. F. Mason, K. Schofield and R. J. Wells, *J. Chem. Soc. (C)*, 1967, 626.

Cyclitols and related compounds. Reviews [9] (see also page A.26)

Class 2a

Structures:

1. (1S,2R,3R,4R,5R,6S)-(−)-cyclohexane 1,2,3,4,5,6-hexaol (inositol)†
2. (+)pinitol
3. (5R,6R)-(−)-2-chlorocyclohexa-1,3-diene-5,6-diol
4. (1S,2S,3R,4S,5R)-(−)-cyclohexane-1,2,3,4,5-pentaol (*gala*-quercitol)
5. (1S,2S,4S,5S)-(+)-cyclohexane-1,2,3,4,5-pentaol (*talo*-quercitol)†
6. (2S,3R,4S,5S,6S)-(−)-2,3,4,5,6-pentahydroxycyclohexanone (*vibo*-inosose)
7. (1R,2S,4S,5R)-(+)-cyclohexane-1,2,3,4,5-pentaol (*proto*-quercitol)
8. (5R,6R)-(−)-cyclohexa-1,3-diene-5,6-diol
9. (R)-(−)-2-hydroxycyclohexanone
10. (2S,3R,4S,5R)-( )-2,3,4,5-tetrahydroxycyclohexanone
11. (1R,2S,3R,4R,5R)-(−)-cyclohexane-1,2,3,4,5-pentaol (viburnitol, *vibo*-quercitol)
12. (2S,3R,5R)-( )-2,3,5-trihydroxyadipic acid
13. (1R,2R)-(−)-cyclohexane-1,2-diol
14. (1R,3R)-(−)-cyclohexane-1,2,3-triol
15. (1R,2S,3S,4R)-(−)-cyclohexane-1,2,3,4-tetraol
16. (1R,2R,3S,4R)-(−)-cyclohexane-1,2,3,4-tetraol
17. D(+)galactose C1, Y21
18. (3R,4R)-(−)-cyclohex-1-ene-3,4-diol
19. (1S,2S,3S,4R)-(+)-cyclohexane-1,2,3,4-tetraol
20. (2R,3S,4R)-( )-2,3,4-trihydroxycyclohexanone
21. (2S,3S,4R)-( )-2,3,4-trihydroxycyclohexanone
22. (3S,4R)-( )-3,4-dihydroxycyclohexane-1,2-dione. Phenylosazone (−)

L(+)mannitol A11.1

(2S,3S)-(−)-2,3-dimethoxysuccinic acid bis-methylamide A3.8

† For application of (R,S) nomenclature to these compounds see R. S. Cahn, C. K. Ingold and V. Prelog, *Angew. Chem. Internat. Edn.*, 1966, 5, 385. The 'fractional' system of stereochemical nomenclature is frequently used in naming cyclitols, and is often more convenient than (R,S) nomenclature; the latter has been used here for consistency. For a description of the fractional systems, see [9].

The following series of compounds closely related to the cyclitols, have also been studied in detail.
Cycloses (ketones derived from cyclitols by oxidation).
Amino analogues (M. Hichens & K. L. Rinehart, *J. Amer. Chem. Soc.*, 1963, 85, 1547).
Halogeno derivatives (G. E. McCasland, S. Furuta and V. Bartuska, *J. Org. Chem.*, 1963, 28, 2096).
Ethers (S. J. Angyal & L. Anderson, *Adv. Carbohydrate Chem.*, 1959, 14, 135).—See also [9].
Many other cyclitols and derived compounds are themselves achiral, but are capable of giving rise to chiral derivatives when suitably substituted. One of the best studied examples is deoxystreptamine, which occurs as substitution products in antibiotics of the Kananycin-neomycin type. See p. C3 and also M. Hichens and K. L. Rinehart, *J. Amer. Chem. Soc.*, 1963, 85, 1547.

1. S. J. Angyal, C. G. MacDonald and N. K. Matheson, *J. Chem. Soc.*, 1953, 3321.
2. A. B. Anderson, D. L. MacDonald and H. O. L. Fischer, *J. Amer. Chem. Soc.*, 1952, 74, 1479.
3. G. E. McCasland, S. Furuta, L. F. Johnson and J. N. Shoolery, *J. Amer. Chem. Soc.*, 1961, 83, 2335.
4. T. Posternak, *Helv. Chim. Acta*, 1932, 15, 948.
5. D. M. Jerina, H. Ziffer and J. W. Daly, *J. Amer. Chem. Soc.*, 1970, 92, 1056.
6. T. Posternak, D. Reymond and H. Friedli, *Helv. Chim. Acta*, 1955, 38, 205.
7. T. Posternak and D. Reymond, *Helv. Chim. Acta*, 1955, 38, 195.
8. T. Posternak, *Helv. Chim. Acta*, 1950, 33, 350, 1594.
9. T. Postenak, 'The Cyclitols', Holden-Day, 1965; S. J. Angyal and A. B. Anderson, *Adv. Carbohydrate Chem.*, 1959, 14, 135.

Class 2a
Proline derivatives

1. (1R,2S)-( )-1-azabicyclo-[3.1.0]hexane. Picrate (+)
2. (S)-(−) 3,4-dehydroproline **A30**
3. (2S,4S)-(−)4-hydroxymethylproline
4. (S)-( )4-methyleneproline. Benzyloxycarbonyl deriv., dicyclohexylammonium salt (−)
5. (2S,4R)-(−)4-hydroxymethylproline
6. (R)-(−)2-methylpyrrolidine
7. (S)-(+)-2-hydroxymethylpyrrolidine (prolinol)
   (S)-(−) proline **A11.14**
8. (S)-(−)4-oxo-proline **Y21**
9. (2S,4S)-(−)4-hydroxypyrrolidine 2-carboxylic acid (*allo*-hydroxyproline)
10. (2S,4S)-(−)4-(methylthio)-pyrrolidine 2-carboxylic acid
11. (2S,4R)-(−)4-hydroxypyrrolidine 2-carboxylic acid (hydroxyproline) Rel. X-ray [6] N-Me (−)
    (R)-(+) methoxysuccinic acid **A2.12**
12. (S)-(−)(2-oxo-4-pyrrolidyl) methyl sulphone
    (S)-(−) methyl thiosuccinic acid **A2.8**
13. (2S,4S)-(−)N-tosyl-2-thia-5-azabicyclo-[2.2.1] heptane
14. (2S,5S)-( )2,5-diazabicyclo[2.2.1] heptane Di-HCl (+)

1. Y. Kaneko, *J. Chem. Soc. Japan*, 1940, **61**, 207.
2. P. Karrer, P. Portmann and M. Suter, *Helv. Chim. Acta*, 1948, **31**, 1617.
3. M. Bethell, G. W. Kenner and R. C. Sheppard, *Nature*, 1962, **194**, 864.
4. A. Neuberger, *J. Chem. Soc.*, 1945, 429.
5. A. A. Patchett and B. Witkop, *J. Amer. Chem. Soc.*, 1957, **79**, 185.
6. J. Zussman, *Acta Cryst.*, 1951, **4**, 72, 493.
7. P. Karrer and K. Erhardt, *Helv. Chim. Acta*, 1951, **34**, 2202.
8. K. G. Untch and G. A. Gibbon, *Tetrahedron Letters*, 1964, 3259.
9. A. V. Robertson and B. Witkop, *J. Amer. Chem. Soc.*, 1962, **84**, 1697.
10. S. Yamada, Y. Murakami and K. Koga, *Tetrahedron Letters*, 1968, 1501.
11. R. Buyle, *Chem. and Ind.*, 1966, 195.
12. P. S. Portoghese and A. A. Mikhail, *J. Org. Chem.*, 1966, **31**, 1059.
13. P. S. Portoghese and V. G. Telang, *Tetrahedron*, 1971, **27**, 1823.

Miscellaneous one- and two-centre compounds

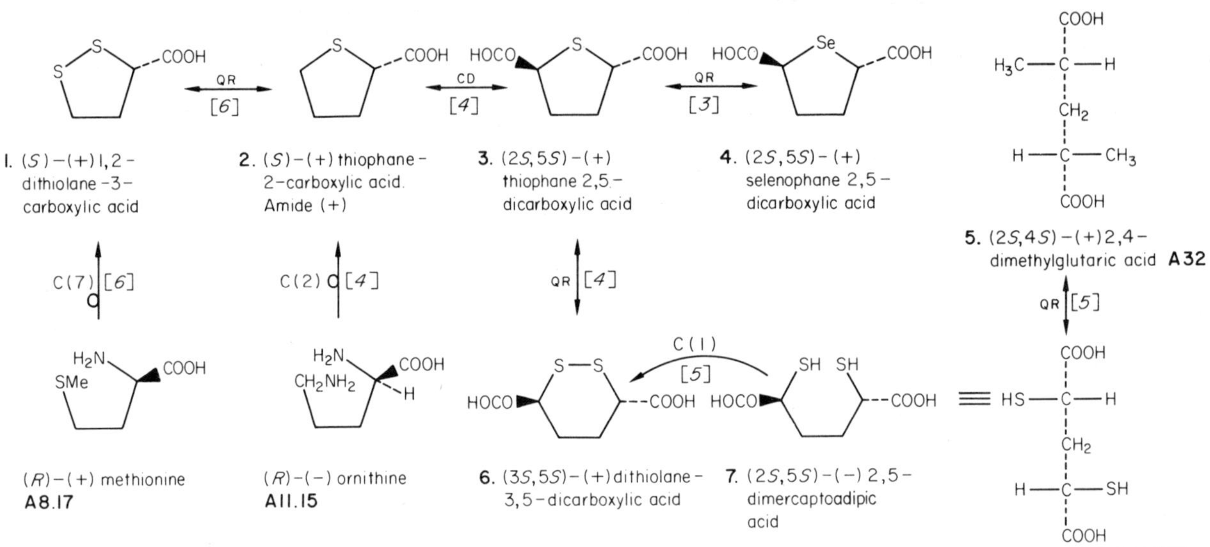

Class 2a

1. (S)-(+)1,2-dithiolane-3-carboxylic acid
2. (S)-(+) thiophane-2-carboxylic acid. Amide (+)
3. (2S,5S)-(+) thiophane 2,5-dicarboxylic acid
4. (2S,5S)-(+) selenophane 2,5-dicarboxylic acid
5. (2S,4S)-(+)2,4-dimethylglutaric acid **A 32**
6. (3S,5S)-(+) dithiolane-3,5-dicarboxylic acid
7. (2S,5S)-(−) 2,5-dimercaptoadipic acid
8. (1R,3S)-(−) 3-methoxycyclohexylamine **A 38** 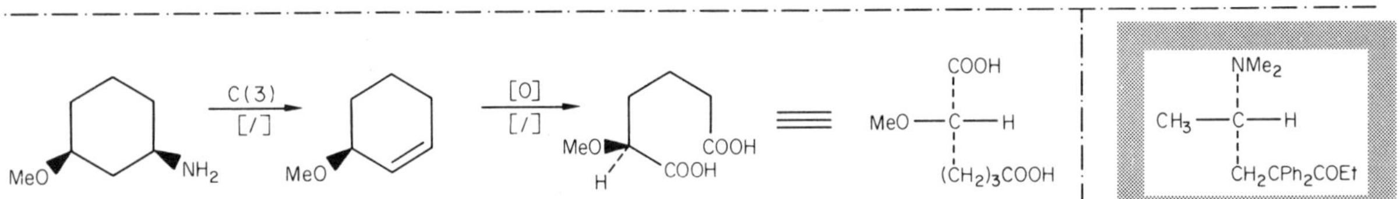
9. (S)-(−) 3-methoxycyclohexene
10. (S)-(−) 2-methoxyadipic acid
11. (S)-(+) 6-dimethylamino-4,4-diphenylheptan-3-one (methadone) Abs. X-ray [2]
12. (R)-(−)3-mercapto-octanedioic acid **A 32** 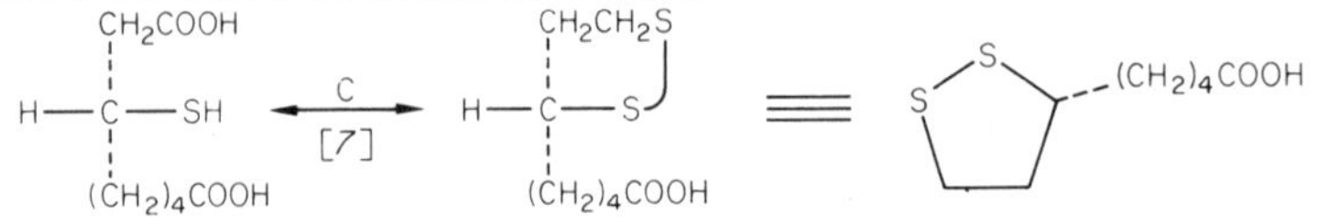
13. (R)-(+) α-lipoic acid
14. (2R,5R)-(−) 2,5-dialkyl tetrahydrofurans (R = Me, Et) 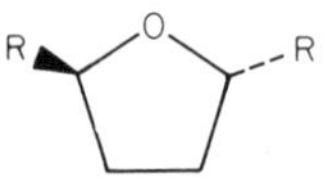
15. (1R,2R)-(−)1,2-dibromocyclohexane 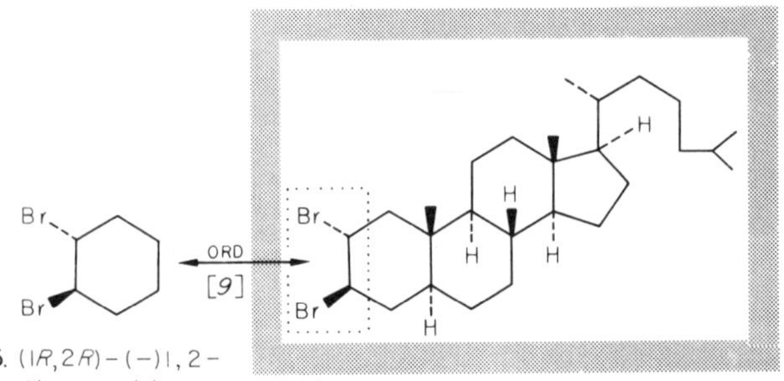
16. 2α,3β-dibromo-5α-cholestane. Abs. X-ray [10] **T 46**
17. (R)-(−) hexane-1,5-diol **Y 14**
18. (R)-(−) hexan-2-ol (R = Me) **A 58**
    (R)-(−) octan-3-ol (R = Et)

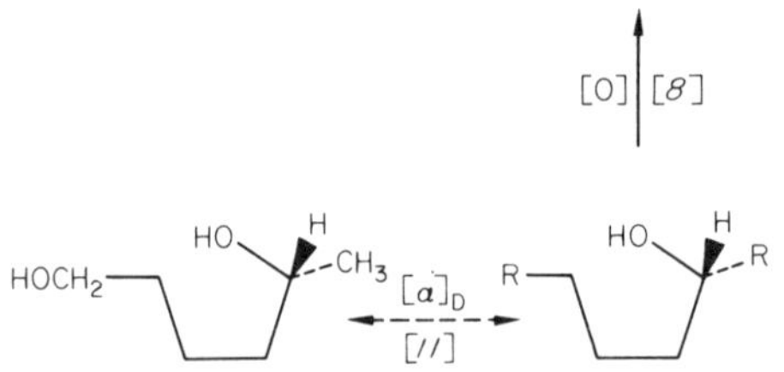

(R)-(+) methionine **A 8.17**
(R)-(−) ornithine **A 11.15**

1. D. S. Noyce and D. B. Denney, *J. Amer. Chem. Soc.*, 1954, **76**, 768.
2. A. W. Hanson and F. R. Ahmed, *Acta Cryst.*, 1958, **11**, 724.
3. A. Fredga, *Tetrahedron*, 1960, **8**, 126.
4. G. Claeson and H-G. Jonsson, *Arkiv Kemi*, 1967, **26**, 247.
5. L. Schotte, *Arkiv Kemi*, 1956, **9**, 441.
6. G. Claeson and J. Pederson, *Tetrahedron Letters*, 1968, 3975.
7. K. Mislow and W. C. Meluch, *J. Amer. Chem. Soc.*, 1956, **78**, 5920.
8. M. L. Mihailović, R. I. Mamuzić, L. Zigić-Mamuzić, J. Bosnjak and Z. Ceković, *Tetrahedron*, 1967, **23**, 215.
9. D. E. Applequist and N. D. Werner, *J. Org. Chem.*, 1963, **28**, 48.
10. H. J. Geise and C. Romers, *Rec. Trav. Chim.*, 1965, **84**, 1626.
11. H. P. Sigg, *Helv. Chim. Acta*, 1964, **47**, 1401.

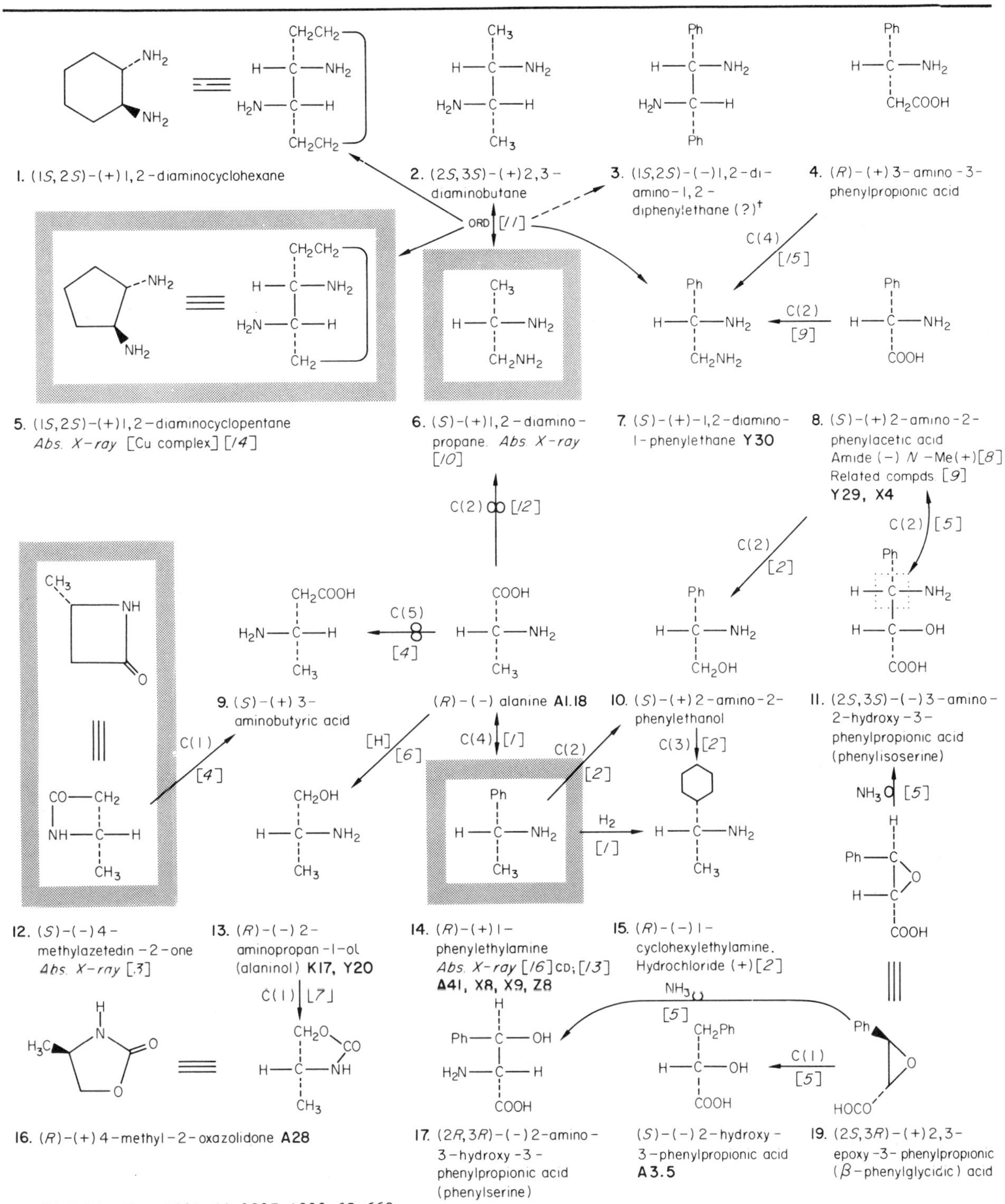

Class 2a, 2b
Aliphatic and aromatic amines and diamines

1. (1S,2S)-(+)1,2-diaminocyclohexane
2. (2S,3S)-(+)2,3-diaminobutane
3. (1S,2S)-(−)1,2-diamino-1,2-diphenylethane (?)†
4. (R)-(+)3-amino-3-phenylpropionic acid
5. (1S,2S)-(+)1,2-diaminocyclopentane Abs. X-ray [Cu complex] [14]
6. (S)-(+)1,2-diaminopropane. Abs. X-ray [10]
7. (S)-(+)-1,2-diamino-1-phenylethane Y30
8. (S)-(+)2-amino-2-phenylacetic acid Amide (−) N−Me(+)[8] Related compds [9] Y29, X4
9. (S)-(+)3-aminobutyric acid
10. (S)-(+)2-amino-2-phenylethanol
11. (2S,3S)-(−)3-amino-2-hydroxy-3-phenylpropionic acid (phenylisoserine)
12. (S)-(−)4-methylazetedin-2-one Abs. X-ray [3]
13. (R)-(−)2-aminopropan-1-ol (alaninol) K17, Y20
14. (R)-(+)1-phenylethylamine Abs. X-ray [16] CD; [13] Δ41, X8, X9, Z8
15. (R)-(−)1-cyclohexylethylamine. Hydrochloride (+)[2]
16. (R)-(+)4-methyl-2-oxazolidone A28
17. (2R,3R)-(−)2-amino-3-hydroxy-3-phenylpropionic acid (phenylserine)
   (S)-(−)2-hydroxy-3-phenylpropionic acid A3.5
19. (2S,3R)-(+)2,3-epoxy-3-phenylpropionic (β-phenylglycidic) acid

1. W. Leithe, *Ber.*, 1931, **64**, 2827; 1932, **65**, 660.
2. G. Ovakimian, M. Kuna and P. A. Levene, *J. Biol. Chem.*, 1940, **135**, 91; 1941, **137**, 337.
3. E. F. Paulus, D. Kobelt and H. Jensen, *Angew. Chem. Internat. Edn.*, 1969, **8**, 990.
4. K. Balenovic, D. Cerar and Z. Fuks, *J. Chem. Soc.*, 1952, 3316.
5. K. Harada, *J. Org. Chem.*, 1966, **31**, 1407.
6. P. Karrer, P. Portmann and M. Suter, *Helv. Chim. Acta*, 1948, **31**, 1617.
7. J. Rétey and F. Lynen, *Biochem. Z.*, 1965, **342**, 256.
8. J. C. Sheehan, H. G. Zachau and W. B. Lawson, *J. Amer. Chem. Soc.*, 1958, **80**, 3349.
9. D. G. Neilson and D. F. Ewing, *J. Chem. Soc. (C)*, 1966, 393.
10. Y. Saito and H. Iwasaki, *Bull. Chem. Soc. Japan*, 1962, **35**, 1131.
11. R. D. Gillard, *Tetrahedron*, 1965, **21**, 503.
12. S. Schnell and P. Karrer, *Helv. Chim. Acta*, 1955, **38**, 2036.
13. J. C. Craig, R. P. K. Chan and S. K. Roy, *Tetrahedron*, 1967, **23**, 3573.
14. M. Itô, F. Marumo and Y. Saito, *Acta Cryst.*, 1971, **B27**, 2187.
15. F. Bergel and J. Butler, *J. Chem. Soc.*, 1961, 4047.
16. M. A. Bush, T. A. Dullforce and G. A. Sim, *Chem. Comm.*, 1969, 1491.

Leucine, histidine, tryptophan and derivatives

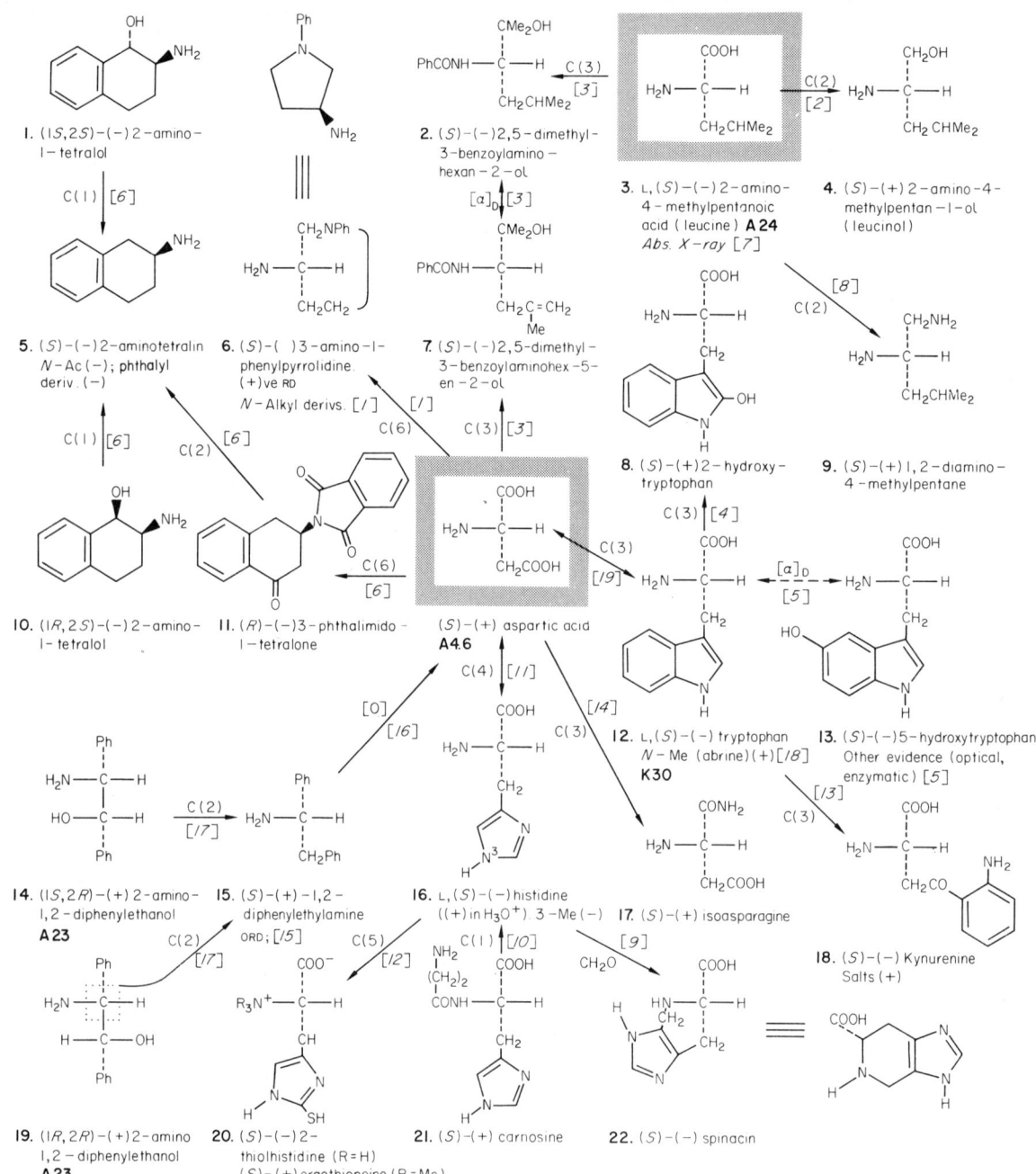

1. D. T. Witiak, Z. Muhi-Eldeen, N. Mahishi, O. P. Sethi and M. C. Gerald, *J. Medicin. Pharmaceut. Chem.*, 1971, **14**, 24.
2. P. Karrer, P. Portmann and M. Suter, *Helv. Chim. Acta*, 1948, **31**, 1617.
3. P. Karrer, W. Jäggi and T. Takahashi, *Helv. Chim. Acta*, 1925, **8**, 360.
4. T. Wieland, O. Weiberg, W. Dilger and E. Fischer, *Ann.*, 1955, **592**, 69.
5. A. J. Morris and M. D. Armstrong, *J. Org. Chem.*, 1957, **22**, 306.
6. F. Zymalkowski and E. Dornhege, *Tetrahedron Letters*, 1968, 5743.
7. M. Mallikarjunan, S. T. Rao, K. Venkatesan and V. R. Sarma, *Acta Cryst.*, 1969, **B25**, 220.
8. S. Schnell and P. Karrer, *Helv. Chim. Acta*, 1955, **38**, 2036.
9. D. Ackermann and S. Skraup, *Z. physiol. Chem.*, 1949, **284**, 129.
10. J. Bernstein, K. A. Losee, C. I. Smith and B. Rubin, *J. Amer. Chem. Soc.*, 1959, **81**, 4433.
11. W. Langenbeck, *Ber.*, 1925, **58**, 227.
12. H. Heath, A. Lawson and C. Rimington, *J. Chem. Soc.*, 1951, 2215.
13. J. L. Warnell and C. P. Berg, *J. Amer. Chem. Soc.*, 1954, **76**, 1708.
14. C. Ressler, *J. Amer. Chem. Soc.*, 1960, **82**, 1641.
15. T. Sasaki, K. Kanematsu, Y. Tsuzuki and K. Tanaka, *J. Medicin. Pharmaceut. Chem.*, 1966, **9**, 847.
16. M. Nakazaki, I. Mita and N. Toshioka, *Bull. Chem. Soc. Japan*, 1963, **36** 161.
17. P. Pratesi, A. La Manna and G. Vitali, *Il farmaco*, 1960, **15**, 387.
18. H. Peter, M. Brugger, J. Schreiber and A. Eschenmoser, *Helv. Chim Acta*, 1963, **46**, 577.
19. E. Hardegger & H. Braunschweiger, *Helv. Chim. Acta*, 1961, **44** 1125.

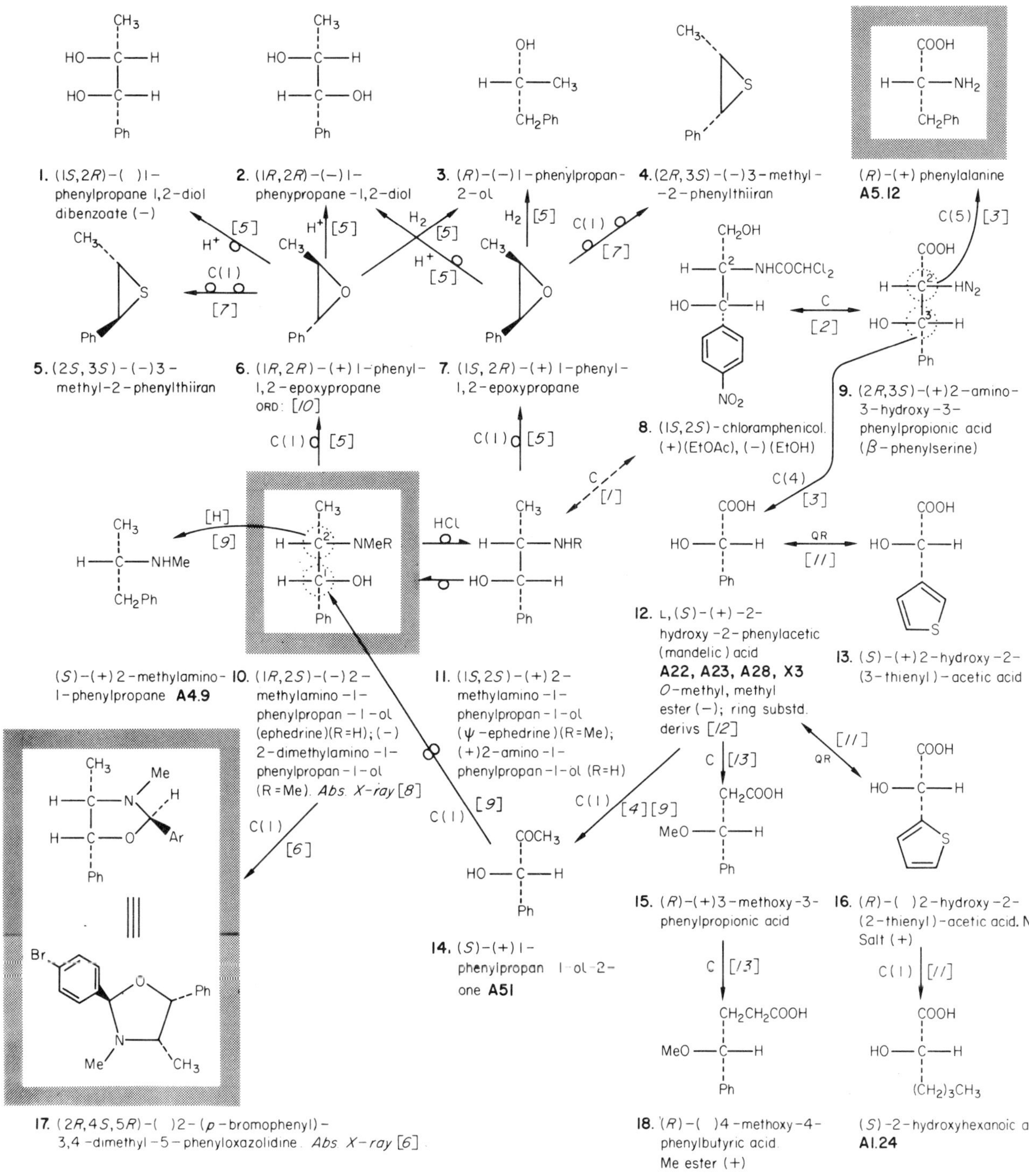

Class 2a, 2b
Aromatic compounds related to ephedrine and mandelic acid

1. (1S,2R)-( )1-phenylpropane 1,2-diol dibenzoate (−)
2. (1R,2R)-(−)1-phenypropane-1,2-diol
3. (R)-(−)1-phenylpropan-2-ol
4. (2R,3S)-(−)3-methyl-2-phenylthiiran
5. (2S,3S)-(−)3-methyl-2-phenylthiiran
6. (1R,2R)-(+)1-phenyl-1,2-epoxypropane ORD: [10]
7. (1S,2R)-(+)1-phenyl-1,2-epoxypropane
8. (1S,2S)-chloramphenicol. (+)(EtOAc), (−)(EtOH)
9. (2R,3S)-(+)2-amino-3-hydroxy-3-phenylpropionic acid (β-phenylserine)
10. (1R,2S)-(−)2-methylamino-1-phenylpropan-1-ol (ephedrine)(R=H); (−)2-dimethylamino-1-phenylpropan-1-ol (R=Me). Abs. X-ray [8]
11. (1S,2S)-(+)2-methylamino-1-phenylpropan-1-ol (ψ-ephedrine)(R=Me); (+)2-amino-1-phenylpropan-1-ol (R=H)
12. L,(S)-(+)-2-hydroxy-2-phenylacetic (mandelic) acid **A22, A23, A28, X3** O-methyl, methyl ester (−); ring substd. derivs [12]
13. (S)-(+)2-hydroxy-2-(3-thienyl)-acetic acid
14. (S)-(+)1-phenylpropan-1-ol-2-one **A51**
15. (R)-(+)3-methoxy-3-phenylpropionic acid
16. (R)-( )2-hydroxy-2-(2-thienyl)-acetic acid. Na Salt (+)
17. (2R,4S,5R)-( )2-(p-bromophenyl)-3,4-dimethyl-5-phenyloxazolidine. Abs. X-ray [6]
18. (R)-( )4-methoxy-4-phenylbutyric acid. Me ester (+)

(S)-(+)2-methylamino-1-phenylpropane **A4.9**

(R)-(+) phenylalanine **A5.12**

(S)-2-hydroxyhexanoic acid **A1.24**

1. G. Fodor, J. Kiss and I. Sallay, *J. Chem. Soc.*, 1951, 1858.
2. M. Honjo, *J. Pharm. Soc. Japan*, 1953, 73, 368.
3. K. Vogler, *Helv. Chim. Acta*, 1950, 33, 2111.
4. J. H. Brewster, *J. Amer. Chem. Soc.*, 1956, 78, 4061.
5. B. Witkop and C. M. Foltz, *J. Amer. Chem. Soc.*, 1957, 79, 197.
6. L. Neelakantan and J. A. Molin-Case, *J. Org. Chem.*, 1971, 36, 2261.
7. I. Moretti, G. Torre and G. Gottarelli, *Tetrahedron Letters*, 1971, 4301.
8. G. N. Ramachandran and S. Raman, *Current Sci. (India)*, 1956, 25, 348.
9. W. Leithe, *Ber.*, 1932, 65, 660.
10. I. Moretti and G. Torre, *Tetrahedron Letters*, 1969, 2717.
11. S. Gronowitz, *Arkiv Kemi*, 1958, 13, 87, 231.
12. P. Pratesi, A. La Manna, A. Campiglio and V. Ghislandi, *J. Chem. Soc.*, 1958, 2069; 1959, 4062.
13. E. W. Yankee and D. J. Cram, *J. Amer. Chem. Soc.*, 1970, 92, 6329.

A22 One-centre compounds related to 3-hydroxybutyric and mandelic acids
Class 2a, 2b

1. (S)-(+) lactic acid **A1.9**
2. (S)-(−)1-(1-cyclohexenyl)1-methoxyethane
3. (S)-(+)1-cyclohexylethanol Me ether (+)
4. (R)-(+)1-halogeno 1-phenylethanes (X = Cl, Br) **A43**
5. (S)-(+)1-amino-propan-2-ol **Y24**
6. (S)-(+) propane-1,2-diol **A14.8**
7. (S)-(+) 2-chloro-2-phenylethanol
8. (S)-(−)1-phenylethanol Me ether (−) Et ether (−) o-amino (−) **A59, Y14, Z1**
9. (R)-(−) 3-chlorobutyric acid
10. (S)-(+) 3-hydroxybutyric acid **A6.4**
11. (S)-(+) phenylthiiran ORD; [7]
12. (R)-(+)1-phenyl-1,2-epoxyethane (phenyloxiran; styrene oxide) **A44**
13. (R)-(−) 2-ethoxy-2-phenylethanol
14. (S)-(+) butane-1,3-diol **A1.12**
15. (S)-(+) heptan-2-ol **K19**
16. (R)-(−)1-phenylethane-1,2-diol
17. (R)-(−) mandelic acid **A21.12**
18. (R)-(−) adrenaline (X = OH, R = Me)
    (R)-(−) synephrine (X = H, R = Me)
    (R)-(−) noradrenaline (X = OH, R = H) **K1**
19. (R)-(−)1,3-dibromobutane
20. (S)-(−) 2-methylthietane-1,1-dioxide
21. (S)-(+) 4-methylthiete-1,1-dioxide
22. (R)-(−)1-phenyl-2,2,2-trifluoroethanol **Z3**

1. P. Pratesi, A. La Manna, A. Campiglio and V. Ghislandi, *J. Chem. Soc.*, 1959, 4062.
2. P. A. Levene and H. L. Haller, *J. Biol. Chem.*, 1926, 67, 329.
3. K. Mislow, *J. Amer. Chem. Soc.*, 1951, 73, 3954.
4. L. A. Paquette and J. P. Freeman, *J. Org. Chem.*, 1970, 35, 2249.
5. P. A. Levene and H. L. Haller, *J. Biol. Chem.*, 1927, 74, 343.
6. E. L. Eliel and D. W. Delmonte, *J. Org. Chem.*, 1956, 21 596.
7. I. Moretti, G. Torre and G. Gottarelli, *Tetrahedron Letters*, 1971, 4301.
8. G. Berti, F. Bottari, P. L. Ferrarini and B. Macchia, *J. Org. Chem.*, 1965, 30, 4091.
9. P. A. Levene and A. Walti, *J. Biol. Chem.*, 1926, 68, 415.
10. R. Lukes, J. Kovar, J. Kloubek and K. Bláha, *Coll. Czech. Chem. Comm.*, 1960, 25, 483.
11. W. A. Cowdrey, E. D. Hughes, C. K. Ingold, S. Masterman and A. D. Scott, *J. Chem. Soc.*, 1937, 1252.
12. H. M. Peters, D. M. Fiegl and H. S. Mosher, *J. Org. Chem.*, 1968, 33, 4245.
13. D. E. Wolf, W. H. Jones, J. Valiant and K. Folkers, *J. Amer. Chem. Soc.*, 1950, 72, 2820.

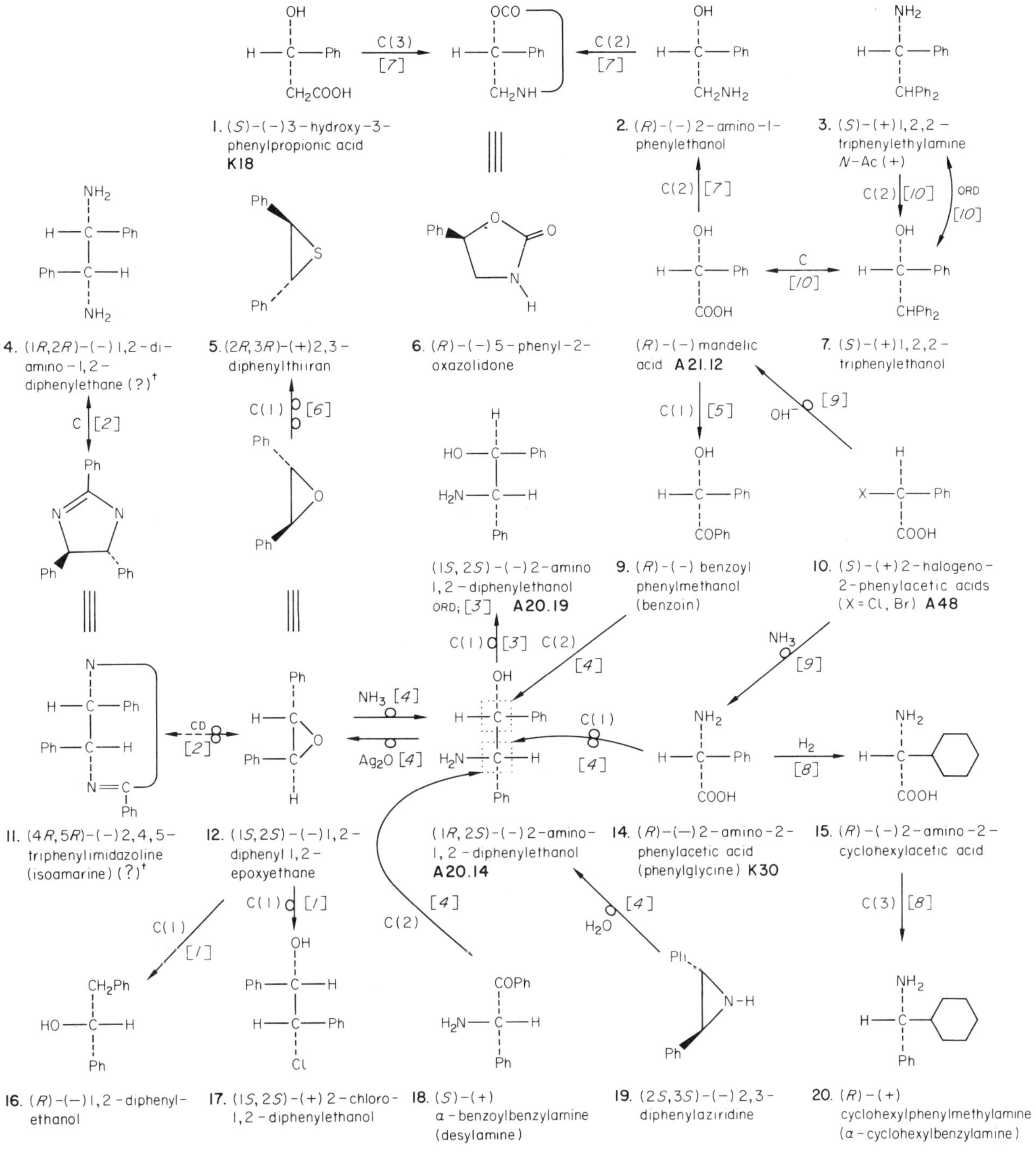

Class 2a, 2b — One- and two-centre compounds related to mandelic acid and phenylglycine

1. (S)-(−)3-hydroxy-3-phenylpropionic acid **K18**
2. (R)-(−)2-amino-1-phenylethanol
3. (S)-(+)1,2,2-triphenylethylamine N-Ac (+)
4. (1R,2R)-(−)1,2-diamino-1,2-diphenylethane (?)†
5. (2R,3R)-(+)2,3-diphenylthiiran
6. (R)-(−)5-phenyl-2-oxazolidone
7. (S)-(+)1,2,2-triphenylethanol
8. (1S,2S)-(−)2-amino-1,2-diphenylethanol ORD; [3] **A20.19**
9. (R)-(−)benzoyl phenylmethanol (benzoin)
10. (S)-(+)2-halogeno-2-phenylacetic acids (X = Cl, Br) **A48**
11. (4R,5R)-(−)2,4,5-triphenylimidazoline (isoamarine) (?)†
12. (1S,2S)-(−)1,2-diphenyl 1,2-epoxyethane
13. (1R,2S)-(−)2-amino-1,2-diphenylethanol **A20.14**
14. (R)-(−)2-amino-2-phenylacetic acid (phenylglycine) **K30**
15. (R)-(−)2-amino-2-cyclohexylacetic acid
16. (R)-(−)1,2-diphenylethanol
17. (1S,2S)-(+)2-chloro-1,2-diphenylethanol
18. (S)-(+) α-benzoylbenzylamine (desylamine)
19. (2S,3S)-(−)2,3-diphenylaziridine
20. (R)-(+) cyclohexylphenylmethylamine (α-cyclohexylbenzylamine)

1. G. Berti, F. Bottari, P. L. Ferrarini and B. Macchia, *J. Org. Chem.*, 1965, **30**, 4091.
2. P. L. Fereday, and S. F. Mason, *Chem. Comm.*, 1971, 1314.
3. G. G. Lyle and W. Lacroix, *J. Org. Chem.*, 1963, **28**, 901.
4. M. B. Watson and G. W. Youngson, *Chem. and Ind.*, 1954, 658.
5. A. McKenzie and H. Wren, *J. Chem. Soc.*, 1908, 309.
6. I. Moretti, G. Torre and G. Gottarelli, *Tetrahedron Letters*, 1971, 4301.
7. C. Schöpf and W. Wüst, *Annalen*, 1959, **626**, 150.
8. V. Ghislandi and D. Vercesi, *Il farmaco*, 1971, **26**, 474.
9. B. C. Hibbin, E. D. Hughes and C. K. Ingold, *Chem. and Ind.*, 1954, 933.
10. C. J. Collins, J. B. Christie and V. F. Raaen, *J. Amer. Chem. Soc.*, 1961, **83**, 4267.

A24 (i) Threonine and related two-centre compounds; (ii) Valine, pinacolylamine and related amino-compounds Class 2a, 2b

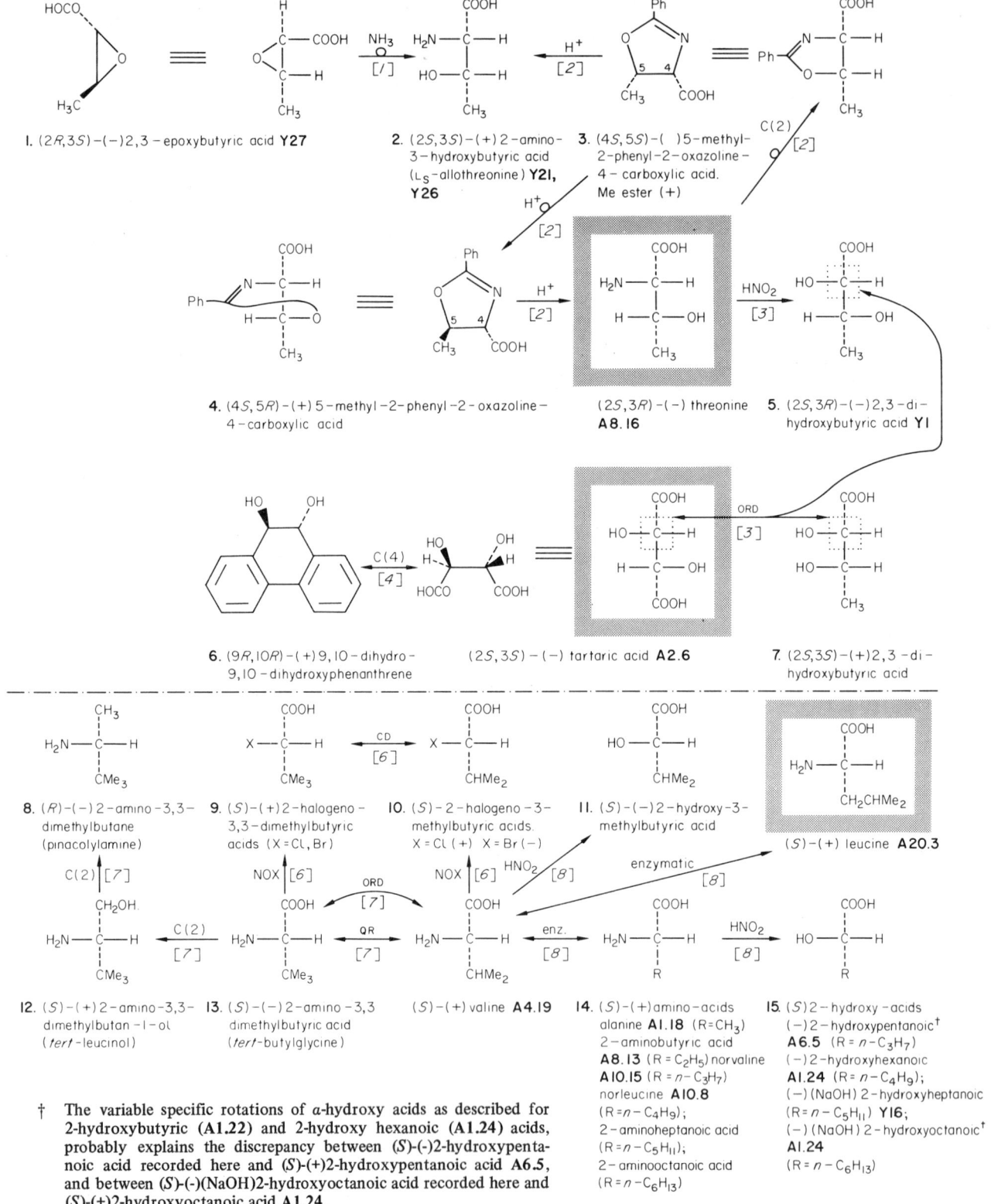

1. (2R,3S)-(−)2,3-epoxybutyric acid Y27
2. (2S,3S)-(+)2-amino-3-hydroxybutyric acid (L_S-allothreonine) Y21, Y26
3. (4S,5S)-( )5-methyl-2-phenyl-2-oxazoline-4-carboxylic acid. Me ester (+)
4. (4S,5R)-(+)5-methyl-2-phenyl-2-oxazoline-4-carboxylic acid
   (2S,3R)-(−) threonine A8.16
5. (2S,3R)-(−)2,3-dihydroxybutyric acid Y1
6. (9R,10R)-(+)9,10-dihydro-9,10-dihydroxyphenanthrene
   (2S,3S)-(−) tartaric acid A2.6
7. (2S,3S)-(+)2,3-dihydroxybutyric acid

8. (R)-(−)2-amino-3,3-dimethylbutane (pinacolylamine)
9. (S)-(+)2-halogeno-3,3-dimethylbutyric acids (X = Cl, Br)
10. (S)-2-halogeno-3-methylbutyric acids. X = Cl (+) X = Br (−)
11. (S)-(−)2-hydroxy-3-methylbutyric acid
    (S)-(+) leucine A20.3
12. (S)-(+)2-amino-3,3-dimethylbutan-1-ol (tert-leucinol)
13. (S)-(−)2-amino-3,3-dimethylbutyric acid (tert-butylglycine)
    (S)-(+) valine A4.19
14. (S)-(+)amino-acids
    alanine A1.18 (R=CH_3)
    2-aminobutyric acid A8.13 (R = C_2H_5) norvaline A10.15 (R = n-C_3H_7) norleucine A10.8 (R = n-C_4H_9);
    2-aminoheptanoic acid (R = n-C_5H_{11});
    2-aminooctanoic acid (R = n-C_6H_{13})
15. (S)2-hydroxy-acids
    (−)2-hydroxypentanoic[†] A6.5 (R = n-C_3H_7)
    (−)2-hydroxyhexanoic A1.24 (R = n-C_4H_9);
    (−)(NaOH)2-hydroxyheptanoic (R = n-C_5H_{11}) Y16;
    (−)(NaOH)2-hydroxyoctanoic[†] A1.24 (R = n-C_6H_{13})

[†] The variable specific rotations of a-hydroxy acids as described for 2-hydroxybutyric (A1.22) and 2-hydroxy hexanoic (A1.24) acids, probably explains the discrepancy between (S)-(−)2-hydroxypentanoic acid recorded here and (S)-(+)2-hydroxypentanoic acid A6.5, and between (S)-(−)(NaOH)2-hydroxyoctanoic acid recorded here and (S)-(+)2-hydroxyoctanoic acid A1.24.

1. K. Harada and J. Oh-hashi, *Bull. Chem. Soc. Japan*, 1966, 39, 2311.
2. D. F. Elliott, *J. Chem. Soc.*, 1950, 62.
3. F. W. Bachelor and G. A. Miana, *Canad. J. Chem.*, 1969, 47, 4089.
4. R. Miura, S. Honmaru and M. Nakazaki, *Tetrahedron Letters*, 1968, 5271.
5. G. Gottarelli and B. Samori, *Tetrahedron Letters*, 1970, 2055.
6. W. G. Galetto and W. Gaffield, *J. Chem. Soc. (C)*, 1969, 2437.
7. H. Pracejus and S. Winter, *Chem. Ber.*, 1964, 97, 3173.
8. C. G. Baker and A. Meister, *J. Amer. Chem. Soc.*, 1951, 73, 1336.

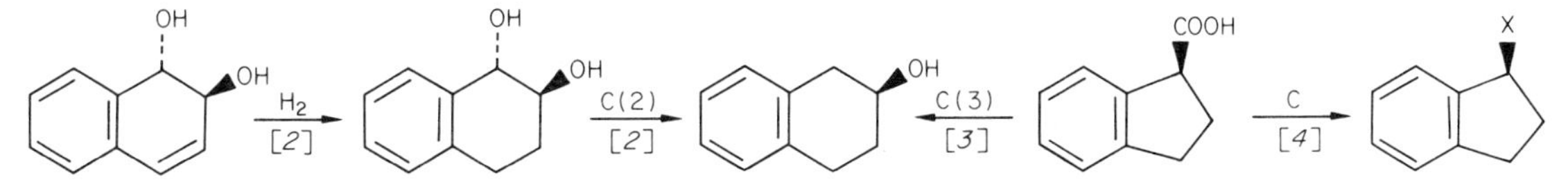

Class 2a, 2b
Miscellaneous compounds; tetralins, indanols etc.

1. (1S,2S)-(+)1,2-dihydro-1,2-dihydroxynaphthalene
2. (1S,2S)-(−)1,2-dihydroxytetralin. di-OAc (+)
3. (S)-(−)1,2,3,4-tetrahydro-2-naphthol. OAc (−)
4. (S)-(−)indane-1-carboxylic acid A50
5. (S)-1-substituted indanes
X = OH (+) X = OAc (−)
X = NH_2 (+) X = NHAc (+)
X = NMe_2 (+)
X = $NMe_3^+ I^-$ (+) T48
6. (1R,2S)-(+) naphthalene 1,2-epoxide 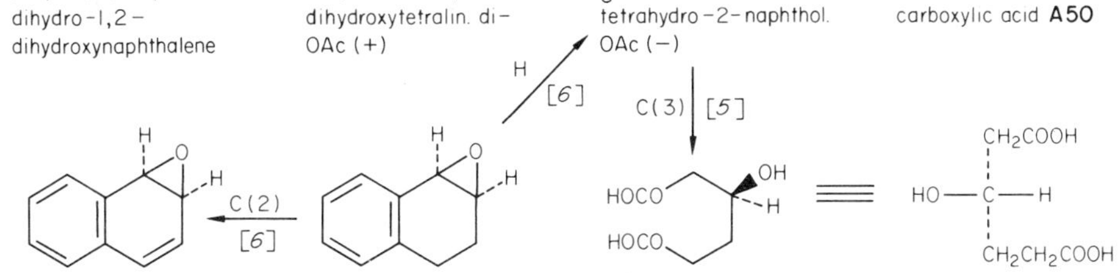
7. (1R,2S)-(+)1,2-epoxytetralin
8. (S)-( ) 3-hydroxyadipic acid. Dihydrazide (+); OMe (−); OMe, di-Me ester (−) K6, K7, Y2
9. (S)-(−) tetralin-1-carboxylic acid A50 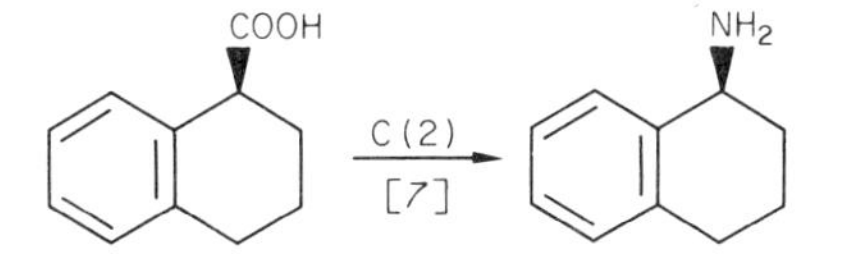
10. (S)-(+) 1,2,3,4-tetrahydro-1-naphthylamine
11. (1S,2S)-(+)2-amino-indan-1-ol (as hydrochloride)
(S)-(−) phenylalanine A5.12
12. (1R,2S)-(+)2-amino-indan-1-ol (as hydrochloride) 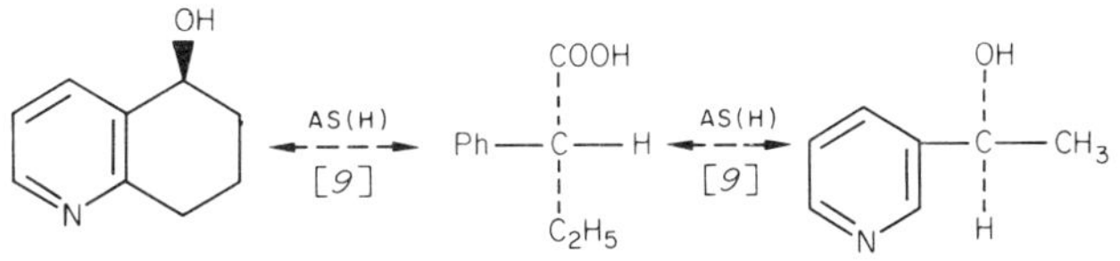
13. (S)-(+) 2-aminoadipic acid Y29
14. (S)-(−) 5-hydroxy-5,6,7,8-tetrahydroquinoline CD; [9]
15. (R)-(−) 2-phenylbutyric acid A41, A61, T30, T31, T38, T57, T58, K5, K16, K24, Y1, Y4, Y14, Y16, Y27, X5, X8, X9
16. (S)-(−)1-(3-pyridyl)-ethanol RD; [9]
17. (R)-(−)3-methyl-2-phenylbutyric acid A41 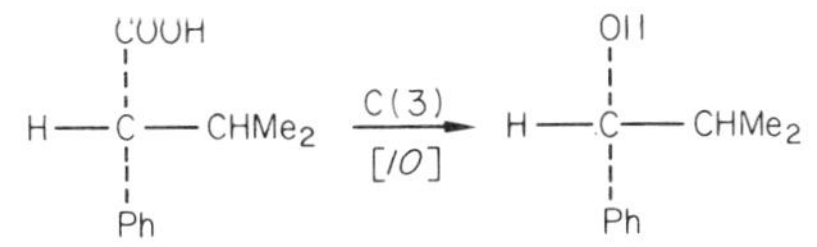
18. (R)-(+) 2-methyl-1-phenylpropan-1-ol
19. (R)-(−) 3,3-dimethyl-2-phenylbutyric acid A41 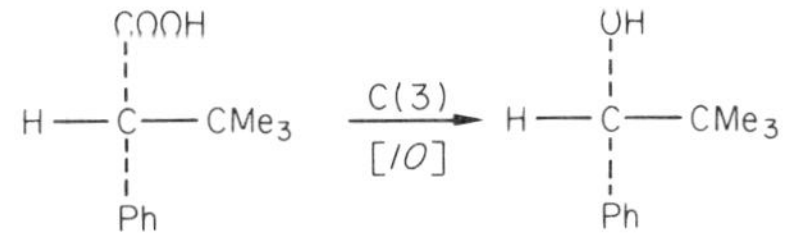
20. (R)-(+) 2,2-dimethyl-1-phenylpropan-1-ol

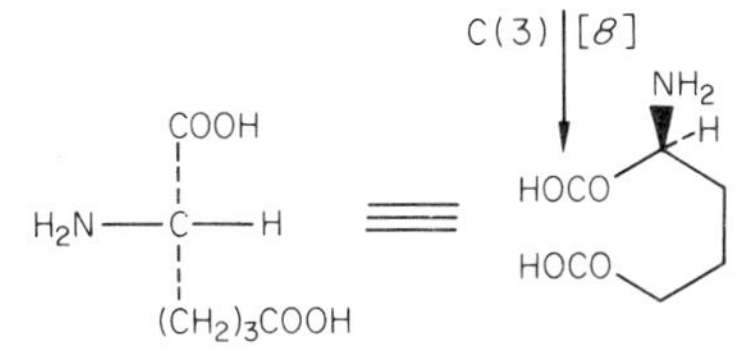

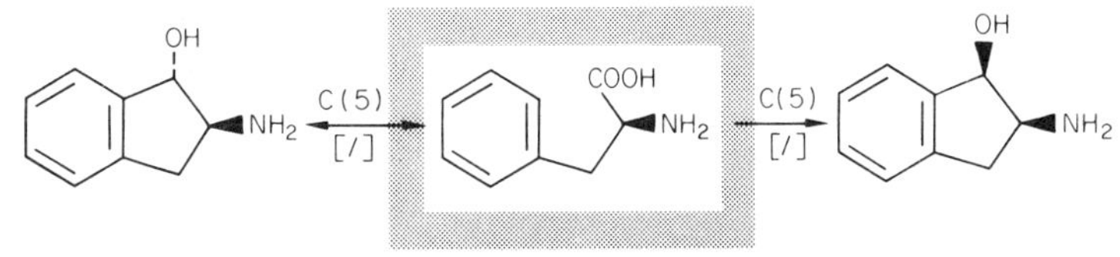

1. E. Dornhege, *Annalen*, 1971, **743**, 42.
2. R. Miura, S. Honmaru and M. Nakazaki, *Tetrahedron Letters*, 1968, 5271.
3. D. Battail-Robert and D. Gagnaire, *Bull. Soc. chim. France*, 1966, 208.
4. J. H. Brewster and J. G. Buta, *J. Amer. Chem. Soc.*, 1966, **88**, 2233.
5. H. Arakawa, N. Torimoto and Y. Masui, *Tetrahedron Letters*, 1968, 4115.
6. D. R. Boyd, D. M. Jerina and J. W. Daly, *J. Org. Chem.*, 1970, **35**, 3170.
7. R. Weidmann and J. P. Guetté, *Compt. rend.*, 1969, **268**, C, 2225.
8. V. Ghislandi and D. Vercesi, *Il farmaco*, 1971, **26**, 474.
9. G. Gottarelli and B. Samori, *Tetrahedron Letters*, 1970, 2055.
10. D. R. Clark and H. S. Mosher, *J. Org. Chem.*, 1970, **35**, 1114.

Quinic acid and other cyclohexane derivatives; piperidines

Class 2a, 3a

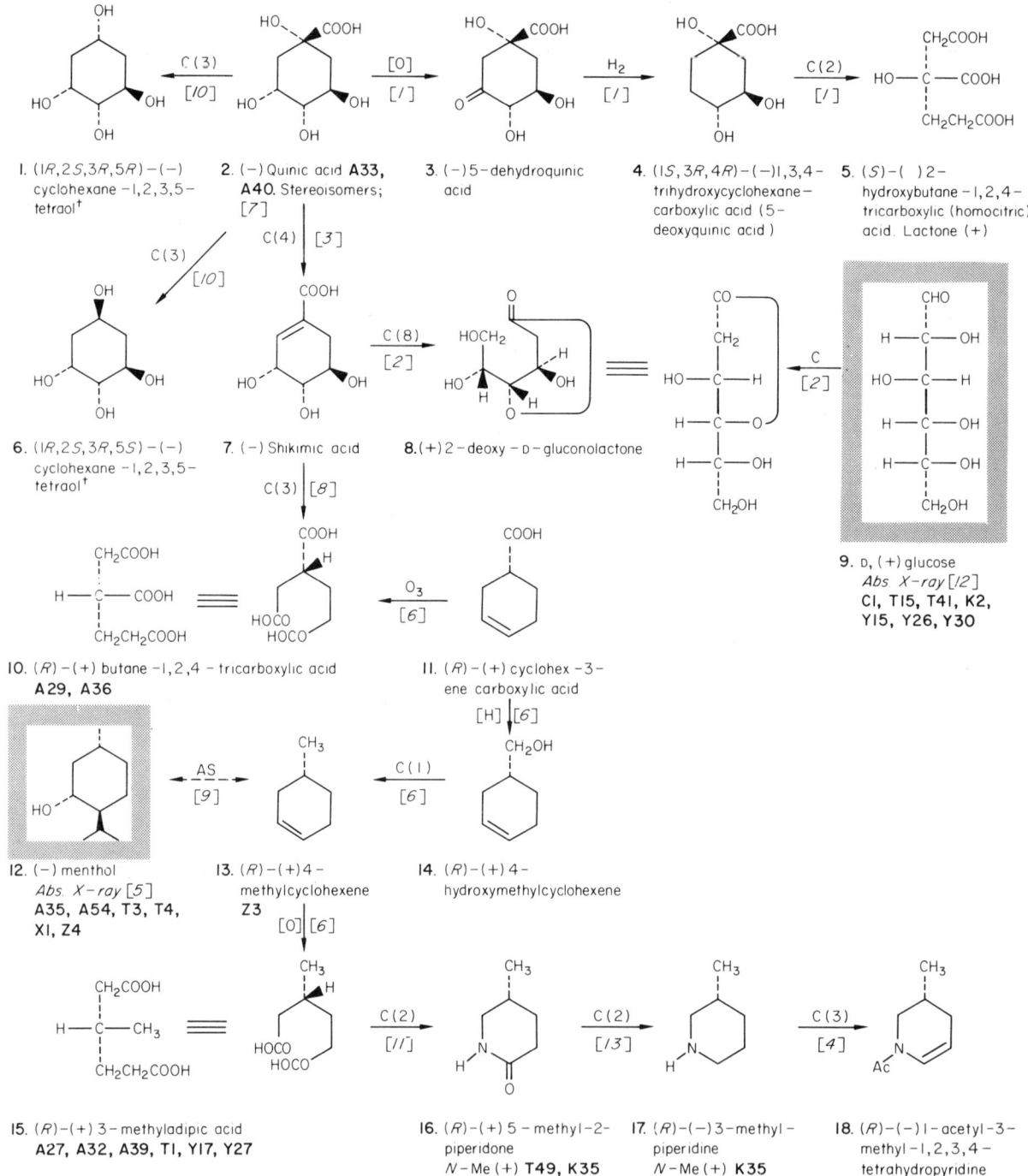

1. (1R,2S,3R,5R)-(−) cyclohexane-1,2,3,5-tetraol†
2. (−) Quinic acid **A33, A40**. Stereoisomers; [7]
3. (−) 5-dehydroquinic acid
4. (1S,3R,4R)-(−)1,3,4-trihydroxycyclohexane-carboxylic acid (5-deoxyquinic acid)
5. (S)-( )2-hydroxybutane-1,2,4-tricarboxylic (homocitric) acid. Lactone (+)
6. (1R,2S,3R,5S)-(−) cyclohexane-1,2,3,5-tetraol†
7. (−) Shikimic acid
8. (+) 2-deoxy-D-gluconolactone
9. D, (+) glucose Abs X-ray [12] C1, T15, T41, K2, Y15, Y26, Y30
10. (R)-(+) butane-1,2,4-tricarboxylic acid **A29, A36**
11. (R)-(+) cyclohex-3-ene carboxylic acid
12. (−) menthol Abs. X-ray [5] **A35, A54, T3, T4, X1, Z4**
13. (R)-(+) 4-methylcyclohexene **Z3**
14. (R)-(+) 4-hydroxymethylcyclohexene
15. (R)-(+) 3-methyladipic acid **A27, A32, A39, T1, Y17, Y27**
16. (R)-(+) 5-methyl-2-piperidone N-Me (+) **T49, K35**
17. (R)-(−) 3-methyl-piperidine N-Me (+) **K35**
18. (R)-(−) 1-acetyl-3-methyl-1,2,3,4-tetrahydropyridine

† Application of (R,S) nomenclature to these compounds; see R. S. Cahn, C. K. Ingold and V. Prelog, *Angew. Chem. Internat. Edn.*, 1966, **5**, 385.

1. U. Thomas, M. G. Kalyanpur and C. M. Stevens, *Biochemistry*, 1966, **5**, 2513.
2. H. O. L. Fischer and G. Dangschat, *Helv. Chim. Acta*, 1937, **20**, 705.
3. G. Dangschat and H. O. L. Fischer, *Biochim. Biophys. Acta*, 1950, **4**, 199.
4. T. Masamune, M. Takasugi and A. Murai, *Tetrahedron*, 1971, **27**, 3369.
5. J. M. Ohrt and R. Parasarathy, *Acta Cryst. (Suppl.)*, 1969, S198.
6. O. Ceder and B. Hansson, *Acta Chem. Scand.*, 1970, **24**, 2693.
7. J. Corse and R. E. Lundin, *J. Org. Chem.*, 1970, **35**, 1904.
8. K. Freudenberg and W. Hohmann, *Annalen*, 1953, **584**, 54.
9. S. I. Goldberg and M. S. Sahli, *J. Org. Chem.*, 1967, **32**, 2059.
10. G. E. McCasland, S. Furuta, L. F. Johnson and J. N. Shoolery, *J. Org. Chem.*, 1964, **29**, 2354.
11. O. Jeger, V. Prelog, E. Sundt, and R. B. Woodward, *Helv. Chim. Acta*, 1954, 37, 2302.
12. S. Neidle and D. Rogers, *Nature*, 1970, **225**, 376.
13. S. Okuda, K. Tsuda and H. Kataoka, *Chem. and Ind.*, 1961, 512.

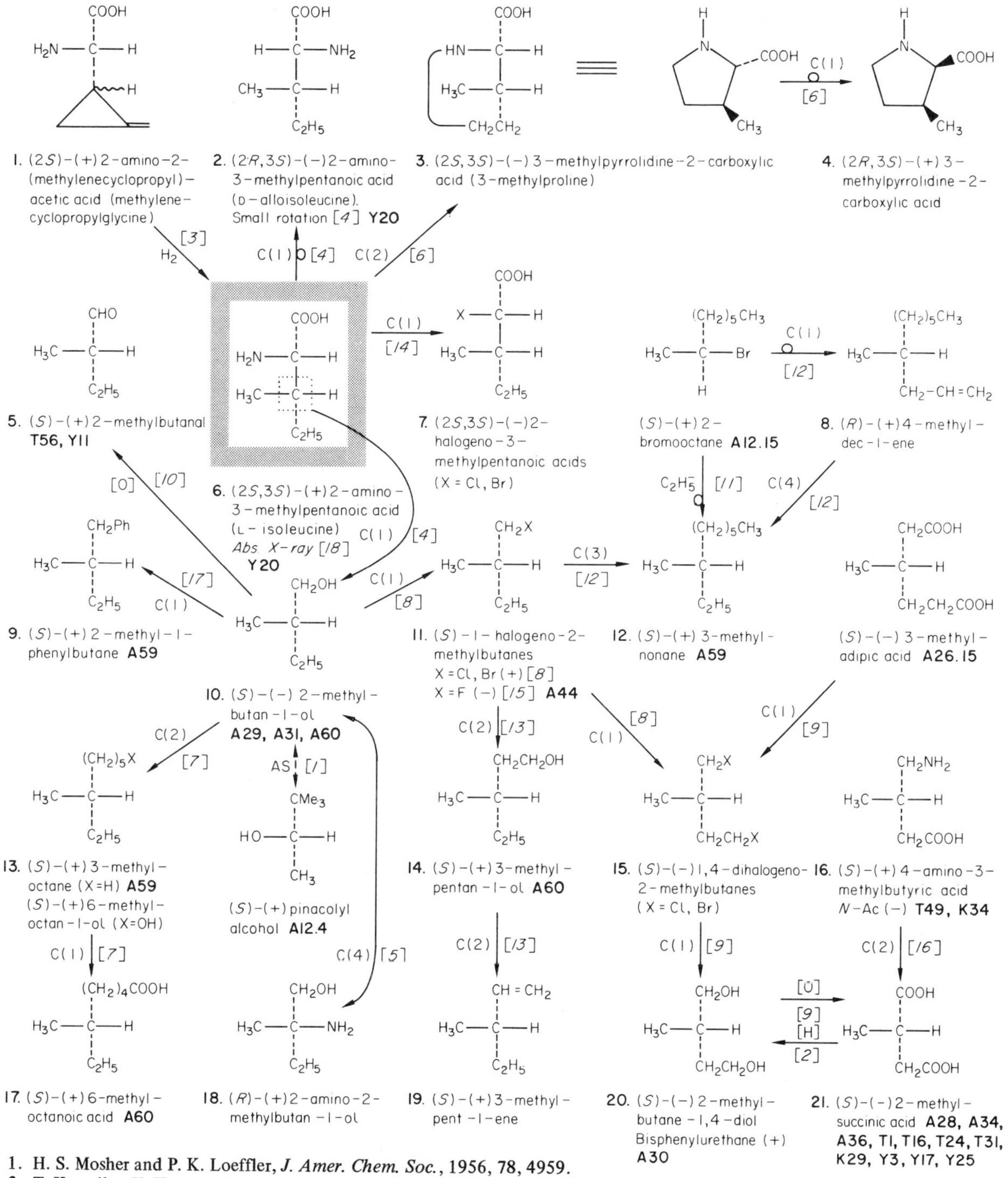

Class 3a   Branched-chain compounds related to isoleucine and 2-methylbutan-1-ol   **A**[27]

1. (2*S*)-(+)2-amino-2-(methylenecyclopropyl)-acetic acid (methylene-cyclopropylglycine)
2. (2*R*,3*S*)-(−)2-amino-3-methylpentanoic acid (D-alloisoleucine). Small rotation [4] **Y20**
3. (2*S*,3*S*)-(−)3-methylpyrrolidine-2-carboxylic acid (3-methylproline)
4. (2*R*,3*S*)-(+)3-methylpyrrolidine-2-carboxylic acid
5. (*S*)-(+)2-methylbutanal **T56, Y11**
6. (2*S*,3*S*)-(+)2-amino-3-methylpentanoic acid (L-isoleucine) Abs X-ray [18] **Y20**
7. (2*S*,3*S*)-(−)2-halogeno-3-methylpentanoic acids (X = Cl, Br)
8. (*R*)-(+)4-methyl-dec-1-ene
9. (*S*)-(+)2-methyl-1-phenylbutane **A59**
10. (*S*)-(−)2-methyl-butan-1-ol **A29, A31, A60**
    (*S*)-(+)pinacolyl alcohol **A12.4**
11. (*S*)-1-halogeno-2-methylbutanes X = Cl, Br (+) [8]; X = F (−) [15] **A44**
12. (*S*)-(+)3-methyl-nonane **A59**
    (*S*)-(−)3-methyl-adipic acid **A26.15**
13. (*S*)-(+)3-methyl-octane (X=H) **A59**
    (*S*)-(+)6-methyl-octan-1-ol (X=OH)
14. (*S*)-(+)3-methyl-pentan-1-ol **A60**
15. (*S*)-(−)1,4-dihalogeno-2-methylbutanes (X = Cl, Br)
16. (*S*)-(+)4-amino-3-methylbutyric acid *N*-Ac (−) **T49, K34**
17. (*S*)-(+)6-methyl-octanoic acid **A60**
18. (*R*)-(+)2-amino-2-methylbutan-1-ol
19. (*S*)-(+)3-methyl-pent-1-ene
20. (*S*)-(−)2-methyl-butane-1,4-diol Bisphenylurethane (+) **A30**
21. (*S*)-(−)2-methyl-succinic acid **A28, A34, A36, T1, T16, T24, T31, K29, Y3, Y17, Y25**

1. H. S. Mosher and P. K. Loeffler, *J. Amer. Chem. Soc.*, 1956, **78**, 4959.
2. T. Kanedko, H. Katsura, H. Asano and K. Wakabayashi, *Chem. and Ind.*, 1960, 1187.
3. D. O. Grey and L. Fowden, *Biochem. J.*, 1962, **82**, 385.
4. F. Ehrlich, *Ber.*, 1907, **40**, 2538.
5. W. Kirmse, H. Arold and B. Kornrumpf, *Chem. Ber.*, 1971, **104**, 1783.
6. J. Kollonitsch, A. N. Scott and G. A. Doldouras, *J. Amer. Chem. Soc.*, 1966, **88**, 3624.
7. L. Crombie and S. H. Harper, *J. Chem. Soc.*, 1950, 2685.
8. H. C. Brown, M. S. Kharash and T. H. Chao, *J. Amer. Chem. Soc.*, 1940, **62**, 3435.
9. J. von Braun and F. Jostes, *Ber.*, 1926, **59**, 1091, 1444.
10. E. J. Badin and E. Pascu, *J. Amer. Chem. Soc.*, 1945, **67**, 1352.
11. S. E. Ulrich, F. H. Gentes, J. F. Lane and E. S. Wallis, *J. Amer. Chem. Soc.*, 1950, **72**, 5127.
12. R. L. Letsinger and J. G. Traynham, *J. Amer. Chem. Soc.*, 1950, **72**, 849.
13. P. Pino, L. Lardicci and L. Centoni, *J. Org. Chem.*, 1959, **24**, 1399.
14. W. Gaffield and W. G. Galetto, *Tetrahedron*, 1971, **27**, 915.
15. D. D. Tanner, H. Tabuuchi and E. V. Blackburn, *J. Amer. Chem. Soc.*, 1971, **93**, 4802.
16. K. Schreiber and H. Ripperger, *Annalen*, 1962, **655**, 114.
17. R. L. Letsinger, *J. Amer. Chem. Soc.*, 1948, **70** 406.
18. J. Trommel and J. M. Bijvoet, *Acta Cryst.*, 1954, **7**, 703.

A²⁸ Alkylsuccinic acids and derived compounds — Class 3a

1. (R)-(+)mercapto-succinic acid **A2.8**

2. (R)-(+)halogeno-succinic acids **A4.1**

(R)-hydroxy acids. (−) lactic **A1.9** (R=Me); (+) malic **A1.25** (R=CH$_2$COOH); (−)mandelic **A21.12** (R=Ph)

1. (R)-(+)3-isopropyl-adipic acid **T2, T6**

2. (S)-(−)2-isopropyl-5-oxohexanoic acid **T10, T25, T41**

3. (R)-(+)2-methyl-4-oxopentanoic acid **Y25**

4. (R)-(+)2-methyl-succinic acid **A27.21**

5. (S)-(+)2-isopropyl-succinic acid **A29, A35, A55, T2, T18**

6. (S)-(−)2-isopropyl-glutaric acid **T3, T10**

7. (S)-( )3-hydroxy-2-methylpropanal

8. (R)-(+)3-methyl-cyclobutene

9. (R)-(+)n-alkyl-succinic acids (R=Et, C$_3$H$_7$, C$_4$H$_9$, C$_5$H$_{11}$, C$_6$H$_{13}$) **A29, A37, A50, A55, T16, T58, Y12, Y15**

10. (2R,3S)-(+)2-bromo-3-methyl succinic acid

11. (2R,3R)-(−)2-amino-3-methylsuccinic (3-methylaspartic) acid

12. (S)-(+)3-hydroxy-2-methylpropionic acid N-phenylcarbamate (+) hydrazide (+)

13. (R)-(+)2-thioalkyl-succinic acids. R=Et, n-C$_3$H$_7$, n-C$_4$H$_9$, n-C$_5$H$_{11}$ ORD; [15]

14. (R)-(+) butane-1,1-2-tricarboxylic acid

15. (S)-(+) 2-(2-thenyl) succinic acid

(R)-(+)4-methyl-2-oxazolidone **A19.16**

16. (S)-( )4-hydroxy-3-methylbutan-2-one **K23**

17. (R)-(−)2-ethyl-3-hydroxymethylbutane-1,4-diol

18. (+) isopilopic acid **K30**

19. (R)-(+)2-benzyl-succinic acid

20. (S)-Methylmalonyl coenzyme A.

1. A. Fredga, *Arkiv Kemi*, 1942, **B15**, no. 23.
2. J. Rétey and F. Lynen, *Biochem. Z.*, 1965, **342**, 256.
3. M. Sprecher and D. B. Sprinson, *J. Biol. Chem.*, 1966, **241**, 868.
4. R. Rossi and P. Diversi, *Tetrahedron*, 1971, **26**, 5033.
5. B. Sjöberg, A. Fredga and C. Djerassi, *J. Amer. Chem. Soc.*, 1959, **81**, 5002.
6. A. Fredga and O. Palm, *Arkiv Kemi*, 1949, **26A**, no. 26.
7. C. Djerassi, O. Halpern, D. I. Wilkinson and E. J. Eisenbraun, *Tetrahedron*, 1958, **4**, 369.
8. R. K. Hill and S. Barcza, *Tetrahedron*, 1966, **22**, 2889.
9. A. Fredga, *Acta Chem. Scand.*, 1949, **3**, 208.
10. A. Fredga and J. K. Miettinen, *Acta Chem. Scand.*, 1947, **1**, 371.
11. H. Minato, *Tetrahedron*, 1962, **18**, 365.
12. M. Matell, *Arkiv Kemi*, 1953, **5**, 17.
13. D. J. Robins and D. H. G. Crout, *J. Chem. Soc. (C)*, 1969, 1386.
14. A. Fredga, *Tetrahedron*, 1960, **8**, 126.
15. A. Fredga, J. P. Jennings, W. Klyne, P. M. Scopes, B. Sjöberg and S. Sjöberg, *J. Chem. Soc.*, 1965, 3928.
16. A. Fredga, *Arkiv Kemi*, 1953, **6**, 277.

Class 3a
Compounds related to 2-methylbutyric acid

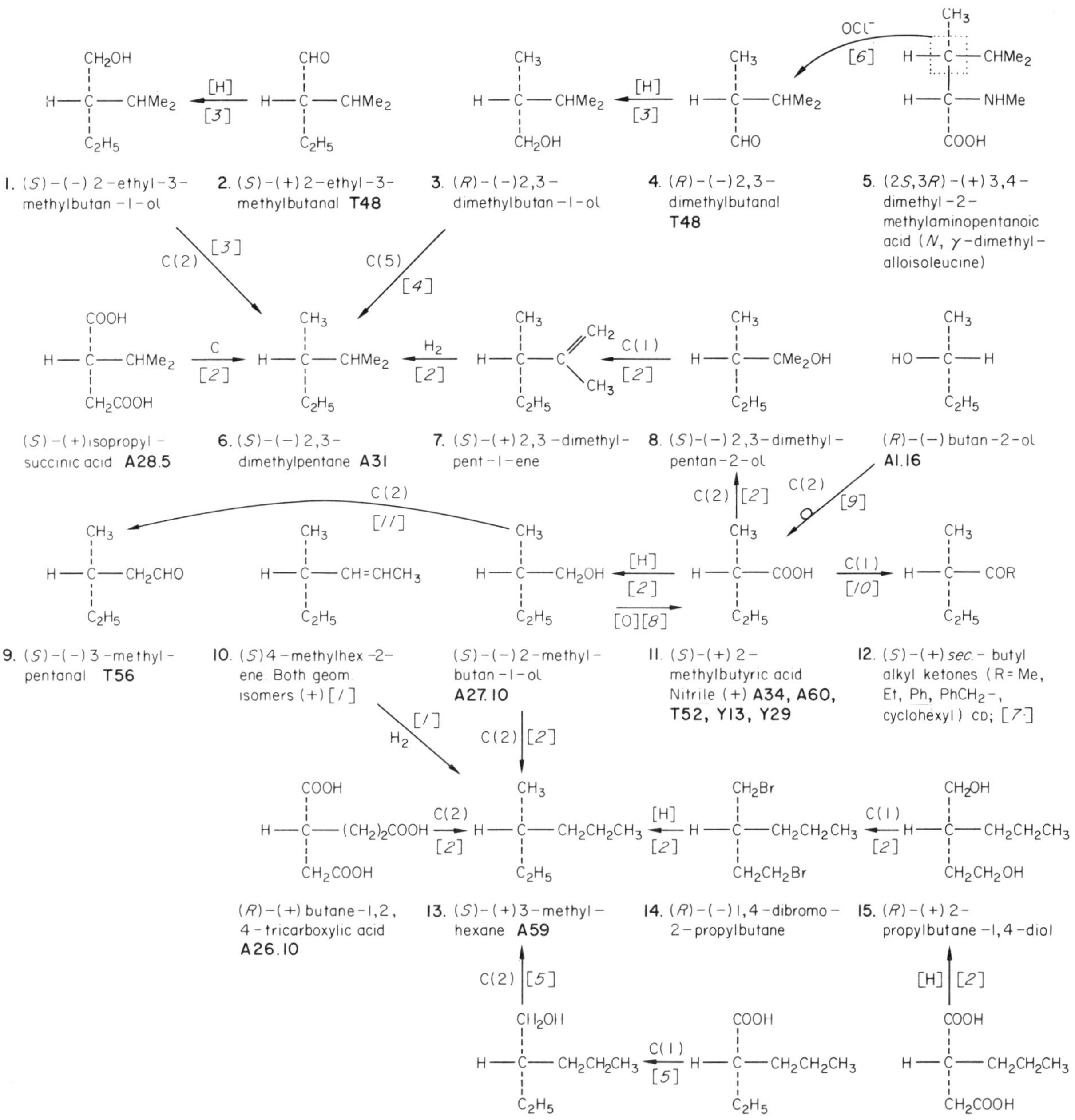

1. (S)-(−) 2-ethyl-3-methylbutan-1-ol

2. (S)-(+) 2-ethyl-3-methylbutanal T48

3. (R)-(−) 2,3-dimethylbutan-1-ol

4. (R)-(−) 2,3-dimethylbutanal T48

5. (2S,3R)-(+) 3,4-dimethyl-2-methylaminopentanoic acid (N, γ-dimethyl-alloisoleucine)

(S)-(+) isopropyl-succinic acid A28.5

6. (S)-(−) 2,3-dimethylpentane A31

7. (S)-(+) 2,3-dimethyl-pent-1-ene

8. (S)-(−) 2,3-dimethylpentan-2-ol

(R)-(−) butan-2-ol A1.16

9. (S)-(−) 3-methyl-pentanal T56

10. (S) 4-methylhex-2-ene. Both geom. isomers (+) [/]

(S)-(−) 2-methyl-butan-1-ol A27.10

11. (S)-(+) 2-methylbutyric acid Nitrile (+) A34, A60, T52, Y13, Y29

12. (S)-(+) sec.-butyl alkyl ketones (R = Me, Et, Ph, PhCH₂-, cyclohexyl) CD; [7]

(R)-(+) butane-1,2,4-tricarboxylic acid A26.10

13. (S)-(+) 3-methylhexane A59

14. (R)-(−) 1,4-dibromo-2-propylbutane

15. (R)-(+) 2-propylbutane-1,4-diol

16. (S)-(+) 2-ethylpentan-1-ol

17. (S)-(+) 2-ethylpentanoic acid A45

(R)-(+) n-propyl-succinic acid A28.9

---

1. R. Rossi, L. Lardicci, and G. Ingrosso, *Tetrahedron*, 1970, **26**, 4067.
2. K. Freudenberg and W. Lwowski, *Annalen*, 1955, **594**, 76, and refs. therein.
3. K. Rsuda, Y. Kishida, and R. Hayatsu, *J. Amer. Chem. Soc.*, 1960, **82**, 3396.
4. P. A. Levene and R. E. Marker, *J. Biol. Chem.*, 1935, **111**, 299.
5. P. A. Levene, A. Rothen and G. M. Meyer, *J. Biol. Chem.*, 1936, **115**, 401.
6. J. C. Sheehan, H. G. Zachau and W. B. Lawson, *J. Amer. Chem. Soc.*, 1958, 80 3349.
7. O. Korver, *Tetrahedron*, 1971, **27**, 4643.
8. W. Marckwald, *Ber.*, 1904, **40**, 2538.
9. J. Kenyon, H. Phillips, and M. P. Pittman, *J. Chem. Soc.*, 1935, 1072.
10. P. D. Bartlett and C. H. Stauffer, *J. Amer. Chem. Soc.*, 1935, **57**, 2580.
11. C. Djerassi and L. E. Geller, *J. Amer. Chem. Soc.*, 1959, **81**, 2789.

A[30] Iso- and hydroxycitric acids; tetrahydrofurans, pyrrolidines Class 3a

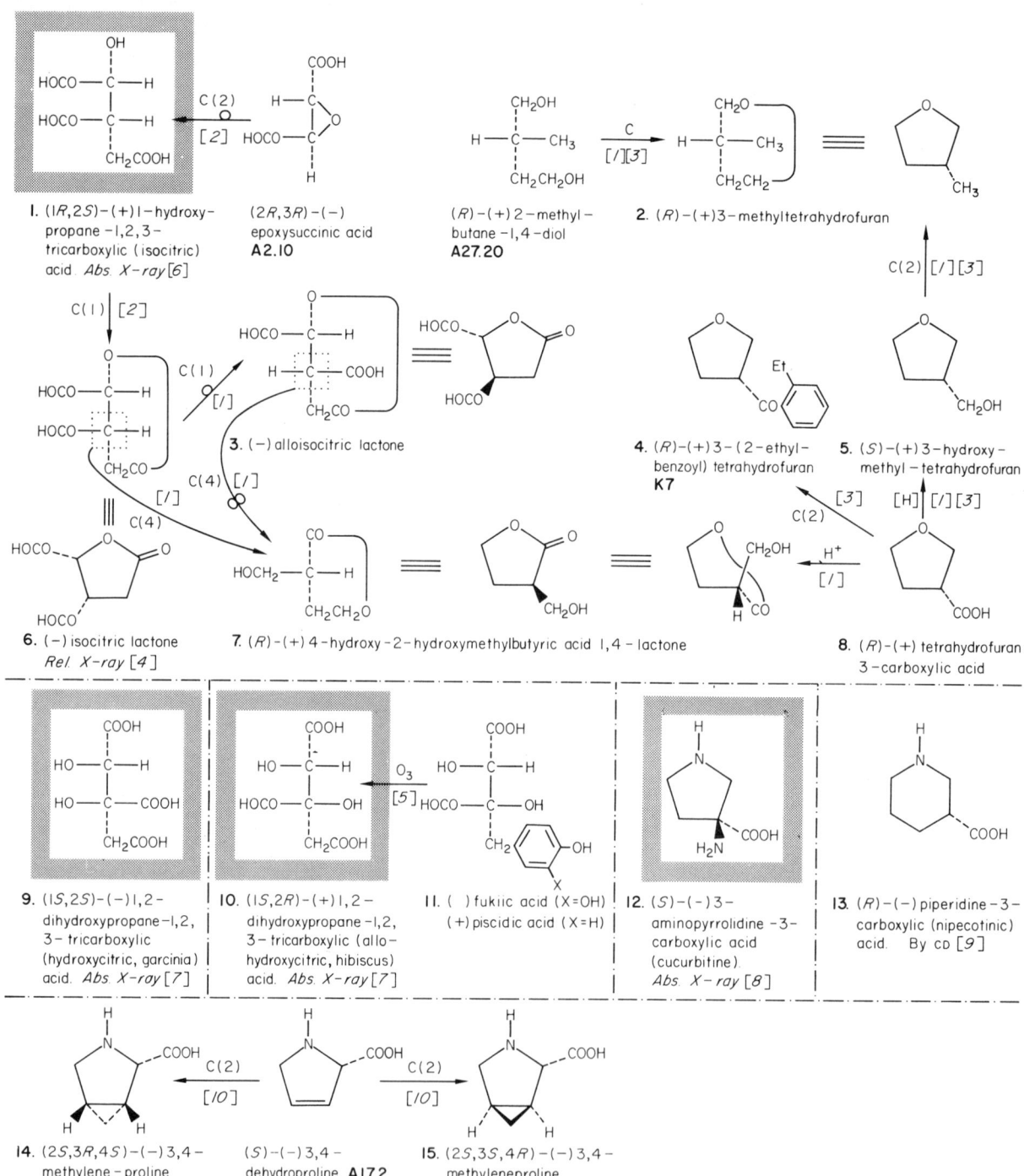

1. (1R,2S)-(+)1-hydroxypropane-1,2,3-tricarboxylic (isocitric) acid. *Abs. X-ray* [6]
   (2R,3R)-(−) epoxysuccinic acid **A2.10**
   (R)-(+)2-methylbutane-1,4-diol **A27.20**
2. (R)-(+)3-methyltetrahydrofuran
3. (−)alloisocitric lactone
4. (R)-(+)3-(2-ethylbenzoyl)tetrahydrofuran **K7**
5. (S)-(+)3-hydroxymethyl-tetrahydrofuran
6. (−)isocitric lactone *Rel. X-ray* [4]
7. (R)-(+)4-hydroxy-2-hydroxymethylbutyric acid 1,4-lactone
8. (R)-(+)tetrahydrofuran 3-carboxylic acid
9. (1S,2S)-(−)1,2-dihydroxypropane-1,2,3-tricarboxylic (hydroxycitric, garcinia) acid. *Abs. X-ray* [7]
10. (1S,2R)-(+)1,2-dihydroxypropane-1,2,3-tricarboxylic (allo-hydroxycitric, hibiscus) acid. *Abs. X-ray* [7]
11. ( )fukiic acid (X=OH) (+)piscidic acid (X=H)
12. (S)-(−)3-aminopyrrolidine-3-carboxylic acid (cucurbitine). *Abs. X-ray* [8]
13. (R)-(−)piperidine-3-carboxylic (nipecotinic) acid. By CD [9]
14. (2S,3R,4S)-(−)3,4-methylene-proline *Rel. X-ray* [10]
    (S)-(−)3,4-dehydroproline **A17.2**
15. (2S,3S,4R)-(−)3,4-methyleneproline *Rel. X-ray* [10]

1. T. Kaneko, H. Katsura, H. Asano and K. Wakabayashi, *Chem. and Ind.*, 1960, 1187.
2. T. Kaneko and H. Katsura, *Chem. and Ind.*, 1960, 1188.
3. R. K. Hill and W. R. Schearer, *J. Org. Chem.*, 1962, 27, 921.
4. J. P. Glusker, A. L. Patterson, S. E. Love and M. L. Dornberg, *Acta Cryst.*, 1963, 16, 1102.
5. T. Yoshihara, A. Ichihara and S. Sakamura, *Tetrahedron Letters*, 1971, 3809.
6. A. L. Patterson, C. K. Johnson, D. van der Helm and J. A. Minkin, *J. Amer. Chem. Soc.*, 1962, 84, 309.
7. J. P. Glusker, J. A. Minkin, C. A. Casciato and F. B. Soule, *Arch. Biochem. Biophys.*, 1969, 13, 573.
8. F. Haï-Fu and L. Cheng-Chiung, *Acta Phys. Sinica*, 1965, 21, 253.
9. H. Ripperger and K. Schreiber, *Chem. Ber.*, 1969, 102, 2864.
10. Y. Fujimoto, F. Irrevere, J. M. Karle, I. L. Karle and B. Witkop, *J. Amer. Chem. Soc.*, 1971, 93, 3471.

Class 3a Branched-chain one- and two-centre compounds including dialkylsuccinic acids A³¹

1. (S)-(+)14-methylhexadecanoic acid
2. (S)-(+)5-methylhept-1-ene
3. (S)-(+)1-chloro-2-methylbutane
4. (2S,3S)-(−)2,3-dimethylbutane-1,4-diol
5. (3S,4S)-(−)3,4-dimethylpyrrolidine
6. (R)-(+)3-dimethylamino-1,1-diphenyl-2-methyl-propan-1-ol (X=OH) (S)-(+)3-dimethylamino-1,1-diphenyl-2-methyl-propane (X=H)
7. (R)-(−)3-amino-2-methylpropionic acid
   (S)-(−)2-methyl-butan-1-ol A27.10
8. (2S,3S)-(−)2,3-dimethylsuccinic acid Anhydride (−); other derivs. [7]
9. (2S,3S)-(+)2,3-dimethyl-3-(3-carboxy-2-pyridyl)-propionic (evoninic) acid
10. (S)-(−)6-dimethylamino-4,4-diphenyl-5-methyl-hexan-3-one (isomethadone)
11. (3S,5S)-(+)6-dimethylamino-4,4-diphenyl-5-methyl-hexan-3-ol (β-methadol)
12. (2S,3S)-(−)2-ethyl-3-methylsuccinic acid. Imide (−) Other derivs [7]
13. (3S,4S)-(−)3,4-dimethylcyclopentanone
14. (3S,4S)-( )3,4-dimethyladipic acid
15. (3R,5S)-(−)6-dimethylamino-4,4-diphenyl-5-methylhexan-3-ol (α-methadol)
16. (2S,3S)-(−)2-ethyl-3-methylbutane-1,4-diol
    (S)-(−)2,3-dimethylpentane A29.6
17. (S)-(−)guaiaretic acid
18. (−)dihydroguaiaretic acid dimethyl ether Y7, Y8
19. (S)-(+)2-hydroxy-2-phenylbutyric (atrolactic) acid A51, T46, K9, K25, Y3, Y13, X3, X6
20. (2S,3R)-(−)2-ethyl-3-methylsuccinic acid
21. (2S,3R)-(−)2-ethyl-3-methylbutane-1,4-diol
22. (R)-(−)3-(3,4-dimethoxyphenyl)-2-methylpropionic acid
    (R)-(−)2-amino-1-(3,4-dimethoxyphenyl)propane A5.6

1. A. W. Schrecker and J. L. Hartwell, *J. Amer. Chem. Soc.*, 1957, **79**, 3827.
2. B. Carnmalm, *Chem. and Ind.*, 1956, 1093.
3. G. E. McCasland and S. Proskow, *J. Amer. Chem. Soc.*, 1956, **78**, 5646.
4. S. F. Velick and J. English, *J. Biol. Chem.*, 1945, **160**, 473.
5. K. Balenović and N. Bregant, *Tetrahedron*, 1959, **5**, 44.
6. A. Fredga and L. Terenius, *Acta Chem. Scand.*, 1964, **18**, 2081.
7. E. Berner and R. Leonardsen, *Annalen*, 1939, **538**, 1.
8. M. Pailer and R. Libiseller, *Monatsh*, 1962, **93**, 511.
9. H. Brockmann and D. Müller-Enoch, *Chem. Ber.*, 1971, **104**, 3704.
10. P. S. Portoghese and D. A. Williams, *Tetrahedron Letters*, 1966, 6299.
11. A. H. Beckett, G. Kirk, and R. Thomas, *J. Chem. Soc.*, 1962, 1386.

Branched chain acids and diacids

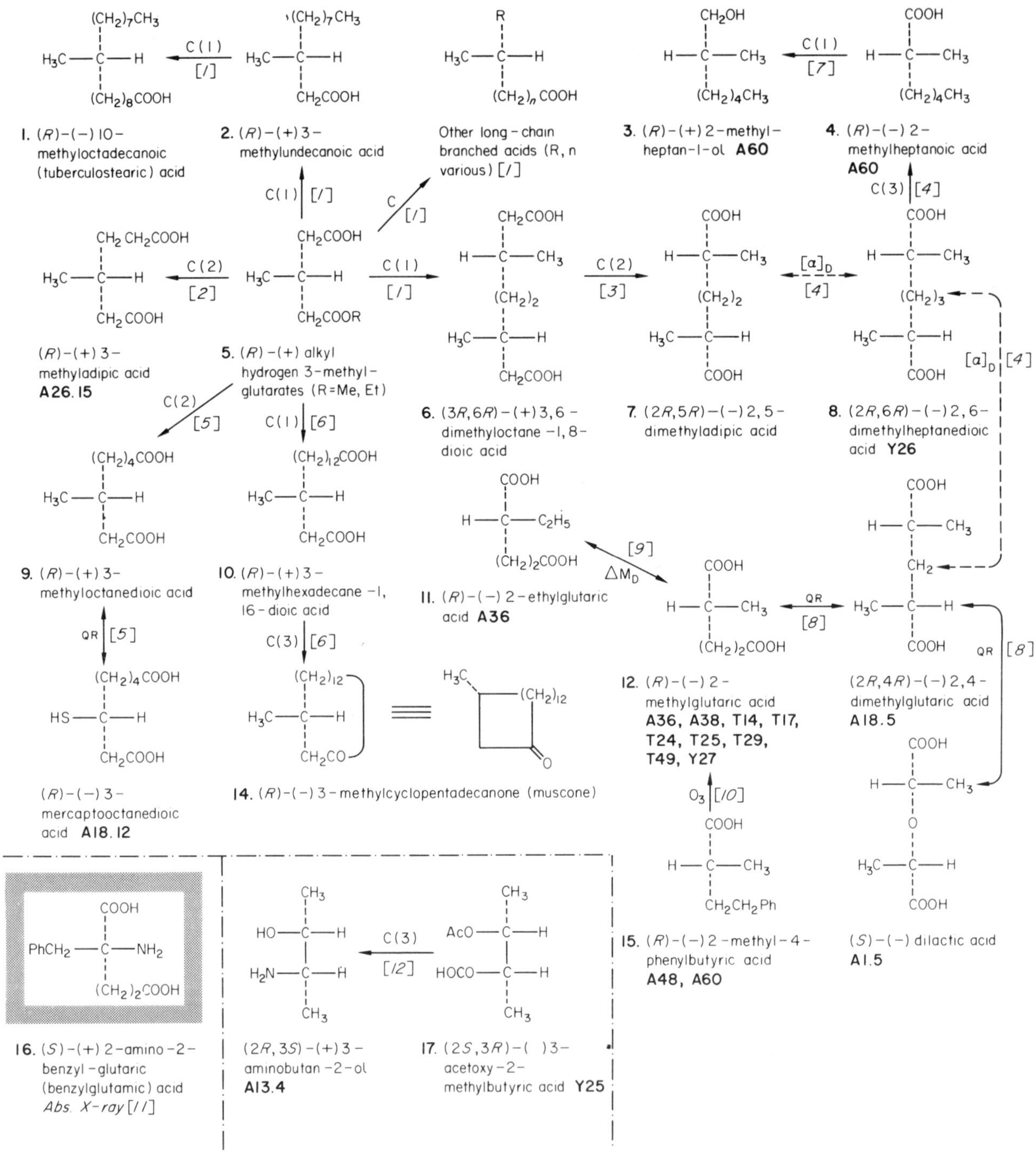

Class 3a

[Scheme of compounds 1–17 with interconversions]

1. (R)-(−)10-methyloctadecanoic (tuberculostearic) acid
2. (R)-(+)3-methylundecanoic acid
3. (R)-(+)2-methyl-heptan-1-ol **A60**
4. (R)-(−)2-methylheptanoic acid **A60**
5. (R)-(+) alkyl hydrogen 3-methyl-glutarates (R=Me, Et)
6. (3R,6R)-(+)3,6-dimethyloctane-1,8-dioic acid
7. (2R,5R)-(−)2,5-dimethyladipic acid
8. (2R,6R)-(−)2,6-dimethylheptanedioic acid **Y26**
9. (R)-(+)3-methyloctanedioic acid
10. (R)-(+)3-methylhexadecane-1,16-dioic acid
11. (R)-(−)2-ethylglutaric acid **A36**
12. (R)-(−)2-methylglutaric acid **A36, A38, T14, T17, T24, T25, T29, T49, Y27**

(R)-(+)3-methyladipic acid **A26.15**

(R)-(−)3-mercaptooctanedioic acid **A18.12**

14. (R)-(−)3-methylcyclopentadecanone (muscone)

(2R,4R)-(−)2,4-dimethylglutaric acid **A18.5**

15. (R)-(−)2-methyl-4-phenylbutyric acid **A48, A60**

(S)-(−) dilactic acid **A1.5**

16. (S)-(+)2-amino-2-benzyl-glutaric (benzylglutamic) acid Abs. X-ray [11]
17. (2S,3R)-(−)3-acetoxy-2-methylbutyric acid **Y25**

(2R,3S)-(+)3-aminobutan-2-ol **A13.4**

---

1. S. Ställberg-Stenhagen, *Arkiv. Kemi*, 1952, **3**, 117, and refs. therein.
2. S. Ställberg-Stenhagen, *Arkiv Kemi*, 1947, **25A**, no. 10.
3. I. Hedlund, *Arkiv Kemi*, 1956, **8**, 89.
4. J. Leitich, W. Oppolzer and V. Prelog, *Experientia*, 1964, **20**, 343.
5. K. Mislow and W. C. Meluch, *J. Amer. Chem. Soc.*, 1956, **78**, 5920.
6. S. Ställberg-Stenhagen, *Arkiv Kemi*, 1951, **3**, 517.
7. P. A. Levene and M. Kuna, *J. Biol. Chem.*, 1941, **140**, 255.
8. A. Fredga, *Arkiv Kemi*, 1947, **24A**, no. 32.
9. K. Mislow and I. V. Steinberg, *J. Amer. Chem. Soc.*, 1955, **77**, 3807.
10. B. Sjöberg, *Arkiv Kemi*, 1958, **12**, 565.
11. T. Ashida, Y. Sasada and M. Kakudo, *Bull. Chem. Soc. Japan*, 1967, **40**, 476.
12. W. D. Celmer, *J. Amer. Chem. Soc.*, 1965, **87**, 1797.

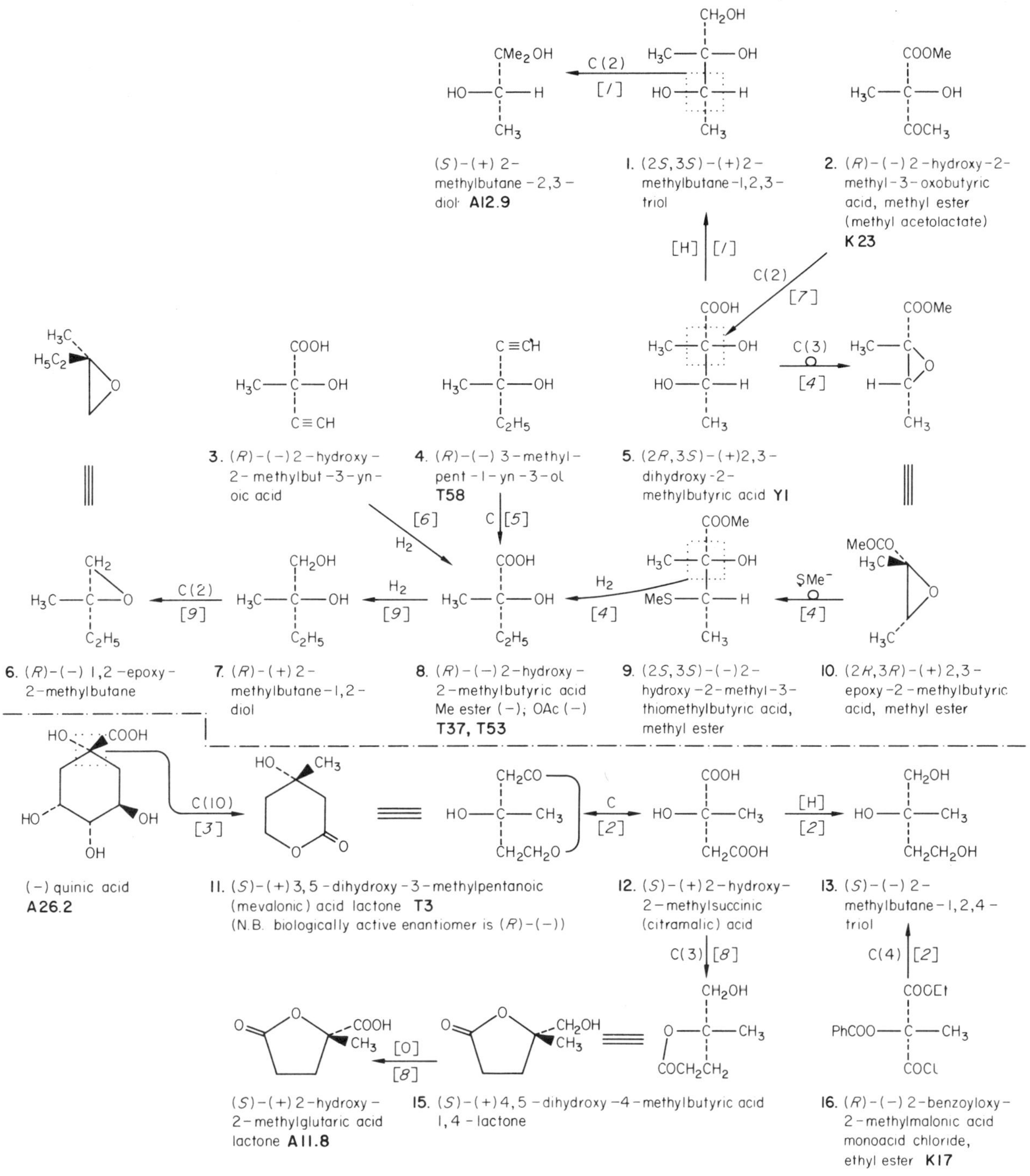

Class 3a
Branched-chain hydroxyacids
A33

(S)-(+) 2-methylbutane-2,3-diol **A12.9**

1. (2S,3S)-(+) 2-methylbutane-1,2,3-triol

2. (R)-(−) 2-hydroxy-2-methyl-3-oxobutyric acid, methyl ester (methyl acetolactate) **K23**

3. (R)-(−) 2-hydroxy-2-methylbut-3-ynoic acid

4. (R)-(−) 3-methylpent-1-yn-3-ol **T58**

5. (2R,3S)-(+) 2,3-dihydroxy-2-methylbutyric acid **Y1**

6. (R)-(−) 1,2-epoxy-2-methylbutane

7. (R)-(+) 2-methylbutane-1,2-diol

8. (R)-(−) 2-hydroxy-2-methylbutyric acid Me ester (−); OAc (−) **T37, T53**

9. (2S,3S)-(−) 2-hydroxy-2-methyl-3-thiomethylbutyric acid, methyl ester

10. (2R,3R)-(+) 2,3-epoxy-2-methylbutyric acid, methyl ester

(−) quinic acid **A26.2**

11. (S)-(+) 3,5-dihydroxy-3-methylpentanoic (mevalonic) acid lactone **T3** (N.B. biologically active enantiomer is (R)-(−))

12. (S)-(+) 2-hydroxy-2-methylsuccinic (citramalic) acid

13. (S)-(−) 2-methylbutane-1,2,4-triol

14. (S)-(+) 2-hydroxy-2-methylglutaric acid lactone **A11.8**

15. (S)-(+) 4,5-dihydroxy-4-methylbutyric acid 1,4-lactone

16. (R)-(−) 2-benzoyloxy-2-methylmalonic acid monoacid chloride, ethyl ester **K17**

1. B. W. Christensen and A. Kjaer, *Proc. Chem. Soc.*, 1962, 307.
2. P. A. Stadler, A. J. Frey, and A. Hofmann, *Helv. Chim. Acta*, 1963, **46**, 2300.
3. M. Eberle and D. Arigoni, *Helv. Chim. Acta*, 1960, **43**, 1508.
4. B. W. Christensen and A. Kjaer, *Acta Chem. Scand.*, 1962, **16**, 2466.
5. D. J. Faulkner and M. R. Petersen, *J. Amer. Chem. Soc.*, 1971, **93**, 3766.
6. D. Dugat, M. Verny and R. Vessière, *Tetrahedron*, 1971, **27**, 1715.
7. D. J. Robins and D. H. G. Crout, *J. Chem. Soc. (C)*, 1970, 1334.
8. H. P. Sigg and H. P. Weber, *Helv. Chim. Acta*, 1968, **51**, 1395.
9. W. Kirmse, H. Arold and B. Kornrumpf, *Chem. Ber.*, 1971, **104**, 1783.

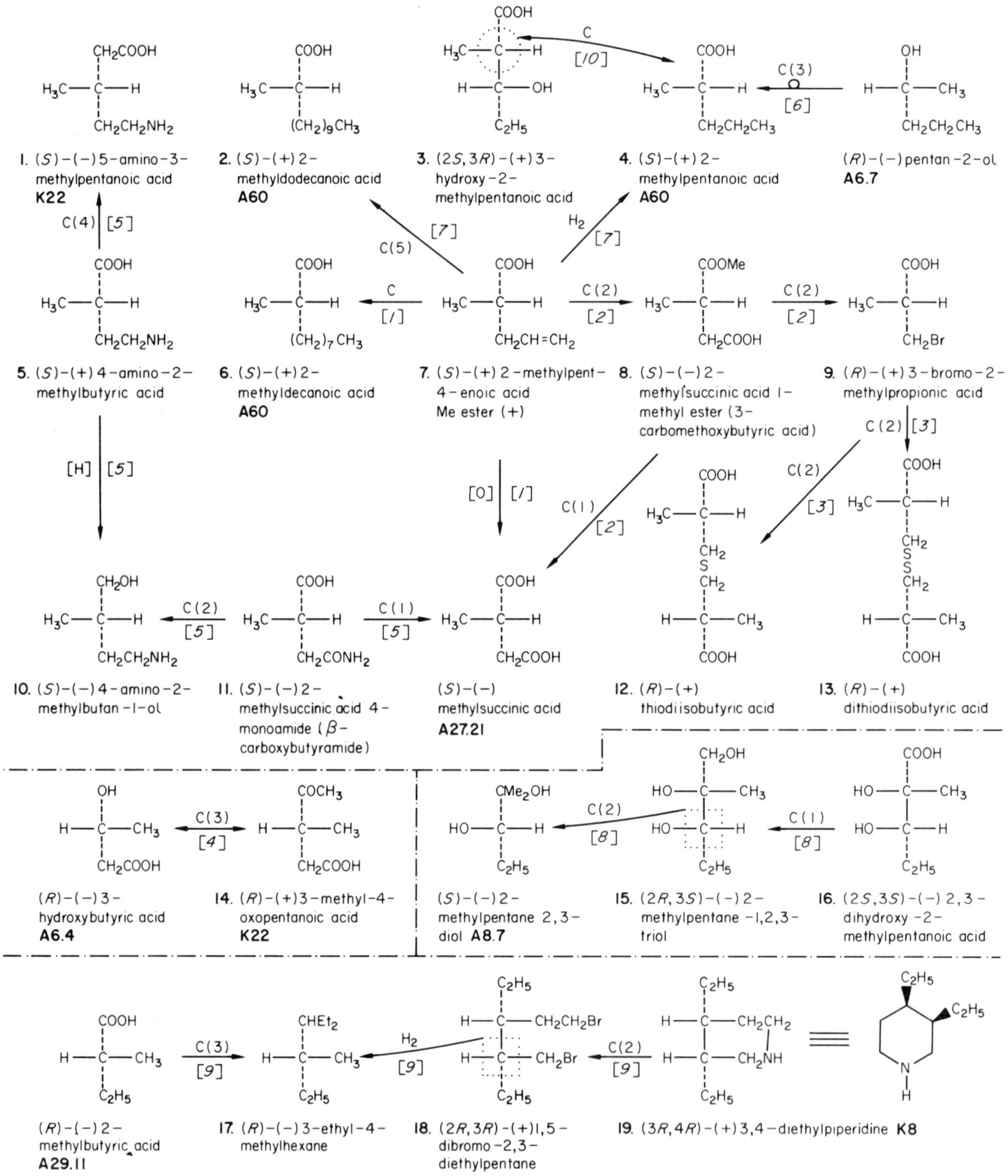

Further compounds derived from branched-chain acids

Class 3a

1. $(S)-(-)$ 5-amino-3-methylpentanoic acid **K22**
2. $(S)-(+)$ 2-methyldodecanoic acid **A60**
3. $(2S,3R)-(+)$ 3-hydroxy-2-methylpentanoic acid
4. $(S)-(+)$ 2-methylpentanoic acid **A60**
    $(R)-(-)$ pentan-2-ol **A6.7**
5. $(S)-(+)$ 4-amino-2-methylbutyric acid
6. $(S)-(+)$ 2-methyldecanoic acid **A60**
7. $(S)-(+)$ 2-methylpent-4-enoic acid Me ester (+)
8. $(S)-(-)$ 2-methylsuccinic acid 1-methyl ester (3-carbomethoxybutyric acid)
9. $(R)-(+)$ 3-bromo-2-methylpropionic acid
10. $(S)-(-)$ 4-amino-2-methylbutan-1-ol
11. $(S)-(-)$ 2-methylsuccinic acid 4-monoamide (β-carboxybutyramide)
    $(S)-(-)$ methylsuccinic acid **A27.21**
12. $(R)-(+)$ thiodiisobutyric acid
13. $(R)-(+)$ dithiodiisobutyric acid
    $(R)-(-)$ 3-hydroxybutyric acid **A6.4**
14. $(R)-(+)$ 3-methyl-4-oxopentanoic acid **K22**
    $(S)-(-)$ 2-methylpentane 2,3-diol **A8.7**
15. $(2R,3S)-(-)$ 2-methylpentane-1,2,3-triol
16. $(2S,3S)-(-)$ 2,3-dihydroxy-2-methylpentanoic acid
    $(R)-(-)$ 2-methylbutyric acid **A29.11**
17. $(R)-(-)$ 3-ethyl-4-methylhexane
18. $(2R,3R)-(+)$ 1,5-dibromo-2,3-diethylpentane
19. $(3R,4R)-(+)$ 3,4-diethylpiperidine **K8**

---

1. S. Ställberg-Stenhagen and E. Stenhagen, *Arkiv Kemi*, 1947, **B24**, no. 9.
2. G. Ställberg, *Acta Chem. Scand.*, 1956, **10**, 1360.
3. G. Ställberg, *Arkiv Kemi*, 1958, **12**, 131.
4. S. Masamune, *J. Amer. Chem. Soc.*, 1960, **82**, 5253.
5. R. Adams and D. Fles, *J. Amer. Chem. Soc.*, 1959, **81**, 4946.
6. H. L. Goering and W. I. Kimoto, *J. Amer. Chem. Soc.*, 1965, **87**, 1748.
7. S. Ställberg-Stenhagen, *Arkiv Kemi*, 1947, **23A**, no. 15.
8. D. G. Manwaring, R. W. Rickards and R. M. Smith, *Tetrahedron Letters*, 1970, 1029.
9. V. Prelog and E. Zalán, *Helv. Chim. Acta*, 1944, **27**, 535, 545.
10. G. A. Snow, *Biochem. J.*, 1965, **94**, 160.

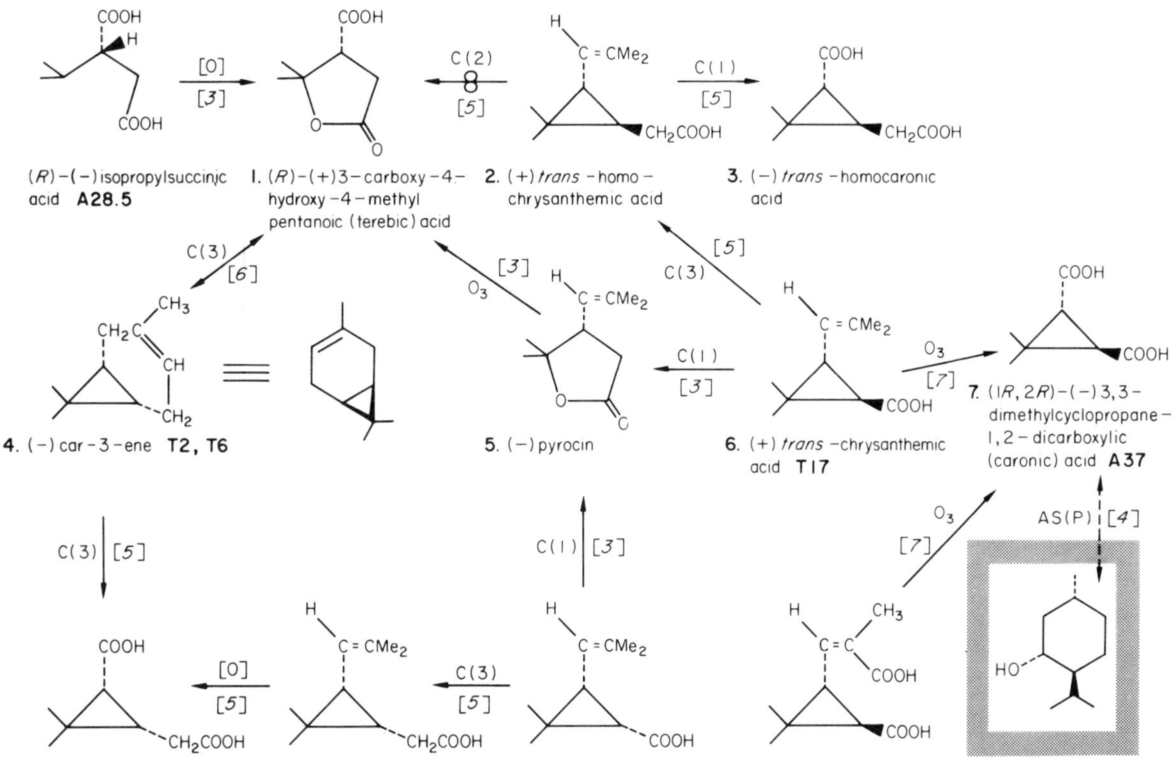

Class 3a
Cyclopropanes, including chrysanthemic acids

1. (R)-(-)isopropylsuccinic acid **A28.5**
2. (R)-(+)3-carboxy-4-hydroxy-4-methyl pentanoic (terebic) acid
3. (+) *trans*-homo-chrysanthemic acid
4. (-) *trans*-homocaronic acid
5. (-) car-3-ene **T2, T6**
6. (-) pyrocin
7. (+) *trans*-chrysanthemic acid **T17**
8. (1R,2R)-(-)3,3-dimethylcyclopropane-1,2-dicarboxylic (caronic) acid **A37**
9. (-) *cis*-homocaronic acid
10. (-) *cis*-homochrysanthemic acid
11. (-) *cis*-chrysanthemic acid
12. (+) *trans*-chrysanthemumdicarboxylic acid. Monomethyl ether = (+) pyrethric acid **T17**

(-) menthol **A26.12**

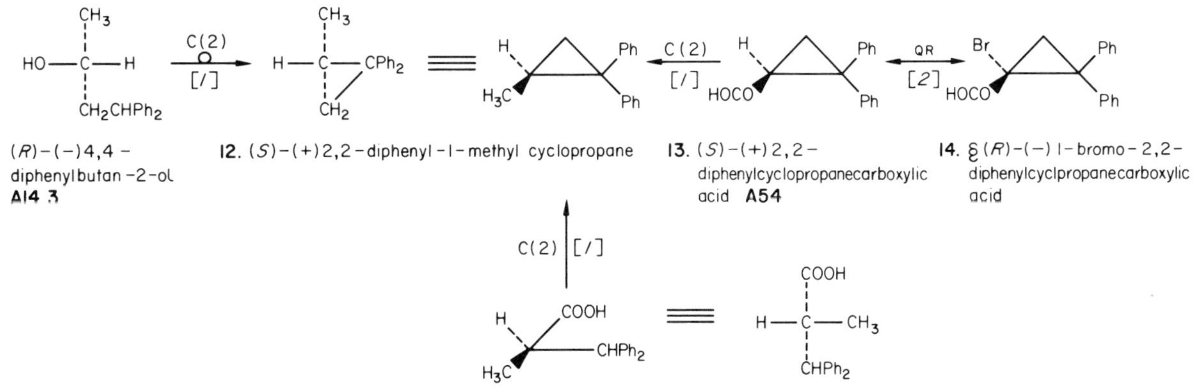

(R)-(-)4,4-diphenylbutan-2-ol **A14.3**

12. (S)-(+)2,2-diphenyl-1-methyl cyclopropane
13. (S)-(+)2,2-diphenylcyclopropanecarboxylic acid **A54**
14. (R)-(-)1-bromo-2,2-diphenylcyclopropanecarboxylic acid

15. (R)-(+)3,3-diphenyl-2-methylpropionic acid
Also by ORD [*1*] Related compds. [*1*]

---

1. H. M. Walborsky and C. G. Pitt, *J. Amer. Chem. Soc.*, 1962, **84**, 4831.
2. H. M. Walborsky, L. Barash, A. E. Young, and F. J. Impasto, *J. Amer. Chem. Soc.*, 1961, **83**, 2517.
3. L. Crombie and S. H. Harper, *J. Chem. Soc.*, 1954, 470.
4. H. M. Walborsky, T. Sugita, M. Ohno and Y. Inouye, *J. Amer. Chem. Soc.*, 1960, **82**, 5255.
5. L. Crombie, J. Crossley and D. A. Mitchard, *J. Chem. Soc.*, 1963, 4957.
6. R. Sandberg, *Arkiv Kemi*, 1960, **16**, 255.
7. H. Staudinger and L. Ruzicka, *Helv. Chim. Acta*, 1924, **7**, 201.

A³⁶ Cyclopentanes — Class 3a

1. (S)-(−) 3-oxocyclopentanecarboxylic acid
2. (R)-(−) 2-methylene-3-oxocyclopentanecarboxylic acid (sarkomycin)
3. (1R,2S)-(+) 2-methyl-3-oxocyclopentanecarboxylic acid
4. (1R,2R)-(−) 1,2-dimethylcyclopentane **A38**
5. (1R,2R)-(−) 2-methylcyclopentanecarboxylic acid **A38** Amide (−)
6. (S)-(−) 3-methylcyclopentanone **T58**
7. (S)-(−) 3-methylcyclopentenyl methyl ketone **T15**
8. (R)-(+) chaulmoogric acid
9. (R)-(−) cyclopent-2-ene acetic acid
10. (S)-(−) 3-oxocyclopentaneacetic acid
11. (1R,4S)-(−) 2-oxobicyclo [2.2.1] heptane (norcamphor, norbornan-2-one) **A46, A47, X3**
12. (S)-(−) 3-ethylcyclopentene
13. (−) methyl dihydrojasmonate
14. ( ) jasmonic acid
15. (1R,2R)-(+) cyclopentane-1,2-diacetic acid **A37**

(R)-(+) butane-1,2,4-tricarboxylic acid **A26.10**
(S)-(−) methylsuccinic acid **A27.21**
(S)-(+) 2-methylglutaric acid **A32.12**
(S)-(+) ethylglutaric acid **A32.11**

16. (S)-(−) 2,2,3-trimethyladipic acid **T54**
17. (S)-(−) 2,2,3-trimethylcyclopentanone
18. (S)-(+) 1-methylene-2,2,3-trimethylcyclopentane
19. (1R,3S)-(+) 1,2,2,3-tetramethylcyclopentylamine
20. (1R,3S)-(+) 1,2,2,3-tetramethylcyclopentanecarboxylic (campholic) acid **T11**

---

1. R. K. Hill and A. G. Edwards, *Tetrahedron*, 1965, **21**, 1501.
2. K. Mislow and I. V. Steinberg, *J. Amer. Chem. Soc.*, 1955, **77**, 3807.
3. S. M. McElvain and E. J. Eisenbraun, *J. Amer. Chem. Soc.*, 1955, **77**, 1599, 3383.
4. D. C. Aldridge, S. Galt, D. Giles and W. B. Turner, *J. Chem. Soc. (C)*, 1971, 1623.
5. V. Rautenstrauch and G. Ohloff, *Helv. Chim. Acta*, 1971, **54**, 1776.
6. Y. Sato, S. Nishioka, O. Yonemitsu and Y. Ban, *Chem. Pharm. Bull. (Japan)*, 1963, **11**, 829.
7. R. K. Hill, P. J. Foley and L. A. Gardella, *J. Org. Chem.*, 1967, **32**, 2330.
8. I. R. Hooper, L. C. Cheney, M. J. Cron, O. B. Fardig, D. A. Johnson, D. L. Johnson, F. M. Palermiti, H. Schmitz and W. B. Wheatley, *Antibiot. Chemotherapy*, 1955, **5**, 585.

Class 3a
Cyclopropanes, cyclohexanes, etc.
A37

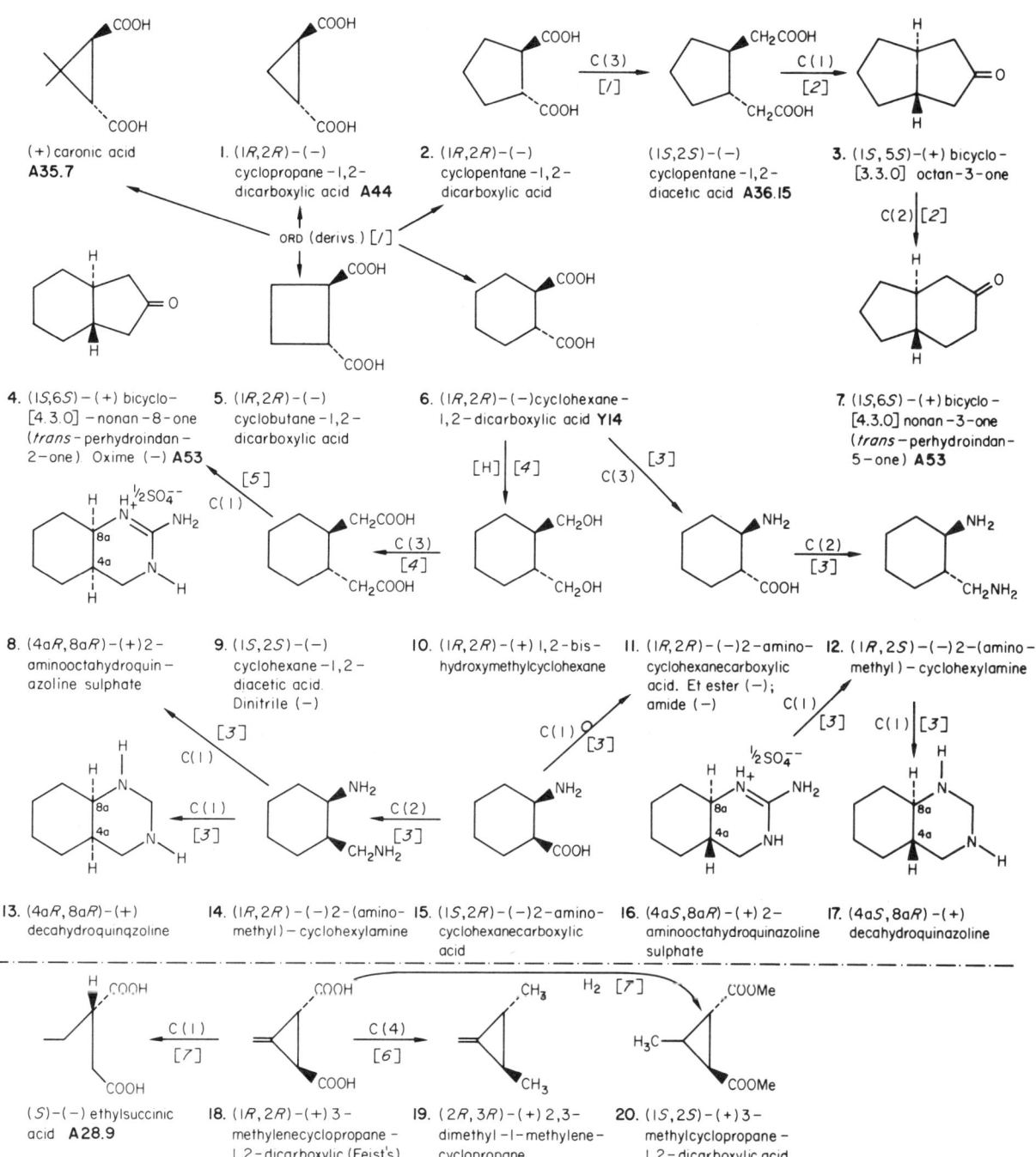

1. Y. Inouye, S. Sawada, M. Ohno and H. M. Walborsky, *Tetrahedron*, 1967, **23**, 3237.
2. P. M. Bourn and W. Klyne, *J. Chem. Soc.*, 1960, 2044.
3. W. L. F. Armarego and T. Kobayashi, *J. Chem. Soc. (C)*, 1969, 1635; 1970, 1597.
4. D. E. Applequist and N. D. Werner, *J. Org. Chem.*, 1963, **28**, 48.
5. W. Hückel, M. Sachs, J. Yantschulewitsch and F. Nerdel, *Ann.*, 1935, **518**, 155.
6. J. J. Gajewski, *J. Amer. Chem. Soc.*, 1971, **93**, 4450.
7. W. von E. Doering and H. D. Roth, *Tetrahedron*, 1970, **26**, 2825.

Cyclohexane derivatives

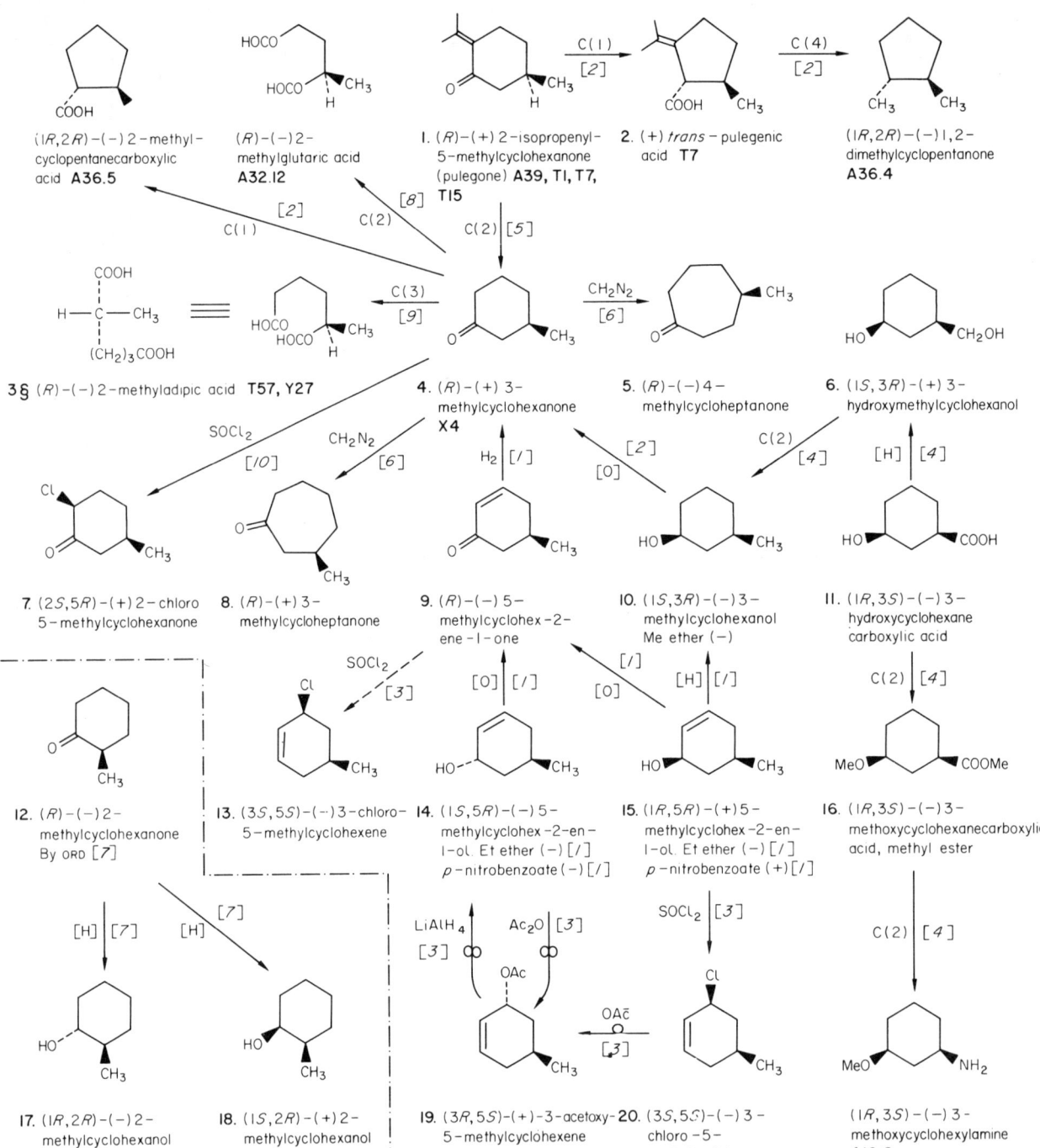

1. H. L. Goering and E. F. Silversmith, *J. Amer. Chem. Soc.*, 1955, 77, 5172.
2. R. K. Hill, P. J. Foley and L. A. Gardello, *J. Org. Chem.*, 1967, 32, 2330.
3. H. L. Goering, T. D. Nevitt and E. F. Silversmith, *J. Amer. Chem. Soc.* 1955, 77, 4042.
4. D. S. Noyce and D. B. Denney, *J. Amer. Chem. Soc.*, 1954, 76, 768.
5. O. Wallach, *Annalen*, 1896, 289, 337.
6. C. Djerassi, B. F. Burrows, C. G. Overberger, T. Takekoshi, C. D. Gutsche and C. T. Chang, *J. Amer. Chem. Soc.*, 1963, 85, 949.
7. C. Beard, C. Djerassi, T. Elliott and R. C. C. Tao, *J. Amer. Chem. Soc.*, 1962, 84, 874.
8. E. J. Eisenbraun and S. M. McElvain, *J. Amer. Chem. Soc.*, 1955, 77, 3383.
9. C. F. Wong, E. Auer and R. T. LaLonde, *J. Org. Chem.*, 1970, 35, 517.
10. C. Djerassi and L. E. Geller, *Tetrahedron*, 1958, 3, 319.

Class 3a
Cyclohexane and decalin derivatives
A³⁹

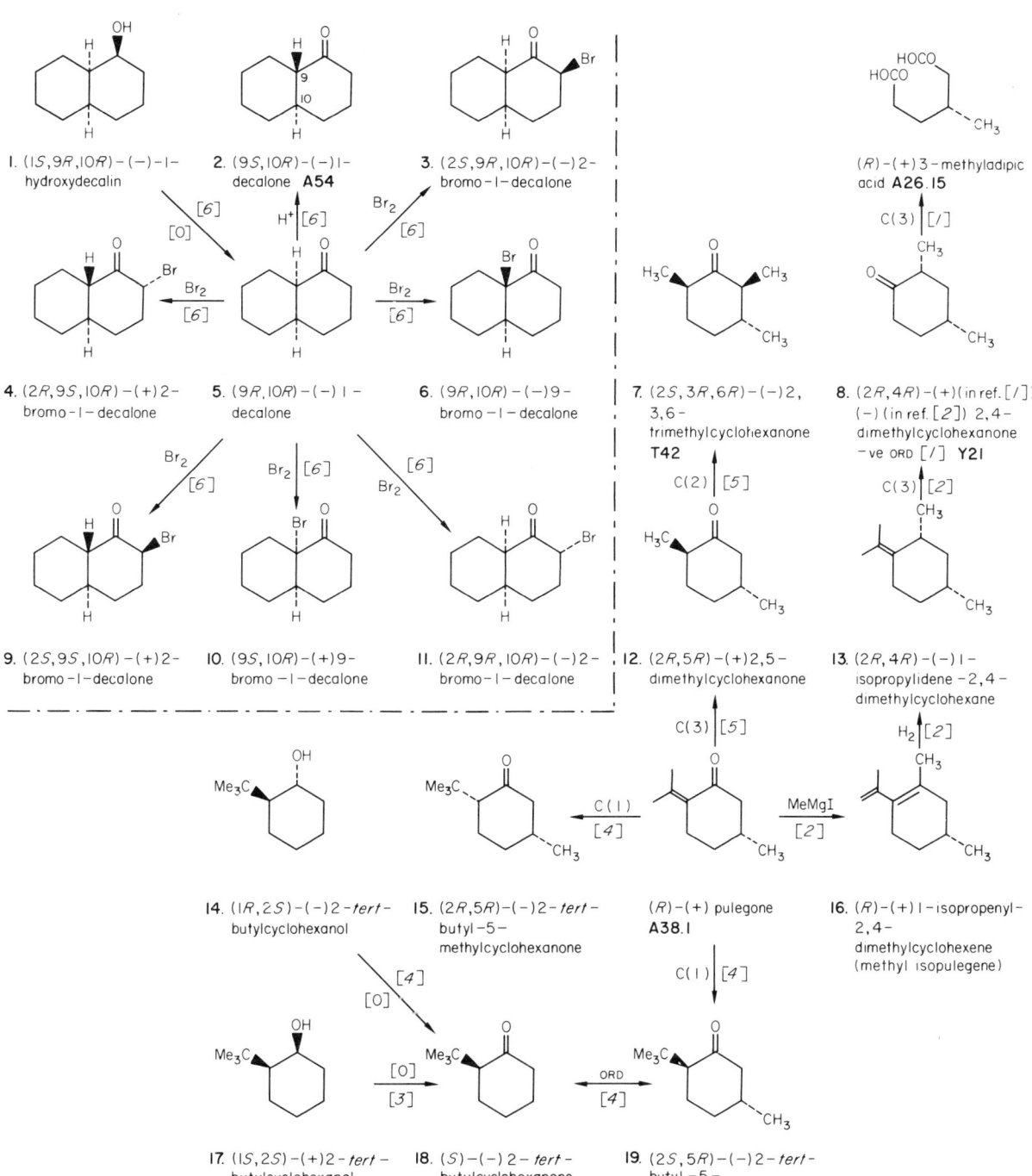

1. (1S,9R,10R)-(−)-1-hydroxydecalin
2. (9S,10R)-(−)-1-decalone **A54**
3. (2S,9R,10R)-(−)-2-bromo-1-decalone
4. (2R,9S,10R)-(+)2-bromo-1-decalone
5. (9R,10R)-(−)-1-decalone
6. (9R,10R)-(−)-9-bromo-1-decalone
7. (2S,3R,6R)-(−)2,3,6-trimethylcyclohexanone **T42**
8. (2R,4R)-(+)(in ref.[1])(−)(in ref.[2]) 2,4-dimethylcyclohexanone −ve ORD [1] **Y21**
9. (2S,9S,10R)-(+)2-bromo-1-decalone
10. (9S,10R)-(+)9-bromo-1-decalone
11. (2R,9R,10R)-(−)2-bromo-1-decalone
12. (2R,5R)-(+)2,5-dimethylcyclohexanone
13. (2R,4R)-(−)1-isopropylidene-2,4-dimethylcyclohexane
14. (1R,2S)-(−)2-tert-butylcyclohexanol
15. (2R,5R)-(−)2-tert-butyl-5-methylcyclohexanone
16. (R)-(+)1-isopropenyl-2,4-dimethylcyclohexene (methyl isopulegene)
17. (1S,2S)-(+)2-tert-butylcyclohexanol
18. (S)-(−)2-tert-butylcyclohexanone
19. (2S,5R)-(−)2-tert-butyl-5-methylcyclohexanone

(R)-(+)3-methyladipic acid **A26.15**

(R)-(+) pulegone **A38.1**

1. E. J. Eisenbraun, J. Osiecki and C. Djerassi, *J. Amer. Chem. Soc.*, 1958, **80**, 1261.
2. J. Wolinsky and D. Chan, *J. Amer. Chem. Soc.*, 1963, **85**, 937.
3. K. L. Cheo, T. H. Elliott and C. C. Tao, *J. Chem. Soc. (C)*, 1966, 1988.
4. C. Djerassi, P. A. Hart and E. J. Warawa, *J. Amer. Chem. Soc.*, 1964, **86**, 78.
5. A. Melera, D. Arigoni, A. Eschenmoser, O. Jeger and L. Ruzicka, *Helv. Chim Acta*, 1956, **39**, 441.
6. C. Djerassi and J. Staunton, *J. Amer. Chem. Soc.*, 1961, **83**, 736.

A40 α-methylaminoacids (e.g. isovaline) and derivatives Class 3a, 3b

1. (R)-( )2-aminobutane-1,2,4-tricarboxylic acid
2. (R)-(+)5-hydroxymethyl-5-(2-hydroxyethyl)-2-pyrrolidone
3. (S)-(−)2-amino-2-methyl-3-phenylpropionic acid (X=H); (S)-(−)α-methyl DOPA (X=OH)
4. (S)-(+)2-amino-2-methylsuccinic (α-methylaspartic) acid
5. (S)-(+)5-ethyl-5-methyl-2-pyrrolidone
6. (R)-(−)2-amino-2-phenylbutyric acid
7. (S)-(−)2-(3,4-dihydroxybenzyl)-2-hydrazinopropionic acid
8. (S)-(+)2-amino-3-hydroxy-2-methyl-propionic acid (α-methylserine)
9. (S)-(+)2-amino-2-methylbutyric acid (isovaline) CD; [4]
10. (S)-(+)2-amino-2-phenylpropionic acid N,N-diMe (+)
11. (R)-(−)2-amino-2-phenylpropan-1-ol A55
12. (R)-(−)2-amino-2-phenylpropionic acid (α-methylphenylglycine) CD;[4]derivs;[12] A55
13. (S)-(−)2-amino-2-phenylbutane. N-Ac (−); N-benzoyl (+) CD;[4]
14. (S)-(+)2-amino-2-methylbut-3-enoic acid
15. (S)-(+)2-amino-2-cyclohexylpropionic acid
16. (R)-(−)2-methyl-2-phenylaziridine
17. (R)-(−)3-amino-3-phenylbutan-2-one
18. (S)-(+)2-anilino-2-phenylpropionic acid
19. (S)-(+)2-(cyclohexylamino)-2-cyclohexylpropionic acid

(−)quinic acid A26.2
(S)-(−)DOPA A5.7

1. S. Yamada and K. Achiwa, *Chem. Pharm. Bull. (Japan)*, 1964, **12**, 1525; 1966, **14**, 537.
2. N. Takamura, S. Terashima, K. Achiwa and S. Yamada, *Chem. Pharm. Bull. (Japan)*, 1967, **15**, 1776.
3. Y. Sugi and S. Mitsui, *Bull. Chem. Soc. Japan*, 1970, **43**, 564.
4. K. Achiwa, S. Terashima, H. Mizuno, N. Takamura, T. Kitagawa, K. Ishikawa and S. Yamada, *Chem. Pharm. Bull. (Japan)*, 1970, **18**, 61.
5. H. Dahn, J. A. Garbarino and C. O'Murchu, *Helv. Chim. Acta*, 1970, **53**, 1370.
6. S. Terashima, K. Achiwa and S. Yamada, *Chem. Pharm. Bull. (Japan)*, 1966, **14**, 572, 579.
7. E. W. Tristram, J. T. Broeke, D. F. Reinhold, M. Sletzinger and D. E. Williams, *J. Org. Chem.*, 1964, **29**, 2053.
8. S. Karady, M. G. Ly, S. H. Pines and M. Sletzinger, *J. Org. Chem.*, 1971, **36**, 1949.
9. S. Terashima, K. Achiwa and S. Yamada, *Chem. Pharm. Bull. (Japan)*, 1966, **14**, 1138.
10. H. Mizuno, S. Terashima, K. Achiwa and S. Yamada, *Chem. Pharm. Bull. (Japan)*, 1967, **15**, 1749.
11. S. Yamada, S. Terashima and K. Achiwa, *Chem. Pharm. Bull. (Japan)*, 1966, **14**, 800.
12. D. J. Cram, L. K. Gaston and H. Jäger, *J. Amer. Chem. Soc.*, 1961, **83**, 2183.
13. J. Knabe and C. Urbahn, *Annalen*, 1971, **750**, 21.

Class 3a, 3b   One-centre compounds related to hydratropic acid   **A⁴¹**

1. (S)-(+) 3-phenylbutan-2-one **A48**
2. (S)-(−) 2,2-dimethyl-3-cyclohexylbutane
3. (R)-(−) 2,2-dimethyl-3-phenylbutane
4. (R)-(+) 3,3-dimethyl-2-phenylbutan-1-ol
5. (R)-(−) 1-amino-2-phenylbutane
6. (R)-(−) 2-methyl-3-phenylbutane
   (R)-(−) 3-methyl-2-phenylbutyric acid **A25.17**
   (R)-(+) 1-phenylethylamine. N-Me (+) [6] **A19.14**
   (R)-(−) 3,3-dimethyl-2-phenylbutyric acid **A25.19**
7. (R)-(−) 2-phenylbutan-1-ol **A42**
8. (S)-(−) 3-hydroxy-2-phenylpropionic (tropic) acid. Amide (−) **K28**
9. (R)-(−) 3-chloro-2-phenylpropionic acid
10. (R)-(−) 2-phenylpropionic (hydratropic) acid. p-NO₂ (+); p-NH₂ (−) **A48, A49, A61, Y5, Y9**
    (R)-(−) 2-phenylbutyric acid **A25.15**
11. (R)-(−) 2-phenylpentanoic acid **A50**
12. (R)-(−) 3,3,4-triphenylhexan-2-one
13. (S) 1,1,2-triphenylbutane. (+)(CHCl₃) (−)(EtOH)
14. (R)-(+) 1,1,2-triphenylbutan-1-ol
15. (R)-(+) 2,2,3-triphenylpentan-1-ol
16. (R)-(+) 2,2,3-triphenylpentanoic acid
17. (R)-(−) 2,2-diphenyl-3-ethylindan-1-one

1. G. Fodor and G. Csepreghy, *J. Chem. Soc.*, 1961, 3222.
2. M-J. Brienne, C. Ouannes and J. Jaques, *Bull. Soc. chim. France*, 1967, 613.
3. D. R. Clark and H. S. Mosher, *J. Org. Chem.*, 1970, **35**, 1114.
4. K. Pettersson, *Arkiv Kemi*, 1956, **10**, 283.
5. J. C. Craig, W. E. Pereira, B. Halpern and J. W. Westley, *Tetrahedron*, 1971, **27**, 1173.
6. S. Fujita, K. Imamura and H. Nozaki, *Bull. Chem. Soc. Japan*, 1971, **44**, 1975.
7. P. A. Levene and R. E. Marker, *J. Biol. Chem.*, 1931, **93**, 749.
8. W. Kirmse and W. Gruber, *Chem. Ber.*, 1971, **104**, 1795.

A⁴² One-centre compounds related to 3-phenylpentanoic acid — Class 3a, 3b

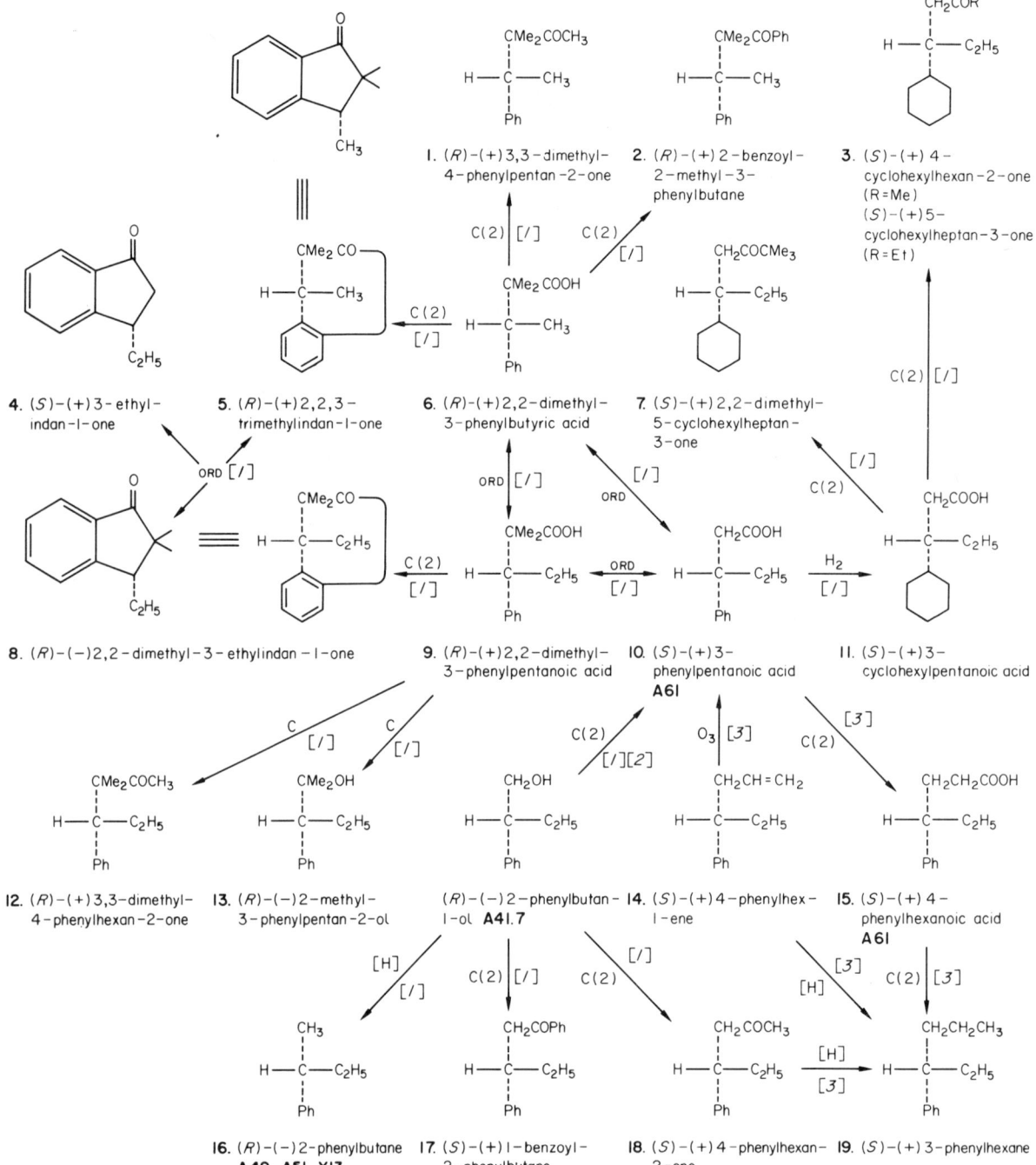

1. M-J. Brienne, C. Ouannes and J. Jacques, *Bull. Soc. chim. France*, 1967, 613.
2. P. A. Levene and R. E. Marker, *J. Biol. Chem.*, 1931, 93, 749.
3. R. Menicagli and L. Lardicci, *Chem. and Ind.*, 1971, 1490.

Class 3a, 3b 2,3-diphenylpropionic acid and related aromatic compounds; tetralones **A⁴³**

1. (1S,2S)-(−)1-amino-1,2-diphenylpropane
2. (2S,3S)-(−)-3-amino-2,3-diphenylpropan-1-ol
3. (2R,3R)-(−)-2,3-diphenylbutane-1,4-diol
4. (2R,3R)-(−)2,3-diphenylbutane
5. (S)-(+)2-benzylhexanoic acid
6. (2S,3S)-( )3-amino-2,3-diphenylpropionic acid.
7. (2R,3R)-(−)-2,3-diphenylsuccinic acid. Ring-substd. derivs. [3]
8. (R)-(−)2,3-diphenylpropan-1-ol
9. (1R,2S)-(−)1-amino-1,2-diphenylpropane
10. (S)-(−)3-phenyl-2-(2-thienyl) propionic acid A50
11. (R)-(−)2,3-diphenylpropionic acid Me ester (−) A48, A50
12. (2S,3R)-( )3-amino-2,3-diphenylpropionic acid
13. (2S,3R)-(−)3-amino-2,3-diphenylpropan-1-ol
14. (S)-(+)2-methyl-1-tetralone. CD; [6]
15. (S)-(−)3-phenyl-2-(2-thienyl) propan-1-ol (X=OH) (R)-(−)1-phenyl-2-(2-thienyl) propane (X=H)
16. (R)-(−)2-methyl-3-phenylpropionic acid A60
    (R)-(−) 2-amino-1-phenylpropane A5.10
17. (1R,2S)-(+)2-methyl-1-tetralol
18. (S)-(−) 2-methyltetralin
19. (S)-(−)2-methyl-1-phenylhexane
20. (R)-(−)2-methyl-1-phenylhexan-3-one
21. (R)-(−) 1-chloro-2-methyl-3-phenylpropane (X=Cl);
    (R)-(+) 2-methyl-3-phenylpropan-1-ol (X=OH)
22. (R)-(+)3-methyl-4-phenylbutyric acid A60
23. (R)-(−)3-methyl-1-tetralone. CD; [6]

1. N. D. Berova and B. J. Kurtev, *Tetrahedron*, 1969, **25**, 2301.
2. R. Buchan and M. B. Watson, *J. Chem. Soc. (C)*, 1968, 2465.
3. D. J. Collins and J. J. Hobbs, *Austral. J. Chem.*, 1970, **23**, 1605.
4. P. E. Verkade, K. S. De Vries and B. M. Wepster, *Rec. Trav. chim.*, 1964, **83**, 1149.
5. A. W. Schrecker, *J. Org. Chem.*, 1957, **22**, 33.
6. J. Barry, H-B. Kagan and G. Snatzke, *Tetrahedron*, 1971, **27**, 4737.
7. K. Pettersson, *Arkiv Kemi*, 1955, **8**, 387.
8. M. B. Watson and G. W. Youngson, *J. Chem. Soc. (C)*, 1968, 258.

Cyclopropanes

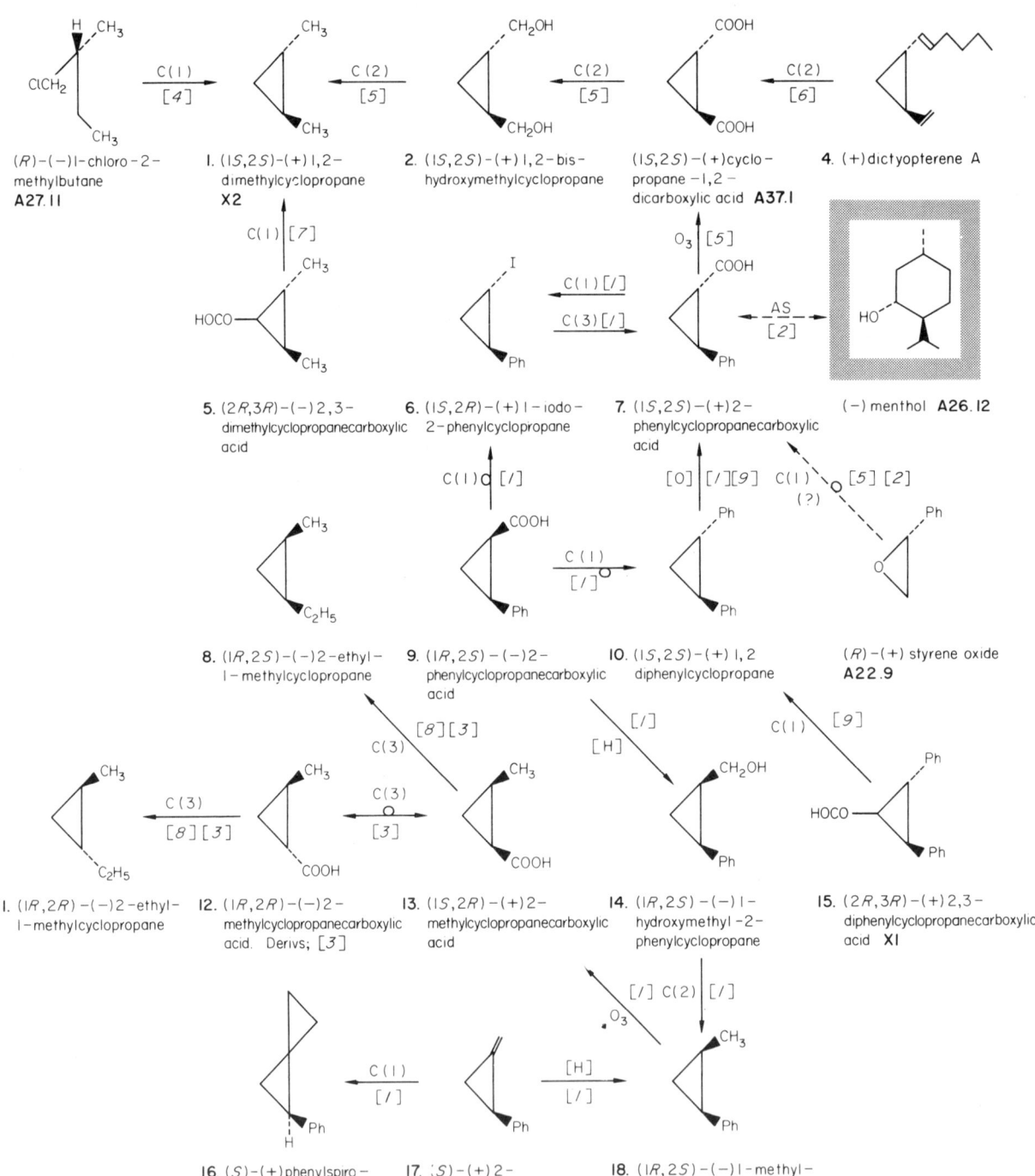

1. T. Aratani, Y. Nakanishi and H. Nozaki, *Tetrahedron*, 1970, **26**, 1675.
2. I. Tömösközi, *Tetrahedron*, 1963, **19**, 1969.
3. R. G. Bergman, *J. Amer. Chem. Soc.*, 1969, **91**, 7405.
4. W. von E. Doering and W. Kirmse, *Tetrahedron*, 1960, **11**, 272.
5. Y. Inouye, T. Sugita and H. M. Walborsky, *Tetrahedron*, 1964, **20**, 1695.
6. R. E. Moore, J. A. Pettus and M. S. Doty, *Tetrahedron Letters*, 1968, 4787.
7. W. M. Jones and J. M. Walbrick, *Tetrahedron Letters*, 1968, 5229.
8. W. L. Carter and R. G. Bergman, *J. Amer. Chem. Soc.*, 1968, **90**, 7344.
9. W. M. Jones and J. W. Wilson, *Tetrahedron Letters*, 1965, 1587.

Class 3a, 3b
Two-centre compounds related to diethylsuccinic acid; adamantanes

A45

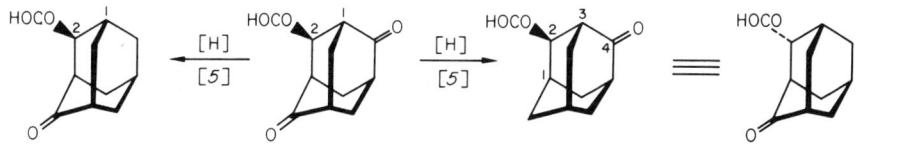

1. (3R,4R)-(+)3,4-diethyladipic acid Dinitrile (+)
2. (3R,4R)-(−)3,4-diethylcyclopentanone
3. (3R,4S)-( )3,4-diethyl-5-hydroxypentanoic acid lactone
4. (2S,3S)-(−)2,3-diethylbutane-1,4-diol
5. (2S,3S)-(−)2,3-diethylsuccinic acid. Anhydride (−); monodiethylamide (−)
6. (S)-(+)2-ethylpentanoic acid dimethylamide
   (S)-(+)2-ethylpentanoic acid **A29.18**
7. (3R,4S)-(−)5-bromo-3,4-diethylpentanoic acid, ethyl ester **K2**
8. (3S,4S)-(−)3,4-bis-(p-hydroxyphenyl)-hexane (hexoestrol)
9. (2S,3R)-(+)2,3-diethylsuccinic acid 1-dimethylamide
10. (2R,3S)-(−)2,3-diethylsuccinic acid 1-methylester
11. (2R,3R)-(−)2,3-diethylglutaric acid **T58**
12. (2S,3S)-(−)2,3-diethyl-4-hydroxybutyric acid lactone **K12, K30**

13. (1R,2S)-( )adamantan-4-one-2-carboxylic acid. By cD [5] (octant rule)
14. (1R,2S)-(+)adamantan-4,8-dione-2-carboxylic acid. By cD [5] (octant rule)
15. (1R,2R)-( )adamantan-4-one-2-carboxylic acid
   For other adamantane derivs., see [5][6]

1. A. R. Battersby, S. W. Breuer and S. Garratt, *J. Chem. Soc. (C)*, 1968, 2467.
2. K. Nagarajan, Ch. Weissmann, H. Schmid and P. Karrer, *Helv. Chim. Acta*, 1963, **46**, 1213.
3. D. H. R. Barton, L. D. S. Godinho and J. K. Sutherland, *J. Chem. Soc.*, 1965, 1779.
4. H. H. Inhoffen, D. Kopp, S. Marić, J. Bekurdts and R. Selimoglu, *Tetrahedron Letters*, 1970, 999.
5. G. Snatzke and G. Eckhardt, *Tetrahedron*, 1968, **24**, 4543.
6. G. Snatzke and G. Eckhardt, *Tetrahedron*, 1970, **26**, 1143.

A⁴⁶ Bicyclo(2,2,1)heptane (norbornane) group — Class 3a, 3b

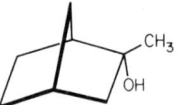

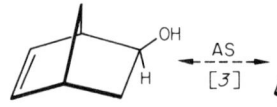
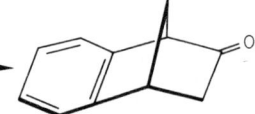

1. (1S,2R,4R)-(+)2-methylnorbornan-2-ol
2. (1R,4R)-(+) norborn-5-en-2-one (dehydronorcamphor)
3. (1R,2S,4R)-(+)norborn-5-en-2-ol (exo-dehydronoborneol)
4. (1R,4R)-(+) benznorbornan-2-one

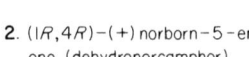

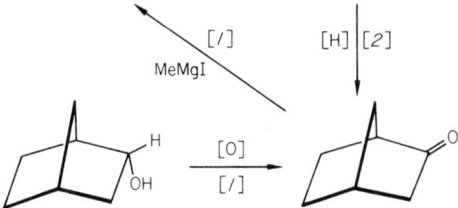
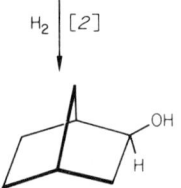
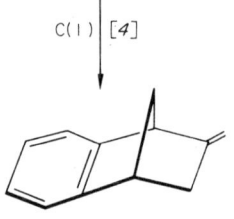

5. (1S,2R,4R)-(+) norbornan-2-ol (endo-norborneol)
   (1S,4R)-(+) norcamphor A36.11
6. (1S,2S,4R)-(−)norbornan-2-ol (exo-norborneol)
7. (1S,4R)-(+)-2-methylenebenznorbornene

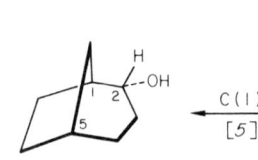

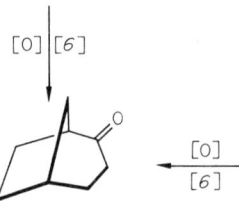
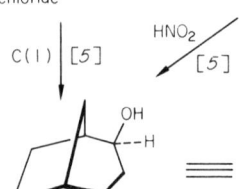

8. (1S,2R,5S)-(+) bicyclo-[3.2.1] octan-2-ol
9. (1S,2R,4R)-(−)2-(aminomethyl)-norbornane hydrochloride
10. (1S,2S,4R)-(+) 2-(aminomethyl)-norbornane hydrochloride

11. (1S,5S)-(+) bicyclo[3.2.1] octan-2-one. ORD; [6]
12. (1S,2S,5S)-(+) (in ref.[6]) (−)(in ref.[5]) bicyclo[3.2.1] octan-2-ol
13. (1R,2R,4S)-(−) bicyclo-[2.2.2] octan-2-ol X10

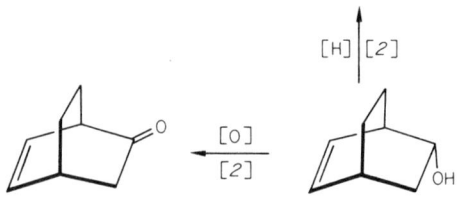

14. (1R,4R)-(+) bicyclo[2.2.2] oct-5-en-2-one
15. (1R,2R,4R)-(+) bicyclo-[2.2.2] oct-5-en-2-ol

1. J. A. Berson, J. S. Walia, A. Remanick, S. Suzuki, P. Reynolds-Warnhoff and D. Willner, *J. Amer. Chem. Soc.*, 1961, **83**, 3986. and references therein.
2. K. Mislow and J. G. Berger, *J. Amer. Chem. Soc.*, 1962, **84**, 1956.
3. D. J. Sandman, K. Mislow, W. P. Giddings, J. Dirlam and G. C. Hanson, *J. Amer. Chem. Soc.*, 1968, **90**, 4877.
4. D. J. Sandman and K. Mislow, *J. Amer. Chem. Soc.*, 1969, **91**, 645.
5. J. A. Berson and D. Willner, *J. Amer. Chem. Soc.*, 1962, **84**, 675.
6. H. M. Walborsky, M. E. Baum and A. A. Youssef, *J. Amer. Chem. Soc.*, 1961, **83**, 988.

Class 3a, 3b — Norbornane group (contd.); bicyclo(2,2,2)octane group A47

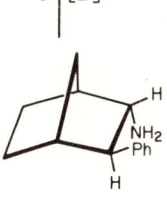
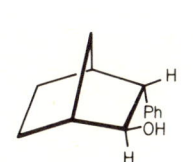

(−)norcamphor A36.11 → **1.** (1R,4S)-(−) 2-phenylnorborn-2-ene → **2.** (1R,3R,4S)-(+)3-phenylnorbornan-2-one ← rearr. [3] **3.** (1R,2R,3R,4S)-(−) 2-phenylnorbornane-1,2-diol

4. (1S,2S,3R,4R)-(+)2-amino-3-phenylnorbornane. Hydrochloride (+). Also by ORD of salicylidene deriv. [3]

5. (1R,3S,4S)-(−)3-phenylnorbornan-2-one

6. (1R,2S,3R,4S)-(+)3-phenylnorbornan-2-ol

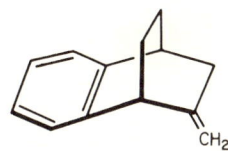
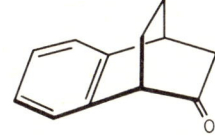
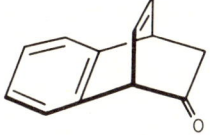

7. (1R,4S)-(−) 2-methylene-5,6-benzobicyclo[2.2.2] octane

8. (1S,4S)-(−)5,6-benzobicyclo [2.2.2] octan-2-one. By CD [1]

9. (1S,4S)-(+) 7,8-benzobicyclo [2.2.2] oct-2-ene-5-one

The following class 3a norbornanes have also been correlated with those on page A46; [5]

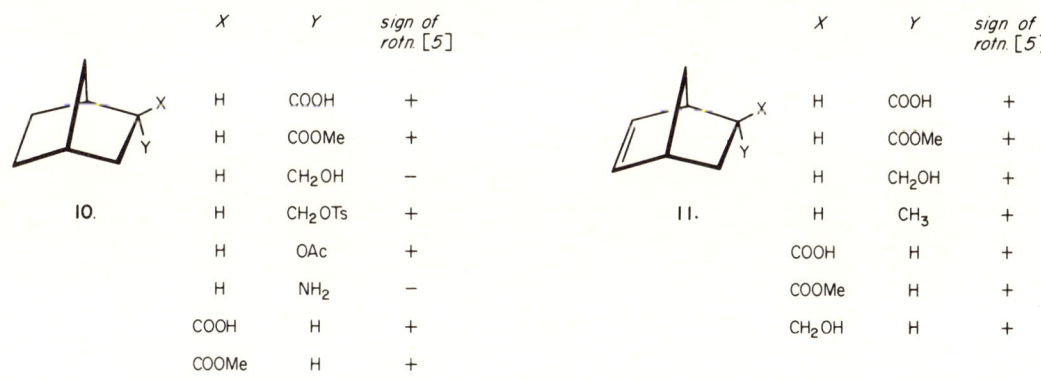

	X	Y	sign of rotn. [5]
10.	H	COOH	+
	H	COOMe	+
	H	CH$_2$OH	−
	H	CH$_2$OTs	+
	H	OAc	+
	H	NH$_2$	−
	COOH	H	+
	COOMe	H	+

	X	Y	sign of rotn. [5]
11.	H	COOH	+
	H	COOMe	+
	H	CH$_2$OH	+
	H	CH$_3$	+
	COOH	H	+
	COOMe	H	+
	CH$_2$OH	H	+

See [5] for many further norbornanes including methyl derivatives providing links with the monoterpenes.

---

1. K. Takeda, S. Hagishita, M. Sugiura, K. Kitahonoki, I. Ban, S. Miyazaki and K. Kuriyama, *Tetrahedron*, 1970, **26**, 1435.
2. C. J. Collins, Z. K. Cheema, R. G. Werth and B. M. Benjamin, *J. Amer. Chem. Soc.*, 1964, **86**, 4913.
3. H. E. Smith and T. C. Willis, *Tetrahedron*, 1970, **26**, 107.
4. H. T. Thomas and K. Mislow, *J. Amer. Chem. Soc.*, 1970, **92**, 6292.
5. J. A. Berson, J. S. Walia, A. Remanick, S. Suzuki, P. Reynolds-Warnhoff and D. Willner, *J. Amer. Chem. Soc.*, 1961, **83**, 3986.

A⁴⁸ One- and two-centre compounds related to 2-phenylpropionic acid — Class 3b

Structures and interconversions:

1. (S)-(+) 2-methyl-4-phenylbutyric acid **A32.15** — PhCH₂CH₂–C(COOH)(CH₃)(H)
   ↑ H₂ [1]
3. (R)-(+) 2-(2-thianaphthenyl)-propionic acid

Y27: (S)-(+) 2-(3-indolyl)-propan-1-ol — indolyl-C(CH₂OH)(CH₃)(H)
   ↑ [H] [2]
4. (S)-(+) 2-(3-indolyl)-propionic acid N–Me (+)

2. (S)-(+) 2-(3-thianaphthenyl)-propionic acid
   ↕ QR [3]
5. (S)-(+) 2-(1-naphthyl)-propionic acid

(S)-(+) 3-phenylbutan-2-one **A41.1** — Ph–C(COCH₃)(CH₃)(H)

ORD and QR [1] ⇌ C(2) [4] / QR [5]

6. (2S,3R)-(−) 2-phenylpentan-3-ol — HO–CH(C₂H₅), CH(CH₃)(Ph)

7. (S)-(+) 2-(2-naphthyl)-propionic acid

8. (S)-(+) 2-phenylpropionic acid **A41.10** — Ph–C(COOH)(CH₃)(H)
   ⇌ QR (amides) [6]
(S)-(+) 2-chloro-2-phenylacetic acid **A23.10** — Ph–C(COOH)(Cl)(H)

(R)-(−) 2,3-diphenylpropionic acid **A43.11** — Ph–C(CH₂Ph)(COOH)(H)
   ↓ [H] [10]
12. (R)-(−) 1,2-diphenylpropane — Ph–C(CH₂Ph)(CH₃)(H)

C(2) [7] ↓
9. (2R,3S)-(+) 3-phenylpentan-2-ol — HO–CH(CH₃), CH(Ph)(C₂H₅)

10. (1R,2S)-(−) 1,2-diphenylpropan-1-ol — Ph–C(OH)(H), C(Ph)(CH₃)(H)
   ← PhMgBr [8]
11. (S)-(+) 1-benzoyl-phenylethane (methyldesoxybenzoin) — Ph–C(COPh)(CH₃)(H)
   → C(2) [9][10] → 12

C(2) [7]
13. (2S,3S)-(+) 3-phenylpentan-2-ol — H–C(CH₃)(OH), H–C(Ph)(C₂H₅)
   → C(2) [7] →
14. (2S,3S)-(+) 2-phenylpentan-3-ol — H–C(C₂H₅)(OH), Ph–C(CH₃)(H)
   → C(3) [7] →
15. (S)-(+) 2-phenylpropanal **A49** — Ph–C(CHO)(CH₃)(H)
   → PhMgBr [8] →
16. (1S,2S)-(−) 1,2-diphenylpropan-1-ol — H–C(Ph)(OH), Ph–C(CH₃)(H)
   ↑ [H] [8] from 11

---

1. B. Sjöberg, *Arkiv Kemi*, 1958, **12**, 565; *Acta Chem. Scand.*, 1960, **14**, 273.
2. T. H. Chan and R. K. Hill, *J. Org. Chem.*, 1970, **35**, 3519.
3. B. Sjöberg, *Arkiv Kemi*, 1957, **11**, 439.
4. K. Mislow and J. Brenner, *J. Amer. Chem. Soc.*, 1953, **75**, 2318.
5. K. Pettersson, *Arkiv Kemi*, 1956, **10**, 297.
6. K. Mislow and M. Heffler, *J. Amer. Chem. Soc.*, 1952, **74**, 3668.
7. D. J. Cram, *J. Amer. Chem. Soc.*, 1952, **74**, 2149 and references therein.
8. F. A. A. Elhafez and D. J. Cram, *J. Amer. Chem. Soc.*, 1952, **74**, 5846.
9. R. A. Barnes and B. R. Juliano, *J. Amer. Chem. Soc.*, 1959, **81**, 6462.
10. M. B. Watson and G. W. Youngson, *J. Chem. Soc. (C)*, 1968, 258.

Class 3b — 2-phenylbutane, 3-phenylbutyric acid and related compounds; indanes — A⁴⁹

1. § (S)-(−)3-tert-butyl-1-methylindene
2. (R)-(−)1-tert-butyl-3-methylindene
3. (R)-(+) (in ref.[6])(−) (in ref.[2]) 3-methylindan-1-one
4. § (S)-(+)3-tert-butylindan-1-one
5. (R)-(+) 3,4-dihydro-4-methylcoumarin
6. (R)-(−) 3-phenylbutan-1-ol
7. (R)-(−) 3-phenylbutyric acid (X=H) p-substituted 3-phenylbutyric acids; X=Me (−) X=COOH (+)[5] A61, T16
   (S)-(+) hydratropic acid A41.10
8. § (S)-(−)4,4-dimethyl-3-phenylpentanoic acid
9. (R)-(−)4-phenylpentanoic acid A61
10. (R)-(−)1-bromo-3-phenylbutane
11. (R)-(−)1-dimethylamino-3-phenylbutane
12. (R)-(−) 3-phenylbut-1-ene
13. (R)-(+) 4-methyl-1-tetralone
14. (2R,3S)-(−) 3-phenylbutan-2-ol
15. (R)-(−) 2-phenylbutane A42.16 (X=Y=H). (R)-(−) 2-(o-hydroxyphenyl)butane (X=OH, Y=H). (R)-(−) Dinoseb (X=OH, Y=NO₂)
16. (2R,3S)-(−)2-amino-3-phenylbutane N-benzoyl (+)
    (S)-(+)2-phenylpropanal A48.15
17. (R)-(−) 1-methyltetralin
18. (2S,3S)-(−) 2-amino-3-phenylbutane N-benzoyl (−)
    (S)-(+) butan-2-ol A1.16
19. (2S,3S)-(−) 3-phenylbutan-2-ol

1. V. Prelog and H. Scherrer, *Helv. Chim. Acta*, 1959, **42**, 2227.
2. J. Almy and D. J. Cram, *J. Amer. Chem. Soc.*, 1969, **91**, 4459.
3. F. A. A. Elhafez and D. J. Cram, *J. Amer. Chem. Soc.*, 1952, **74**, 5846.
4. D. J. Cram, *J. Amer. Chem. Soc.*, 1952, **74**, 2149 and references therein.
5. V. K. Honwad and A. S. Rao, *Tetrahedron*, 1965, **21**, 2592.
6. J. Grimshaw and P. G. Millar, *J. Chem. Soc. (C)*, 1970, 2324.
7. D. J. Cram, *J. Amer. Chem. Soc.*, 1952, **74**, 2137.
8. D. J. Cram and J. E. McCarty, *J. Amer. Chem. Soc.*, 1954, **76**, 5740.
9. J. Barry, H-B. Kagan and G. Snatzke, *Tetrahedron*, 1971, **27**, 4737.

A⁵⁰ One-centre aromatic acids, tetralin derivatives and indanes   Class 3b

1. (S)-(+) 2-phenylhexanoic acid
2. (S)-(+) 2-phenylheptanoic acid
3. (S)-(+) 2-phenyl-3-(2-thienyl)-propionic acid
   (S)-(+) 2,3-diphenyl-propionic acid **A43.11**
4. (S)-(+) 2-phenylpentan-1-ol
5. (S)-(+) 2-phenylpentane
6. (S)-(+) 2-phenylpentane-1,5-diol
   (S)-(+) 2-phenylpentanoic acid **A41.11**
   (R)-(+) 3-phenyl-2-(2-thienyl)-propionic acid **A43.10**
7. (R)-(+) butylsuccinic acid **A28.9**
8. (S)-(+) 2-phenyl-pent-4-enoic acid
9. (S)-(+) 5-bromo-2-phenylpentanoic acid **A52**
10. (S)-(+) 2-phenylglutaric acid. Anhydride (+)
11. (R)-(+) 2-(2-thienyl)-succinic acid
12. (S)-(+) 2-phenylsuccinic acid **Y6**
    (S)-(−) indane-1-carboxylic acid **A25.4**
    (S)-(−) tetralin-1-carboxylic acid **A25.9**
13. (S)-(−) 4-oxotetralin-1-carboxylic acid
14. (S)-(+) 1-methylindane
15. (S)-(−) 1-hydroxymethylindane
16. (S)-(+) 1-acetylindane
17. (S)-(−) 1-acetyltetralin

1. J. H. Brewster and J. G. Buta, *J. Amer. Chem. Soc.*, 1966, **88**, 2233.
2. A. Fredga, *Chem. Ber.*, 1956, **89**, 322.
3. K. Pettersson, *Arkiv Kemi*, 1956, **9**, 509.
4. R. Weidmann and J. P. Guetté, *Compt. rend.*, 1969, **268C**, 2225.
5. L. Westman, *Arkiv. Kemi*, 1957, **11**, 431.
6. K. Pettersson, *Arkiv. Kemi*, 1954, **7**, 39, 347.
7. L. Westman, *Arkiv. Kemi*, 1957, **12**, 161.
8. B. Sjöberg, *Acta Chem. Scand.*, 1960, **14**, 273.
9. K. Pettersson, *Arkiv. Kemi*, 1955, **8**, 387.
10. K. Kawazu, T. Fujita and T. Mitsui, *J. Amer. Chem. Soc.*, 1959, **81**, 933.
11. A. Fredga and L. Westman, *Arkiv. Kemi*, 1954, **7**, 193.

Class 3b — One-centre compounds related to atrolactic acid — A[51]

1. (S)-(+)2-hydroxy-3-methyl-2-phenylbutyric acid **X3**
2. (S)-(+)3-hydroxy-3-phenylbutan-2-one
3. (R)-(−)2-halogeno-2-phenylpropionic acids (X = Cl, Br)
4. (S)-(−)3-methyl-7-methoxyphthalide-3-carboxylic acids (X = H, Cl) **Y28**
5. (S)-(+)2-hydroxy-2-phenylpentan-3-one
   (S)-(+) atrolactic acid **A31.19**
6. (S)-( )2-hydroxy-2-(2-acetylaminophenyl)-propionic acid. Cinchonidine salt (−)
7. (S)-(−)3-methylphthalide-3-carboxylic acid
   (S)-(+)1-phenylpropan-1-ol-2-one **A21.14**
8. (1S,2S)-(−)1,2-diphenylpropane-1,2-diol
9. (S)-(−)1-benzoyl-1-phenylethanol (methylbenzoin)
10. (S)-( )1-amino-2-phenylpropan-2-ol. Hydrochloride (+)
11. (R)-(−)3-hydroxy-3-phenylbutyric acid
12. (R)-(+) 3-phenylbutane-1,3-diol
13. (R)-( )1-amino-2-phenylbutan-2-ol. Hydrochloride (−)
14. (R)-(−)2-hydroxy-2-phenylbutyric acid **X3**
15. (R)-(+)2-phenylbutane-1,2-diol
16. (R)-(+)2-phenylbutan-2-ol O-Ac (+) **A56**
    (S)-(+)2-phenylbutane **A42.16**
17. (S)-(+)3-hydroxy-3-phenylpentanoic acid
18. (S)-(−)3-phenylpentane-1,3-diol
19. (R)-(+)3,3-dimethyl-2-hydroxy-2-phenylbutyric acid **X3**

1. D. J. Cram, K. R. Kopecky, F. Hauk and A. Langemann, *J. Amer. Chem. Soc.*, 1959, 81, 5754.
2. S. Mitsui and Y. Kudo, *Chem. & Ind.*, 1965, 381.
3. S. Mitsui, S. Imaizumi, Y. Senda and K. Konno, *Chem. & Ind.*, 1964, 233.
4. V. N. Dobrynin, A. I. Gurevich, M. G. Karapetyan, M. N. Kolosov and M. M. Shemyakin, *Tetrahedron Letters*, 1962, 901.
5. D. J. Cram and J. Allinger, *J. Amer. Chem. Soc.*, 1954, 76, 4516.
6. J. H. Brewster, *J. Amer. Chem. Soc.*, 1956, 78, 4061.
7. W. A. Cowdrey, E. D. Hughes, C. K. Ingold, S. Masterman and A. D. Scott, *J. Chem. Soc.*, 1937, 1252.
8. H. Mizuno, S. Terashima and S. Yamada, *Chem. Pharm. Bull. (Japan)*, 1971, 19, 227.
9. K. Shingu, S. Hagishita and M. Nakagawa, *Tetrahedron Letters*, 1967, 4371.

A52 Aromatic cyclohexane derivatives — Class 3b

1. § (S)-(+) 2-dimethylamino-2-phenylcyclohexanone [1]
2. (1R,2S)-(−) 2-dimethylamino-2-phenylcyclohexanol
3. (1R,2R)-(+) 1,2-epoxy-1-phenylcyclohexane
4. (1R,2S)-(+) 2-dimethylamino-1-phenylcyclohexanol
5. (S)-(−) 2-phenylcyclohexanone
6. (1R,2S)-(−) 2-phenylcyclohexanol
7. (1R,2R)-(+) 1-phenylcyclohexane-1,2-diol
8. (1R,2S)-(+) 1-phenylcyclohexane-1,2-diol
9. (1S,2S)-(+) 2-phenylcyclohexanol

(S)-(+) 5-bromo-2-phenylpentanoic acid **A50.9**

10. (S)-(+) 2-phenyladipic acid
11. (R)-(+) 3-phenylcyclohex-1-ene

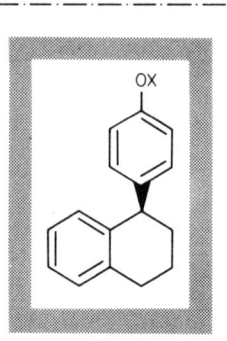

12. (R)-(−) 1-(p-hydroxyphenyl)-tetralin (X=H) Abs. X-ray [5]
(R)-(−) nafenopin (X=CMe₂COOH) [5]

1. G. Berti, personal communication.
2. G. Berti, B. Macchia, F. Macchia and L. Monti, *J. Org. Chem.*, 1968, **33**, 4045.
3. L. Westman, *Arkiv Kemi*, 1958, **12**, 167.
4. G. Berti, B. Macchia, F. Macchia and L. Monti, *J. Chem. Soc. (C)*, 1971, 3371.
5. W. L. Bencze, B. Kisis, R. T. Puckett and N. Finch, *Tetrahedron*, 1970, **26**, 5407.

Class 4a
Decalins and perhydroindane derivatives
A53

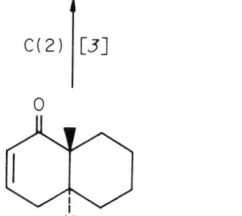
1. (9R,10S)-(+)9-methyldecalin-1,2-dione

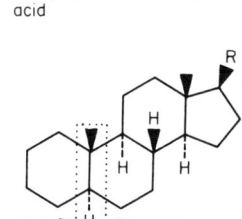
2. (1S,2R)-(−)2-carboxy-2-methylcyclohexane propionic acid

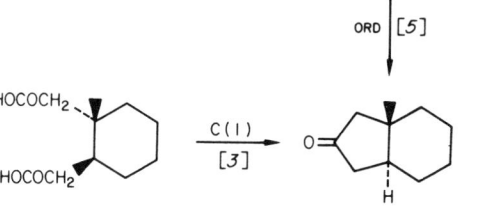
3. (8R,9S)-(−)8-methylperhydroindan-1-one

(+) trans−perhydroindan-2-one **A37.4**

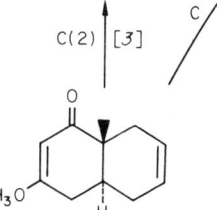
4. (9R,10S)-(+)9-methyl-Δ² -1-octalone

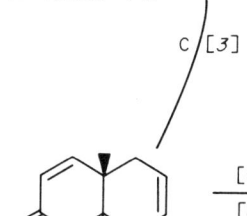
5. 5α-steroids **T47**

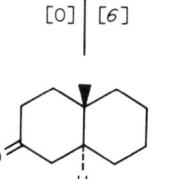
6. (1S,2S)-(−)1-methylcyclohexane-1,2-diacetic acid

7. (8S,9S)-(+)8-methylperhydroindan-2-one

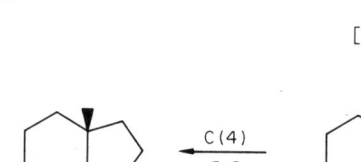
8. (9R,10S)-(−)3-methoxy-9-methyl-Δ²,⁶-1-hexalone **T23**

9. (9R,10S)-(−)10-methyl-Δ³,⁶-2-hexalone

10. (9S,10S)-(+)10-methyl-2-decalone **A54, T19**

11. (1S,2R)-(−)2-carboxy-2-methylcyclohexane acetic acid **T34**

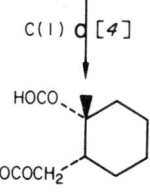

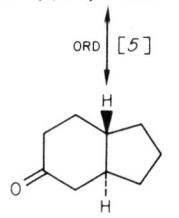
12. (8S,9S)-(+)8-methylperhydroindan-5-one

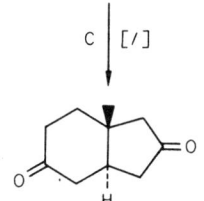
13. (9S,10R)-(−)10-methyl-Δ⁶-2-octalone

14. (1R,2R)-(+)2-carboxy-2-methylcyclohexane acetic acid **K4**

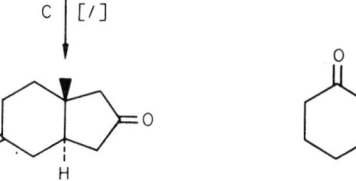

(+) trans−perhydroindan-5-one **A37.7**

15. (8R,9R)-(−)8-methylperhydroindane-2,5-dione

16. (9R,10R)-(+)9-methyl-1-decalone **T38, K35**

17. (1R,2R)-(+)2-carboxy-2-methylcyclohexanepropionic acid

---

1. C. Djerassi, D. Marshall and T. Nakano, *J. Amer. Chem. Soc.*, 1958, 80, 4853.
2. L. B. Barkley, M. W. Farrar, W. S. Knowles, H. Raffelson and Q. E. Thompson, *J. Amer. Chem. Soc.*, 1954, 76, 5014.
3. C. Djerassi, R. Riniker and B. Riniker, *J. Amer. Chem. Soc.*, 1956, 78, 6362.
4. F. Gautschi, O. Jeger, V. Prelog and R. B. Woodward, *Helv. Chim. Acta*, 1955, 38, 296.
5. P. M. Bourn and W. Klyne, *J. Chem. Soc.*, 1960, 2044.
6. B. Riniker, J. Kalvoda, D. Arigoni, A. Fürst, O. Jeger, A. M. Gold and R. B. Woodward, *J. Amer. Chem. Soc.*, 1954, 76, 312.

A⁵⁴ Further decalins; cyclopropane derivatives — Class 4a

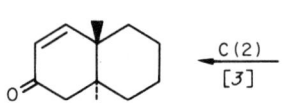
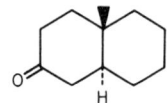

1. (9S,10R)-(−)10-methyl-Δ³-2-octalone

(9S,10S)-(+)10-methyl-2-decalone **A53.10**

2. (S)-(+)9-methyl-Δ⁴-3-octalone

3. (9R,10S)-(+)10-methyl-2-decalone

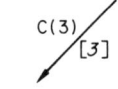

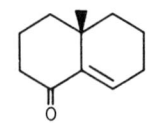
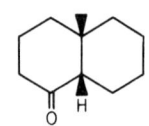
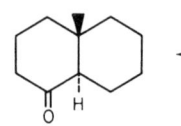
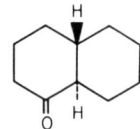

4. (S)-(−)10-methyl-Δ⁸-1-octalone

5. (9S,10S)-(+)10-methyl-1-decalone

6. (9R,10S)-(+)10-methyl-1-decalone

(9R,10S)-(+)1-decalone **A39.2**

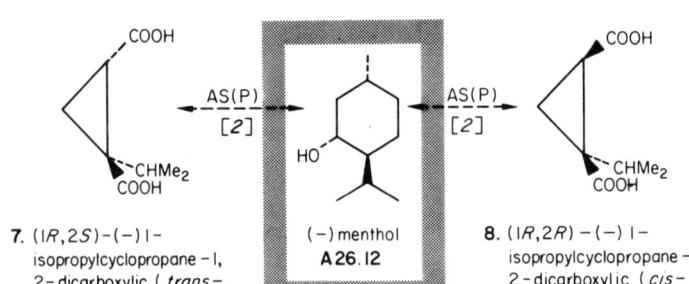

7. (1R,2S)-(−)1-isopropylcyclopropane-1,2-dicarboxylic (*trans*-umbellularic) acid

(−)menthol **A26.12**

8. (1R,2R)-(−)1-isopropylcyclopropane-1,2-dicarboxylic (*cis*-umbellularic) acid **T7**

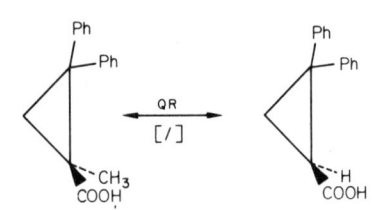

9. (S)-(−)1-methyl-2,2-diphenylcyclopropanecarboxylic acid

10. (S)-(+)2,2-diphenylcyclopropanecarboxylic acid **A35.13**

---

1. H. M. Walborsky, L. Barash, A. E. Young and F. J. Impastato, *J. Amer. Chem. Soc.*, 1961, **83**, 2517.
2. H. M. Walborsky, T. Sugita, M. Ohno and Y. Inouye, *J. Amer. Chem. Soc.*, 1960, **82**, 5255.
3. C. Djerassi and D. Marshall, *J. Amer. Chem. Soc.*, 1958, **80**, 3986, and references therein.
4. C. Djerassi and J. Staunton, *J. Amer. Chem. Soc.*, 1961, **83**, 736.

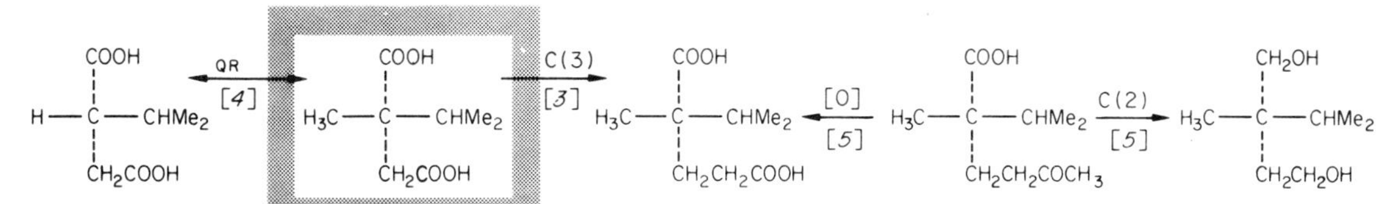

Class 4a, 4b
Quaternary acids and related compounds

1. (S)-(+) 2-isopropyl-2-methylsuccinic acid. Abs. X-ray[3] **T7**
2. (S)-(+) 2-isopropyl-2-methylglutaric acid
3. (S)-(+) 2-isopropyl-2-methyl-5-oxo-hexanoic acid **T7**
4. (S)-(−) 2-isopropyl-2-methylbutane-1,4-diol (small rotation)

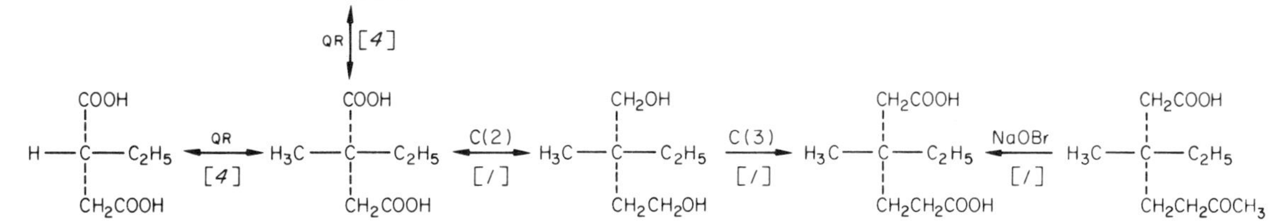

(S)-(+) isopropylsuccinic acid **A28.5**

(R)-(+) ethylsuccinic acid **A28.9**

5. (R)-(+) 2-ethyl-2-methylsuccinic acid **X4**
6. (R)-(−) 2-ethyl-2-methylbutane-1,4-diol
7. (R)-(+) 3-ethyl-3-methyladipic acid

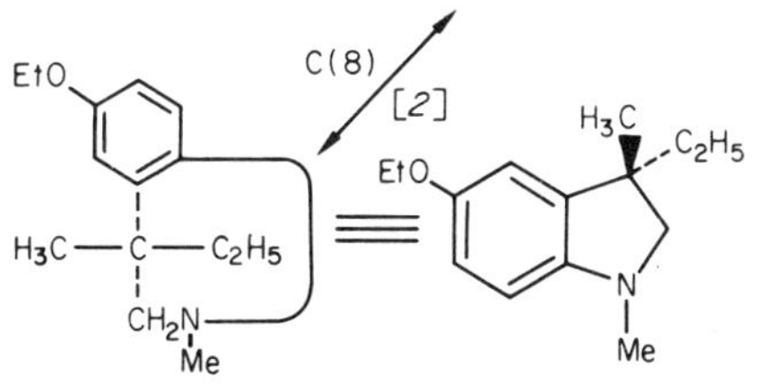

8. (S)-( ) 1,3-dimethyl-5-ethoxy-3-ethyl-2,3-dihydroindole **K30** 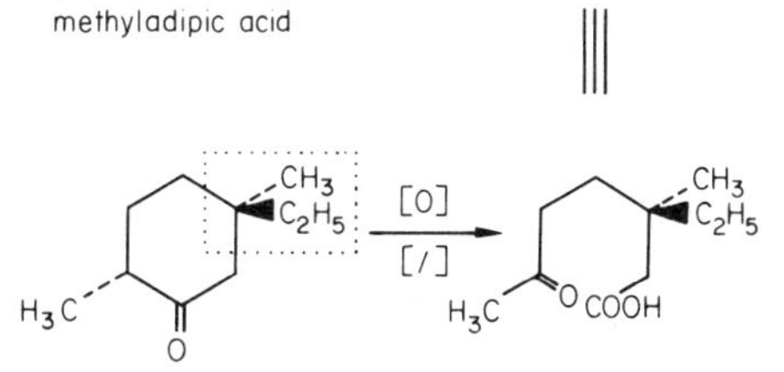
9. (2R,5R)-(+) 2,5-dimethyl-5-ethyl-cyclohexanone [8] **T34**
10. (R)-(+) 3-ethyl-3-methyl-6-oxoheptanoic acid

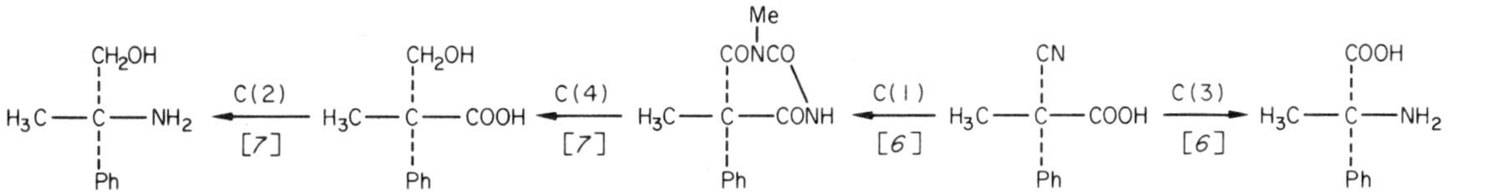

(R)-(−) 2-amino-2-phenylpropan-1-ol **A40.11**

11. (R)-(+) 3-hydroxy-2-methyl-2-phenylpropionic (α-methyltropic) acid.
12. (R)-(−) 2-cyano-2-phenylpropionic acid

(R)-(−) 2-amino-2-phenylpropionic acid **A40.12**

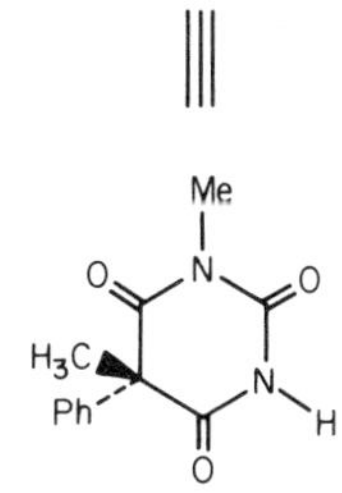

13. (S)-(+) N-methyleudan. Other barbiturates [6]

---

1. M. R. Cox, G. A. Ellestad, A. J. Hannaford, I. R. Wallwork, W. B. Whalley and B. Sjöberg, *J. Chem. Soc.*, 1965, 7257.
2. R. B. Longmore and B. Robinson, *Chem. & Ind.*, 1969, 622.
3. M. R. Cox, H. P. Koch, W. B. Whalley, M. B. Hursthouse and D. Rogers, *Chem. Comm.*, 1967, 212.
4. J. Porath, *Arkiv Kemi*, 1949, 1, 385; 1951, 3, 163.
5. E. J. Eisenbraun, F. Burian, J. Osiecki and C. Djerassi, *J. Amer. Chem. Soc.*, 1960, 82, 3476.
6. J. Knabe and C. Urbahn, *Annalen*, 1971, 750, 21.
7. J. Knabe, H. Junginger and W. Geismar, *Arch. Pharm.*, 1971, 304, 1.
8. C. Djerassi, B. Green, W. B. Whalley and C. G. De Grazia, *J. Chem. Soc. (C)*, 1966, 624.

A56 Quaternary aromatic compounds derived from 2-methyl-2-phenylbutyric acid

Class 4a, 4b

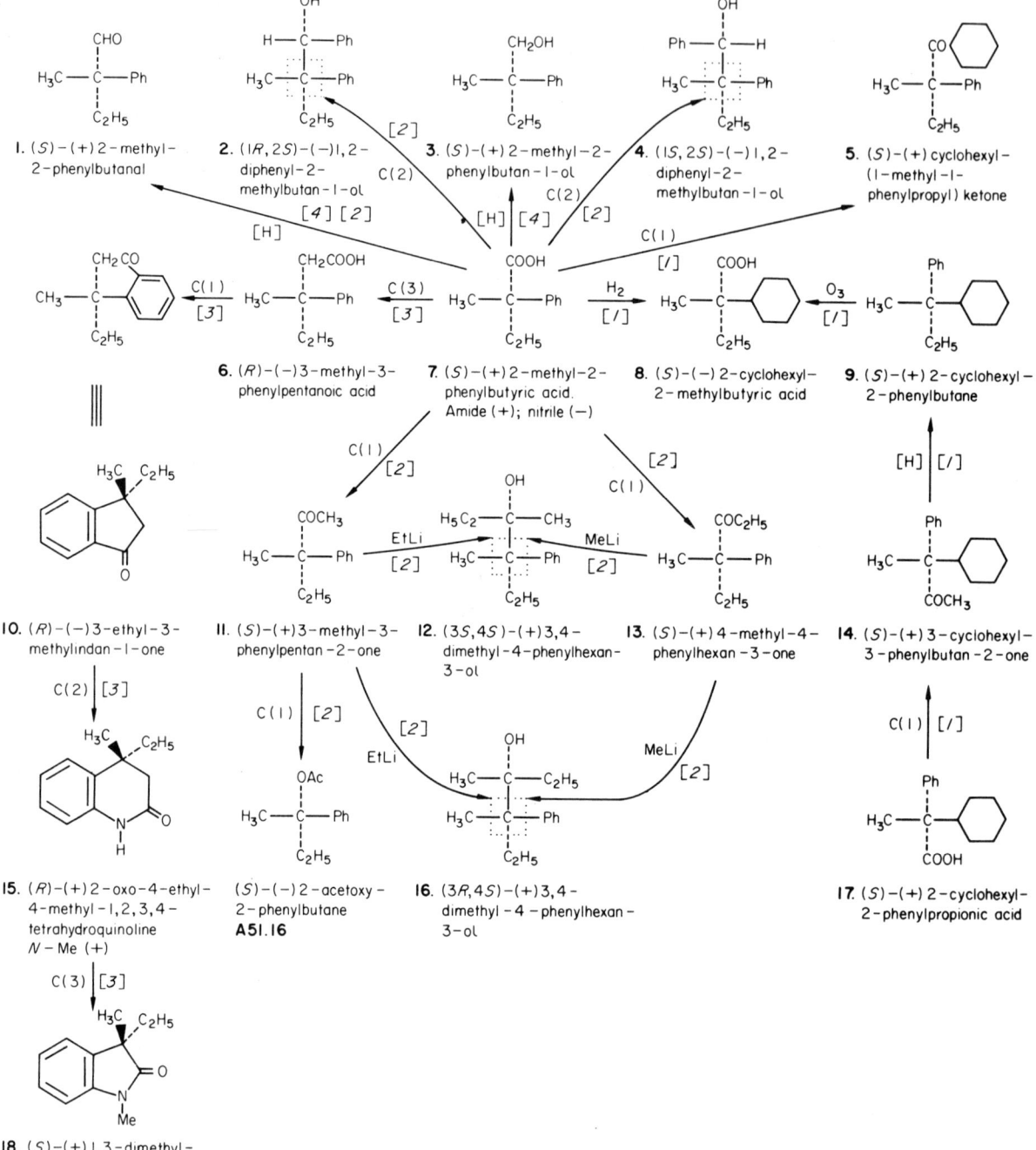

1. (S)-(+) 2-methyl-2-phenylbutanal
2. (1R,2S)-(−) 1,2-diphenyl-2-methylbutan-1-ol
3. (S)-(+) 2-methyl-2-phenylbutan-1-ol
4. (1S,2S)-(−) 1,2-diphenyl-2-methylbutan-1-ol
5. (S)-(+) cyclohexyl-(1-methyl-1-phenylpropyl) ketone
6. (R)-(−) 3-methyl-3-phenylpentanoic acid
7. (S)-(+) 2-methyl-2-phenylbutyric acid. Amide (+); nitrile (−)
8. (S)-(−) 2-cyclohexyl-2-methylbutyric acid
9. (S)-(+) 2-cyclohexyl-2-phenylbutane
10. (R)-(−) 3-ethyl-3-methylindan-1-one
11. (S)-(+) 3-methyl-3-phenylpentan-2-one
12. (3S,4S)-(+) 3,4-dimethyl-4-phenylhexan-3-ol
13. (S)-(+) 4-methyl-4-phenylhexan-3-one
14. (S)-(+) 3-cyclohexyl-3-phenylbutan-2-one
15. (R)-(+) 2-oxo-4-ethyl-4-methyl-1,2,3,4-tetrahydroquinoline N−Me (+)
16. (3R,4S)-(+) 3,4-dimethyl-4-phenylhexan-3-ol
17. (S)-(+) 2-cyclohexyl-2-phenylpropionic acid
18. (S)-(+) 1,3-dimethyl-3-ethyloxindole K30

(S)-(−) 2-acetoxy-2-phenylbutane A51.16

1. B. Calas and L. Giral, *Bull. Soc. chim. France*, 1971, 2629.
2. D. J. Cram and J. Allinger, *J. Amer. Chem. Soc.*, 1954, 76, 4516, and references therein.
3. R. K. Hill and G. R. Newkome, *Tetrahedron*, 1969, 25, 1249.
4. H. M. Walborsky and L. E. Allen, *J. Amer. Chem. Soc.*, 1971, 93, 5465.

Supplementary tables of one-centre compounds

Mills and Klyne [1] have made a number of generalizations relating the absolute configurations of certain classes of one-centre compounds to their signs of molecular rotation at the sodium D line.
P. A. Levene and his co-workers [2], [3], [4], [5] interrelated a large number of 1-centre compounds by chemical reactions not involving the asymmetric centre. Some of these correlations have been incorporated in the main body of Section A, and the results of many others are tabulated here. Where the same compounds have been prepared by Levene *et al.* and by other workers the agreement is good, but caution should be exercised in taking absolute configurations determined in this work, and not elsewhere, completely without reservation, due to the following factors:

(1) One or two cases of sign contradiction between different papers in the Levene series have been noted.

(2) Many of the reactions were carried out with material of low optical rotation and/or low optical purity.

A non-standard form of Fischer projection was used by Levene, in common with several other authors at about this time [6]. Bonds which should be interpreted as lying above the plane of the paper are shown as dotted lines, i.e. the configurations shown, if interpreted literally, are the converse of the Fischer convention prevailing at the time.

$$
\begin{array}{ccc}
\begin{array}{c} d \\ | \\ a \cdots C \cdots e \\ | \\ b \end{array}
& \equiv &
\begin{array}{c} d \\ \vdots \\ a - C - e \\ \vdots \\ b \end{array}
\end{array}
$$

Projection formula as used by Levene *et al.*   Usual Fischer projection.

References for pages 61–67

1. J. A. Mills and W. Klyne, *Progr. Stereochem.*, 1954, **1**, 177.
2. P. A. Levene and A. Rothen, *J. Org. Chem.*, 1936, **1**, 76.
3. P. A. Levene and R. E. Marker, *J. Biol. Chem.*, 1931, **91**, 405.
4. P. A. Levene and R. E. Marker, *J. Biol. Chem.*, 1931, **91**, 77.
5. P. A. Levene and M. Kuna, *J. Biol. Chem.*, 1941, **140**, 255.
6. A. Fredga and J. K. Miettinen, *Acta Chem. Scand.*, 1947, **1**, 371.

Class 2a

Secondary alcohols

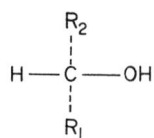

($R_1 > R_2$; (S) chirality); these are normally dextrorotatory when R_1 and R_2 are n-alkyl groups [1]

	R_1	R_2	Name	$[a]_D$	ref.
A1.16	C_2H_5	CH_3	butan-2-ol	+	[2]
1.	C_6H_{11}	C_3H_7-n	1-cyclohexylbutan-1-ol	−	[2]
2.	C_6H_{11}	CH_3	1-cyclohexylethanol	+	[2]
3.	$(CH_2)_2C_6H_{11}$	C_2H_5	6-cyclohexylhexan-3-ol	+	[2]
4.	C_6H_{11}	C_4H_9-n	1-cyclohexylpentan-1-ol	−	[2]
5.	$CH_2-C_6H_{11}$	C_2H_5	1-cyclohexylpentan-3-ol	+	[2]
6.	C_6H_{11}	C_2H_5	1-cyclohexylpropan-1-ol	−	[2]
7.	$C_7H_{15}-n$	C_2H_5	decan-3-ol	+	[2]
8.	$C_6H_{13}-n$	C_3H_7-n	decan-4-ol	+	[2]
9.	$C_5H_{11}-n$	C_4H_9-n	decan-5-ol	+	[2]
10.	C_4H_9-n	C_2H_5	heptan-3-ol	+	[2]
11.	$-CH=CH_2$	C_4H_9-n	hept-1-en-3-ol	+	[2]
A18.18	C_4H_9-n	CH_3	hexan-2-ol	+	[2]
A6.12	C_3H_7-n	C_2H_5	hexan-3-ol	+	[2]
A6.11	$-CH=CH_2$	C_3H_7-n	hex-1-en-3-ol	+	[2]
12.	$(CH_2)_2CH=CH_2$	CH_3	hex-5-en-2-ol	+	[2]
13.	C_3H_7-i	CH_3	3-methylbutan-2-ol	+	[2]
14.	C_3H_7-i	C_4H_9-n	2-methylheptan-3-ol	−	[2]
15.	$CH_2-C_3H_7-i$	C_3H_7-n	2-methylheptan-4-ol	+	[2]
16.	C_3H_7-i	C_3H_7-n	2-methylhexan-3-ol	−	[2]
17.	$CH_2-C_3H_7-i$	C_2H_5	5-methylhexan-3-ol	+	[2]
18.	$CH_2-C_3H_7-i$	$C_5H_{11}-n$	2-methylnonan-5-ol	+	[2]
19.	C_3H_7-i	$C_5H_{11}-n$	2-methyloctan-3-ol	−	[2]
20.	$CH_2-C_3H_7-i$	C_4H_9-n	2-methyloctan-4-ol	+	[2]
A8.6	C_3H_7-i	C_2H_5	2-methylpentan-3-ol	−	[2]
21.	$CH_2-C_3H_7-i$	CH_3	4-methylpentan-2-ol	+	[2]
22.	$C_6H_{13}-n$	C_2H_5	nonan-3-ol	+	[2]
23.	$C_5H_{11}-n$	C_3H_7-n	nonan-4-ol	+	[2]
A12.16	$C_6H_{13}-n$	CH_3	octan-2-ol	+	[2]
24.	C_4H_9-n	C_3H_7-n	octan-4-ol	+	[2]
25.	$-CH=CHCH_3$	C_4H_9-n	oct-2-en-4-ol	+	[2]
A6.7	C_3H_7-n	CH_3	pentan-2-ol	+	[2]
A12.11	$-CH=CHCH_3$	CH_3	pent-3-en-2-ol	−	[2]
A6.8	$CH_2-CH=CH_2$	CH_3	pent-4-en-2-ol	+	[2]
26.	$(CH_2)_2Ph$	C_2H_5	6-phenylhexan-3-ol	+	[2]
27.	CH_2Ph	C_2H_5	1-phenylpentan-3-ol	+	[2]
28.	$C_8H_{17}-n$	C_2H_5	undecan-3-ol	+	[2]

References on page 61

Class 2b
Secondary alcohols

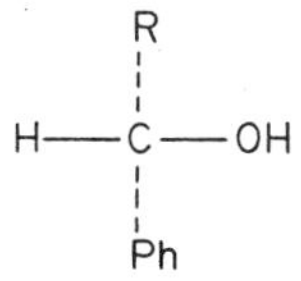

$$\begin{array}{c} R \\ | \\ H-C-OH \\ | \\ Ph \end{array}$$

	R	Name	$[\alpha]_D$	ref
1.	C_3H_7-n	1-phenylbutan-1-ol	–	[2]
2.	C_4H_9-n	1-phenylpentan-1-ol	–	[2]
A22.6	CH_3	1-phenylethanol	–	[2]
3.	C_2H_5	1-phenylpropan-1-ol	–	[2]

Class 3a
Hydrocarbons

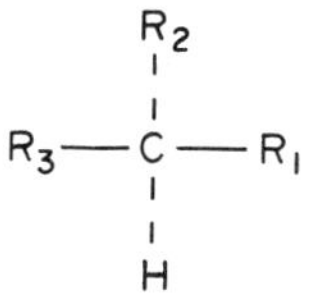

$$\begin{array}{c} R_2 \\ | \\ R_3-C-R_1 \\ | \\ H \end{array}$$

($R_1 > R_2 > R_3$) (*S*) chirality); these are normally dextrorotatory when R_1, R_2, R_3 are *n*-alkyl groups [*1*].

All of these examples have $R_3=CH_3$

	R_1	R_2	Name	$[\alpha]_D$	ref.
4.	C_6H_{11}	C_2H_5	2-cyclohexylbutane	–	[2]
5.	$CH_2C_6H_{11}$	C_2H_5	1-cyclohexyl-2-methylbutane	+	[2]
6.	$(CH_2)_2C_6H_{11}$	C_2H_5	1-cyclohexyl-3-methylpentane	+	[2]
7.	$CH_2C_3H_7-i$	C_2H_5	2,4-dimethylhexane	+	[3]
8.	$(CH_2)_2C_3H_7-i$	C_2H_5	2,5-dimethylheptane	+	[4]
9.	$C_7H_{15}-n$	C_2H_5	3-methyldecane	+	[2]
10.	$C_5H_{11}-n$	C_4H_9-n	5-methyldecane	+	[2]
11.	C_4H_9-n	C_2H_5	3-methylheptane **K22**	+	[2]
A29.13	C_3H_7-n	C_2H_5	3-methylhexane	+	[2]
A27.12	$C_6H_{13}-n$	C_2H_5	3-methylnonane	+	[2]
12.	$C_5H_{11}-n$	C_3H_7-n	4-methylnonane **K21**	+	[2]
A27.13	$C_5H_{11}-n$	C_2H_5	3-methyloctane	+	[2]
13.	C_4H_9-n	C_3H_7-n	4-methyloctane	+	[2]
A27.9	CH_2Ph	C_2H_5	2-methyl-1-phenylbutane	+	[2]
14.	$(CH_2)_2Ph$	C_2H_5	3-methyl-1-phenylpentane	+	[2]

References on page 61

Class 3a

Alcohols

(S) chirality

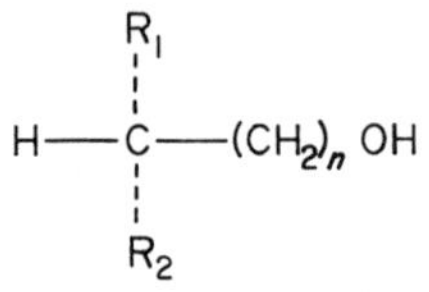

H—C(R₁)(R₂)—(CH₂)ₙOH

	R₁	R₂	n	Name	[α]_D	ref.
1.	CH₃	C₃H₇–i	1	3,4-dimethylpentan-1-ol	–	[4]
A27.10	CH₃	C₂H₅	1	2-methylbutan-1-ol	–	[2]
A32.3	CH₃	C₅H₁₁–n	1	2-methylheptan-1-ol	–	[2]
2.	CH₃	C₄H₉–n	2	3-methylheptan-1-ol	–	[2]
3.	CH₃	C₃H₇–n	3	4-methylheptan-1-ol	±0	[2]
4.	CH₃	C₂H₅	4	5-methylheptan-1-ol	+	[2]
5.	CH₃	C₄H₉–n	1	2-methylhexan-1-ol	–	[2]
6.	CH₃	C₃H₇–n	2	3-methylhexan-1-ol	–	[2]
7.	CH₃	C₂H₅	3	4-methylhexan-1-ol	+	[2]
8.	CH₃	C₅H₁₁–n	3	4-methylnonan-1-ol	–	[2]
9.	CH₃	C₄H₉–n	4	5-methylnonan-1-ol	±0	[2]
10.	CH₃	C₅H₁₁–n	2	3-methyloctan-1-ol	–	[2]
11.	CH₃	C₄H₉–n	3	4-methyloctan-1-ol	–	[2]
12.	CH₃	C₃H₇–n	4	5-methyloctan-1-ol	+	[2]
13.	CH₃	C₃H₇–n	1	2-methylpentan-1-ol	–	[2]
A27.14	CH₃	C₂H₅	2	3-methylpentan-1-ol	+	[2]

Class 3a

Amines

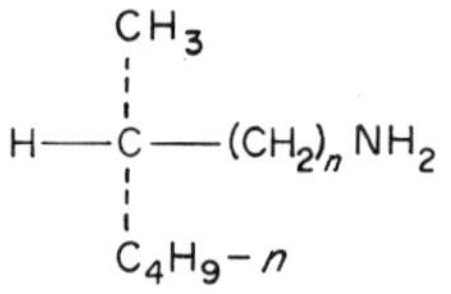

H—C(CH₃)(C₄H₉–n)—(CH₂)ₙNH₂

	n	Name	[α]_D	ref.
14.	5	1-amino-6-methyldecane	+	[2]
15.	2	1-amino-3-methylheptane	–	[2]
16.	1	1-amino-2-methylhexane	–	[2]
17.	4	1-amino-5-methylnonane	+	[2]
18.	3	1-amino-4-methyloctane	–	[2]

References on page 61

Class 3a

Carboxylic acids

The following generalizations have been made: [1]

COOH H$_3$C—C—H (CH$_2$)$_n$CH$_3$	(CH$_2$)$_n$COOH H$_3$C—C—H C$_2$H$_5$	CH$_2$COOH H$_3$C—C—H (CH$_2$)$_n$CH$_3$
($n > 1$)	($n > 0$)	($n > 1$)
(S) chirality dextrorotatory [1]	(S) chirality dextrorotatory [1]	(S) chirality laevorotatory [1]

$$R_2 - \underset{R_1}{\overset{H}{C}} - (CH_2)_n COOH \;;\; (S) \text{ chirality}$$

	R$_1$	R$_2$	n	name	$[\alpha]_D$	ref.
1.	CH$_3$	C$_3$H$_7$–i	0	2,3-dimethylbutyric	+	[4]
2.	CH$_3$	C$_3$H$_7$–i	1	3,4-dimethylpentanoic	–	[4]
3.	C$_2$H$_5$	C$_6$H$_{13}$–n	0	2-ethyloctanoic	+	[2]
A29.11	CH$_3$	C$_2$H$_5$	0	2-methylbutyric	+	[2]
A34.6	CH$_3$	C$_8$H$_{17}$–n	0	2-methyldecanoic	+	[2]
4.	CH$_3$	C$_5$H$_{11}$–n	3	5-methyldecanoic	+	[2]
A34.2	CH$_3$	C$_{10}$H$_{21}$–n	0	2-methyldodecanoic	+	[2]
A32.4	CH$_3$	C$_5$H$_{11}$–n	0	2-methylheptanoic	+	[5]
5.	CH$_3$	C$_4$H$_9$–n	1	3-methylheptanoic	–	[2]
6.	CH$_3$	C$_3$H$_7$–n	2	4-methylheptanoic	+	[2]
7.	CH$_3$	C$_2$H$_5$	3	5-methylheptanoic	+	[2]
8.	CH$_3$	C$_4$H$_9$–n	0	2-methylhexanoic	+	[2]
9.	CH$_3$	C$_3$H$_7$–n	1	3-methylhexanoic	–	[2]
10.	CH$_3$	C$_2$H$_5$	2	4-methylhexanoic	+	[2]
11.	CH$_3$	C$_7$H$_{15}$–n	0	2-methylnonanoic	+	[2]
12.	CH$_3$	C$_5$H$_{11}$–n	2	4-methylnonanoic	+	[2]
13.	CH$_3$	C$_4$H$_9$–n	3	5-methylnonanoic	+	[2]
14.	CH$_3$	C$_6$H$_{13}$–n	0	2-methyloctanoic	+	[2]
15.	CH$_3$	C$_5$H$_{11}$–n	1	3-methyloctanoic	–	[2]
16.	CH$_3$	C$_4$H$_9$–n	2	4-methyloctanoic	+	[2]
17.	CH$_3$	C$_3$H$_7$–n	3	5-methyloctanoic	+	[2]
A27.17	CH$_3$	C$_2$H$_5$	4	6-methyloctanoic	+	[2]
A32.15	CH$_3$	(CH$_2$)$_2$Ph	0	2-methyl-4-phenylbutyric	+	[2]
A43.22	CH$_3$	CH$_2$Ph	1	3-methyl-4-phenylbutyric	–	[2]
18.	(CH$_2$)$_2$Ph	CH$_3$	3	5-methyl-7-phenylheptanoic	+	[2]
19	CH$_3$	(CH$_2$)$_2$Ph	1	4-methyl-6-phenylhexanoic	–	[2]
20	CH$_3$	(CH$_2$)$_2$Ph	1	3-methyl-5-phenylpentanoic	–	[2]
A43.16	CH$_3$	CH$_2$Ph	0	2-methyl-3-phenylpropionic	+	[2]
A34.4	CH$_3$	C$_3$H$_7$–n	0	2-methylpentanoic	+	[2]
21.	CH$_3$	C$_2$H$_5$	1	3-methylpentanoic	+	[2]

References on page 61

Class 3a
Bromoalkanes
(S) chirality

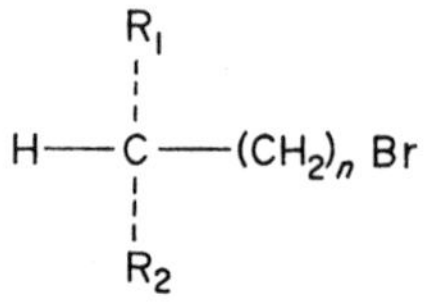

$$H-\underset{R_2}{\overset{R_1}{\underset{|}{\overset{|}{C}}}}-(CH_2)_n\,Br$$

	R_1	R_2	n	name	$[\alpha]_D$	ref.
1.	CH_3	C_3H_7-i	1	1-bromo-2,3-dimethylbutane	+	[4]
2.	C_3H_7-i	CH_3	2	1-bromo-2,3-dimethylpentane	−	[4]
3.	CH_3	C_2H_5	1	1-bromo-2-methylbutane	+	[2]
4.	CH_3	$C_5H_{11}-n$	4	1-bromo-5-methyldecane	+	[2]
5.	CH_3	$C_5H_{11}-n$	1	1-bromo-2-methylheptane	+	[2]
6.	CH_3	C_4H_9-n	2	1-bromo-3-methylheptane	+	[2]
7.	CH_3	C_3H_7-n	3	1-bromo-4-methylheptane	+	[2]
8.	CH_3	C_4H_9-n	1	1-bromo-2-methylhexane	+	[2]
9.	CH_3	C_3H_7-n	2	1-bromo-3-methylhexane	+	[2]
10.	CH_3	C_2H_5	3	1-bromo-4-methylhexane	+	[2]
11.	CH_3	C_2H_5	4	1-bromo-5-methylhexane	+	[2]
12.	CH_3	$C_5H_{11}-n$	3	1-bromo-4-methylnonane	+	[2]
13.	CH_3	C_4H_9-n	4	1-bromo-5-methylnonane	+	[2]
14.	CH_3	$C_5H_{11}-n$	2	1-bromo-3-methyloctane	+	[2]
15.	CH_3	C_4H_9-n	3	1-bromo-4-methyloctane	+	[2]
16.	CH_3	C_3H_7-n	4	1-bromo-5-methyloctane	+	[2]
17.	CH_3	C_2H_5	5	1-bromo-6-methyloctane	+	[2]
18.	CH_3	C_3H_7-n	1	1-bromo-2-methylpentane	+	[2]
19.	CH_3	C_2H_5	2	1-bromo-3-methylpentane	+	[2]

Class 3b
Alcohols

$$H-\underset{R_2}{\overset{R_1}{\underset{|}{\overset{|}{C}}}}-(CH_2)_n\,OH$$

(S) chirality

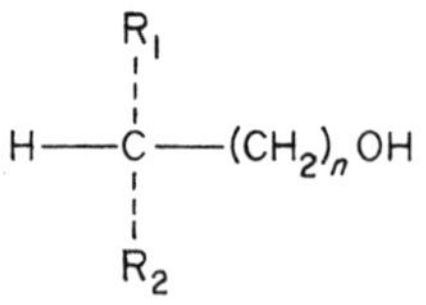

	R_1	R_2	n	name	$[\alpha]_D$	ref.
20.	Ph	CH_3	2	3-phenylbutan-1-ol	+	[2]
21.	Ph	C_2H_5	4	5-phenylheptan-1-ol	+	[2]
22.	Ph	C_2H_5	3	4-phenylhexan-1-ol	−	[2]
23.	Ph	CH_3	4	5-phenylhexan-1-ol	+	[2]
24.	Ph	C_2H_5	2	3-phenylpentan-1-ol	+	[2]
25.	Ph	CH_3	3	4-phenylpentan-1-ol	+	[2]
26.	CH_3	Ph	1	2-phenylpropan-1-ol	−	[2]

References on page 61

A63

Class 3b
Carboxylic acids

(S) chirality

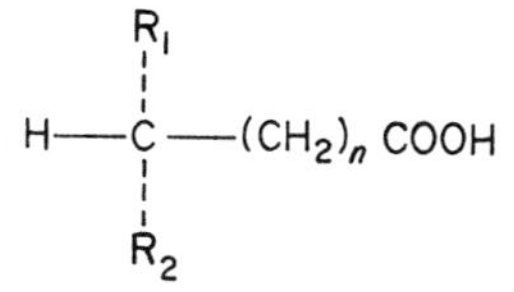

$$H-\overset{R_1}{\underset{R_2}{C}}-(CH_2)_n\,COOH$$

	R_1	R_2	n	name	$[\alpha]_D$	ref.
A25.15	C_2H_5	Ph	0	2-phenylbutyric	+	[2]
A49.7	Ph	CH_3	1	3-phenylbutyric	+	[2]
1.	Ph	C_2H_5	3	5-phenylheptanoic	+	[2]
A42.15	Ph	C_2H_5	2	4-phenylhexanoic	+	[2]
2.	Ph	CH_3	3	5-phenylhexanoic	+	[2]
A42.10	Ph	C_2H_5	1	3-phenylpentanoic	+	[2]
A49.9	Ph	CH_3	2	4-phenylpentanoic	+	[2]
A41.10	CH_3	Ph	0	2-phenylpropionic	+	[2]

References on page 61

-C-

Carbohydrates

The treatment of carbohydrate stereoisomerism given here is very much an outline, and standard texts [1] [2] [20] should be consulted for detailed accounts of intraseries correlations. Carbohydrates shown in other sections of the 'Atlas' and not specifically included in this section have been chemically correlated by unambiguous methods with one or more compounds included here.

Scope[†]

(1) The parent trioses, tetroses, pentoses and hexoses which are fundamental to carbohydrate stereochemical nomenclature are shown. Derivatives of these, such as deoxy-, aminodeoxy- etc. derivatives are not included.
(2) All carbohydrates for which Bijvoet X-rays are available are included.
(3) Di- and polysaccharides are not included as their stereochemistry follows fairly readily from that of monosaccharides.
(4) Nucleosides likewise may be considered as simple substitution products of the sugars and require no further description within the context of the 'Atlas'.

Tetroses, Pentoses and Hexoses

Nomenclature

According to the traditional system of carbohydrate nomenclature, the configuration of any monosaccharide having up to six carbon atoms is completely defined by a prefix D or L followed by a syllable 'gluc-', 'mann-' etc. This latter specifies both the number of carbon atoms in the sugar and the relative stereochemistry of all chiral centres, while the D or L prefix designates the enantiomeric series to which the compound belongs.

Sugars having the D prefix are considered to be derived from D $(R-)$ glyceraldehyde by the chain-lengthening process represented on p. C1, in which the asymmetric carbon atom of the glyceraldehyde becomes the penultimate carbon atom (ultimate chiral carbon atom) of the chain, i.e. C_3 of a tetrose, C_4 of a pentose and C_5 of a hexose. Sugars derived in this way from L $(S-)$ glyceraldehyde belong to the antipodal series.

In many cases the homologation steps represented by arrows on page C1 have been realised in practice, by sequences such as R–CHO → RCH(OH)CN → RCH(OH)CHO (Fischer-Kiliani reaction) but even in cases where they have not, the configurations of the individual monosaccharides are supported by a wealth of other evidence [1] [2].

For a table of compounds to which the homologation reaction has been applied, see reference [2], p. 144.

Specification of chirality by the (R,S) system is superfluous for carbohydrates and their simple derivatives to which the traditional system is applicable, but it may be invoked to specify the chirality at additional centres not covered by the carbohydrate 'local' convention [3].

[†] We thank Dr. R. D. Guthrie for helpful discussion.
References on page 72

Aldoheptoses, -octoses and higher sugars

Nomenclature

Definition of the stereochemistry of heptoses, octoses and nonoses requires two prefixes, one defining the configuration at C_2–C_5 as in a hexose, and the other, which appears first in the name, defining the configuration at the remaining chiral centre(s). The known aldoheptoses, aldoöctoses, octuloses and nonuloses are tabulated in Reference [1], pp. 245-50.

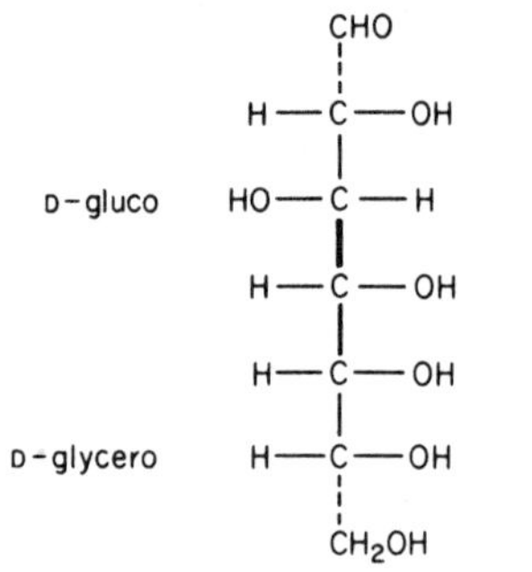

D-glycero-D-glucoheptose.

Anomerism of monosaccharides

The formation of five- and six-membered rings by sugars is denoted by the insertion of 'furano-' and 'pyrano-' respectively into the name. The configuration at the new C_1 chiral centre thus formed is described by the additional symbol α or β [1] [2].

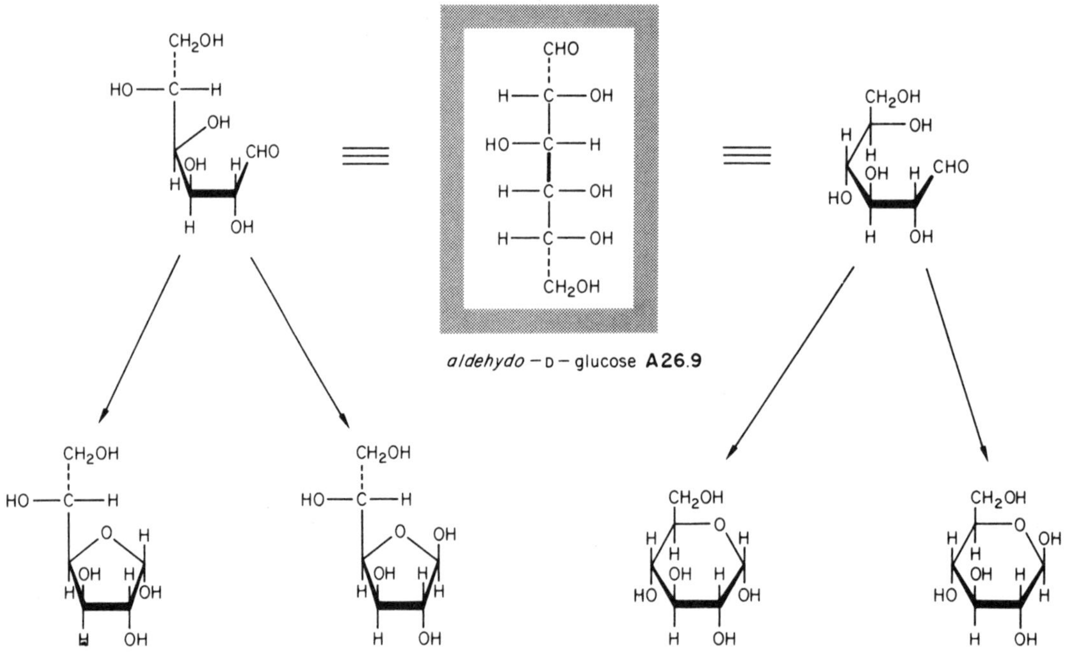

aldehydo-D-glucose A26.9

1. α-D-glucofuranose   2. β-D-glucofuranose   3. α-D-glucopyranose   4. β-D-glucopyranose

Fundamental aldoses and alditols.

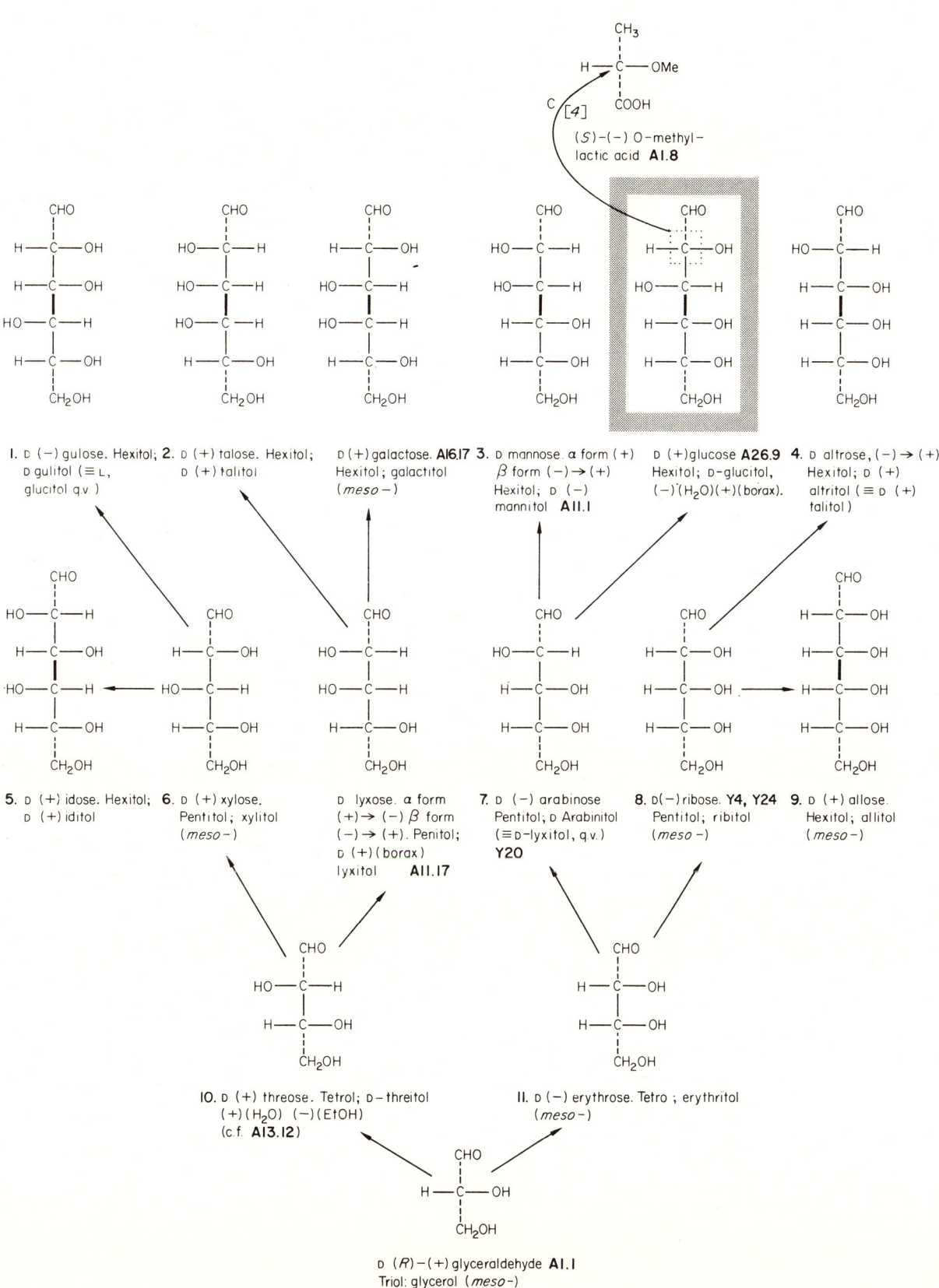

C² Ketoses.

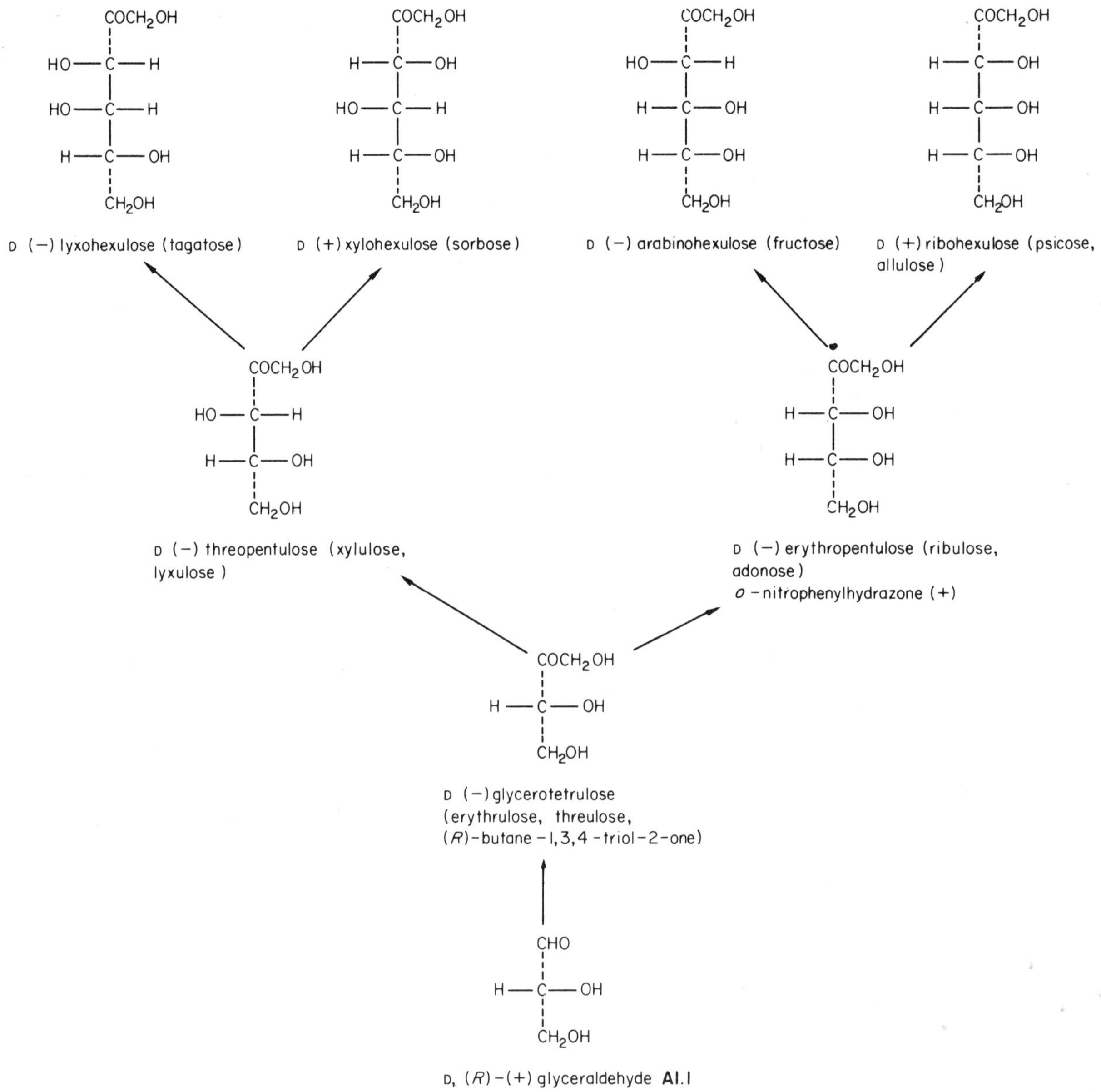

$$\text{D (−) lyxohexulose (tagatose)}$$
$$\text{D (+) xylohexulose (sorbose)}$$
$$\text{D (−) arabinohexulose (fructose)}$$
$$\text{D (+) ribohexulose (psicose, allulose)}$$

D (−) threopentulose (xylulose, lyxulose)

D (−) erythropentulose (ribulose, adonose)
o −nitrophenylhydrazone (+)

D (−) glycerotetrulose (erythrulose, threulose, (R)-butane−1,3,4−triol−2−one)

D, (R)−(+) glyceraldehyde **Al.1**

References for pages 71-73
1. L. Hough and A. C. Richardson in Rodd's 'Chemistry of Carbon Compounds', 2nd edn., volume IF, 1967.
2. J. Stanek, M. Cerný, J. Kocourek and J. Pacák, 'The Monosaccharides', Academic Press, 1963.
3. R. S. Cahn, C. K. Ingold and V. Prelog, *Angew. Chem. (Internat. Ed.)*, 1966, **5**, 385.
4. M. L. Wolfrom, R. U. Lemieux, S. M. Olin and D. I. Weisblat, *J. Amer. Chem. Soc.*, 1949, **71**, 4057.
5. G. N. Ramachandran and R. Chandrasekaran, *Biochim. Biophys. Acta*, 1967, **148**, 317.
6. R. Parthasarathy and R. E. Davis, *Acta Cryst.*, 1967, **23**, 1049.
7. R. Hoge and J. Trotter, *J. Chem. Soc. (A)*, 1968, 267.
8. H. Shimanouchi, N. Saito and Y. Sasada, *Bull. Chem. Soc. Japan*, 1969, **42**, 1239.
9. Y. Tsukuda, Y. Nakagawa, H. Kano, T. Sato, M. Shiro and H. Koyama, *Chem. Comm.*, 1967, 975.
10. E. Shefter and K. N. Trueblood, *Acta Cryst.*, 1965, **18**, 1067.
11. G. Koyama, K. Maeda, H. Umezawa and Y. Iitaka, *Tetrahedron Letters*, 1966, 597.
12. E. Subramanian and D. J. Hunt, *Acta Cryst.*, 1970, **B26**, 303.
13. J. W. Moncrief and S. P. Sims, *Chem. Comm.*, 1969, 914.
14. R. E. Marsh and J. Waser, *Acta Cryst.*, 1971, **B26**, 1030.
15. P. F. Wiley, D. J. Duchamp, V. Hsiung and C. C. Chidester, *J. Org. Chem.*, 1971, **36**, 2670.
16. T. Ikekawa, H. Umezawa and Y. Iitaka, *J. Antibiotics (Tokyo)*, 1966, **19A**, 49.
17. B. E. Davidson and A. T. McPhail, *J. Chem. Soc. (B)*, 1970, 660.
18. G. Koyama, Y. Iitaka, K. Maeda and H. Umezawa, *Tetrahedron Letters*, 1968, 1875.
19. S. Neidle, D. Rogers and M. B. Hursthouse, *Tetrahedron Letters*, 1968, 4725.
20. J. F. Stoddart, 'Stereochemistry of Carbohydrates', Wiley-Interscience, 1971.

Further carbohydrates for which anomalous dispersion X-ray data are available. C³

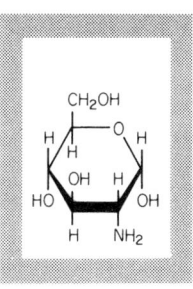
1. (+)2-amino-2-deoxy-α-D-glucopyranose. *Abs. X-ray* [5]

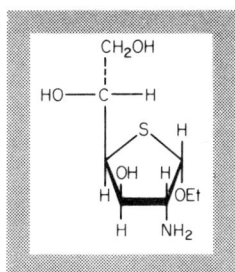
2. (+) ethyl 1-thio α-D glucofuranoside. *Abs. X-ray* [6]

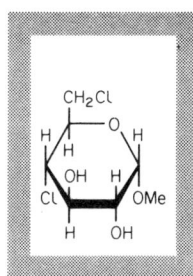
3. Methyl 4,6-dichloro-4,6-dideoxy α-D-glucopyranoside. *Abs. X-ray* [7]

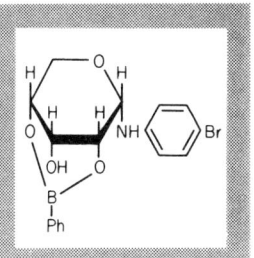
4. N-(p-bromophenyl) α-D ribopyranosylamine 2,4-phenylboronate. *Abs. X-ray* [8]

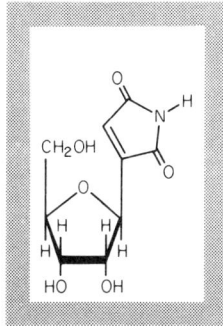
5. Showdomycin (3-β-D-ribofuranosyl maleimide). *Abs. X-ray* [9]

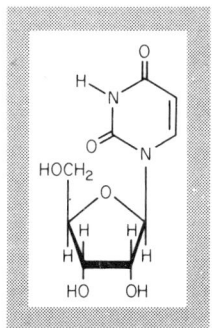
6. Uridine. *Abs. X-ray* [10]

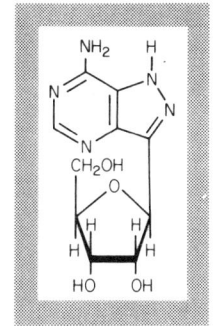
7. Formycin. *Abs. X-ray* [11]

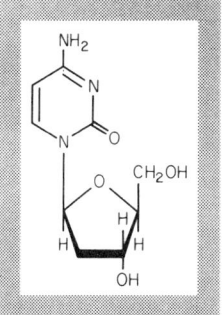
8. 2'-deoxycytidine. *Abs. X-ray* [12]

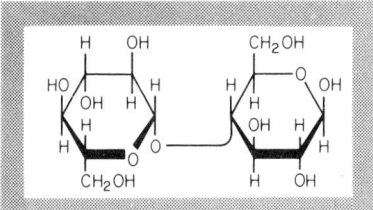
9. Cellobiose. *Abs. X-ray* [13]

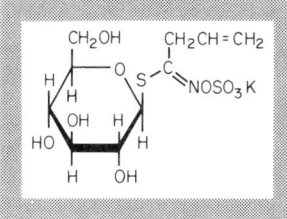
10. Sinigrin. *Abs. X-ray* [14]

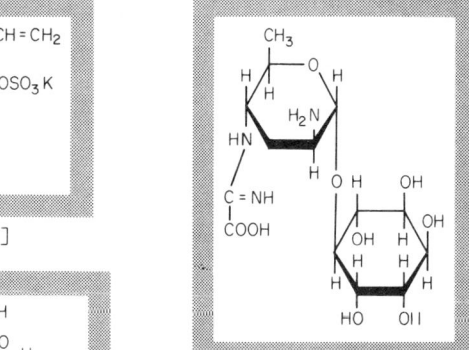
11. Kasugamycin. *Abs. X-ray* [16]

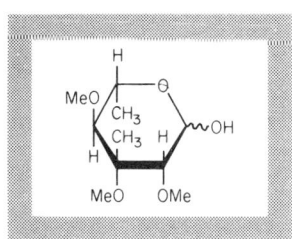
12. L-nogalose (3-C-methyl-2,3,4,-tri-O-methyl-6-deoxy-L-mannopyranose). *Abs. X-ray* [15]

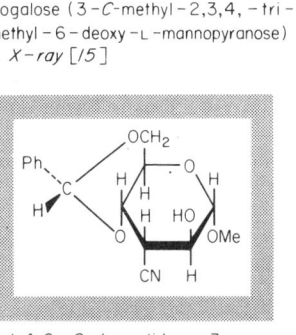
13. Methyl 4,6-O-benzylidene-3-cyano-3-deoxy α-D-altropyranoside. *Abs. X-ray* [17]

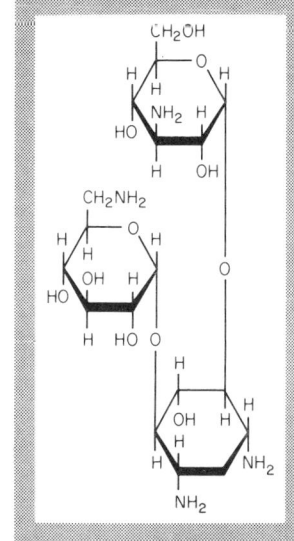
14. Kanamycin A. *Abs. X-ray* [18]

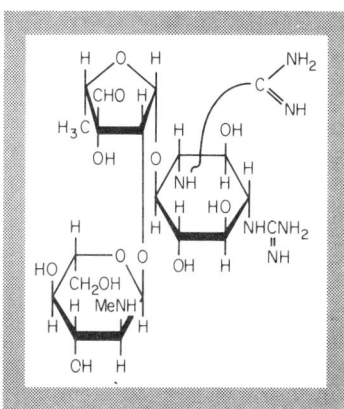
15. Streptomycin. *Abs. X-ray* [19]

References on page 72

-T-
Terpenes (including Steroids)

Introductory Notes to Chapter T

Scope

The term 'terpene' has been given its widest interpretation here, and modified terpene types such as furanoterpenes and norterpenes (including steroids) are included. The arrangement of material is basically in order of increasing molecular weight, i.e. from mono- to tetra-terpenes, and within each subdivision the simple types, as far as practicable, precede the more complex.

Arrangement

Monoterpenes	T1–T17
Sesquiterpenes	T1–3, T6, T11–12, T16–32
Diterpenes	T32–T41
Sesterterpenes	T41
Triterpenes	T42–T51
Steroids	T46–T50
Nortriterpenoids (Limonoids, etc.)	T52–T53
Tetraterpenes (including ionone group)	T54–T55
Phytol-tocopherol-phylloquinone group	T56
Miscellaneous modified terpenes	T57–T58

The section on steroids is considerably shorter than would be justified on the basis of their importance as natural products, but they fall into the category of a 'stereochemically homogenous' series of compounds (see Introduction), for which extensive further description in the 'Atlas' would be superfluous.

Citronellal and related terpenes.

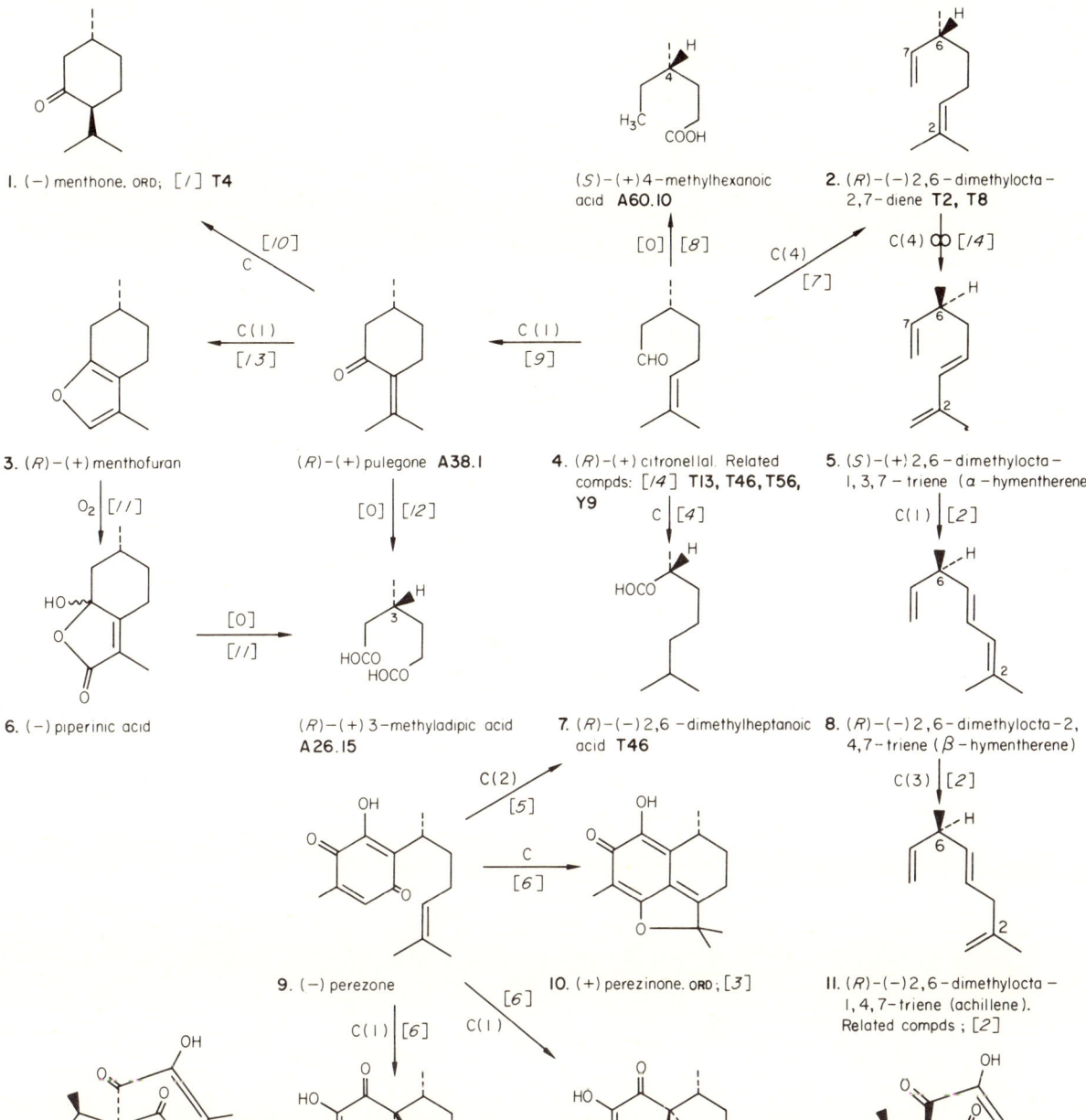

1. C. Djerassi, R. Riniker and B. Riniker, *J. Amer. Chem. Soc.*, 1956, **78**, 6377.
2. K. H. Schulte-Elte and M. Gadola, *Helv. Chim. Acta*, 1971, **54**, 1095.
3. P. Joseph-Nathan and M. P. Gonzales, *Canad. J. Chem.*, 1969, **47**, 2465.
4. J. v. Braun and W. Teuffert, *Ber.*, 1929, **62**, 235.
5. F. Kögl and A. G. Boer, *Rec. Trav. Chim.*, 1935, **54**, 779.
6. F. Walls, J. Padilla, P. Joseph-Nathan, F. Giral, M. Escobar and J. Romo, *Tetrahedron*, 1966, **22**, 2387.
7. D. Arigoni and O. Jeger, *Helv. Chim. Acta*, 1954, **37**, 881.
8. N. Kishner, *J. Russ. Phys-Chem. Soc.*, 1911, **43**, 951.
9. W. Kühn and H. Schinz, *Helv. Chim. Acta*, 1953, **36**, 161.
10. A. J. Birch, *Ann. Reps. Chem. Soc.*, 1950, **47**, 190.
11. R. B. Woodward and R. H. Eastman, *J. Amer. Chem. Soc.*, 1950, **72**, 399.
12. F. W. Semmler, *Ber.*, 1892, **25**, 3512.
13. W. Treibs, *Ber.*, 1937, **70**, 85.
14. E. Klein and W. Rojahn, *Chem. Ber.*, 1964, **97**, 2700.

T 2
Limonene, phellandrene and related monoterpenes.

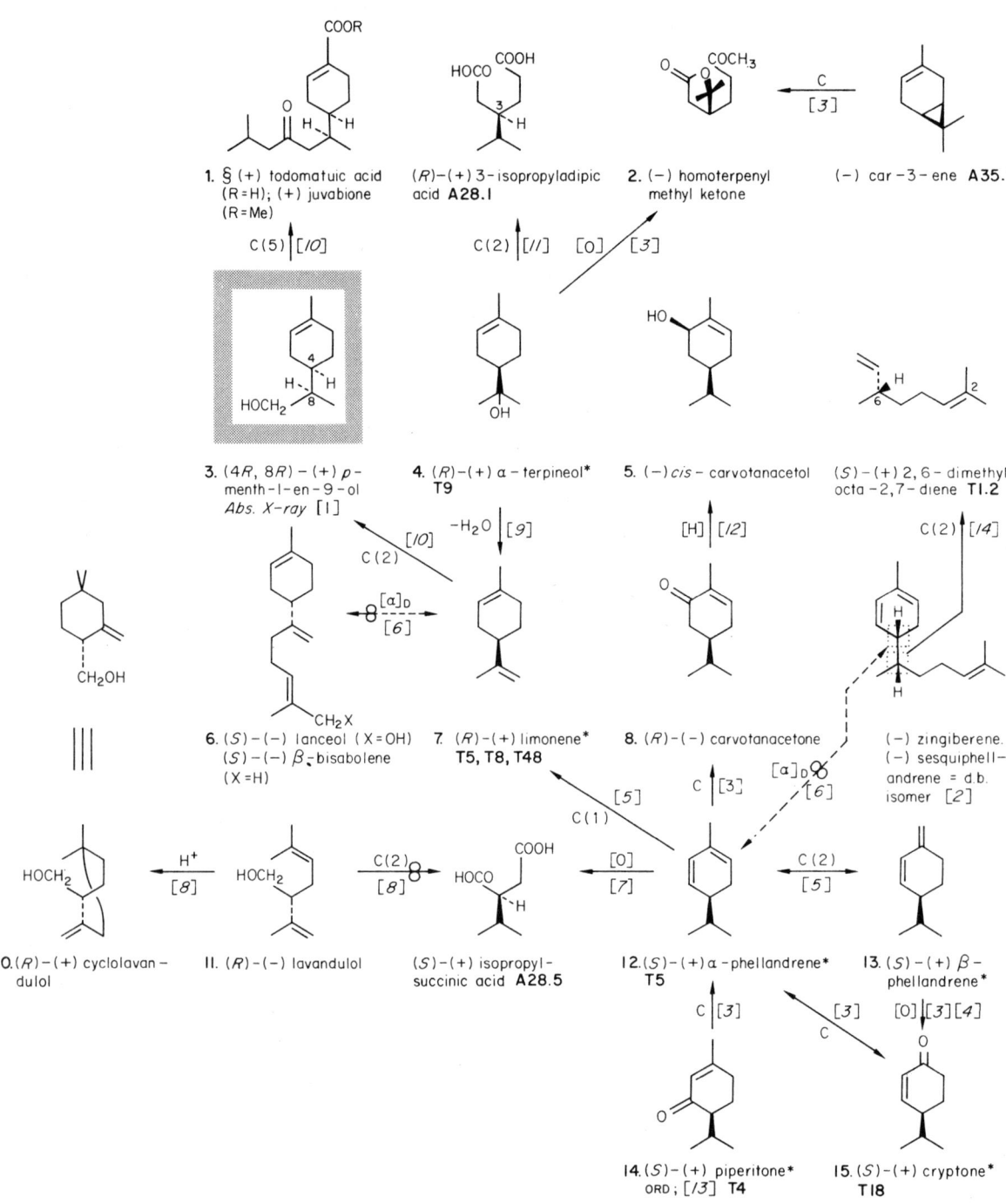

1. § (+) todomatuic acid (R=H); (+) juvabione (R=Me)
2. (−) homoterpenyl methyl ketone
3. (4R, 8R)−(+) p-menth-1-en-9-ol Abs. X-ray [1]
4. (R)−(+) α-terpineol* T9
5. (−) cis-carvotanacetol
6. (S)−(−) lanceol (X=OH); (S)−(−) β-bisabolene (X=H)
7. (R)−(+) limonene* T5, T8, T48
8. (R)−(−) carvotanacetone
10. (R)−(+) cyclolavandulol
11. (R)−(−) lavandulol
12. (S)−(+) α-phellandrene* T5
13. (S)−(+) β-phellandrene*
14. (S)−(+) piperitone* ORD; [13] T4
15. (S)−(+) cryptone* T18

(R)−(+) 3-isopropyladipic acid A28.1
(−) car-3-ene A35.4
(S)−(+) 2,6-dimethyl-octa-2,7-diene T1.2
(−) zingiberene. (−) sesquiphellandrene = d.b. isomer [2]
(S)−(+) isopropylsuccinic acid A28.5

1. J. F. Blount, B. A. Pawson, and G. Saucy, *Chem. Comm.*, 1969, 715.
2. D. W. Connell and M. D. Sutherland, *Austral. J. Chem.*, 1966, **19**, 283.
3. A. J. Birch, *Ann. Reps. Chem. Soc.*, 1950, **47**, 190.
4. O. Wallach, *Annalen,* 1905, **343**, 28.
5. G. G. Acheson and T. F. West, *J. Chem. Soc.*, 1949, 812.
6. J. A. Mills, *J. Chem. Soc.*, 1952, 4976.
7. T. A. Henry and H. Paget, *J. Chem. Soc.*, 1928, 70.
8. M. Soucek and L. Dolejs, *Coll. Czech. Chem. Comm.*, 1959, **24**, 3802.
9. G. Wagner, *Ber.*, 1894, **27**, 1636, 2270.
10. B. A. Pawson, H.-C. Cheung, S. Gurbaxani, and G. Saucy, *Chem. Comm.*, 1968, 1057.
11. K. Freudenberg and W. Lwowski, *Annalen*, 1955, **594**, 76.
12. A. S. Hallsworth, H. B. Henbest, and T. I. Wrigley, *J. Chem. Soc.*, 1957, 1969.
13. C. Djerassi, R. Riniker, and B. Riniker, *J. Amer. Chem. Soc.*, 1956, **78**, 6377.
14. D. Arigoni and O. Jeger, *Helv. Chim. Acta*, 1954, **37**, 881.

Linalool and related monoterpenes.

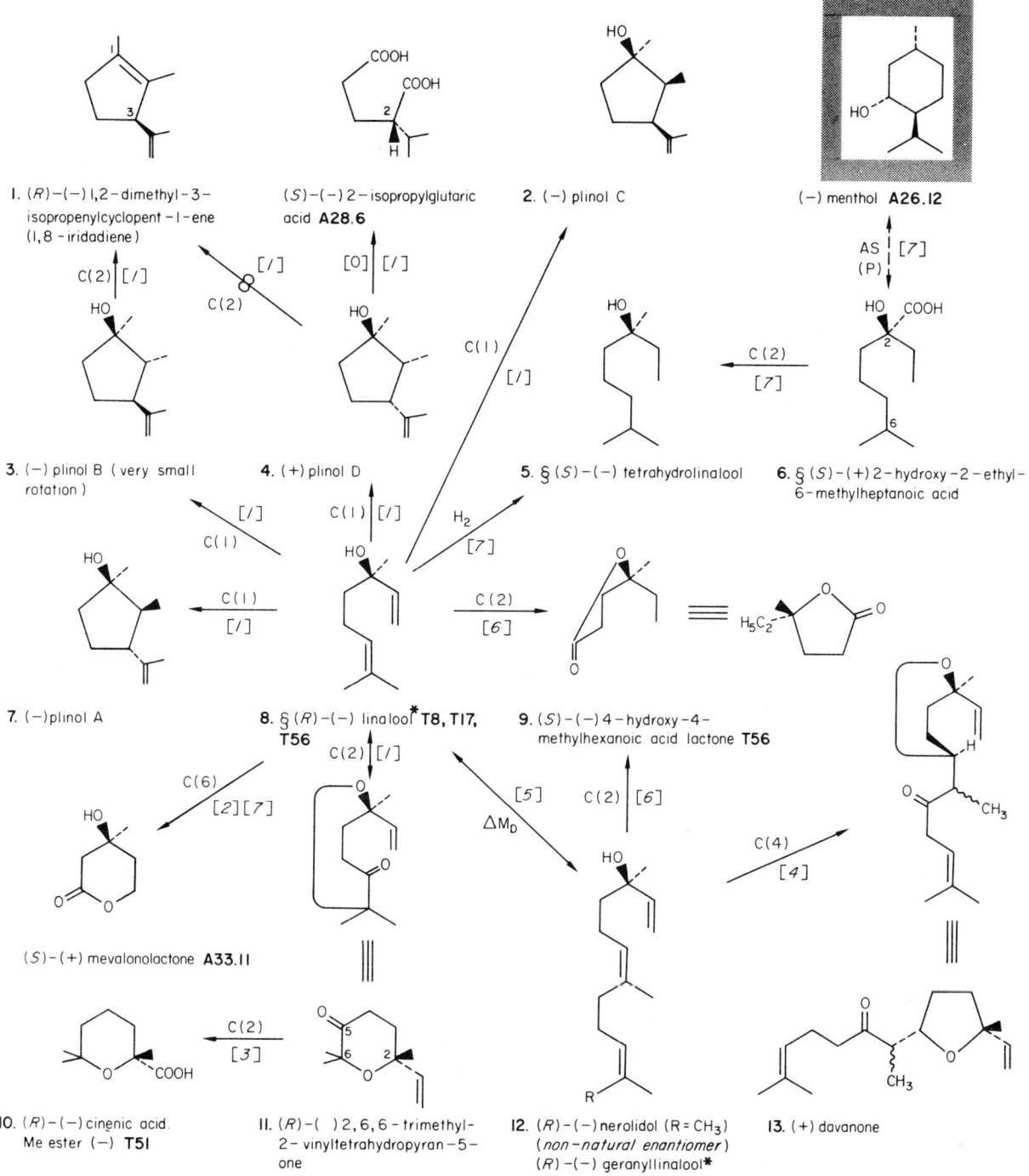

1. (R)-(-)1,2-dimethyl-3-isopropenylcyclopent-1-ene (1,8-iridadiene)
2. (-) plinol C
3. (-) plinol B (very small rotation)
4. (+) plinol D
5. § (S)-(-) tetrahydrolinalool
6. § (S)-(+) 2-hydroxy-2-ethyl-6-methylheptanoic acid
7. (-) plinol A
8. § (R)-(-) linalool* **T8, T17, T56**
9. (S)-(-) 4-hydroxy-4-methylhexanoic acid lactone **T56**
10. (R)-(-) cinenic acid. Me ester (-) **T51**
11. (R)-( ) 2,6,6-trimethyl-2-vinyltetrahydropyran-5-one
12. (R)-(-) nerolidol (R = CH₃) (non-natural enantiomer) (R)-(-) geranyllinalool* (R = CH₂CH₂CH = CMe₂)
13. (+) davanone

(S)-(-) 2-isopropylglutaric acid **A28.6**

(-) menthol **A26.12**

(S)-(+) mevalonolactone **A33.11**

1. H. Strickler, G. Ohloff and K. Kováts, *Tetrahedron Letters*, 1964, 649.
2. R. H. Cornforth, J. W. Cornforth and G. Popják, *Tetrahedron*, 1962, **18**, 1351.
3. M. Nagai, O. Tanaka and S. Shibata, *Tetrahedron Letters*, 1966, 4797.
4. G. Ohloff and W. Giersch, *Helv. Chim. Acta,* 1970, **53**, 841.
5. B. Kimland and T. Norin, *Acta. Chem. Scand.*, 1967, **21**, 825.
6. P. Vlad and M. Soucek, *Coll. Czech. Chem. Comm.*, 1962, **27**, 1726.
7. V. Prelog and E. Watanabe, *Annalen*, 1957, **603**, 1; R. H. Cornforth, J. W. Cornforth and V. Prelog, *Annalen*, 1960, **634**, 197.

T⁴ *p*-menthane group of monoterpenes; menthols, menthylamines etc.

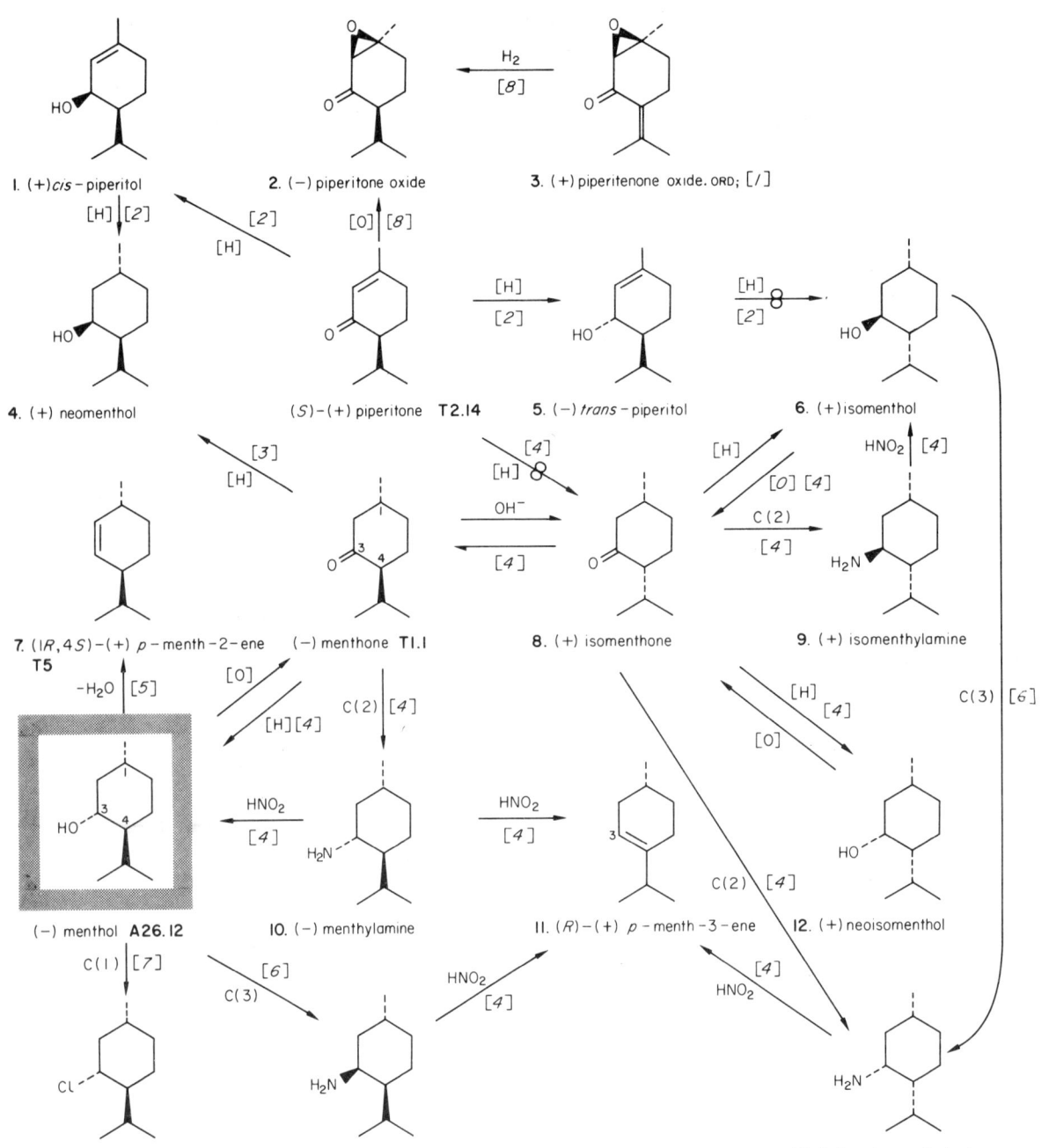

1. (+)*cis*-piperitol
2. (−) piperitone oxide
3. (+) piperitenone oxide. ORD; [*1*]
4. (+) neomenthol
5. (−) *trans*-piperitol
6. (+) isomenthol
7. (1R,4S)-(+) *p*-menth-2-ene T5
8. (+) isomenthone
9. (+) isomenthylamine
10. (−) menthylamine
11. (R)-(+) *p*-menth-3-ene
12. (+) neoisomenthol
13. (+) menthyl chloride T5
14. (+) neomenthylamine
15. (+) neoisomenthylamine

(−) menthol A26.12
(S)-(+) piperitone T2.14
(−) menthone T1.1

1. S. Shimizu, J. Katsuhara and Y. Inouye, *Agr. Biol. Chem. (Japan)*, 1966, **30**, 89.
2. A. K. Macbeth and J. S. Shannon, *J. Chem. Soc.*, 1952, 2852.
3. O. Zeitschel and H. Schmidt, *Ber.*, 1926, **59**, 2298.
4. J. Read, *Chem. Revs.*, 1930, **7**, 1 and references therein.
5. J. Read and J. A. Hendry, *Ber.*, 1938, **71**, 2544.
6. A. K. Bose, J. F. Kistner and L. Farber, *J. Org. Chem.*, 1962, **27**, 2925.
7. R. Mechoulam and Y. Gaoni, *Tetrahedron Letters*, 1967, 1109.
8. E. Klein and G. Ohloff, *Tetrahedron*, 1963, **19**, 1091.

p-menthane group (contd.); carvone and related compds. T5

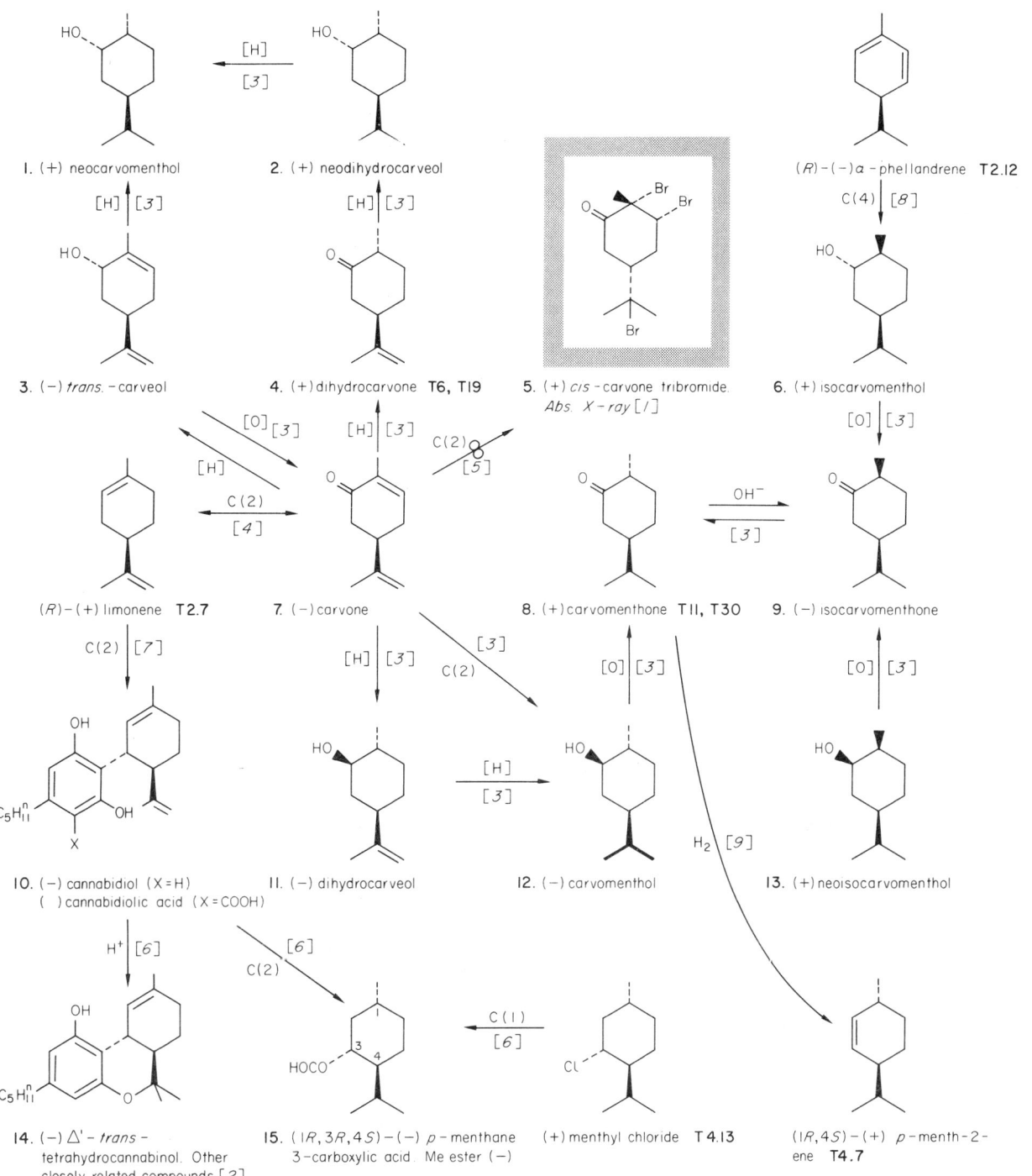

1. (+) neocarvomenthol
2. (+) neodihydrocarveol
3. (−) trans-carveol
4. (+) dihydrocarvone T6, T19
5. (+) cis-carvone tribromide. Abs. X-ray [1]
6. (+) isocarvomenthol
7. (−) carvone
8. (+) carvomenthone T11, T30
9. (−) isocarvomenthone
10. (−) cannabidiol (X=H)
    ( ) cannabidiolic acid (X=COOH)
11. (−) dihydrocarveol
12. (−) carvomenthol
13. (+) neoisocarvomenthol
14. (−) Δ¹-trans-tetrahydrocannabinol. Other closely related compounds [2]
15. (1R,3R,4S)-(−) p-menthane 3-carboxylic acid. Me ester (−)

(+) menthyl chloride T4.13
(R)-(−)α-phellandrene T2.12
(R)-(+) limonene T2.7
(1R,4S)-(+) p-menth-2-ene T4.7

1. R. W. Schevitz and M. G. Rossmann, *Chem. Comm.*, 1969, 711.
2. Y. Gaoni and R. Mechoulam, *J. Amer. Chem. Soc.*, 1971, 93, 217.
3. S. H. Schroeter and E. L. Eliel, *J. Org. Chem.*, 1965, 30, 1 and refs. therein.
4. F. W. Semmler and J. Feldstein, *Ber.*, 1914, 47, 384.
5. J. Wolinsky, J. J. Hamsher and R. O. Hutchins, *J. Org. Chem.*, 1970, 35, 207.
6. R. Mechoulam and Y. Gaoni, *Tetrahedron Letters*, 1967, 1109.
7. T. Petrzilka, W. Haefliger, C. Sikemeier, G. Ohloff and A. Eschenmoser, *Helv. Chim. Acta*, 1967, 50, 719.
8. A. Blumann, E. W. Della, C. A. Henrick, J. Hodgkin and P. R. Jefferies, *Austral. J. Chem.*, 1962, 15, 290.
9. N. L. McNiven and J. Read, *J. Chem. Soc.*, 1952, 159.

T6 Monoterpenes; carane and m-menthane groups.

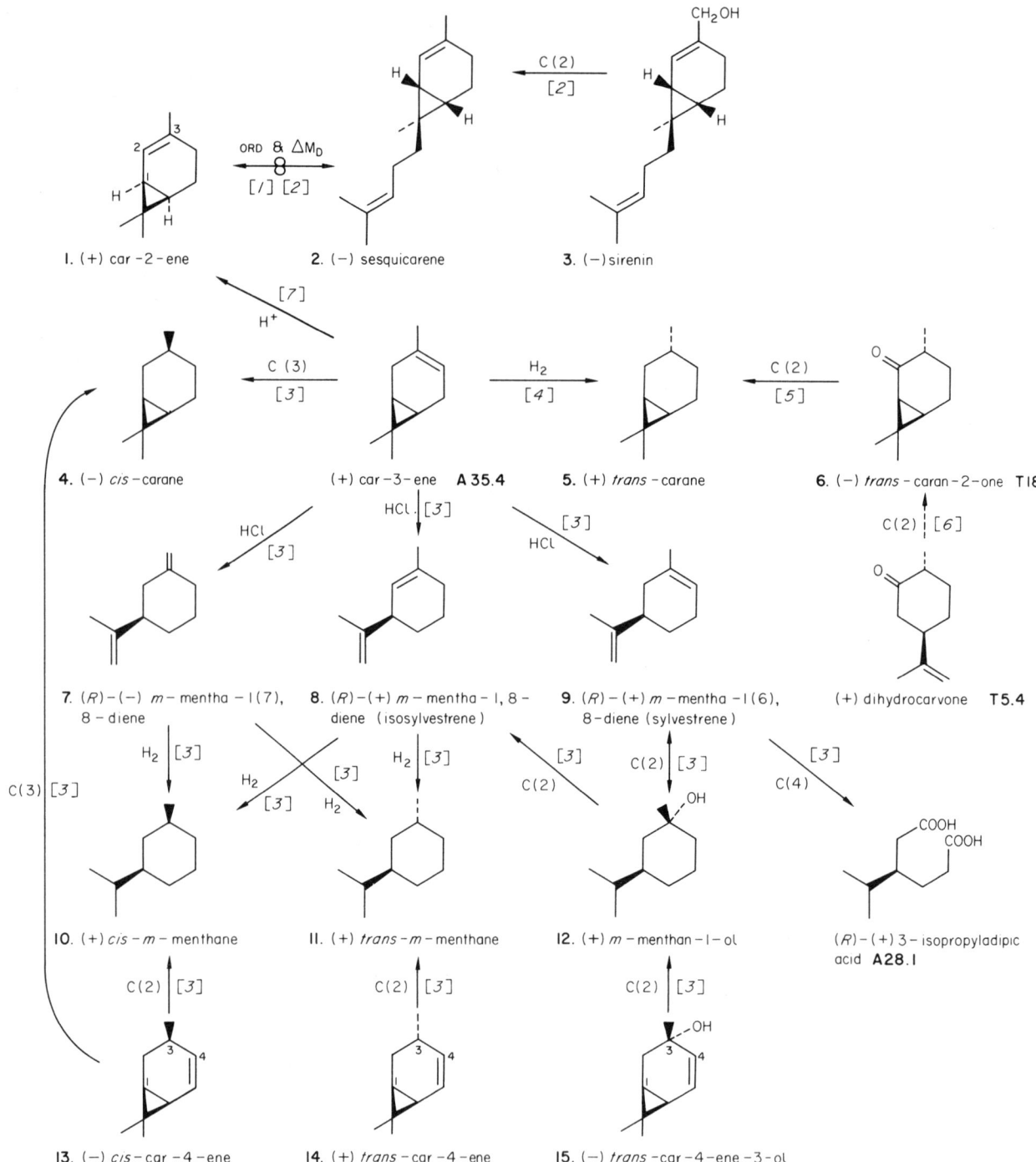

1. Y. Ohta and Y. Hirose, *Tetrahedron Letters*, 1968, 1251.
2. J. J. Plattner and H. Rapoport, *J. Amer. Chem. Soc.*, 1971, **93**, 1758.
3. K. Gollnick and G. Schade, *Tetrahedron Letters*, 1966, 5157 and refs. therein.
4. V. Krestinski and F. Solodki, *J. prakt. Chem.*, 1930, **126**, 14.
5. F. W. Semmler and J. Feldstein, *Ber.*, 1914, **47**, 384.
6. A. von Baeyer, *Ber.*, 1894, **27**, 1915.
7. J. Verghese, *J. Indian Chem. Soc.*, 1959, **36**, 151.

Thujane (bicyclo(3,1,0)hexane) group. T7

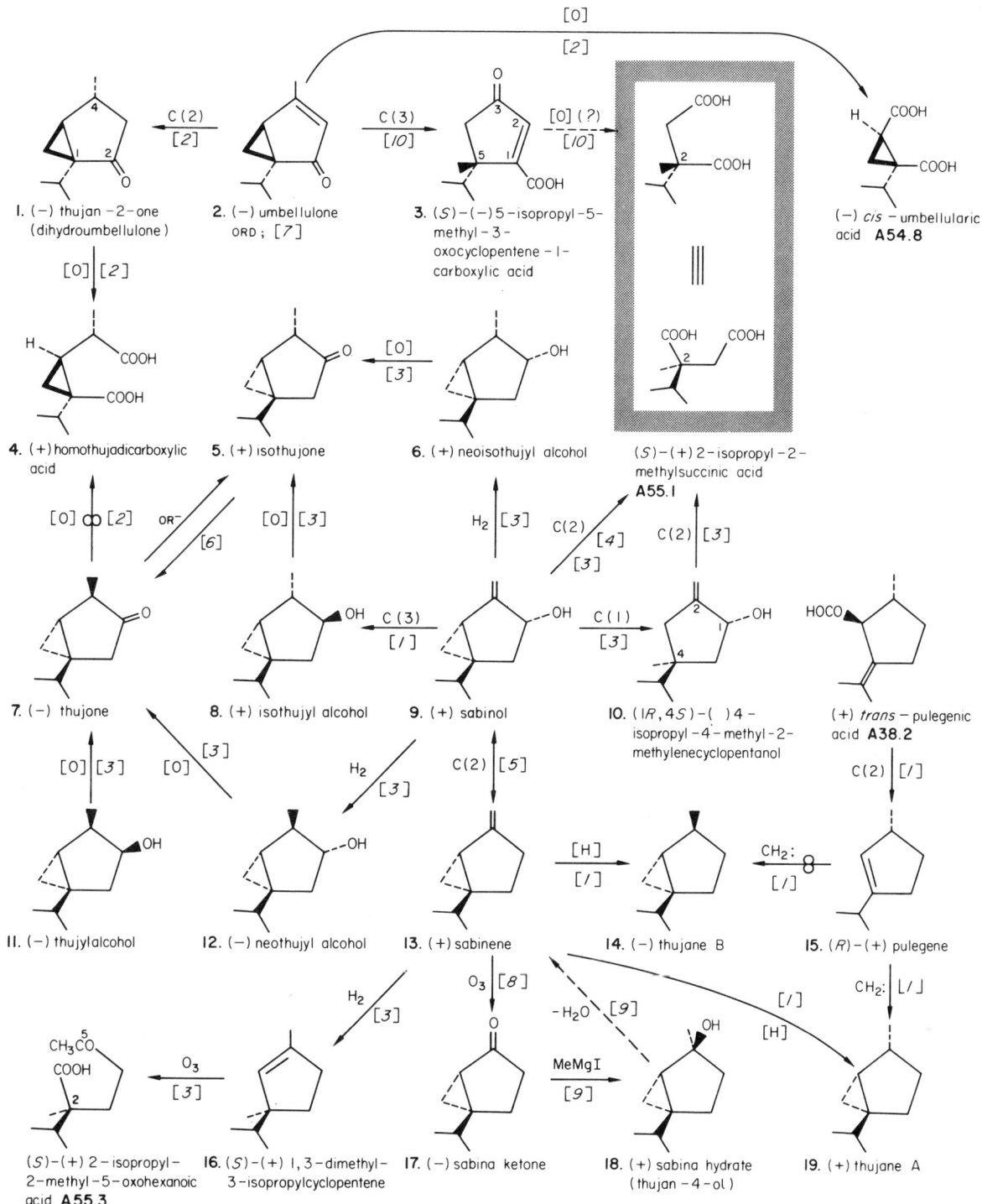

1. G. Ohloff, G. Uhde, A. F. Thomas and E. Kovats, *Tetrahedron*, 1966, **22**, 309.
2. F. W. Semmler, *Ber.*, 1907, **40**, 5019; 1908, **41**, 3988; H. N. Rydon, *J. Chem. Soc.*, 1936, 829.
3. T. Norin, *Acta Chem. Scand.*, 1962, **16**, 640.
4. J. D. Edwards and N. Ichikawa, *J. Org. Chem.*, 1964, **29**, 503.
5. F. W. Semmler, *Ber.*, 1902, **35**, 2047.
6. R. H. Eastman and A. V. Winn, *J. Amer. Chem. Soc.*, 1960, **82**, 5908.
7. C. Djerassi, R. Riniker and B. Riniker, *J. Amer. Chem. Soc.*, 1956, **78**, 6377.
8. H. Schmidt, *Z. angew. Chem.*, 1929, **42**, 126.
9. J. W. Daly, F. C. Green and R. H. Eastman, *J. Amer. Chem. Soc.*, 1958, **80**, 6330.
10. R. H. Eastman and A. Oken, *J. Amer. Chem. Soc.*, 1953, **75**, 1029.

T⁸ Pinane (bicyclo(3.1.1)heptane) group.

Review; [3]

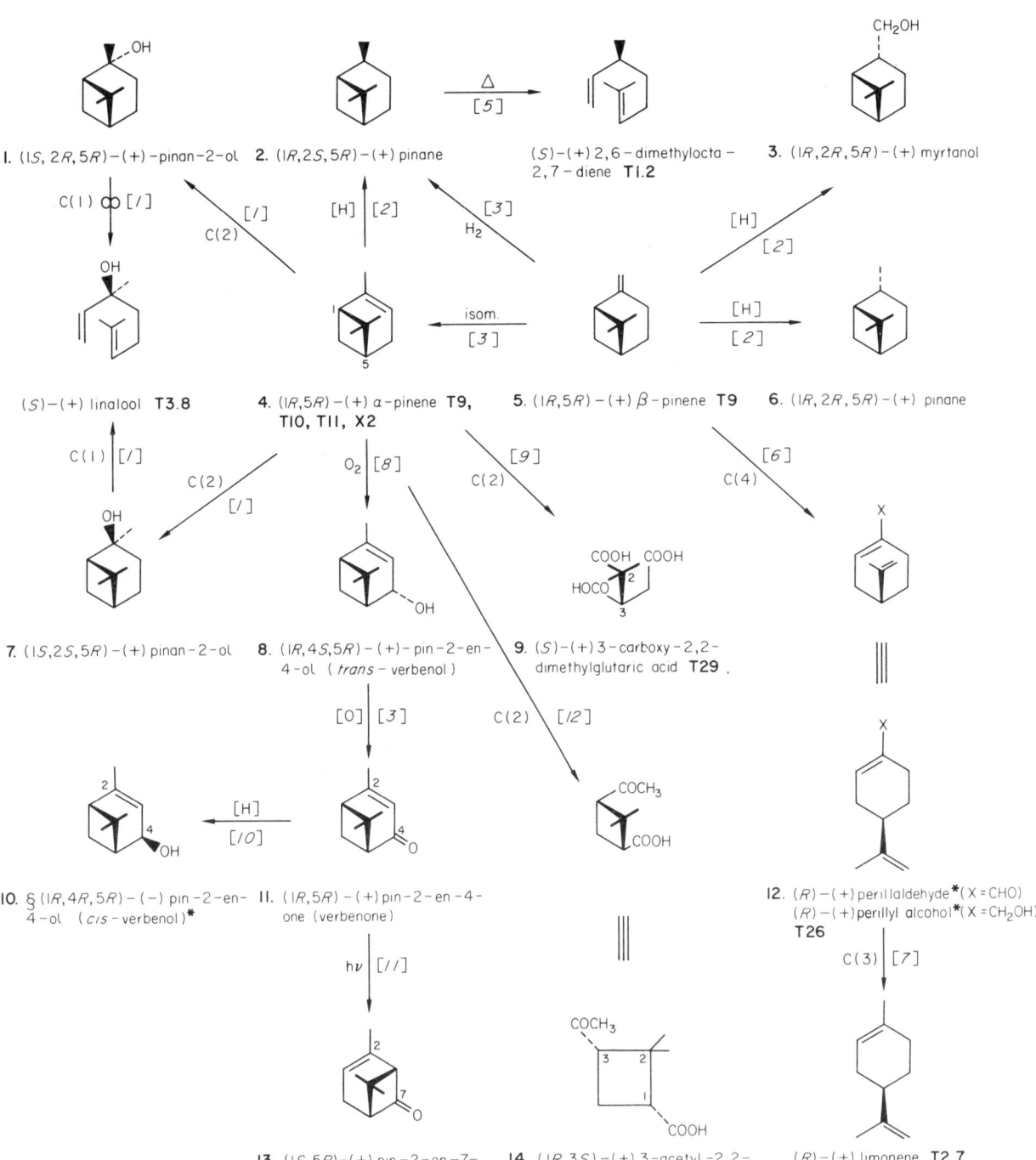

1. (1S, 2R, 5R)-(+)-pinan-2-ol
2. (1R, 2S, 5R)-(+) pinane
   (S)-(+) 2,6-dimethylocta-2,7-diene **T1.2**
3. (1R, 2R, 5R)-(+) myrtanol
4. (1R, 5R)-(+) α-pinene **T9, T10, T11, X2**
5. (1R, 5R)-(+) β-pinene **T9**
6. (1R, 2R, 5R)-(+) pinane
7. (1S, 2S, 5R)-(+) pinan-2-ol
8. (1R, 4S, 5R)-(+)-pin-2-en-4-ol (*trans*-verbenol)
9. (S)-(+) 3-carboxy-2,2-dimethylglutaric acid **T29**
10. § (1R, 4R, 5R)-(−) pin-2-en-4-ol (*cis*-verbenol)*
11. (1R, 5R)-(+) pin-2-en-4-one (verbenone)
12. (R)-(+) perillaldehyde*(X = CHO)
    (R)-(+) perillyl alcohol*(X = CH₂OH) **T26**
13. (1S, 5R)-(+) pin-2-en-7-one (chrysanthenone). ORD [4]
14. (1R, 3S)-(+) 3-acetyl-2,2-dimethylcyclobutanecarboxylic (pinononic) acid.
    (R)-(+) limonene **T2.7**
    (S)-(+) linalool **T3.8**

1. G. Ohloff and E. Klein, *Tetrahedron*, 1962, **18**, 37.
2. H. C. Brown and G. Zweifel, *J. Amer. Chem. Soc.*, 1964, **86**, 393.
3. D. V. Banthorpe and D. Whittaker, *Chem. Revs.*, 1966, **66**, 643.
4. A. Moscowitz, K. Mislow, M. A. W. Glass and C. Djerassi, *J. Amer. Chem. Soc.*, 1962, **84**, 1945.
5. R. Reinäcker and G. Ohloff, *Angew. Chem.*, 1961, **73**, 240.
6. G. Büchi, W. Hofheinz and J. V. Paukstelis, *J. Amer. Chem. Soc.*, 1969, **91**, 6473.
7. F. W. Semmler and B. Zaar, *Ber.*, 1911, **44**, 52.
8. G. Whitham, *J. Chem. Soc.*, 1961, 2232.
9. P. A. Plattner and H. Kläui, *Helv. Chim. Acta*, 1943, **26**, 1553.
10. C. A. Reece, J. O. Rodin, R. G. Brownlee, W. G. Duncan and R. M. Silverstein, *Tetrahedron*, 1968, **24**, 4249.
11. J. J. Hurst and G. H. Whitham, *J. Chem. Soc.*, 1960, 2864.
12. M. Harispe, D. Mea and A. Horeau, *Bull. Soc. Chim. France*, 1964, 1035.

Pinane group (contd.).

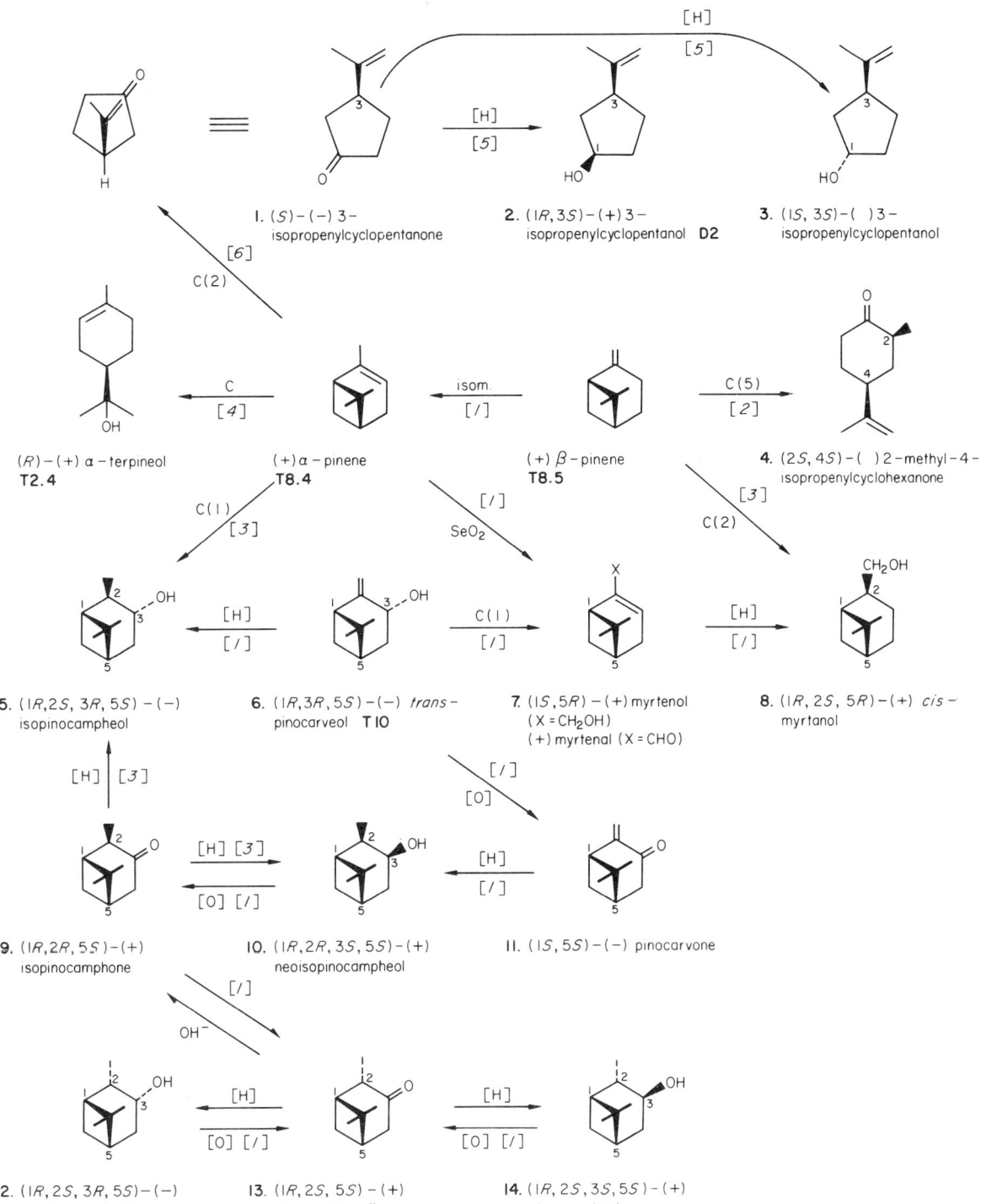

1. D. V. Banthorpe and D. Whittaker, *Chem. Revs.*, 1966, **66**, 643.
2. A. Van der Gen, L. M. Van der Linde, J. G. Witteveen and H. Boelens, *Rec. Trav. Chim.*, 1971, **90**, 1034.
3. H. C. Brown and G. Zwiefel, *J. Amer. Chem. Soc.*, 1964, **86**, 393.
4. M. Delepine, *Bull. Soc. chim. France*, 1924, **35**, 1655.
5. C. Djerassi and B. Tursch, *J. Amer. Chem. Soc.*, 1961, **83**, 4609.
6. M. Harispe, A. Boime and R. Charronat, *Bull. Soc. chim. France*, 1958, 481; Y-R. Naves, ibid., 1372.

T10 Bicyclo(2.2.1)heptane types; camphor and related terpenes.

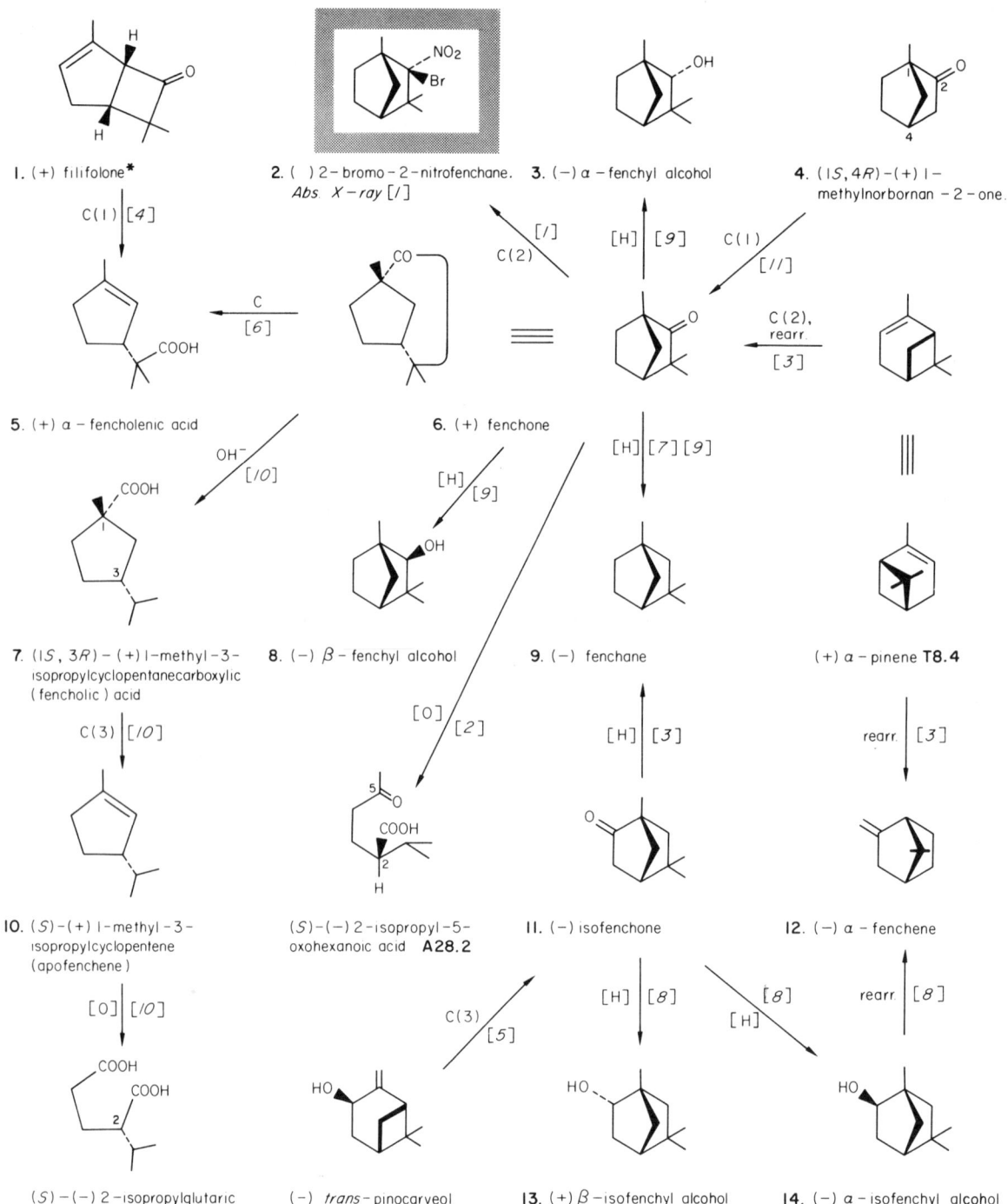

1. (+) filifolone*
2. ( ) 2-bromo-2-nitrofenchane. Abs. X-ray [1]
3. (−) α-fenchyl alcohol
4. (1S,4R)-(+) 1-methylnorbornan-2-one.
5. (+) α-fencholenic acid
6. (+) fenchone
7. (1S,3R)-(+)1-methyl-3-isopropylcyclopentanecarboxylic (fencholic) acid
8. (−) β-fenchyl alcohol
9. (−) fenchane
   (+) α-pinene T8.4
10. (S)-(+) 1-methyl-3-isopropylcyclopentene (apofenchene)
    (S)-(−) 2-isopropyl-5-oxohexanoic acid A28.2
11. (−) isofenchone
12. (−) α-fenchene
    (S)-(−) 2-isopropylglutaric acid A28.6
    (−) trans-pinocarveol T9.5
13. (+) β-isofenchyl alcohol
14. (−) α-isofenchyl alcohol

1. M. C. Rerat, *Compt. Rend.*, 1968, **266(C)**, 612.
2. O. Wallach, *Annalen*, 1911, 379, 182.
3. A. J. Birch, *Ann. Reps. Chem. Soc.*, 1950, 47, 190 and references therein.
4. R. B. Bates, M. J. Onore, S. K. Paknikar, C. Steelink and E. P. Blanchard, *Chem. Comm.*, 1967, 1037.
5. M. P. Hartshorn and A. F. A. Wallis, *J. Chem. Soc.*, 1964, 5254.
6. G. B. Cockburn, *J. Chem. Soc.*, 1899, 501.
7. L. Wolff, *Annalen*, 1911, 394, 97.
8. W. Hückel, *Bull. Soc. Chim. belges*, 1962, 71, 473 and references therein; W. Hückel and H. J. Kern, *Annalen*, 1965, 687, 40 and references therein.
9. J. Kenyon and H. E. M. Priston, *J. Chem. Soc.*, 1925, 1472.
10. O. Wallach, *Annalen*, 1911, 379, 182.
11. J. A. Berson, J. S. Walia, A. Remanick, S. Suzuki, P. Reynolds-Warnhoff and D. Willner, *J. Amer. Chem. Soc.*, 1961, 83, 3986.

Bicyclo(2.2.1)heptane types; fenchane group. T11

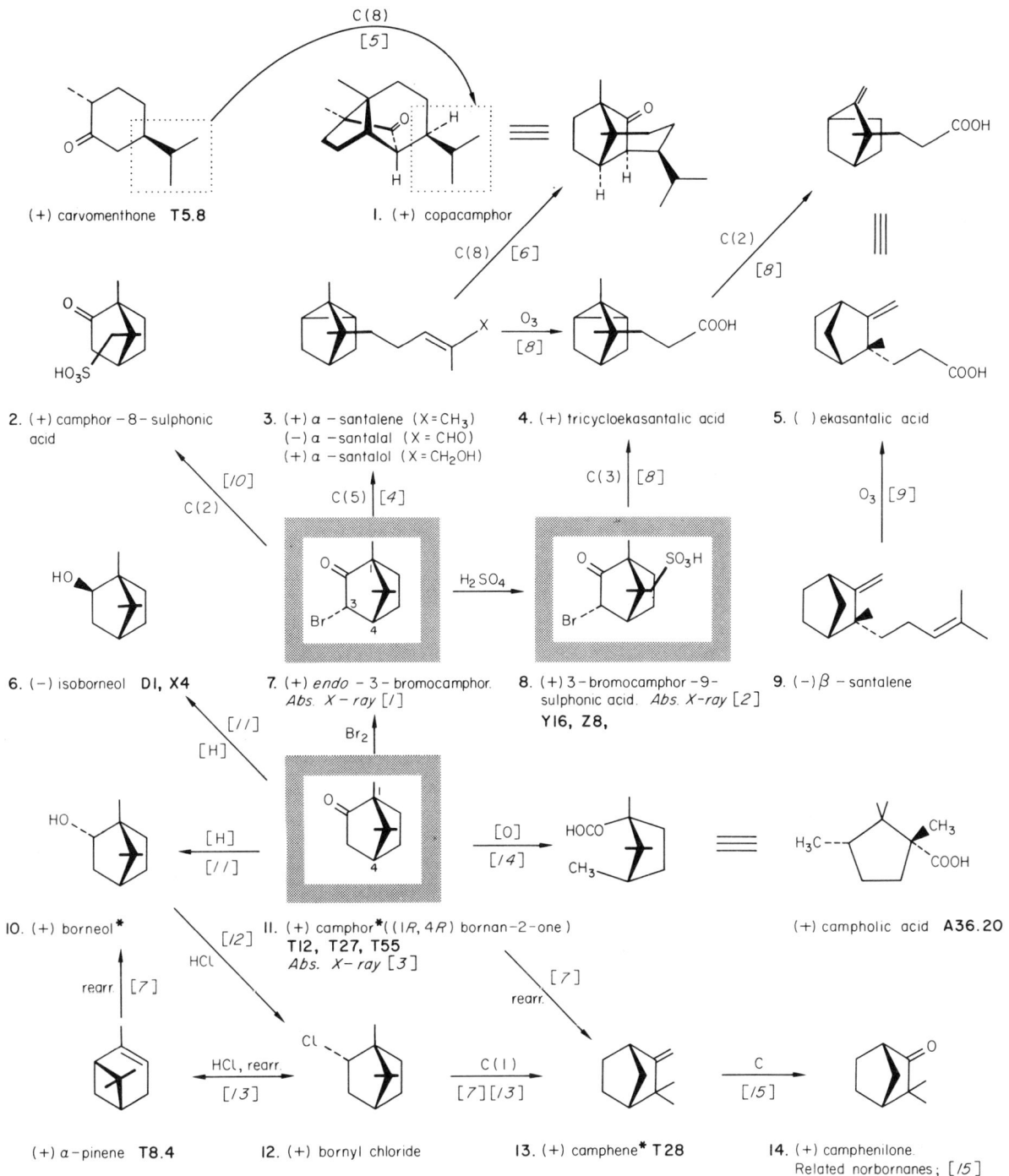

1. M. G. Northolt and J. H. Palm, *Rec. Trav. Chim.*, 1966, **85**, 143; F. H. Allen and D. Rogers, *Chem. Comm.*, 1966, 836.
2. J. A. Wunderlich, *Acta. Cryst.*, 1967, **23**, 846.
3. H. A. J. Oonk, Ph.D. Thesis, Utrecht, 1965.
4. E. J. Corey, S. W. Chow and R. A. Scherrer, *J. Amer. Chem. Soc.*, 1952, **79**, 5773.
5. E. Piers, R. W. Britton, R. J. Keziere and R. D. Smillie, *Canad. J. Chem.*, 1971, **49**, 2620.
6. M. Kolbe-Haugwitz and L. Westfelt, *Acta. Chem. Scand.*, 1970, **24**, 1623.
7. A. J. Birch, *Ann. Reps. Chem. Soc.*, 1950, **47**, 190.
8. P. C. Guha and S. C. Bhattacharrya, *J. Indian Chem. Soc.*, 1944, **21**, 271.
9. L. Ruzicka and G. Thomann, *Helv. Chim. Acta*, 1935, **18**, 355.
10. F. W. Semmler and K. Bode, *Ber.*, 1907, **40**, 1137.
11. F. Ullmann and A. Schmid, *Ber.*, 1910, **43**, 3202.
12. G. Wagner and W. Brickner, *Ber.*, 1899, **32**, 2302.
13. P. D. Bartlett and J. D. Gill, *J. Amer. Chem. Soc.*, 1941, **63**, 1273.
14. H. Rupe and C. A. Kloppenburg, *Helv. Chim. Acta*, 1919, **2**, 363.
15. J. A. Berson, J. S. Walia, A. Remanick, S. Suzuki, P. Reynolds-Warnhoff and D. Willner, *J. Amer. Chem. Soc.*, 1961, **83**, 3986 and references therein.

T 12 Further mono- and sesquiterpenes related to camphor.

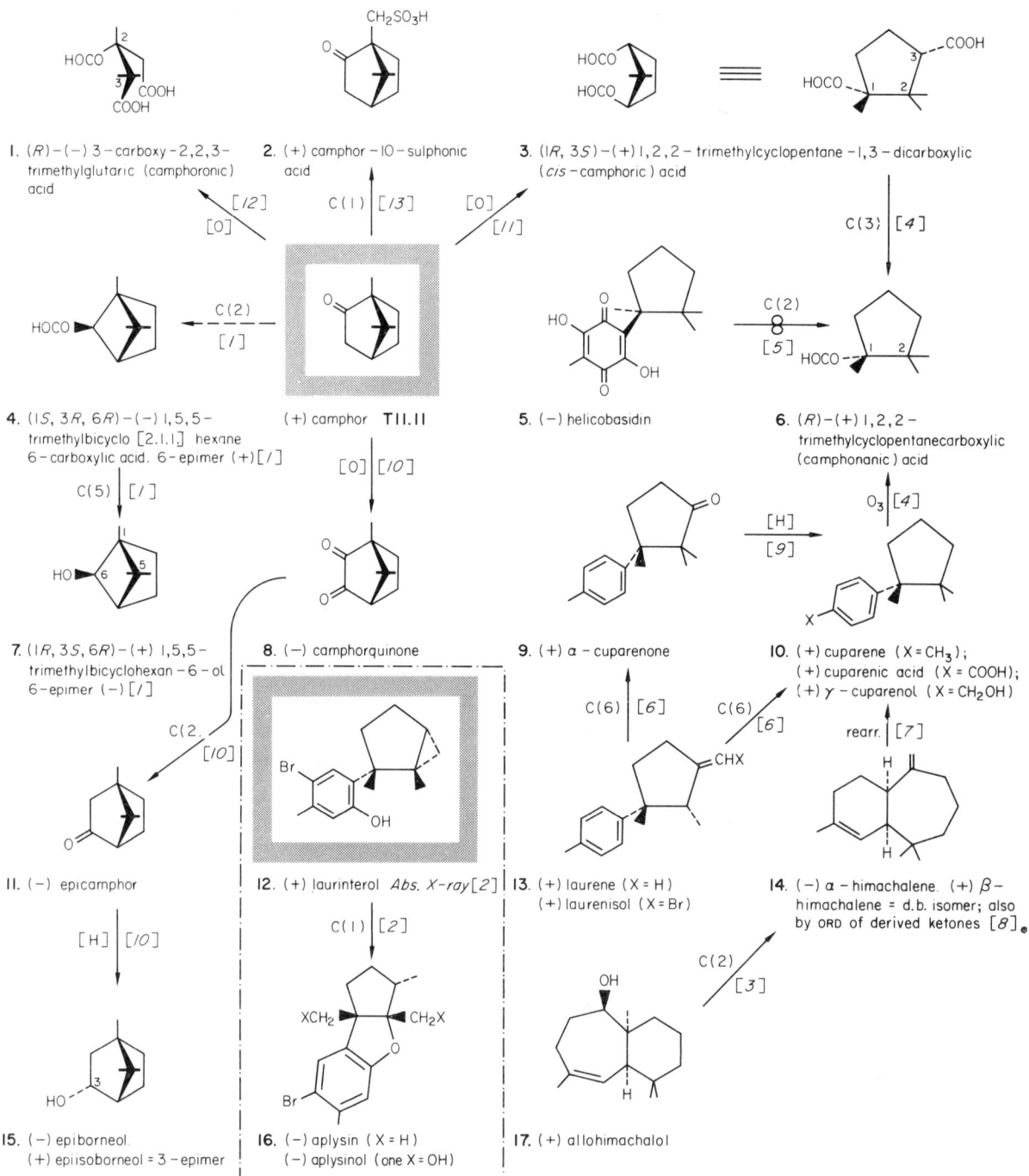

1. J. Meinwald, A. Lewis and P. G. Gassman, *J. Amer. Chem. Soc.*, 1962, **84**, 977.
2. A. F. Cameron, G. Ferguson and J. M. Robertson, *Chem. Comm.*, 1967, 271.
3. S. C. Bisarya and S. Dev, *Tetrahedron*, 1968, **24**, 3869.
4. C. Enzell and H. Erdtman, *Tetrahedron*, 1958, **4**, 361.
5. S. Natori, H. Ogawa, K. Yamaguchi and H. Nishikawa, *Chem. Pharm. Bull. Japan*, 1963, **11**, 1343.
6. T. Irie, T. Suzuki, Y. Yasunari, E. Kurosawa and T. Masamune, *Tetrahedron*, 1969, **25**, 459.
7. H. S. Subba Rao, N. P. Damodaran and S. Dev, *Tetrahedron Letters*, 1968, 2213.
8. T. C. Joseph and S. Dev, *Tetrahedron*, 1968, **24**, 3841.
9. G. L. Chetty and S. Dev, *Tetrahedron Letters*, 1964, 73.
10. W. Hückel and O. Fechtig, *Annalen*, 1962, **652**, 81.
11. O. Aschan, *Annalen*, 1901, **316**, 192.
12. J. Bredt, *Ber.*, 1893, **26**, 3047.
13. E. Wedekin, D. Schenk and R. Stüsser, *Ber.*, 1923, **56**, 633.

T13
Cyclopentanoid monoterpenes. Iridanes and secoiridanes; loganin.

Review; [13].

1. (−) asperuloside. **T15**
2. (1R,2R,3S)-(−)2-formyl-3-methylcyclopentylacetaldehyde. Bis-DNP (−)
3. (+) genepin
4. (1R,2R,3R)-(−)3-methylcyclopentane-1,2-dicarboxylic (trans-cis-nepetic) acid **T14**
5. (−) loganin. Abs. X-ray [1]
6. (−) sweroside
7. (−) menthiafolin
8. (1R,2R,3S)-(−)3-methylcyclopentane-1,2-dicarboxylic (cis-cis-nepetic) acid.
9. (−) verbenalin
    (S)-(−) citronellal **T14**
10. (−) secologanin **K2**
11. (−) foliamenthin
12. (+) iridomyrmecin. Rel. X-ray [2]
13. (−) iridodial
14. (+) δ-nepetalinic acid **T14**
15. (−) isoiridomyrmecin
16. (+) α-nepetalinic acid **T14, T25**
17. (−) actinidine. Related alkaloids [7]

1. P. J. Lentz and M. G. Rossman, *Chem. Comm.*, 1969, 1269.
2. J. F. McConnell, A. McL. Mathieson and B. P. Schoenborn, *Tetrahedron Letters*, 1962, 445.
3. H. Inouye, T. Yoshida and S. Tobita, *Tetrahedron Letters*, 1968, 2945.
4. A. R. Battersby, R. S. Kapil and R. Southgate, *Chem. Comm.*, 1968, 131.
5. T. Sakan, A. Fujino, F. Murai, Y. Butsugan and A. Suzui, *Bull. Chem. Soc. Japan,* 1959, 32, 315.
6. G. Büchi and R. E. Manning, *Tetrahedron*, 1962, 18, 1049.
7. G. W. K. Cavill and A. Zeitlin, *Austral. J. Chem.*, 1962, 20, 349.
8. K. J. Clark, G. I. Fray, R. H. Jaeger and R. Robinson, *Tetrahedron*, 1959, 6, 217.
9. L. H. Briggs, B. F. Cain, P. W. LeQuesne and J. N. Shoolery, *J. Chem. Soc.*, 1965, 2595.
10. A. R. Battersby, A. R. Burnett, G. D. Knowles and P. G. Parsons, *Chem. Comm.*, 1968, 1277, 1280.
11. H. Inouye, T. Yoshida, Y. Nakamura and S. Tobita, *Chem. Pharm. Bull. Japan*, 1970, 18, 1889.
12. C. Djerassi, T. Nakano, A. N. James, L. H. Zalkow, E. J. Eisenbraun and J. N. Shoolery, *J. Org. Chem.*, 1961, 26, 1192.
13. 'Cyclopentanoid Terpene Derivatives' (W. I. Taylor and A. R. Battersby, Eds.), Marcel Dekker, Inc., 1969.
14. G. W. K. Cavill and D. L. Ford, *Austral. J. Chem.*, 1960, 13, 296.
15. P. Lowe, Ch. v. Szczepanski, C. J. Coscia and D. Arigoni, *Chem. Comm.*, 1968, 1276.

T14 Cyclopentanoid monoterpenes (contd.); nepetic and nepatalinic acids.

1. (+) neomatabiol
   ( ) isoneomatabiol = epimer [4]

   (R)-(−)2-methylglutaric acid A32.12

2. (−) furopelargone A (furopelargones B,C and D closely related) [2]

3. (−) matatabiether

4. (+) isodihydronepetalactone

5. (+) dihydronepetalactone

   (+) trans−cis−nepetic acid T13.4 (X=OH)

6. (+) trans−cis−nepetonic acid (X=Me)

7. ( ) neonepetalactone

8. (1S,2R,3S)-(+)3-methylcylopentane-1,2-dicarboxylic (cis−trans-nepetic) acid (X=OH). ( ) cis−trans-nepetonic acid (X=Me) T15

9. (+) nepetalactone

10. ( ) isonepetalactone

11. (+) γ-nepatalinic acid. See also [/]

12. ( ) α-nepetalic acid (X=CHO)
    (+) α-nepatalinic acid T13.15 (X=COOH)

13. (+) β-nepatalinic acid

14. ( ) δ-nepetalic acid (X=CHO)
    (+) δ-nepatalinic acid T13.14 (X=COOH)

15. (+) γ-skytanthine

16. (+) α-skytanthine

17. (+) β-skytanthine

18. (+) δ-skytanthine

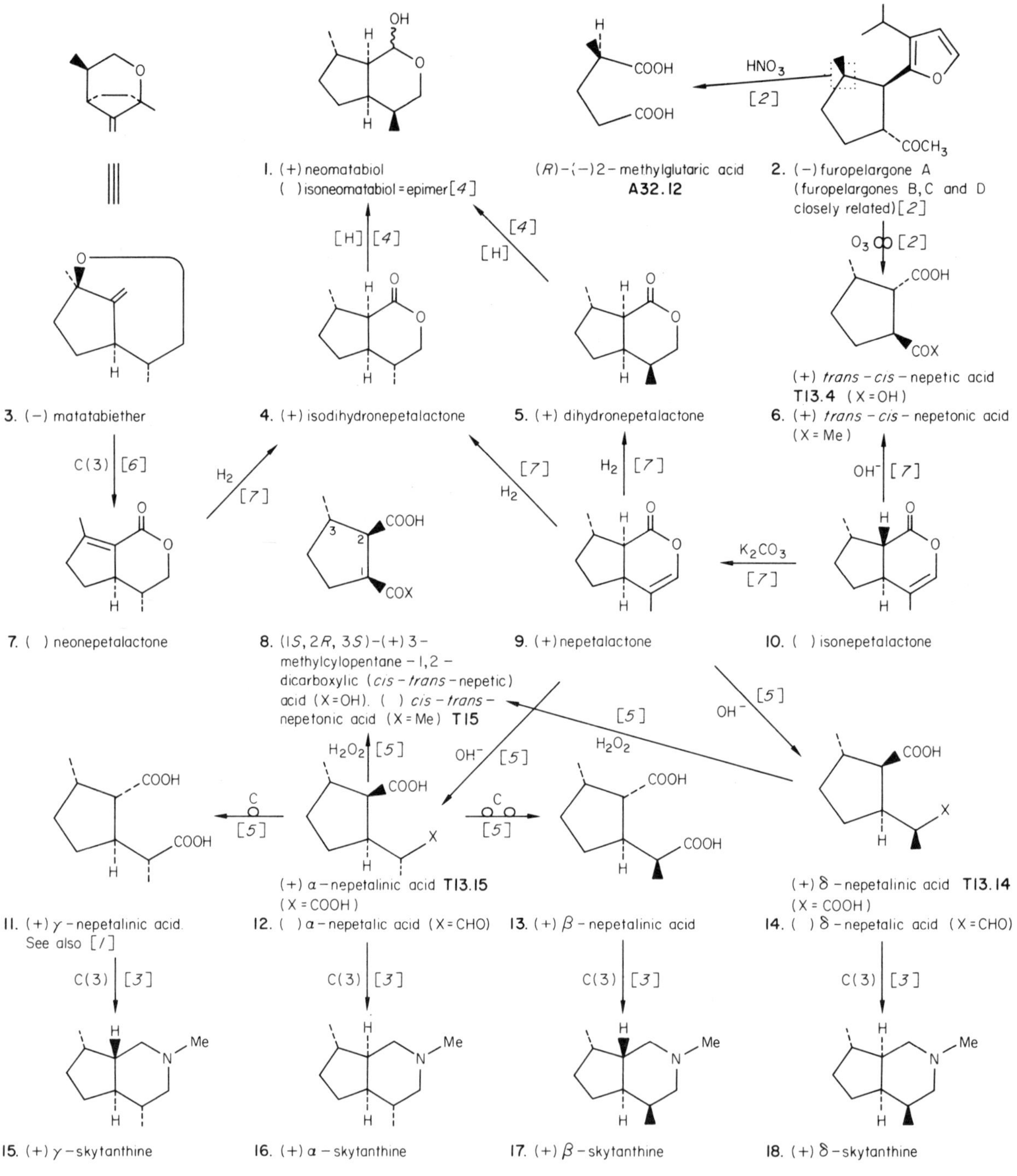

1. R. Trave, *Gazzetta*, 1970, **100**, 1061.
2. M. Romanuk. V. Herout, F. Sorm, Y. R. Naves, P. Tullen, R. B. Bates and C. W. Sigel, *Coll. Czech. Chem. Comm.*, 1964, **29**, 1048.
3. E. J. Eisenbraun, A. Bright and H. H. Appel, *Chem. and Ind.*, 1962, 1242.
4. S. B. Hyeon, S. Isoe and T. Sakan, *Tetrahedron Letters*, 1968, 5325.
5. R. B. Bates, E. J. Eisenbraun and S. M. McElvain, *J. Amer. Chem. Soc.*, 1958, **80**, 3420 and references therein.
6. S. Izoe, T. Ono, S. B. Hyeon and T. Sakan, *Tetrahedron Letters*, 1968, 5319.
7. T. Sakan, S. Isoe, S. B. Hyeon, R. Katsumura, T. Maeda, J. Wolinsky, D. Dickerson, M. Slabaugh and D. Nelson, *Tetrahedron Letters*, 1965, 4097, and refs. therein.

Cyclopentanoid monoterpenes (contd.).

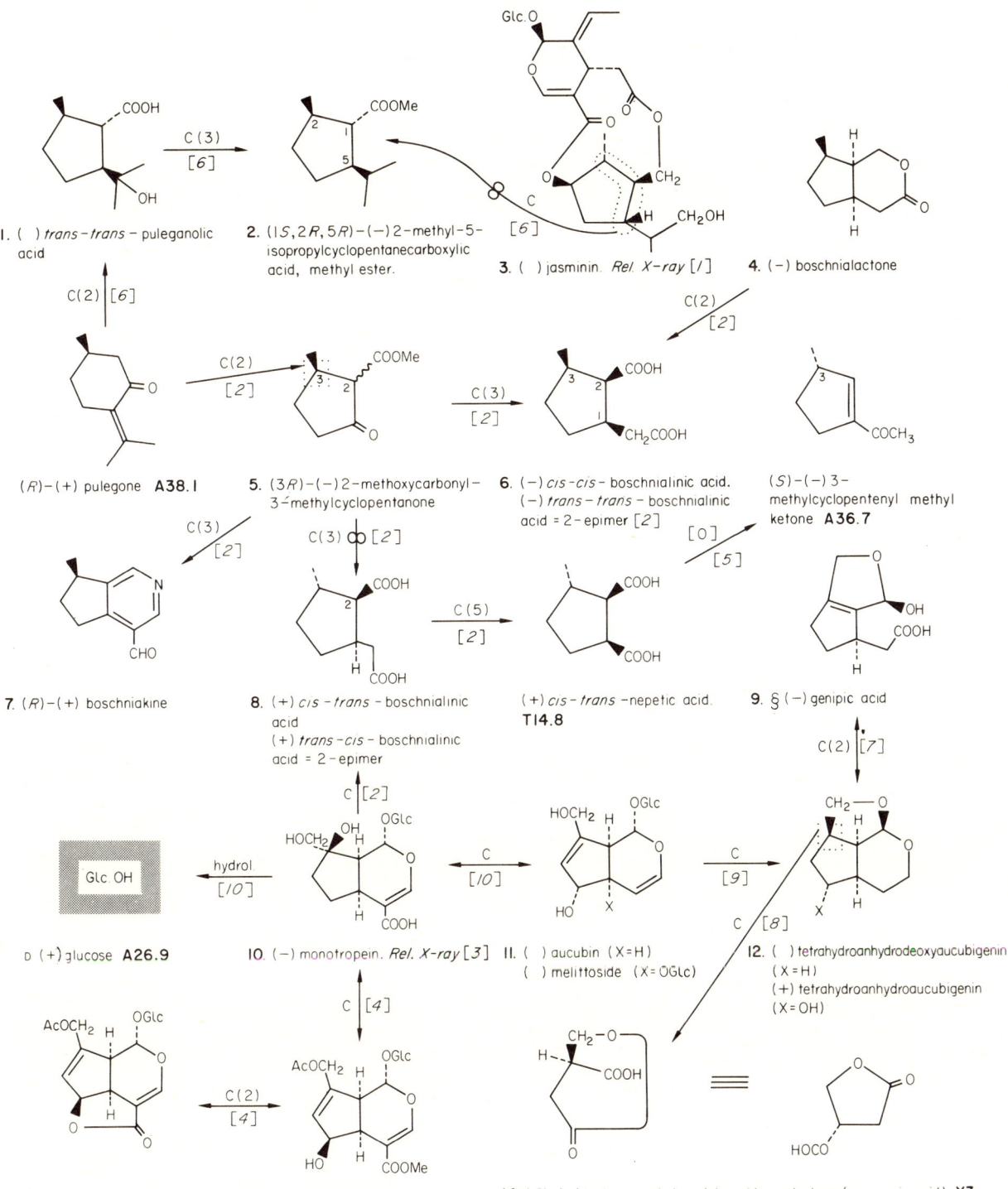

1. ( ) *trans-trans* − puleganolic acid
2. (1*S*,2*R*,5*R*)−(−)2-methyl-5-isopropylcyclopentanecarboxylic acid, methyl ester.
3. ( ) jasminin. *Rel. X-ray* [1]
4. (−) boschnialactone
5. (3*R*)−(−)2-methoxycarbonyl-3-methylcyclopentanone
6. (−) *cis-cis*- boschnialinic acid. (−) *trans-trans* − boschnialinic acid = 2-epimer [2]
7. (*R*)−(+) boschniakine
8. (+) *cis-trans*- boschnialinic acid (+) *trans-cis*- boschnialinic acid = 2-epimer
9. § (−) genipic acid
10. (−) monotropein. *Rel. X-ray* [3]
11. ( ) aucubin (X=H); ( ) melittoside (X=OGlc)
12. ( ) tetrahydroanhydrodeoxyaucubigenin (X=H); (+) tetrahydroanhydroaucubigenin (X=OH)
13. ( ) daphylloside
14. (*S*)−(−) hydroxymethylsuccinic acid γ − lactone (paraconic acid) Y3

(*R*)−(+) pulegone **A38.1**

(+) *cis-trans* −nepetic acid. **T14.8**

(*S*)−(−) 3-methylcyclopentenyl methyl ketone **A36.7**

D (+) glucose **A26.9**

(−) asperuloside **T13.1**

1. A. Shimada and M. Fukuyo, *Abstr. 22nd Japanese Chem. Soc. Meeting*, 1969, 60.
2. T. Sakan, F. Murai, Y. Hayashi, Y. Honda, T. Shono, M. Nakajima and M. Kato, *Tetrahedron*, 1967, **23**, 4635.
3. N. Masaki, M. Hirabayashi, K. Fuji, K. Osaki and H. Inouye, *Tetrahedron Letters*, 1967, 2367.
4. H. Inouye, S. Ueda, M. Hirabayashi and N. Shimokawa, *Yakagaku Zasshi*, 1966, **86**, 943.
5. S. M. McElvain and E. J. Eisenbraun, *J. Amer. Chem. Soc.*, 1955, **77**, 1599.
6. T. Kamikawa, K. Inouye, T. Kubota and M. C. Woods, *Tetrahedron*, 1970, **26**, 4561.
7. W. H. Tallent, *Tetrahedron*, 1964, **20**, 1781.
8. H. Uda, M. Maruyama, K. Kabuki and S. Fujise, *Nippon Kagaku Zasshi*, 1964, **85**, 279.
9. P. Karrer and H. Schmid, *Helv. Chim. Acta*, 1946, **29**, 525.
10. H. Inouye and K. Fuji, *Chem. Pharm. Bull. (Japan)*, 1964, **12**, 901 and references therein.

T 16 Miscellaneous mono- and sesquiterpenes.

1. (−) plumieride
2. ( )
3. (+) plumericin, isoplumericin = geom. isomer [2]
4. (S)-(+) calacone
5. (R)-(−) α-curcumene
6. (S)-(+) ar-turmerone
7. (S)-(+) 4-(p-tolyl)-pentanoic acid
8. (S)-(+) γ-curcumene*
9. ( ) carquejol
10. ( ) dihydroabscisin II.[†] AC by Mills' rule for cyclohex-1-en-3-ol system [7]
11. (S)-(+) abscisin II (dormin)[†]
12. ( ) dihydroabscisin II.[†] AC by Mills' rule for cyclohex-1-en-3-ol system [7]
13. (+) carquejanone (ortho-menthone; (2R, 3S)-3-methyl-2-isopropycyclohexanone. By CD [5] (octant rule) stereoisomers; [6]
14. ( ) cryptomerone. By CD. [8] (αβ-unsaturated ketone chromophore)

(S)-(+) lactic acid A1.9
(R)-(+) ethylsuccinic acid A28.9
(S)-(−) methylsuccinic acid A27.21
(S)-(+) 3-(p-tolyl)-butyric acid A49.7

1. O. Halpern and H. Schmid, *Helv. Chim. Acta*, 1958, **41**, 1109.
2. G. Albers-Schönberg and H. Schmid, *Helv. Chim. Acta*, 1961, **44**, 1447.
3. J. Vrkoc, V. Herout and F. Sorm, *Coll. Czech. Chem. Comm.*, 1961, **26**, 1343.
4. V. K. Honwad and A. S. Rao, *Tetrahedron*, 1965, **21**, 2593.
5. A. F. Thomas, *Helv. Chim. Acta*, 1967, **50**, 963.
6. M.-G. Ferretti-Alloise, A. Jacot-Guillarmod and Y.-R. Naves, *Helv. Chim. Acta*, 1970, **53**, 551.
7. J. W. Cornforth, W. Draber, B. V. Milborrow and G. Ryback, *Chem. Comm.*, 1967, 114.
8. S. Itô, M. Kodama, H. Nishiya and S. Narita, *Tetrahedron Letters*, 1969, 3185.

† Note added in proof; ACs revised by T. Oritani and Yamashita, *Tetrahedron Letters*, 1972, 2521.

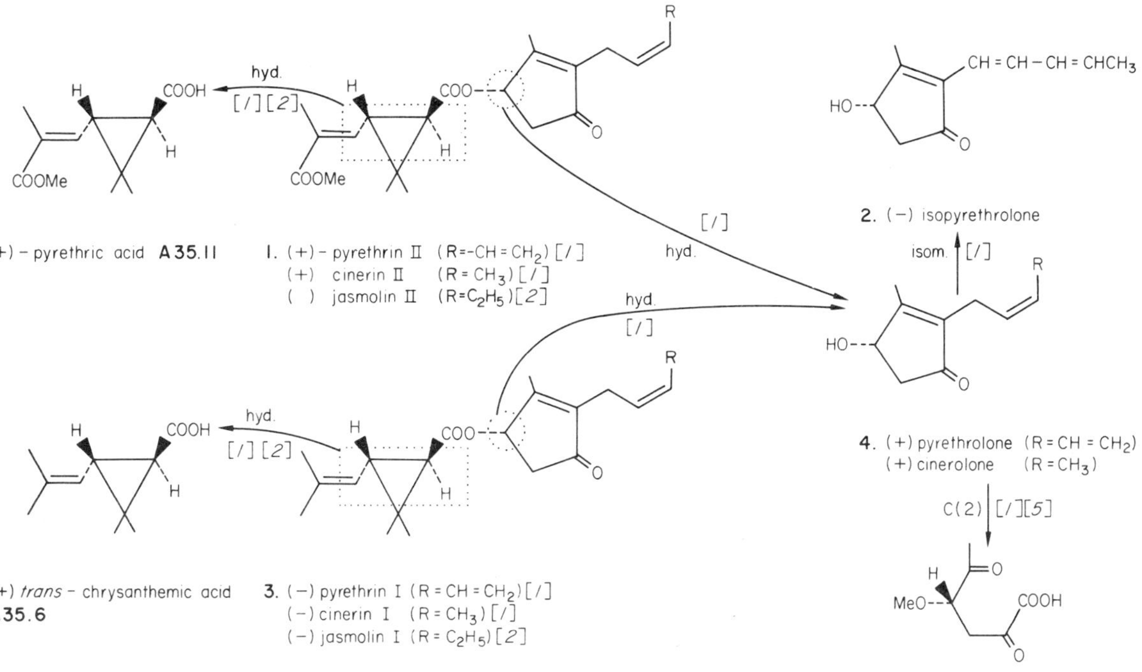

Pyrethroids, furanoterpenes (ngaione etc.).

(+)-pyrethric acid **A35.11**

1. (+)-pyrethrin II (R=-CH=CH$_2$) [/]
   (+) cinerin II (R=CH$_3$) [/]
   ( ) jasmolin II (R=C$_2$H$_5$) [2]

2. (−) isopyrethrolone

(+) *trans*-chrysanthemic acid **A35.6**

3. (−) pyrethrin I (R=CH=CH$_2$) [/]
   (−) cinerin I (R=CH$_3$) [/]
   (−) jasmolin I (R=C$_2$H$_5$) [2]

4. (+) pyrethrolone (R=CH=CH$_2$)
   (+) cinerolone (R=CH$_3$)

5. (S)-(−) 2,5-dioxo-4-methoxyhexanoic acid

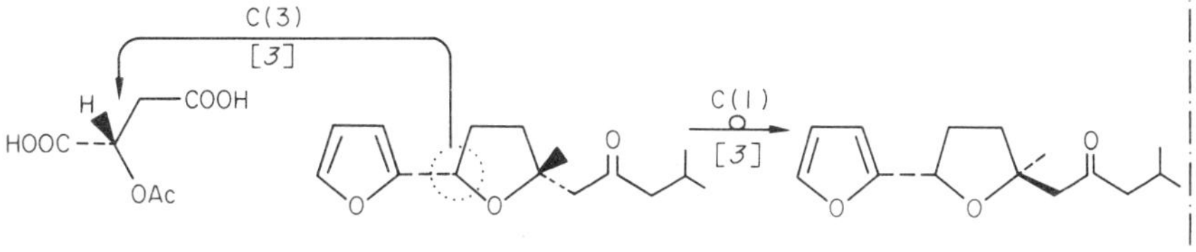

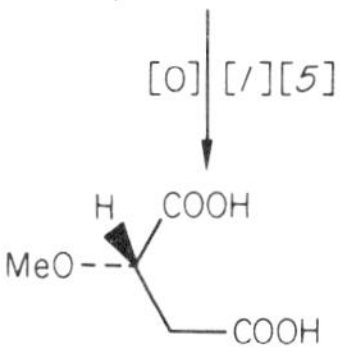

(S)-(−) acetylmalic acid **A1.25**

6. (−) ngaione (ipomeamarone)*

7. (−) epingaione

(S)-(−) methoxysuccinic acid **A2.12**

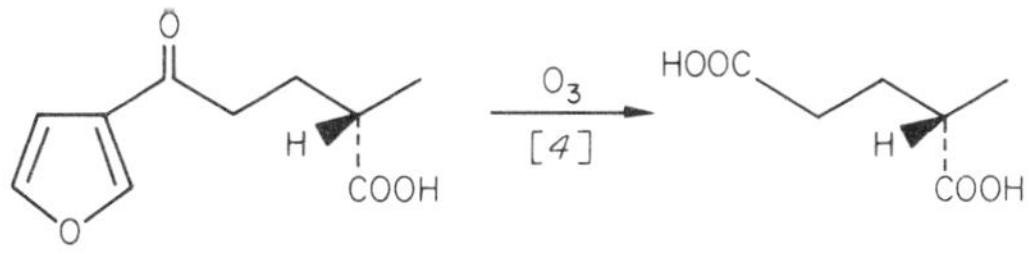

8. (+) batatic acid

(S)-(+) 2-methylglutaric acid **A32.12**

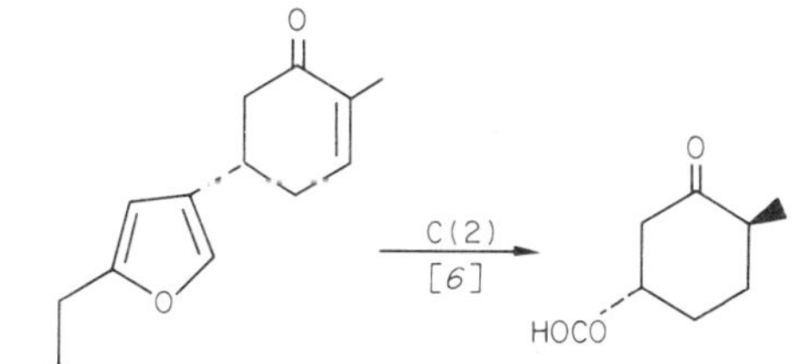

9. (+) bilobanone

10. (1S,4S)-(−) 4-methyl-3-oxocyclohexanecarboxylic acid. By CD (octant rule; −ve C.E.) [5]

1. L. Crombie and M. Elliott, *Fortschr. Chem. org. Naturstoffe*, 1961, **19**, 120 and references therein.
2. P. J. Godin, R. J. Sleeman, M. Snarey and E. M. Thain, *J. Chem. Soc. (C)*, 1966, 332.
3. B. F. Hegarty, J. R. Kelly, R. J. Park and M. D. Sutherland, *Austral. J. Chem.*, 1970, **23**, 107.
4. T. Kubota and K. Naya, *Chem. and Ind.*, 1954, 1427.
5. Y. Katsuda, T. Chikamoto and Y. Inouye, *Bull. Agric. Chem. Soc. Japan*, 1958, **22**, 427; 1959, **23**, 174.
6. H. Irie, H. Kimura, N. Otani, K. Ueda and S. Uyeo, *Chem. Comm.*, 1967, 678.

T18 Cadinane group and related compounds. Discussion of nomenclature and stereochemistry: [11]

1. (−) copaene. Related compd.; (−) mustakone (= oxocopaene) [1]
2. (−) cubeb camphor (cubebol)
   (−) trans-caran-2-one T6.6
3. (+) ε-muurolene (ε-cadinene) [13]
4. (−) cadinene dihydrohalides (X = Cl, Br). Rel. X-ray [2]
5. (−) torreyol (δ-cadinol, cedrelanol)
6. (−) oplopanone
7. (−) cadinane
8. (−) epi-cubenol
9. (−) β-cadinene
10. (+) γ-cadinene
11. (−) cubenol
12. (−) chiloscyphone. By CD [3]
13. (+) germacrene D. (germacrenes B and C = d.b. isomers [4]
14. (−) β-bourbonene. Also by ORD of derivs [5]. (+) α-bourbonene = d.b. isomer.
15. (−) ε-bulgarene. stereoisomers: [11]
16. (+) bulgarene dihydrobromide Rel X-ray [6]

(S)-(+) cryptone T2.15

(S)-(+) isopropylsuccinic acid A28.5

1. V. H. Kapadia, B. A. Nagasampagi, V. G. Naik and S. Dev, *Tetrahedron*, 1965, **21**, 607; P. de Mayo, R. E. Williams, G. Büchi and S. H. Feairheller, *Tetrahedron*, 1965, **21**, 619.
2. F. Hanic, *Chem. Listy*, 1958, **52**, 165.
3. A. Matsuo and S. Hayashi, *Tetrahedron Letters*, 1970, 1289.
4. K. Yoshihara, Y. Ohta, T. Sakai and Y. Hirose, *Tetrahedron Letters*, 1969, 2263.
5. J. Krepinský, Z. Samek and F. Sorm, *Tetrahedron Letters*, 1966, 3209.
6. A. Línek, R. Vlahov, M. Holub and V. Herout, *Tetrahedron Letters*, 1968, 23.
7. A. Tanaka, H. Uda and A. Yoshikoshi, *Chem. Comm.*, 1969, 308.
8. V. Sýkora, V. Herout and F. Sorm, *Coll. Czech. Chem. Comm.*, 1958, **23**, 2181.
9. L. Westfelt, *Acta Chem. Scand.*, 1970, **24**, 1618.
10. K. Takeda, H. Minato and M. Ishikawa, *Tetrahedron*, 1966, 7th Suppl., 219.
11. R. Vlahov, M. Holub and V. Herout, *Coll. Czech. Chem. Comm.*, 1967, **32**, 822.
12. F. Vonásek, V. Herout and F. Sorm, *Coll. Czech. Chem. Comm.*, 1960, **25**, 919.
13. L. Westfelt, *Acta Chem. Scand.*, 1964, **18**, 572.
14. M. D. Soffer, G. E. Günay, O. Korman and M. B. Adams, *Tetrahedron Letters*, 1963, 389 and references therein.
15. Y. Ohta and Y. Hirose, *Tetrahedron Letters*, 1967, 2073.

Eudesmane and maaliol groups of sesquiterpenes.

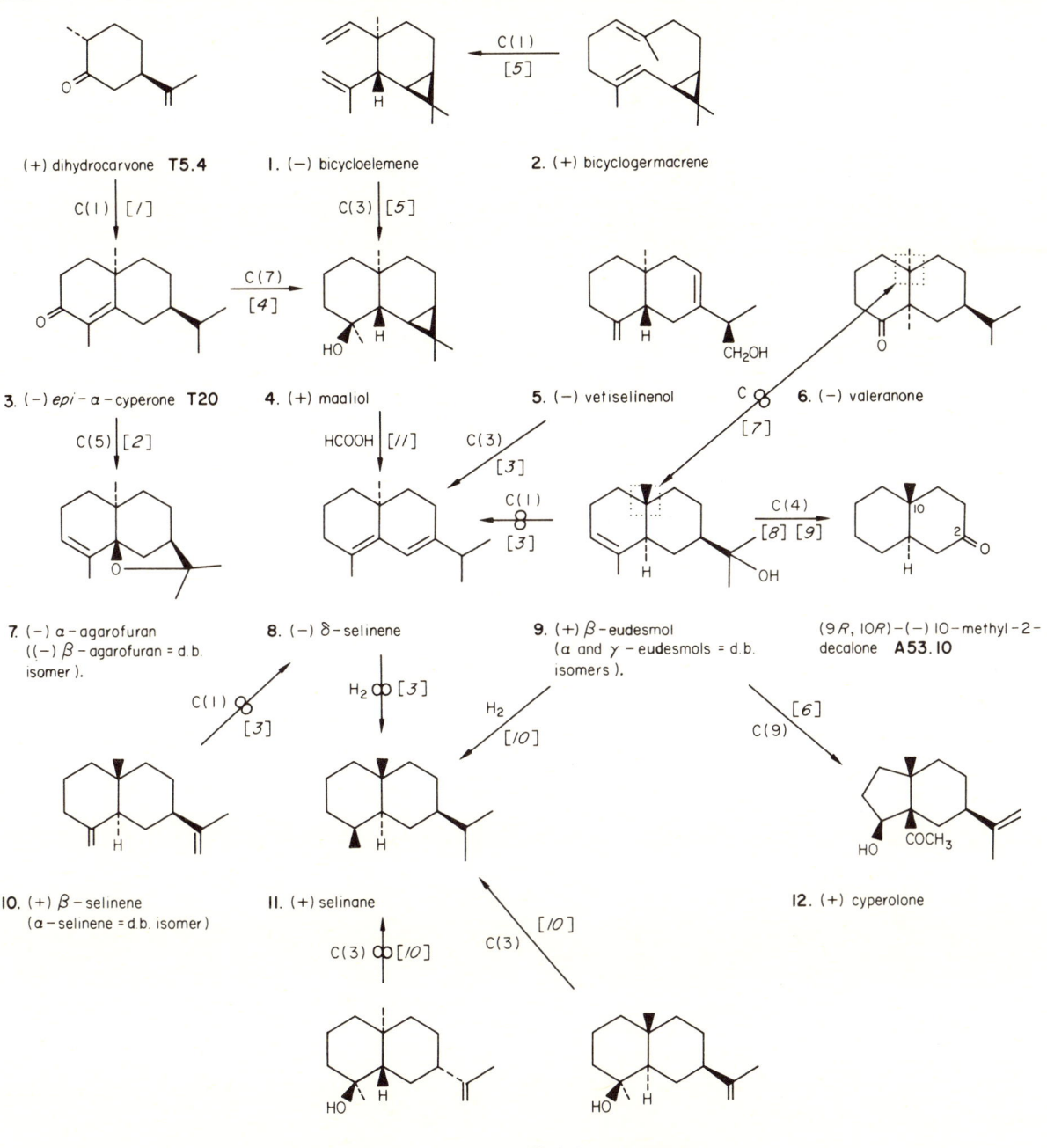

1. T. G. Halsall, D. W. Theobald and K. B. Walshaw, *J. Chem. Soc.*, 1964, 1029.
2. H. C. Barrett and G. Büchi, *J. Amer. Chem. Soc.*, 1967, 89, 5665.
3. N. H. Andersen, *Tetrahedron Letters*, 1970, 1755.
4. R. B. Bates, G. Büchi, T. Matsuura and R. R. Schaffer, *J. Amer. Chem. Soc.*, 1960, 82, 2327.
5. K. Takeda, I. Horibe and H. Minato, *Chem. Comm.*, 1971, 308.
6. H. Hiniko, K. Aota, Y. Maebayashi and T. Takemoto, *Chem. Pharm. Bull. Japan*, 1966, 14, 1441.
7. W. Klyne, S. C. Bhattacharyya, S. K. Paknikar, C. S. Narayanan, K. S. Kulkarni, J. Krepinský, M. Ramanuk, V. Herout and F. Sorm, *Tetrahedron Letters*, 1964, 1443.
8. B. Riniker, J. Kalvoda, D. Arigoni, A. Fürst, O. Jeger, A. M. Gold and R. B. Woodward, *J. Amer. Chem. Soc.*, 1954, 76, 313.
9. W. Cocker and T. B. H. McMurry, *Tetrahedron*, 1960, 8, 181.
10. L. H. Zalkow, V. B. Zalkow and D. R. Brannon, *Chem. and Ind.*, 1963, 38; V. B. Zalkow, A. M. Shaligram and L. H. Zalkow, ibid., 1964, 194.
11. G. Büchi, M. S. von Wittenau and D. M. White, *J. Amer. Chem. Soc.*, 1959, 81, 1968.

T 20
Elemane group and related sesquiterpenes.

1. (+) β-cyperone T21
2. (+) α-cyperone
   (+) epi-α-cyperone T19.3
3. (−) hexahydrochamaecynone
4. (+) tetrahydrosaussurea lactone T21
5. (−) elemol. p-nitrobenzoate (−)
6. (+) tetrahydro-epi-α-cyperone
7. (−) chamaecynone
8. (−) occidenol
9. §(+) hedycarol
10. (−) eleman-8-one
11. (+) chamaecynenol. Abs. X-ray [1]
12. (+) linderalactone.
13. (−) isolinderalactone
14. (−) isofuranogermacrene
15. ( ) α-rotunol (OH α)
    ( ) β-rotunol (OH β)
16. (+) zeylanine
17. (−) shiromodiol acetate. Abs. X-ray [2]
18. 17β acetoxy − 5α − androst − 3 − en − 2 − one
    5β − spirost − 3 − en − 2 − one T47

(R)-(−) 2-phenylbutyric acid A 25.15

1. K. Takase, S. Ibe, T. Asao, T. Nozoe, H. Shimanouchi and Y. Sasada, *Chem. and Ind.*, 1968, 1638.
2. R. J. McClure, G. A. Sim, P. Coggon and A. T. McPhail, *Chem. Comm.*, 1970, 128.
3. A. E. Bradfield, B. H. Hegde, B. S. Rao, J. L. Simonsen and A. E. Gillam, *J. Chem. Soc.*, 1936, 667.
4. T. G. Halsall, D. W. Theobald and K. B. Walshaw, *J. Chem. Soc.*, 1964, 1029.
5. T. Nozoe, Y. S. Cheng and T. Toda, *Tetrahedron Letters*, 1966, 3663.
6. K. Takeda, I. Horibe, M. Teraoka and H. Minato, *J. Chem. Soc. (C)*, 1969, 1786 and refs. therein.
7. K. Takeda, I. Horibe, M. Teraoka and H. Minato, *J. Chem. Soc. (C)*, 1970, 973.
8. A. D. Wagh, S. K. Paknikar and S. C. Bhattacharyya, *Tetrahedron*, 1964, 20, 2647.
9. B. Tomita and Y. Hirose, *Tetrahedron Letters*, 1970, 235.
10. H. Hiniko, K. Aota, D. Kuwano and T. Takemoto, *Tetrahedron Letters*, 1969, 2741.
11. K. Takeda, I. Horibe, M. Teraoka and H. Minato, *J. Chem. Soc. (C)*, 1969, 1491.
12. R. Howe and F. J. McQuillin, *J. Chem. Soc.*, 1958, 1194.
13. D. H. R. Barton and E. J. Tarlton, *J. Chem. Soc.*, 1954, 3492.
14. R. V. H. Jones and M. D. Sutherland, *Chem. Comm.*, 1968, 1229.

13. H. Bruderer, D. Arigoni and O. Jeger, *Helv. Chim. Acta*, 1956, 39, 858.
14. M. Sumi, W. G. Dauben and W. K. Hayes, *J. Amer. Chem. Soc.*, 1958, 80, 5704.
15. S. M. Kupchan and J. E. Kelsey, *Tetrahedron Letters*, 1967, 2863.
16. M. Nakazaki, H. Chikamatsu and M. Maeda, *Tetrahedron Letters*, 1966, 4499.
17. M. Nakazaki, *Chem. and Ind.*, 1962, 413.

Sesquiterpenes related to santonin.

1. ( ) bromogeigerin acetate. Rel. X-ray [1]
2. (−) geigerin. Related compd; (−) torilin [16]
3. ( ) isophotoartemisin acetate
4. (+) eupatoriopicrin [15]
5. (+) asperilin. Representative eudesmanolide lactone.
6. ( ) tetrahydroalantolactone
7. (−) artemisin
8. § (+) balchanolide. Representative germacranolide lactone [15]
9. (−) lindenene (X=H); ( ) linderene (X=OH). Also by ORD of derivs. [6]
10. § (+) 8α-hydroxy-4,5,11α (H) eudesman-13-oic acid
11. § (−) ψ-santonin
12. (−) pyrethrosin Rel. X-ray [3]
13. ( ) octahydrodehydroxylinderene
14. (−) 2-bromo-β-desmotroposantonin. Abs. X-ray [4]
15. (−) α-santonin. Rel. X-ray [5] T22
   (S)-(+) alanine A1.18
16. (−) lindestrene. (+) atractylon = dihydro-deriv.
17. (+) occidol
   (+) tetrahydrosaussurea lactone T20.4
   (+) β-cyperone T20.1

1. A. T. McPhail, *Proc. Chem. Soc.*, 1960, 278.
2. K. Takeda and M. Ikuta, *Tetrahedron Letters*, 1964, 277.
3. E. J. Gabe, S. Neidle, D. Rogers and C. E. Nordman, *Chem. Comm.*, 1971, 559.
4. A. T. McPhail, B. Rimmer, J. M. Robertson and G. A. Sim, *J. Chem. Soc. (B)*, 1967, 101.
5. J. D. M. Asher and G. A. Sim, *Proc. Chem. Soc.*, 1962, 335.
6. K. Takeda, H. Minato, M. Ishikawa and M. Miyawaki, *Tetrahedron*, 1964, 20, 1655 and refs. therein.
7. M. Nakazaki and H. Arakawa, *Proc. Chem. Soc.*, 1962, 151.
8. D. M. Simonović, A. S. Rao and S. C. Bhattacharyya, *Tetrahedron*, 1963, 19, 1061.
9. W. Cocker and M. A. Nisbet, *J. Chem. Soc.*, 1963, 534.
10. D. H. R. Barton, O. C. Böckman and P. de Mayo, *J. Chem. Soc.*, 1960, 2263.
11. V. Herout, M. Suchý and F. Sorm, *Coll. Czech. Chem. Comm.*, 1961, 26, 2612.
12. D. H. R. Barton and J. T. Pinhey, *Proc. Chem. Soc.*, 1960, 279.

References continued on page 96

T22 Santanolide and guaianolide lactones etc.

ORD/CD of santanolides and guaianolides: [14]

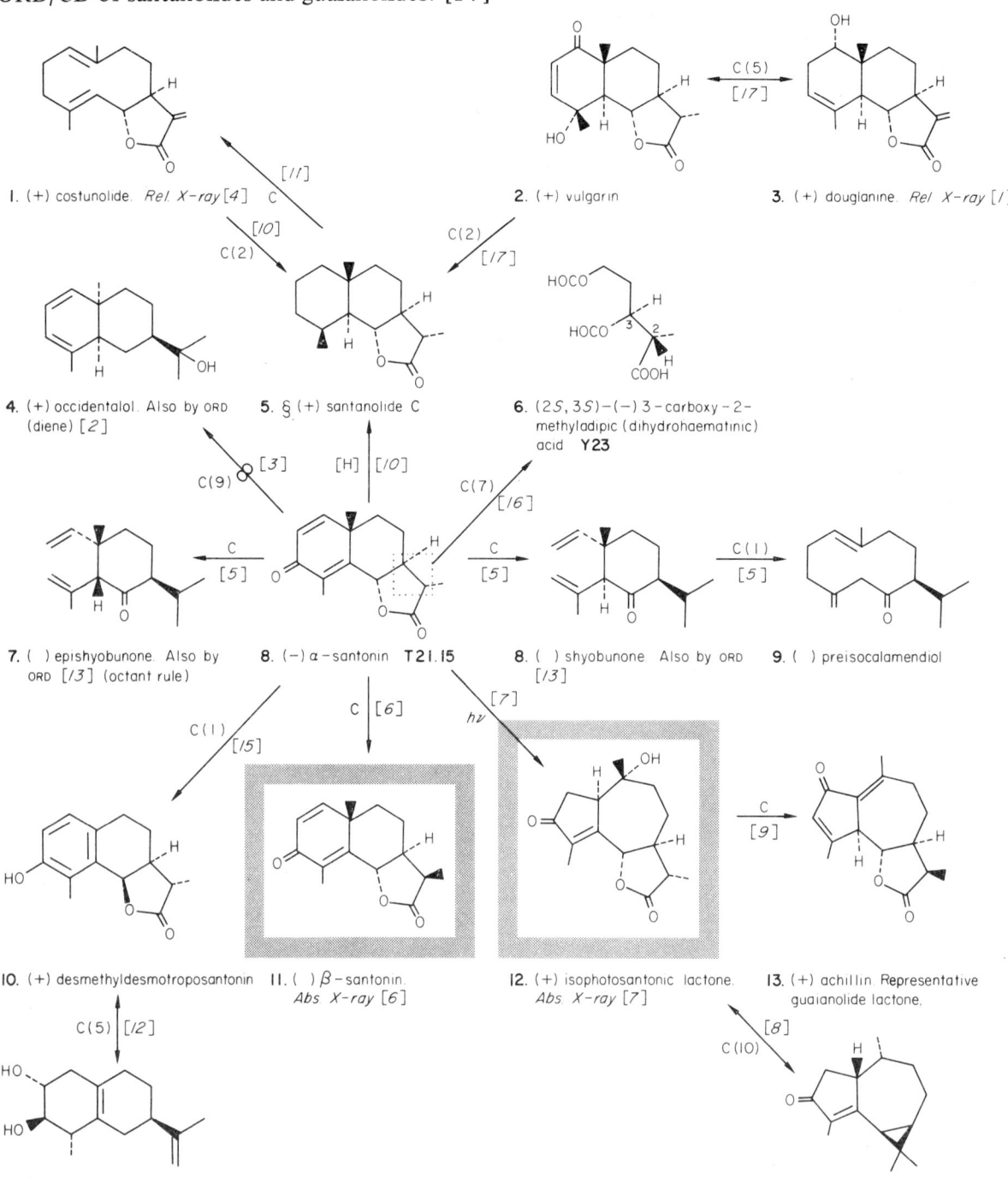

1. M. ul-Haque, C. N. Caughan, M. T. Emerson, T. A. Geissman and S. Matsueda, *J. Chem. Soc. (B)*, 1970, 598.
2. A. G. Hortmann and J. B. de Roos, *J. Org. Chem.*, 1969, **34**, 736.
3. M. Ando, K. Nanaumi, T. Nakagawa, T. Asao and K. Takase, *Tetrahedron Letters*, 1970, 3891.
4. F. Sorm, M. Suchý, M. Holub, A. Línek, I. Hadinec and C. Novák, *Tetrahedron Letters*, 1970, 1893.
5. K. Kato, Y. Hirata and S. Yamamura, *Chem. Comm.*, 1970, 1324.
6. P. Coggon and G. A. Sim, *J. Chem. Soc. (B)*, 1969, 237.
7. J. D. M. Asher and G. A. Sim, *Proc. Chem. Soc.*, 1961, 111.
8. G. Büchi and H. J. Loewenthal, *Proc. Chem. Soc.*, 1962, 280.
9. J. N. Marx and E. H. White, *Tetrahedron*, 1969, **25**, 2117.
10. V. Herout, M. Suchý and F. Sorm, *Coll. Czech. Chem. Comm.*, 1961, **26**, 2612.
11. E. J. Corey and A. G. Hortmann, *J. Amer. Chem. Soc.*, 1963, **85**, 4033.
12. N. Katsui, A. Murai, M. Takasugi, K. Imaizumi, T. Masamune and K. Tomiyama, *Chem. Comm.*, 1968, 43.
13. M. Iguchi, A. Nishiyama, H. Koyama, S. Yamamura and Y. Hirata, *Tetrahedron Letters*, 1968, 5315.
14. W. Stöcklin, T. G. Waddell and T. A. Geissman, *Tetrahedron*, 1970, **26**, 2397.
15. S. M. Sharif, S. Nozoe, K. Tsuda and N. Ikekawa, *J. Org. Chem.*, 1963, **28**, 793.
16. I. Fleming, *Nature*, 1967, **216**, 151.
17. S. Matsueda and T. A. Geissman, *Tetrahedron Letters*, 1967, 2159, and references therein.

Eremophilane and related sesquiterpene types

1. (−) fukinolidiol Abs X-ray[1]
   Related compds[14], (+)
   fukinanolide, (−) fukinolide etc.

2. (+) fukinan-8-ol

3. (+) ligularenolide

4. (+) furanoligularenone

5. (−) petasalbine. Representative furanoeremophilane

6. (−) eremophilone

7. (+) fukinone

8. (3S, 5S, 9S, 10R)-(−)5,10-dimethyl-3-isopropyl-2-decalone. Stereoisomers; [3][12]

9. (+) nardostachone [4]

10. (+) nootkatane

11. ( ) hydroxydihydroeremophilone. Rel X-ray [5]

(9S, 10R)-(+) 3-methoxy-9-methyl-Δ^{2,6}-1-hexalone
A53.8

12. (−) isoishwarone

13. (+) nootkatone. Isonootkatone = d.b. isomer

14. (−) petasin

15. (1R, 9S, 10S)-(−) 1,9-dimethyldecalin-2,6-dione

16. (+) ishwarone

17. cholest-4-en-3-one T47

18. (+) warburgiadione

1. C. Katayama, A. Furusaki, I. Nitta, M. Hayashi and K. Naya, *Bull. Chem. Soc. Japan*, 1970, **43**, 1976.
2. K. Naya, I. Takagi. Y. Kawaguchi, Y. Asada, Y. Hirose and N. Shinoda, *Tetrahedron*, 1968, **24**, 5871.
3. F. Patil, G. Ourisson, Y. Tanahashi, M. Wada and T. Takahashi, *Bull. Soc. chim. France*, 1968, 1047.
4. A. R. Pinder, *Tetrahedron Letters*, 1970, 413, and references therein.
5. D. F. Grant, *Acta Cryst.*, 1957, **10**, 498.
6. T. Takahashi, Y. Ishizaki, T. Takahashi and K. Tori, *Tetrahedron Letters*, 1968, 3739.
7. T. R. Govindachari, K. Nagarajan and P. C. Parthasarathy, *Chem. Comm.*, 1969, 823.
8. W. D. McLeod, *Tetrahedron Letters*, 1965, 4779.
9. S. D. Sastry, M. L. Mahes-wari, K. K. Chakravarti and S. C. Bhattacharyya, *Tetrahedron*, 1967, **23**, 2491.
10. L. Novotny, V. Herout and F. Sorm, *Coll. Czech. Chem. Comm.*, 1964, **29**, 2189.
11. C. Djerassi, R. Mauli and L. H. Zalkow, *J. Amer. Chem. Soc.*, 1959, **81**, 3424.
12. L. H. Zalkow, A. M. Shaligram, S. Hu and C. Djerassi, *Tetrahedron*, 1966, **22**, 337.
13. D. Herbst and C. Djerassi, *J. Amer. Chem. Soc.*, 1960, **82**, 4337.
14. K. Naya, I. Takagi, M. Hayashi, S. Nakamura, M. Kobayashi and S. Katsumura, *Chem. and Ind.*, 1968, 318.
15. L. H. Zalkow, F. X. Markley and C. Djerassi, *J. Amer. Chem. Soc.*, 1960, **82**, 6354.
16. C. J. W. Brooks and G. H. Draffan, *Tetrahedron*, 1969, **25**, 2865.

T24 Pseudoguaianolide and xanthanolide lactones.

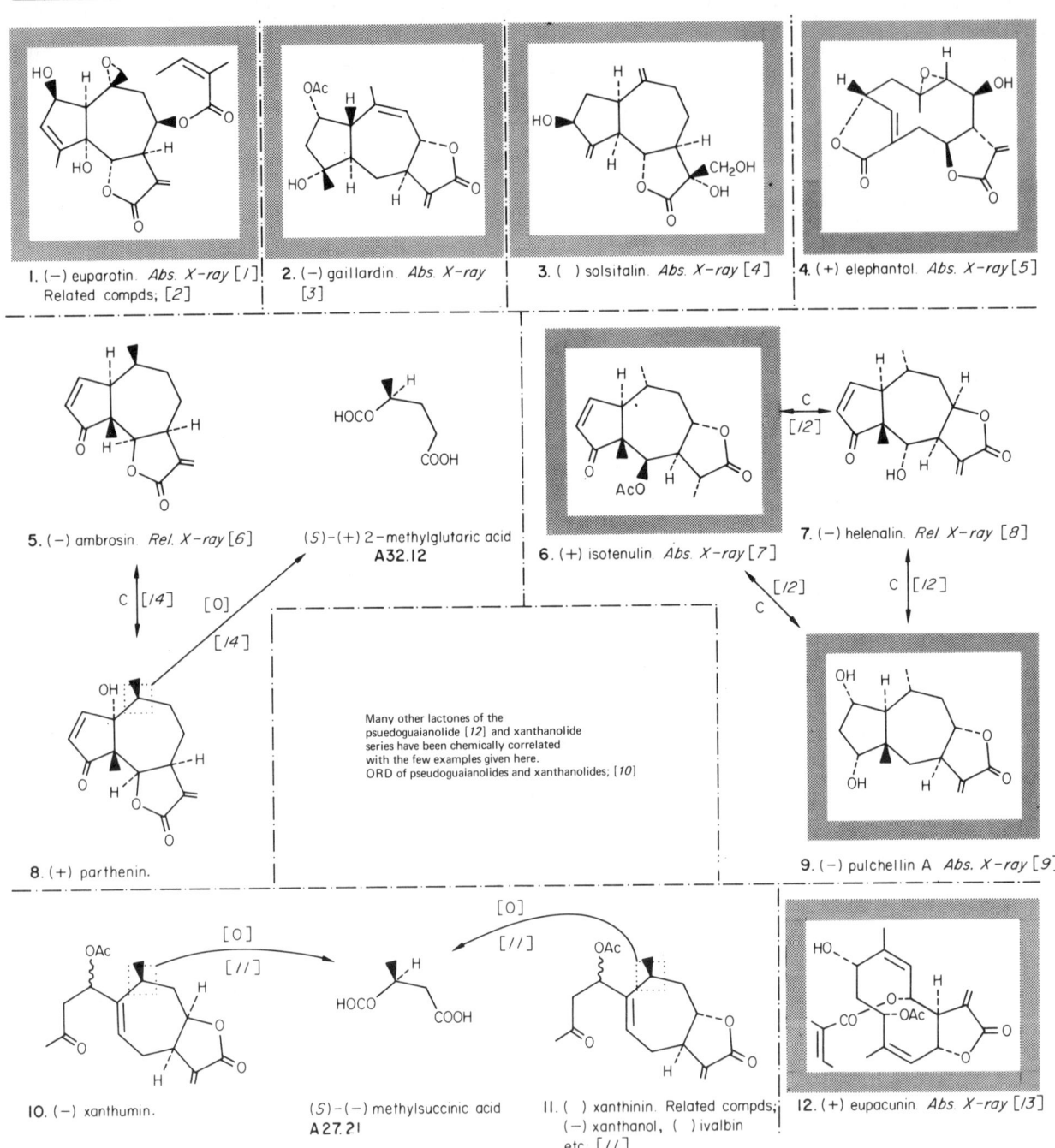

1. (−) euparotin. *Abs. X-ray* [1] Related compds; [2]
2. (−) gaillardin. *Abs. X-ray* [3]
3. ( ) solsitalin. *Abs. X-ray* [4]
4. (+) elephantol. *Abs. X-ray* [5]
5. (−) ambrosin. *Rel. X-ray* [6]
   (S)-(+) 2-methylglutaric acid A32.12
6. (+) isotenulin. *Abs. X-ray* [7]
7. (−) helenalin. *Rel. X-ray* [8]
8. (+) parthenin.
9. (−) pulchellin A *Abs. X-ray* [9]
10. (−) xanthumin.
    (S)-(−) methylsuccinic acid A27.21
11. ( ) xanthinin. Related compds; (−) xanthanol, ( ) ivalbin etc. [11]
12. (+) eupacunin. *Abs. X-ray* [13]

Many other lactones of the psuedoguaianolide [12] and xanthanolide series have been chemically correlated with the few examples given here.
ORD of pseudoguaianolides and xanthanolides; [10]

1. S. M. Kupchan, J. C. Hemingway, J. M. Cassady, J. R. Knox, A. T. McPhail and G. A. Sim, *J. Amer. Chem. Soc.*, 1967, **89**, 465.
2. S. M. Kupchan, J. E. Kelsey, M. Maruyama and J. M. Cassady, *Tetrahedron Letters*, 1968, 3517.
3. T. A. Dullforce, G. A. Sim, D. N. J. White, J. E. Kelsey and S. M. Kupchan, *Tetrahedron Letters*, 1969, 973.
4. W. E. Thiessen and H. Hope, *Acta Cryst.*, 1970, **B26**, 554.
5. S. M. Kupchan, Y. Aynehchi, J. M. Cassady, A. T. McPhail, G. A. Sim, H. K. Schnoes and A. L. Burlinghame, *J. Amer. Chem. Soc.*, 1966, **88**, 3674.
6. M. T. Emerson, W. Herz, C. N. Caughlan and R. W. Witters, *Tetrahedron Letters*, 1966, 6151.
7. D. Rogers and M. ul-Haque, *Proc. Chem. Soc.*, 1963, 942.
8. M. ul-Haque and C. N. Caughlan, *J. Chem. Soc. (B)*, 1969, 956.
9. T. Sekita, S. Inayama and Y. Iitaka, *Acta Cryst.*, 1971, **B27**, 877.
10. W. Stöcklin, T. G. Waddell and T. A. Geissman, *Tetrahedron*, 1970, **26**, 2397.
11. T. E. Winters, T. A. Geissman and D. Safir, *J. Org. Chem.*, 1969, **34**, 153.
12. W. Herz, K. Aota, M. Holub and Z. Samek, *J. Org. Chem.*, 1970, **35**, 2611, and references therein.
13. S. M. Kupchan, M. Maruyama, R. J. Hemingway, J. C. Hemingway, S. Shibuya, T. Fujita, P. D. Cradwick, A. D. U. Hardy and G. A. Sim, *J. Amer. Chem. Soc.*, 1971, **93**, 4914.
14. W. Herz, H. Watanabe, M. Miyazaki and Y. Kishida, *J. Amer. Chem. Soc.*, 1962, **84**, 2601.

T25
Bicyclo [5.3.0]-decane types: guaiol and related sesquiterpenes.

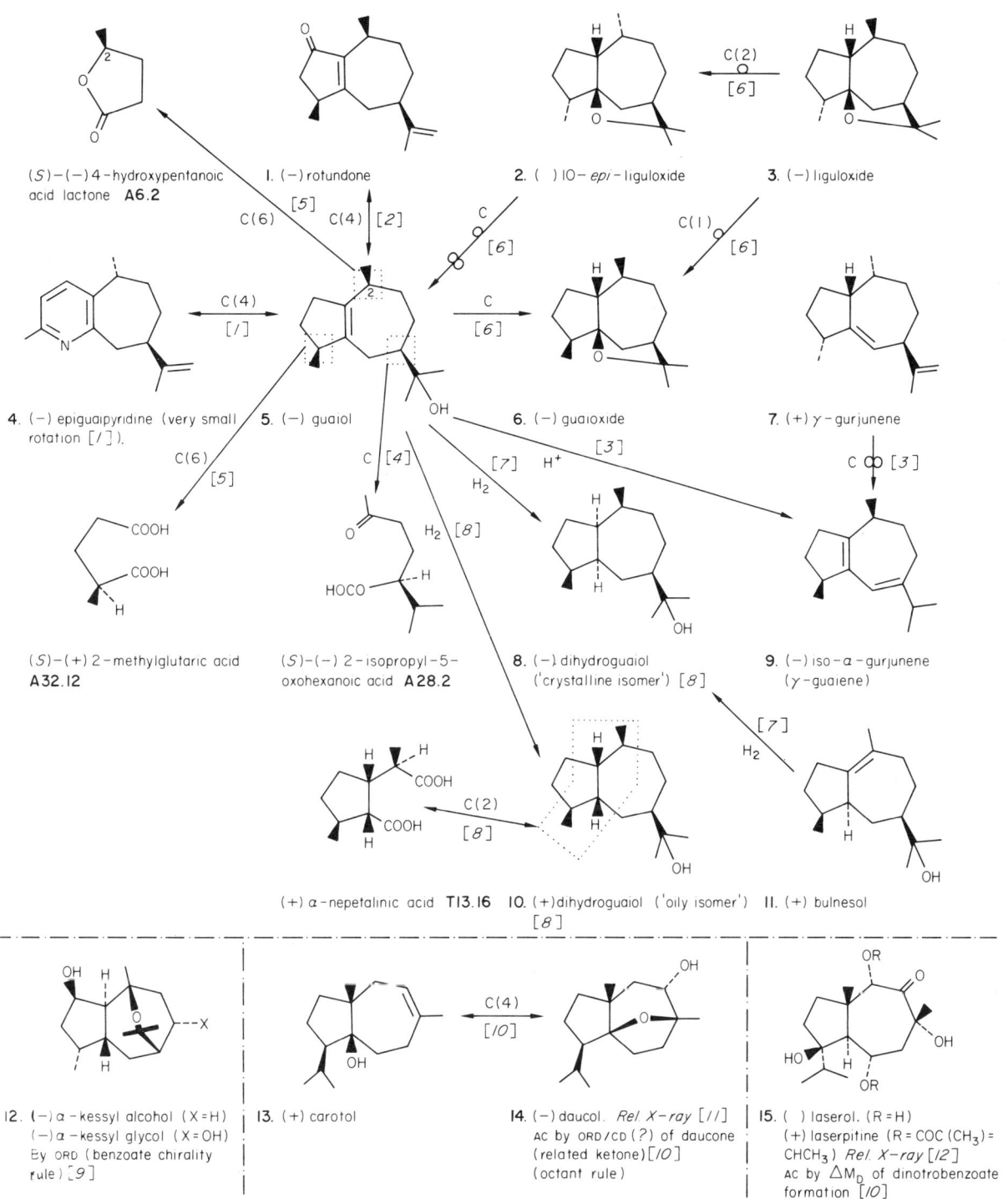

1. G. Büchi, I. M. Goldman and D. W. Mayo, *J. Amer. Chem. Soc.*, 1966, **88**, 3109.
2. V. H. Kapadia, V. G. Naik, M. S. Wadia and S. Dev, *Tetrahedron Letters*, 1967, 4661.
3. C. Ehret and G. Ourisson, *Tetrahedron*, 1969, **25**, 1785.
4. H. Minato, *Tetrahedron*, 1962, **18**, 365.
5. K. T. Akeda and H. Minato, *Tetrahedron Letters*, 1960, no. 22, 33.
6. H. Ishii, T. Tozyo, M. Nakamura and H. Minato, *Tetrahedron*, 1970, **26**, 2911.
7. L. Dolejs, A. Mironov and F. Sorm, *Coll. Czech. Chem. Comm.*, 1961, **26**, 1015.
8. E. J. Eisenbraun, T. George, B. Riniker and C. Djerassi, *J. Amer. Chem. Soc.*, 1960, **82**, 3648.
9. S. Itô, M. Kodama, T. Nozoe, H. Hiniko, Y. Hiniko, Y. Takeshita and T. Takemoto, *Tetrahedron*, 1967, **23**, 553.
10. M. Holub, J. Tax, P. Sedmera and F. Sorm, *Coll. Czech. Chem. Comm.*, 1970, **35**, 3597 and references therein.
11. R. B. Bates, C. D. Green and T. C. Sneath, *Tetrahedron Letters*, 1969, 3461.
12. M. Holub, V. Herout, F. Sorm and A. Linek, *Tetrahedron Letters*, 1965, 1441.

T 26 Further bicyclo[5.3.0] decanes; decalins, especially cyclopropyl types

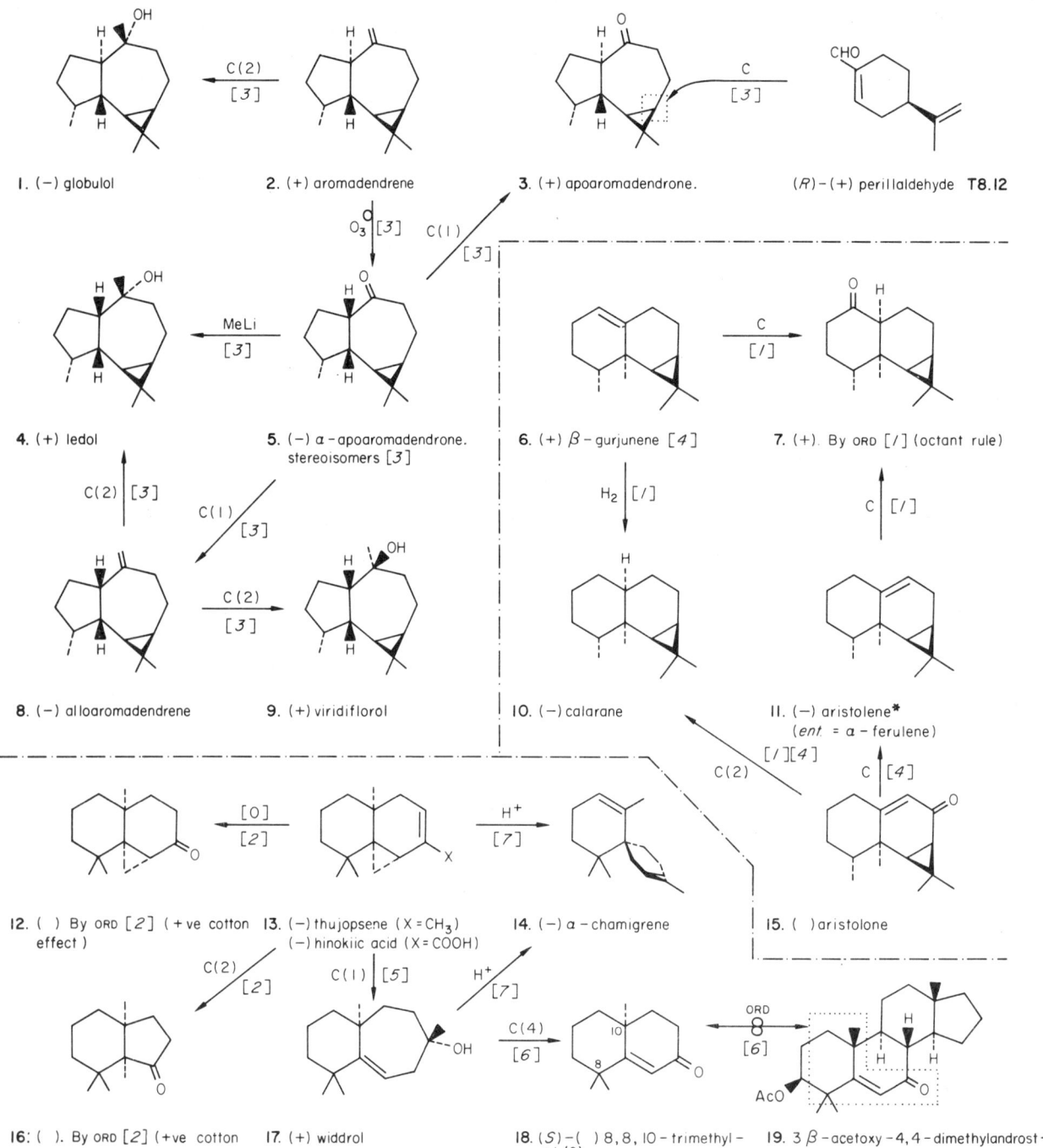

1. G. Büchi, F. Greuter and T. Tokoroyama, *Tetrahedron Letters*, 1962, 827.
2. T. Norin, *Acta Chem. Scand.*, 1963, **17**, 738.
3. G. Büchi, W. Hofheinz and J. V. Paukstelis, *J. Amer. Chem. Soc.*, 1969, **91**, 6473 and references therein.
4. J. Vrcoc, J. Krepinský, V. Herout and F. Sorm, *Tetrahedron Letters*, 1963, 225.
5. S. Nagahama, *Bull. Chem. Soc. Japan*, 1960, **33**, 1467.
6. C. Enzell, *Acta Chem. Scand.*, 1962, **16**, 1553.
7. S. Itô, K. Endo, T. Yoshida, M. Yatagai and M. Kodama, *Chem. Comm.*, 1967, 186.

Bridged-ring sesquiterpenes

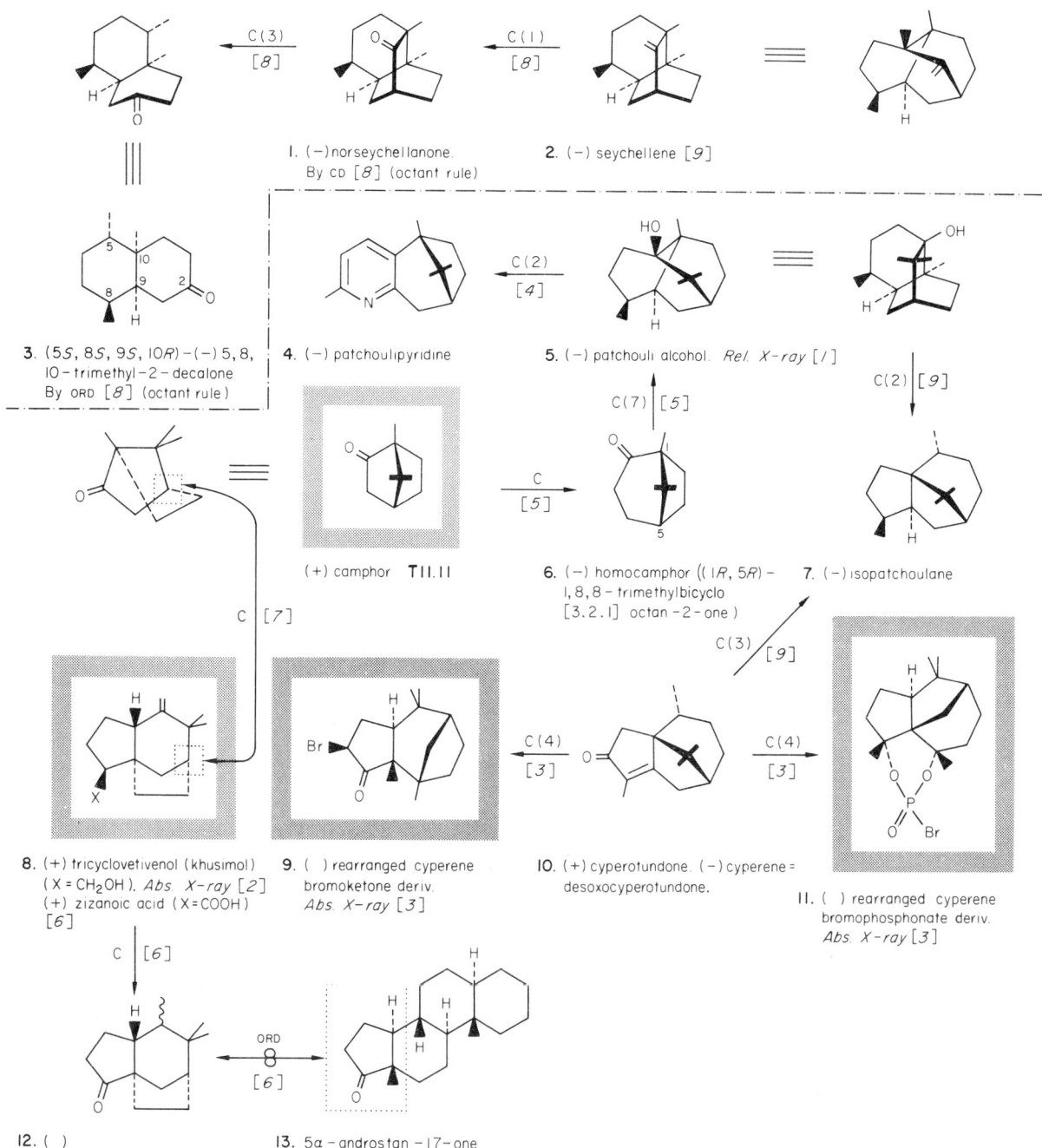

1. (−) norseychellanone. By CD [8] (octant rule)
2. (−) seychellene [9]
3. (5S, 8S, 9S, 10R)−(−)5,8,10-trimethyl-2-decalone By ORD [8] (octant rule)
4. (−) patchoulipyridine
5. (−) patchouli alcohol. Rel. X-ray [1]
6. (−) homocamphor ((1R, 5R)-1,8,8-trimethylbicyclo[3.2.1]octan-2-one)
7. (−) isopatchoulane
8. (+) tricyclovetivenol (khusimol) (X = CH₂OH). Abs. X-ray [2] (+) zizanoic acid (X=COOH) [6]
9. ( ) rearranged cyperene bromoketone deriv. Abs. X-ray [3]
10. (+) cyperotundone. (−) cyperene = desoxycyperotundone.
11. ( ) rearranged cyperene bromophosphonate deriv. Abs. X-ray [3]
12. ( )
13. 5α-androstan-17-one **T37, T47**

---

1. M. Dobler, J. D. Dunitz, B. Gubler, H. P. Weber, G. Büchi and J. Padilla, *Proc. Chem. Soc.*, 1963, 383.
2. R. M. Coates, R. F. Farney, S. M. Johnson and I. C. Paul, *Chem. Comm.*, 1969, 999.
3. H. Dreyfus, J. C. Thierry, R. Weiss, O. Kennard, W. D. S. Motherwell, J. C. Coppola and D. G. Watson, *Tetrahedron Letters*, 1969, 3757.
4. G. Büchi, I. M. Goldman and D. W. Mayo, *J. Amer. Chem. Soc.*, 1966, **88**, 3109.
5. G. Büchi, W. D. MacLeod and J. Padilla, *J. Amer. Chem. Soc.*, 1964, **86**, 4438.
6. F. Kido, H. Uda and A. Yoshikoshi, *Tetrahedron Letters*, 1968, 1247.
7. D. F. MacSweeney and R. Ramage, *Tetrahedron*, 1971, **27**, 1481.
8. G. Wolff and G. Ourisson, *Tetrahedron Letters*, 1968, 3849.
9. H. Hiniko, K. Aota and T. Takemoto, *Chem. Pharm. Bull. (Japan)*, 1966, **14**, 890.

T 28
Longifolene and caryophyllene (bicyclo-[7.2.0]-undecane) groups.

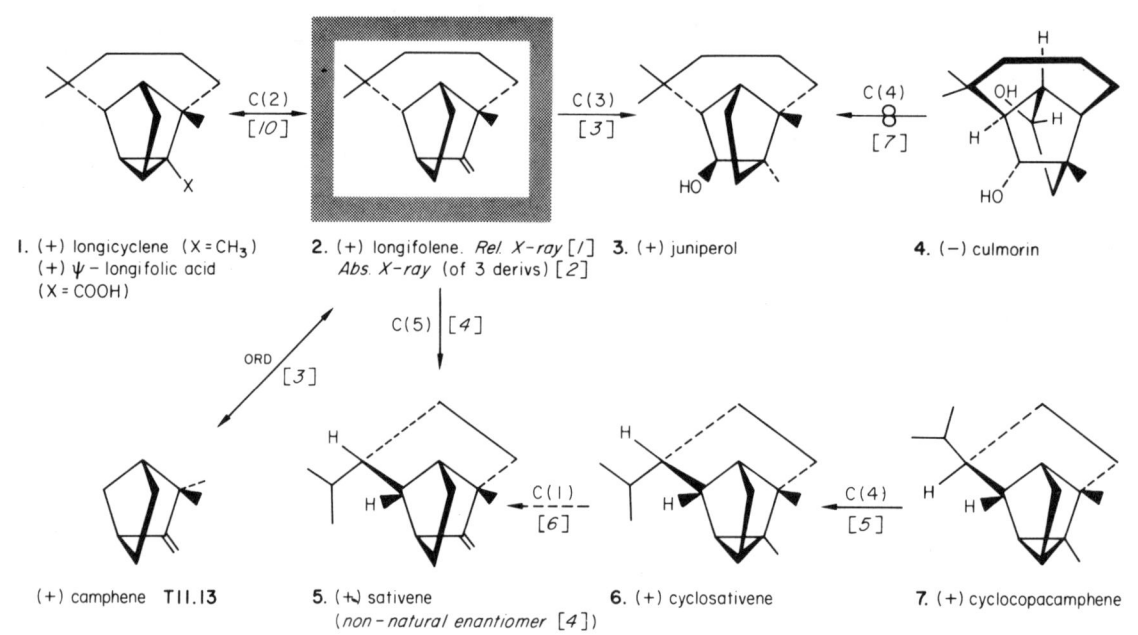

1. (+) longicyclene (X = CH₃)
   (+) ψ - longifolic acid (X = COOH)
2. (+) longifolene. *Rel. X-ray* [1] *Abs. X-ray* (of 3 derivs) [2]
3. (+) juniperol
4. (−) culmorin

(+) camphene T11.13

5. (+) sativene (*non-natural enantiomer* [4])
6. (+) cyclosativene
7. (+) cyclocopacamphene

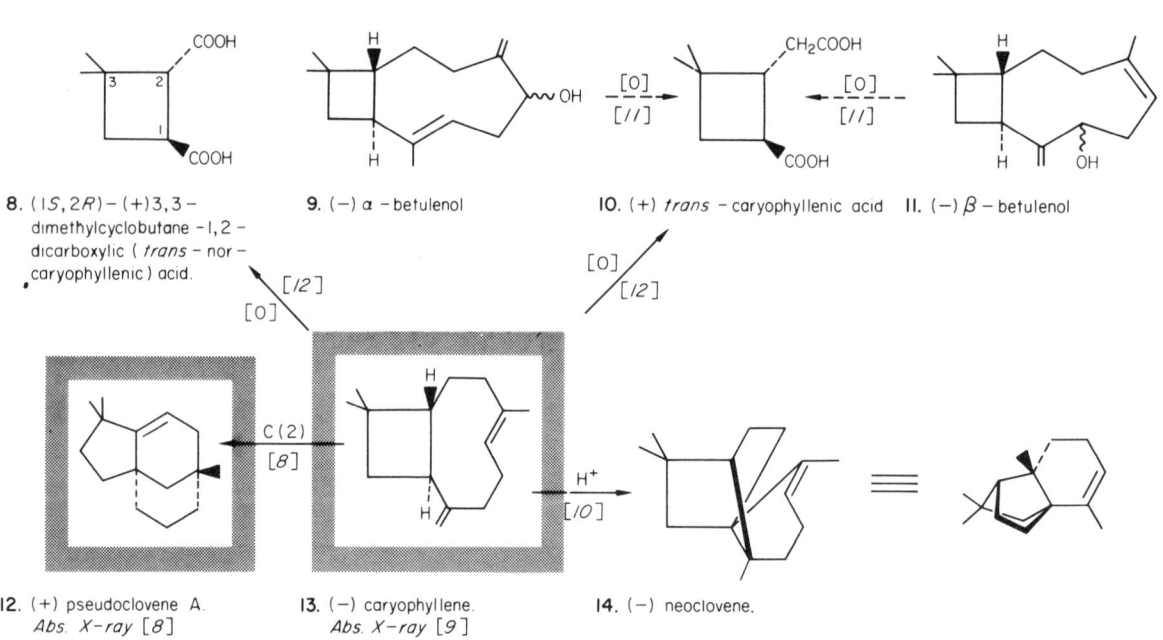

8. (1S,2R)-(+)3,3-dimethylcyclobutane-1,2-dicarboxylic (*trans*-nor-caryophyllenic) acid.
9. (−) α - betulenol
10. (+) *trans* - caryophyllenic acid
11. (−) β - betulenol
12. (+) pseudoclovene A. *Abs. X-ray* [8]
13. (−) caryophyllene. *Abs. X-ray* [9]
14. (−) neoclovene.

1. R. H. Moffett and D. Rogers, *Chem. and Ind.*, 1953, 916.
2. J.-C. Thierry and R. Weiss, *Tetrahedron Letters*, 1969, 2663.
3. G. Jacob, G. Ourisson and A. Rassat, *Bull. Soc. Chim. France*, 1959, 1374.
4. P. de Mayo and R. E. Williams, *J. Amer. Chem. Soc.*, 1965, 87, 3275.
5. J. E. McMurry, *J. Org. Chem.*, 1971, 36, 2826.
6. F. Kido, R. Sakuma, H. Uda and A. Yoshikoshi, *Tetrahedron Letters*, 1969, 3169.
7. D. H. R. Barton and N. H. Werstiuk, *Chem. Comm.*, 1967, 30.
8. D. M. Hawley, G. Ferguson, T. F. W. McKillop and J. M. Robertson, *J. Chem. Soc. (B)*, 1969, 599.
9. D. M. Hawley, J. S. Roberts, G. Ferguson and A. L. Porte, *Chem. Comm.*, 1967, 942.
10. U. R. Nayak and S. Dev, *Tetrahedron*, 1968, 24, 4099.
11. M. Holub, V. Herout, M. Horák and F. Sorm, *Coll. Czech. Chem. Comm.*, 1959, 24, 3730.
12. W. C. Evans, G. R. Ramage and J. L. Simonsen, *J. Chem. Soc.*, 1934, 1806.

Cedrene and acorane (spiro-[5,6]-decane) groups.

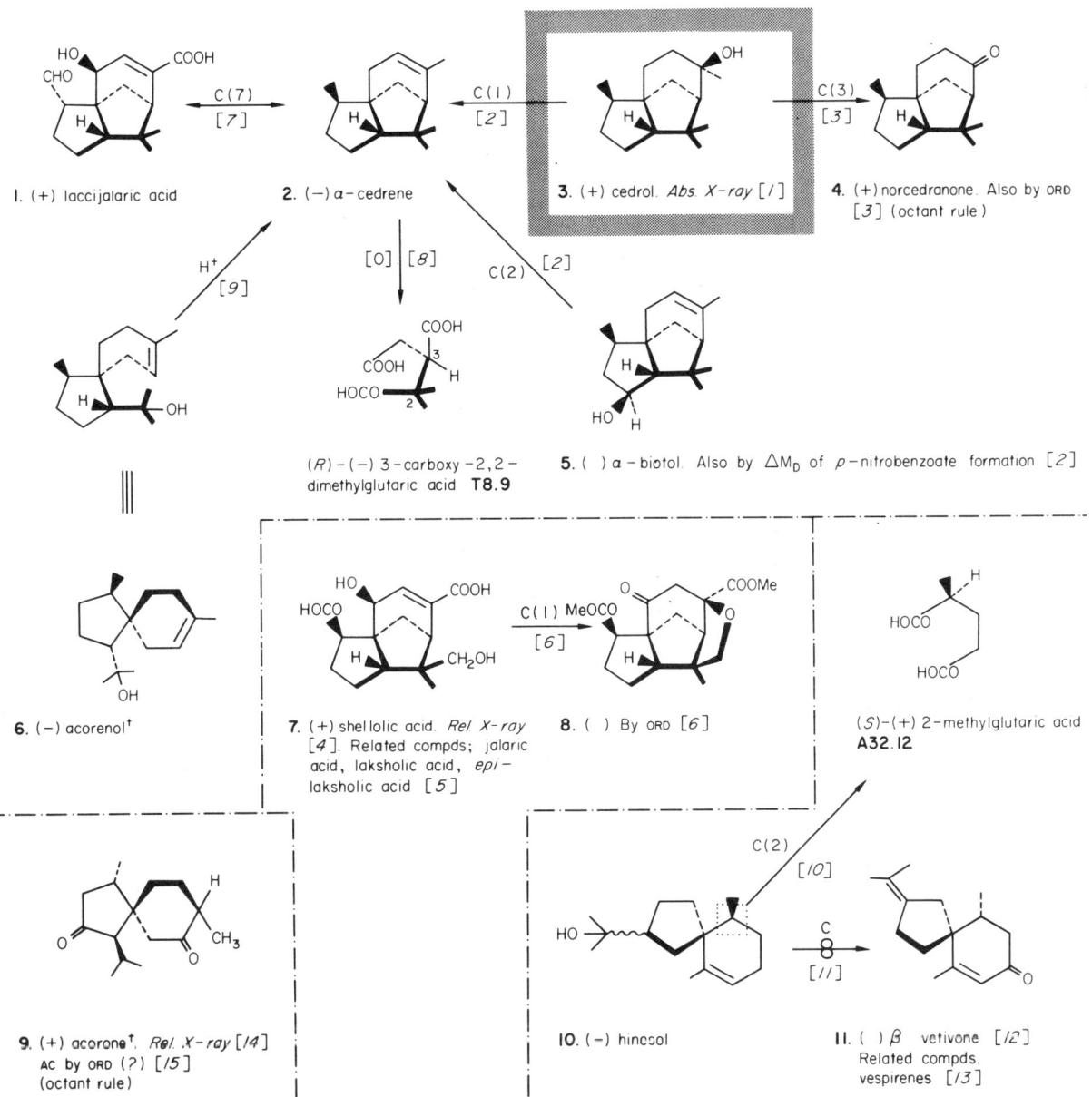

1. V. Amirthalingam, D. F. Grant and A. Senol, *Acta Cryst.*, 1969 (Suppl.), S129.
2. B. Tomita, Y. Hirose and T. Nakatsuka, *Tetrahedron Letters*, 1968, 843.
3. G. Stork and F. H. Clarke, *J. Amer. Chem. Soc.*, 1961, 83, 3114.
4. E. J. Gabe, *Acta Cryst.*, 1962, 15, 759.
5. M. S. Wadia, V. V. Mhaskar and S. Dev, *Tetrahedron Letters*, 1963, 513; R. G. Khurana, M. S. Wadia, V. V. Mhaskar and S. Dev, *Tetrahedron Letters*, 1964, 1537.
6. R. C. Cookson, N. Lewin and A. Morrison, *Tetrahedron*, 1962, 18, 547.
7. A. N. Singh, A. B. Upadhye, M. S. Wadia, V. V. Mhaskar and S. Dev, *Tetrahedron*, 1969, 25, 3855.
8. P. A. Plattner and H. Kläui, *Helv. Chim. Acta*, 1943, 26, 1553.
9. B. Tomita and Y. Hirose, *Tetrahedron Letters*, 1970, 143.
10. W. Z. Chow, O. Motl and F. Sorm, *Coll. Czech. Chem. Comm.*, 1962, 27, 1914.
11. I. Yosioka and T. Kimura, *Chem. Pharm. Bull. Japan*, 1965, 13, 1430.
12. J. A. Marshall and P. C. Johnson, *J. Amer. Chem. Soc.*, 1967, 89, 2750.
13. N. H. Andersen, M. S. Falcone and D. D. Syrdal, *Tetrahedron Letters*, 1970, 1759.
14. C. E. McEachan, A. T. McPhail and G. A. Sim, *J. Chem. Soc. (C)*, 1966, 579.
15. J. Vrkoc, J. Jonás, V. Herout and F. Sorm, *Coll. Czech. Chem. Comm.*, 1964, 29, 539.

† See also N. H. Anderson and D. D. Syrdal, *Tetrahedron Letters*, 1972, 899.

T 30 Humulene (cycloundecane) type; helminthosporal; illudin and miscellaneous types.

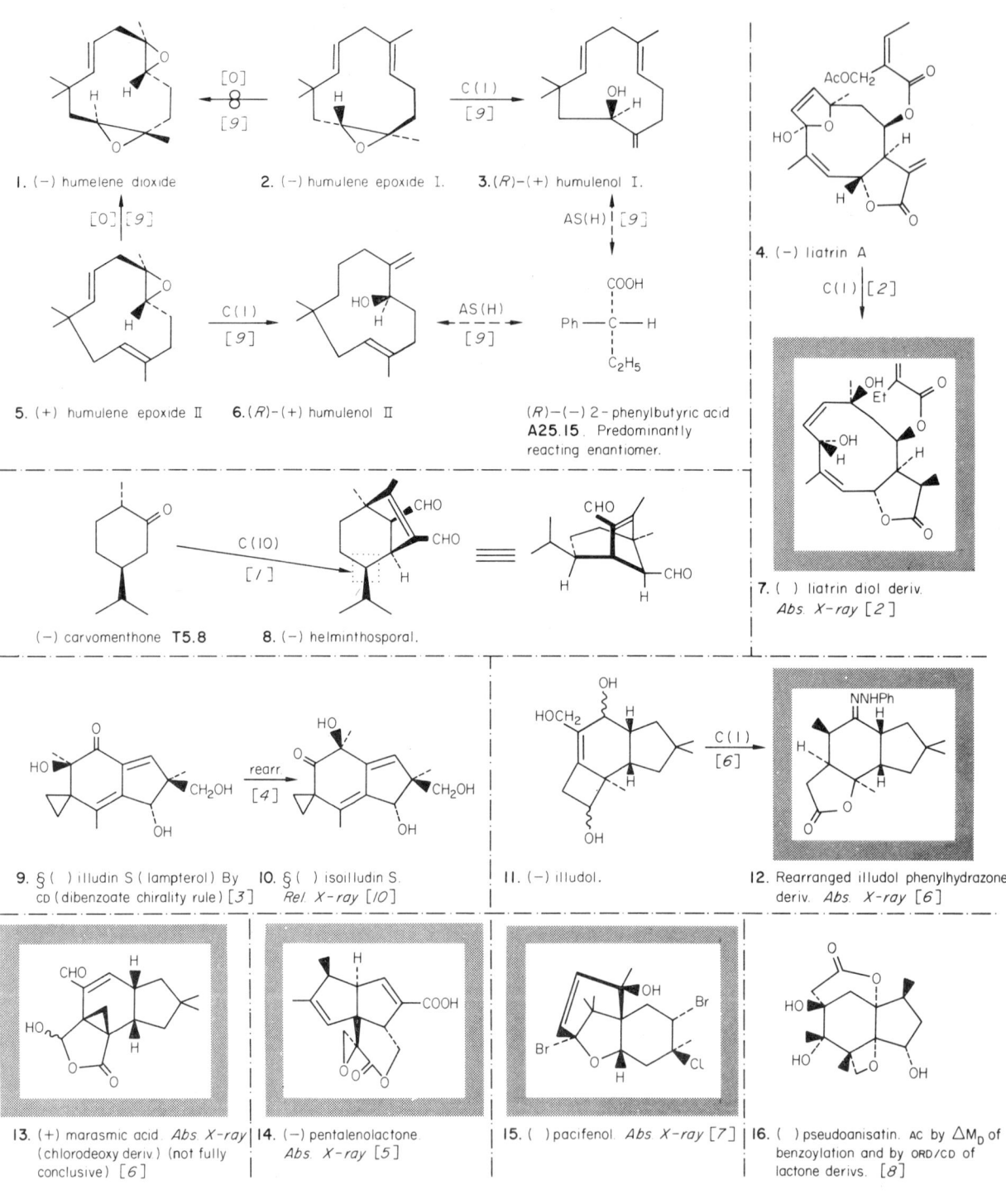

1. (−) humelene dioxide
2. (−) humelene epoxide I.
3. (R)-(+) humulenol I.
4. (−) liatrin A
5. (+) humulene epoxide II
6. (R)-(+) humulenol II

(R)-(−) 2-phenylbutyric acid **A25.15**. Predominantly reacting enantiomer.

(−) carvomenthone **T5.8**
8. (−) helminthosporal.
7. ( ) liatrin diol deriv. *Abs X-ray* [2]

9. § ( ) illudin S (lampterol) By CD (dibenzoate chirality rule) [3]
10. § ( ) isoilludin S. *Rel. X-ray* [10]
11. (−) illudol.
12. Rearranged illudol phenylhydrazone deriv. *Abs X-ray* [6]

13. (+) marasmic acid. *Abs X-ray* (chlorodeoxy deriv.) (not fully conclusive) [6]
14. (−) pentalenolactone *Abs X-ray* [5]
15. ( ) pacifenol. *Abs X-ray* [7]
16. ( ) pseudoanisatin. AC by ΔM_D of benzoylation and by ORD/CD of lactone derivs. [8]

1. E. J. Corey and S. Nozoe, *J. Amer. Chem. Soc.*, 1963, **85**, 3527.
2. S. M. Kupchan, V. H. Davies, T. Fujita, M. R. Cox and R. F. Bryan, *J. Amer. Chem. Soc.*, 1971, **93**, 4916.
3. N. Harada and K. Nakanishi, *Chem. Comm.*, 1970, 310.
4. K. Nakanishi, M. Ohashi, M. Tada and Y. Yamada, *Tetrahedron*, 1965, **21**, 1231.
5. D. G. Martin, G. Slomp, S. Mizsak, D. J. Duchamp and C. G. Chidester, *Tetrahedron Letters*, 1970, 4901.
6. P. D. Cradwick and G. A. Sim, *Chem. Comm.*, 1971, 431.
7. J. J. Sims, W. Fenical, R. M. Wing and P. Radlick, *J. Amer. Chem. Soc.*, 1971, **93**, 3774.
8. M. Okigawa and N. Kawano, *Tetrahedron Letters*, 1971, 75.
9. N. P. Damodaran and S. Dev, *Tetrahedron*, 1968, **24**, 4123, 4133.
10. Y. Saito, quoted in ref [4] above.

9. A. J. Weinheimer, P. H. Washecheck, D. van der Helm and M. B. Hossain, *Chem. Comm.*, 1968, 1070.
10. J. Gutzwiller and Ch. Tamm, *Helv. Chim. Acta*, 1965, **48**, 157.
11. W. O. Gotfredsen, J. F. Grove and Ch. Tamm, *Helv. Chim. Acta*, 1967, **50**, 1666 and references therein.

Verrucarin, ovalicin and picrotoxinin groups. T31

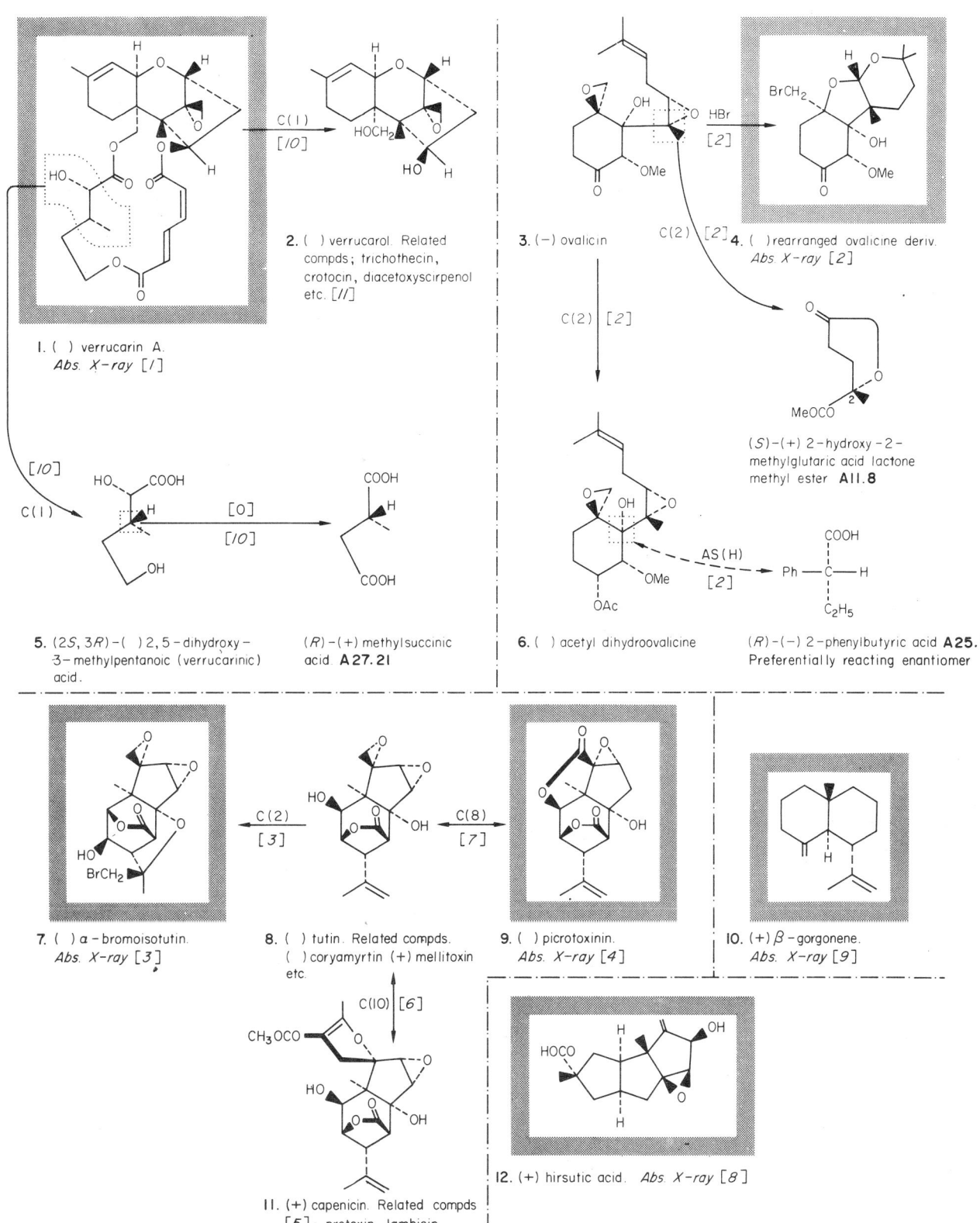

1. ( ) verrucarin A. Abs. X-ray [1]
2. ( ) verrucarol. Related compds; trichothecin, crotocin, diacetoxyscirpenol etc. [11]
3. (−) ovalicin
4. ( ) rearranged ovalicine deriv. Abs X-ray [2]
5. (2S, 3R)-( ) 2,5-dihydroxy-3-methylpentanoic (verrucarinic) acid.
   (R)-(+) methylsuccinic acid. A27.21
6. ( ) acetyl dihydroovalicine
   (S)-(+) 2-hydroxy-2-methylglutaric acid lactone methyl ester A11.8
   (R)-(−) 2-phenylbutyric acid A25.15 Preferentially reacting enantiomer
7. ( ) α-bromoisotutin. Abs. X-ray [3]
8. ( ) tutin. Related compds. ( ) coryamyrtin (+) mellitoxin etc.
9. ( ) picrotoxinin. Abs. X-ray [4]
10. (+) β-gorgonene. Abs. X-ray [9]
11. (+) capenicin. Related compds [5]; pretoxin, lambicin
12. (+) hirsutic acid. Abs. X-ray [8]

1. A. T. McPhail and G. A. Sim, *J. Chem. Soc. (C)*, 1966, 1394.
2. H. P. Sigg and H. P. Weber, *Helv. Chim. Acta*, 1968, **51**, 1395.
3. B. M. Craven, *Acta Cryst.*, 1964, **17**, 396.
4. B. M. Craven, *Acta Cryst.*, 1962, **15**, 387.
5. A. Corbella, G. Jommi, B. Rindone and C. Scolastico, *Tetrahedron*, 1969, **25**, 4835.
6. A. Corbella, G. Jommi and C. Scolastico, *Tetrahedron Letters*, 1966, 4819.
7. A. Corbella, G. Jommi, B. Rindone and C. Scolastico, *Ann. Chim. (Rome)*, 1967, **57**, 758.
8. F. W. Comer and J. Trotter, *J. Chem. Soc. (B)*, 1966, 11.

References continued on page 106

T 32 Drimane-type sesquiterpenes; abietane type diterpenes.

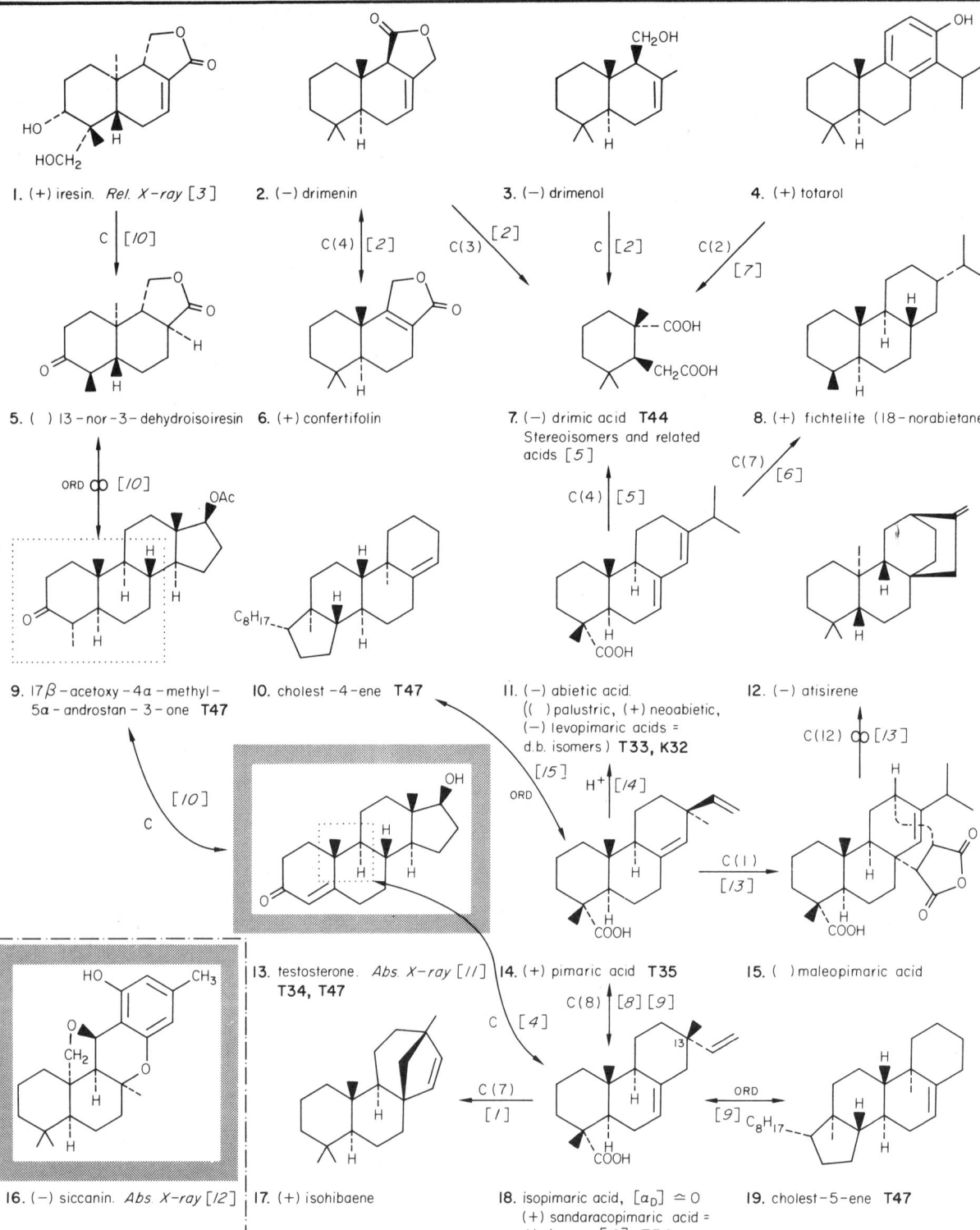

1. E. Wenkert, P. W. Jeffs and J. R. Mahajan, *J. Amer. Chem. Soc.*, 1964, **86**, 2218.
2. H. H. Appel, J. D. Connolly, K. H. Overton and R. P. M. Bond, *J. Chem. Soc.*, 1960, 4685.
3. M. G. Rossmann and W. N. Lipscomb, *J. Amer. Chem. Soc.*, 1958, **80**, 2592.
4. A. Afonso, *J. Org. Chem.*, 1970, **35**, 1949 and references therein.
5. K. Schaffner, R. Viterbo, D. Arigoni and O. Jeger, *Helv. Chim. Acta*, 1956, **39**, 174.
6. A. W. Burgstahler and J. N. Marx, *J. Org. Chem.*, 1969, **34**, 1562.
7. Y-L. Chow and H. Erdtman, *Acta Chem. Scand.*, 1962, **16**, 1305.
8. V. Galik, J. Kuthan and F. Petru, *Chem. and Ind.*, 1960, 722.
9. W. Antkowiak, O. E. Edwards, R. Howe and J. W. ApSimon, *Canad. J. Chem.*, 1965, **43**, 1257.
10. C. Djerassi and S. Burstein, *Tetrahedron*, 1959, 7, 37.
11. A. Cooper, E. M. Gopalakrishna and D. A. Norton, *Acta Cryst.*, 1968, **B24**, 935.
12. K. Hirai, S. Nozoe, K. Tsuda, Y. Iitaka, K. Ishibashi and M. Shirasaka, *Tetrahedron Letters*, 1967, 2177.
13. L. H. Zalkow and N. N. Girotra, *J. Org. Chem.*, 1964, **29**, 1299; A. H. Kapadi, R. R. Sobti and S. Dev, *Tetrahedron Letters*, 1965, 2729.
14. E. Wenkert and J. W. Chamberlin, *J. Amer. Chem. Soc.*, 1959, **81**, 688.
15. A. K. Bose and W. A. Struck, *Chem. and Ind.*, 1959, 1628.

T33

Diterpenes- abietane and podocarpane types.

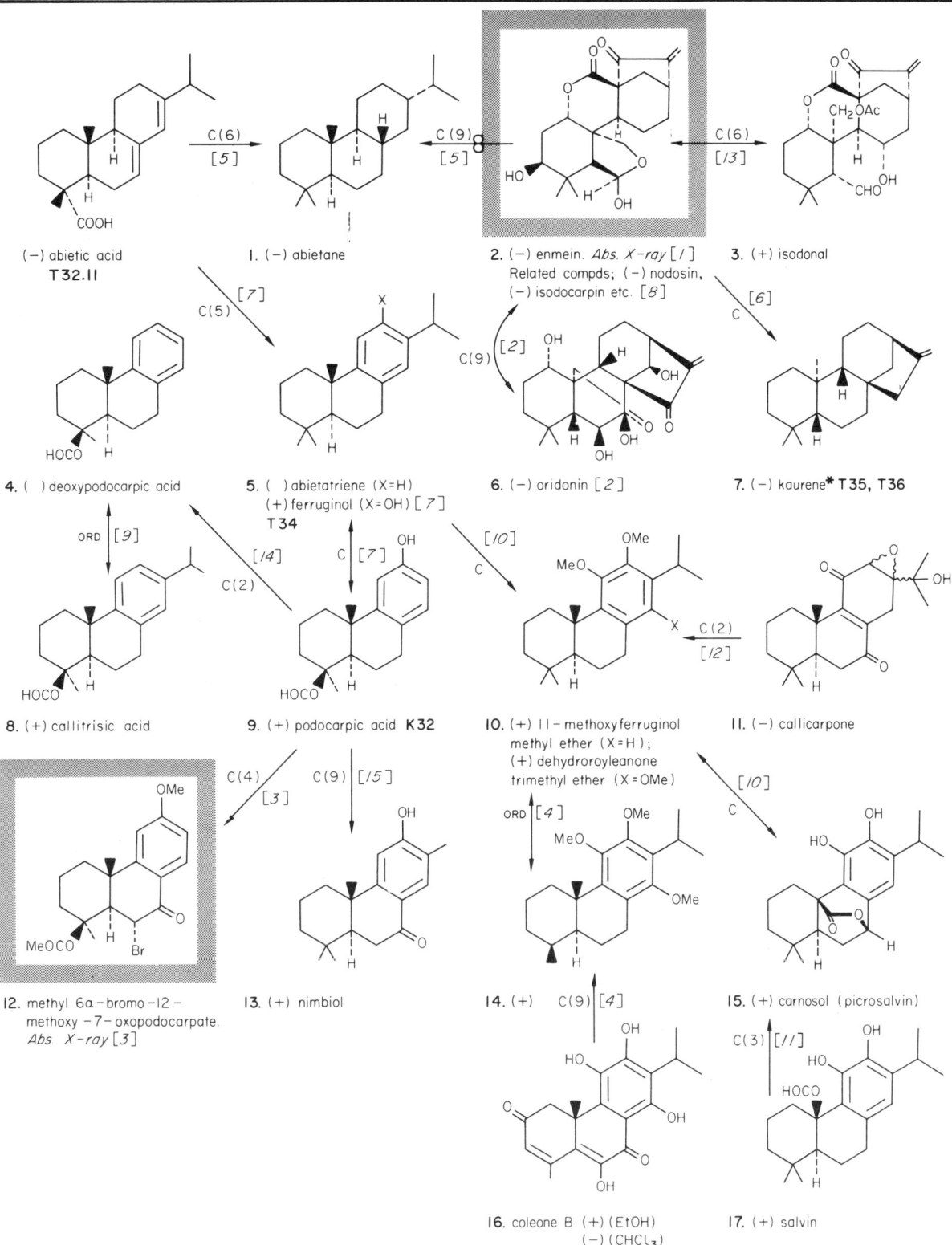

1. (−) abietic acid **T32.11**
2. 1. (−) abietane
3. 2. (−) enmein. *Abs. X-ray* [1] Related compds; (−) nodosin, (−) isodocarpin etc. [8]
4. 3. (+) isodonal
5. 4. ( ) deoxypodocarpic acid
6. 5. ( ) abietatriene (X=H) (+) ferruginol (X=OH) [7] **T34**
7. 6. (−) oridonin [2]
8. 7. (−) kaurene* T35, T36
9. 8. (+) callitrisic acid
10. 9. (+) podocarpic acid K32
11. 10. (+) 11-methoxyferruginol methyl ether (X=H); (+) dehydroroyleanone trimethyl ether (X=OMe)
12. 11. (−) callicarpone
13. 12. methyl 6α-bromo-12-methoxy-7-oxopodocarpate. *Abs. X-ray* [3]
14. 13. (+) nimbiol
15. 14. (+)
16. 15. (+) carnosol (picrosalvin)
17. 16. coleone B (+) (EtOH) (−) (CHCl$_3$)
18. 17. (+) salvin

1. N. Natsume and Y. Iitaka, *Acta Cryst.*, 1966, **20**, 197.
2. E. Fujita, T. Fujita, H. Katayama and M. Shibuya, *Chem. Comm.*, 1967, 252.
3. G. R. Clark and T. N. Waters, *J. Chem. Soc. (C)*, 1970, 887.
4. M. Ribi, A. Chang Sin-Ren, H. P. Küng and C. H. Eugster, *Helv. Chim. Acta*, 1969, **52**, 1685.
5. E. Fujita, T. Fujita and H. Katayama, *Chem. Comm.*, 1967, 968.
6. E. Fujita, T. Fujita and H. Katayama, *Tetrahedron*, 1970, **26**, 1009.
7. W. P. Campbell and D. Todd, *J. Amer. Chem. Soc.*, 1942, **64**, 928.
8. E. Fujita, T. Fujita and M. Shibuya, *Tetrahedron Letters*, 1966, 3153.
9. L. J. Gough, *Tetrahedron Letters*, 1968, 295.
10. C. H. Brieskorn, A. Fuchs, J. B. Bredenberg, J. D. McChesney and E. Wenkert, *J. Org. Chem.*, 1964, **29**, 2293.
11. C. R. Narayanan and H. Linde, *Tetrahedron Letters*, 1965, 3647.
12. K. Kawazu and T. Mitsui, *Tetrahedron Letters*, 1966, 3519.
13. T. Kubota and I. Kubo, *Tetrahedron Letters*, 1967, 3781.
14. E. Wenkert and B. G. Jackson, *J. Amer. Chem. Soc.*, 1958, **80**, 217.
15. R. H. Bible, *Tetrahedron*, 1960, **11**, 22.

T34
Labdane series; manool, sclareol and ambreinolide group.

1. (+) norambreinolide **T42**
2. (+) manoyl oxide†
3. (+) colensenone
   (+) abietatriene **T33.5**
4. (−) sclareol† **T36**
5. ( )
6. (+) manool† **T36**
7. (+) 13-epimanool*†
8. (+) ambreinolide **T37, T42**
9. (+) torulosol†
10. ( ) isocupressic acid†

testosterone **T32.13**

(+) sandaracopimaric acid **T32.18**

11. (+) rimuene. Rel. X-ray [1]
12. (−) rosenonolactone. Rel. X-ray [2]. AC by ORD [3]

(2R,5R)-(+) 2,5-dimethyl-5-ethylcyclohexanone **A55.9**

13. (−) dolabradiene.
14. (+) allodevadarool (erythroxydiol Y)
15. (+)

(1S,2R)-(−)-2-carboxy-2-methylcyclohexaneacetic acid **A53.11**

†see footnote to page T36

1. B. F. Anderson, D. Hally and T. N. Waters, *Acta Cryst.*, 1970, **26B**, 882.
2. A. I. Scott, S. A. Sutherland, D. W. Young, L. Guglielmetti, D. Arigoni and G. A. Sim, *Proc. Chem. Soc.*, 1964, 19.
3. C. Djerassi, B. Green, W. B. Whalley and C. G. De Grazia, *J. Chem. Soc. (C)*, 1966, 624.
4. J. D. Connolly, Y. Kitahara, K. H. Overton and A. Yoshikoshi, *Chem. Pharm. Bull. Japan*, 1965, **13**, 603.
5. J. D. Connolly, R. McCrindle, R. D. H. Murray, A. J. Renfrew, K. H. Overton and A. Melera, *J. Chem. Soc. (C)*, 1966, 268.
6. G. A. Ellestad, B. Green, A. Harris, W. B. Whalley and H. Smith, *J. Chem. Soc.*, 1965, 7246.
7. T. McCreadie, K. H. Overton and A. J. Allison, *J. Chem. Soc. (C)*, 1971, 317.
8. J. D. Connolly, R. McCrindle, R. D. H. Murray and K. H. Overton, *J. Chem. Soc. (C)*, 1966, 273.
9. A. K. Bose and S. Harrison, *Chem. and Ind.*, 1963, 254.
10. V. Galik, J. Kuthan and F. Petru, *Chem. and Ind.*, 1960, 722.
11. C. Enzell, *Acta Chem. Scand.*, 1961, **15**, 1303.
12. J. R. Hosking and C. W. Brandt, *Ber.*, 1935, **68**, 1311.
13. J. W. Rowe and J. H. Scroggins, *J. Org. Chem.*, 1964, **29**, 1554.
14. H. R. Schenk, H. Gutmann, O. Jeger and L. Ruzicka, *Helv. Chim. Acta*, 1952, **35**, 817.
15. L. Ruzicka, C. F. Seidel and L. L. Engel, *Helv. Chim. Acta*, 1942, **25**, 621.
16. E. Lederer and M. Stoll, *Helv. Chim. Acta*, 1950, **33**, 1345.
17. P. K. Grant and R. M. Carman, *J. Chem. Soc.*, 1962, 3740.

References continued on page 111

Pimarane, Kaurane and gibbane series.

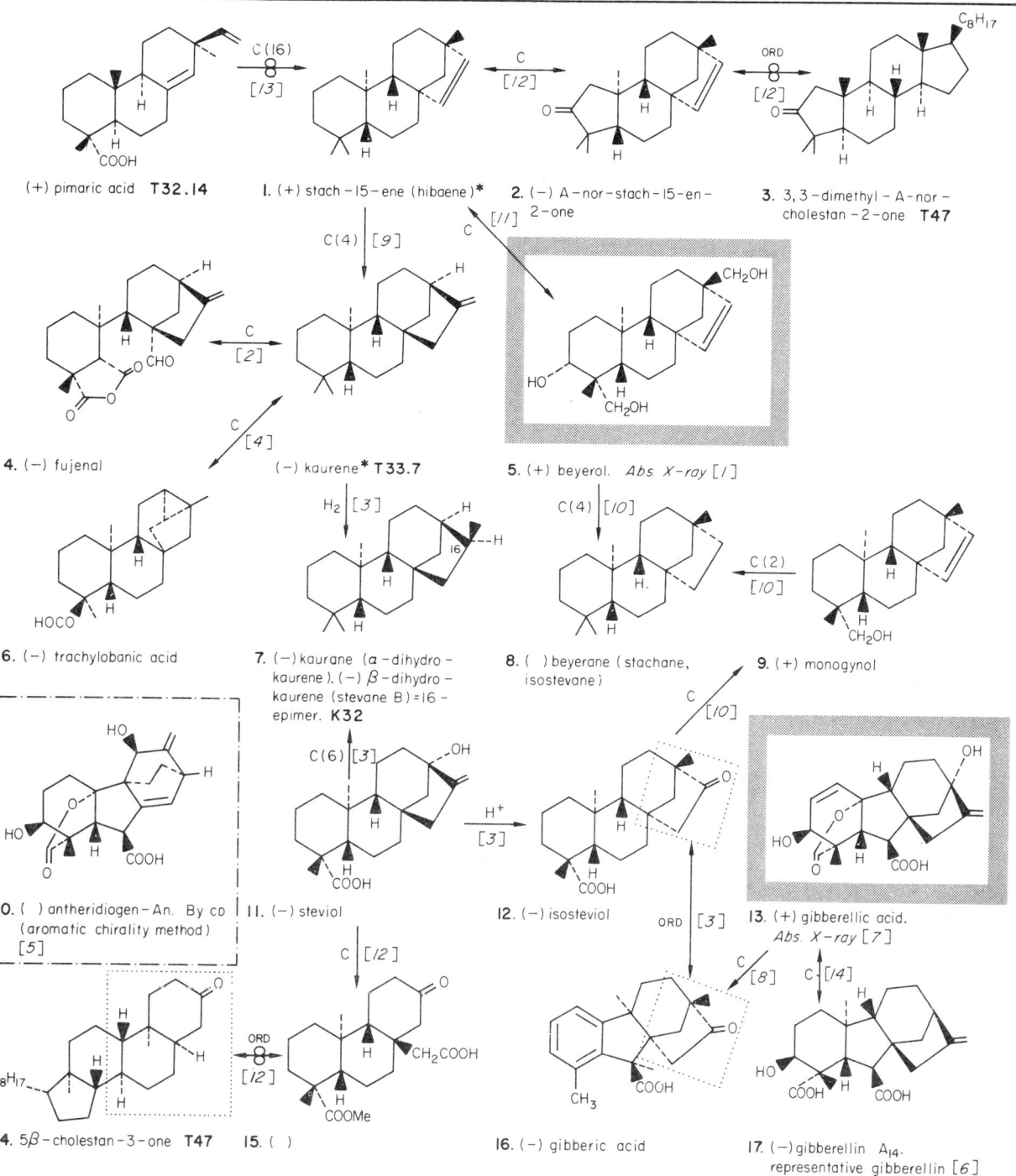

1. (+) pimaric acid T32.14
1. (+) stach-15-ene (hibaene)*
2. (−) A-nor-stach-15-en-2-one
3. 3,3-dimethyl-A-nor-cholestan-2-one T47
4. (−) fujenal
5. (+) beyerol. Abs. X-ray [1]
6. (−) trachylobanic acid
7. (−) kaurane (α-dihydro-kaurene). (−) β-dihydro-kaurene (stevane B) = 16-epimer. K32
8. ( ) beyerane (stachane, isostevane)
9. (+) monogynol
10. ( ) antheridiogen-An. By CD (aromatic chirality method) [5]
11. (−) steviol
12. (−) isosteviol
13. (+) gibberellic acid. Abs. X-ray [7]
14. 5β-cholestan-3-one T47
15. ( )
16. (−) gibberic acid
17. (−) gibberellin A₁₄. representative gibberellin [6]

(−) kaurene* T33.7

1. A. M. O'Connell and E. N. Maslen, *Acta Cryst.*, 1966, **21**, 744.
2. B. E. Cross, R. H. B. Galt and J. R. Hanson, *J. Chem. Soc.*, 1963, 5052.
3. F. Dolder, H. Lichti, E. Mosettig and P. Quitt, *J. Amer. Chem. Soc.*, 1960, **82**, 246.
4. G. Hugel, L. Lods, J. M. Mellor, D. W. Theobald and G. Ourisson, *Bull. Soc. chim. France*, 1963, 1974.
5. K. Nakanishi, M. Endo, U. Näf and L. F. Johnson, *J. Amer. Chem. Soc.*, 1971, **93**, 5579.
6. J. F. Grove, *Quart. Rev.*, 1961, **15**, 56.
7. F. McCapra, A. T. McPhail, A. I. Scott, G. A. Sim and D. W. Young, *J. Chem. Soc. (C)*, 1966, 1577.
8. G. Stork and H. Newman, *J. Amer. Chem. Soc.*, 1959, **81**, 3168.
9. A. H. Kapadi and S. Dev, *Tetrahedron Letters*, 1965, 1255.
10. A. H. Kapadi and S. Dev, *Tetrahedron Letters*, 1964, 2751.
11. P. R. Jefferies, R. S. Rosich and D. E. White, *Tetrahedron Letters*, 1963, 1793.
12. C. Djerassi, P. Quitt, E. Mosettig, R. C. Cambie, P. S. Rutledge and L. H. Briggs, *J. Amer. Chem. Soc.*, 1961, **83**, 3720 and references therein.
13. W. Herz, A. K. Pinder and R. N. Mirrington, *J. Org. Chem.*, 1966, **31**, 2257.
14. B. E. Cross, *J. Chem. Soc. (C)*, 1966, 501.

◀18. M. Mangoni and L. Bellardini, *Gazzetta*, 1964, **94**, 1108.
19. O. Jeger, O. Dürst and G. Büchi, *Helv. Chim. Acta*, 1947, **30**, 1853.

T 36
Labdane group; cafestol, phyllocladene and related compounds.

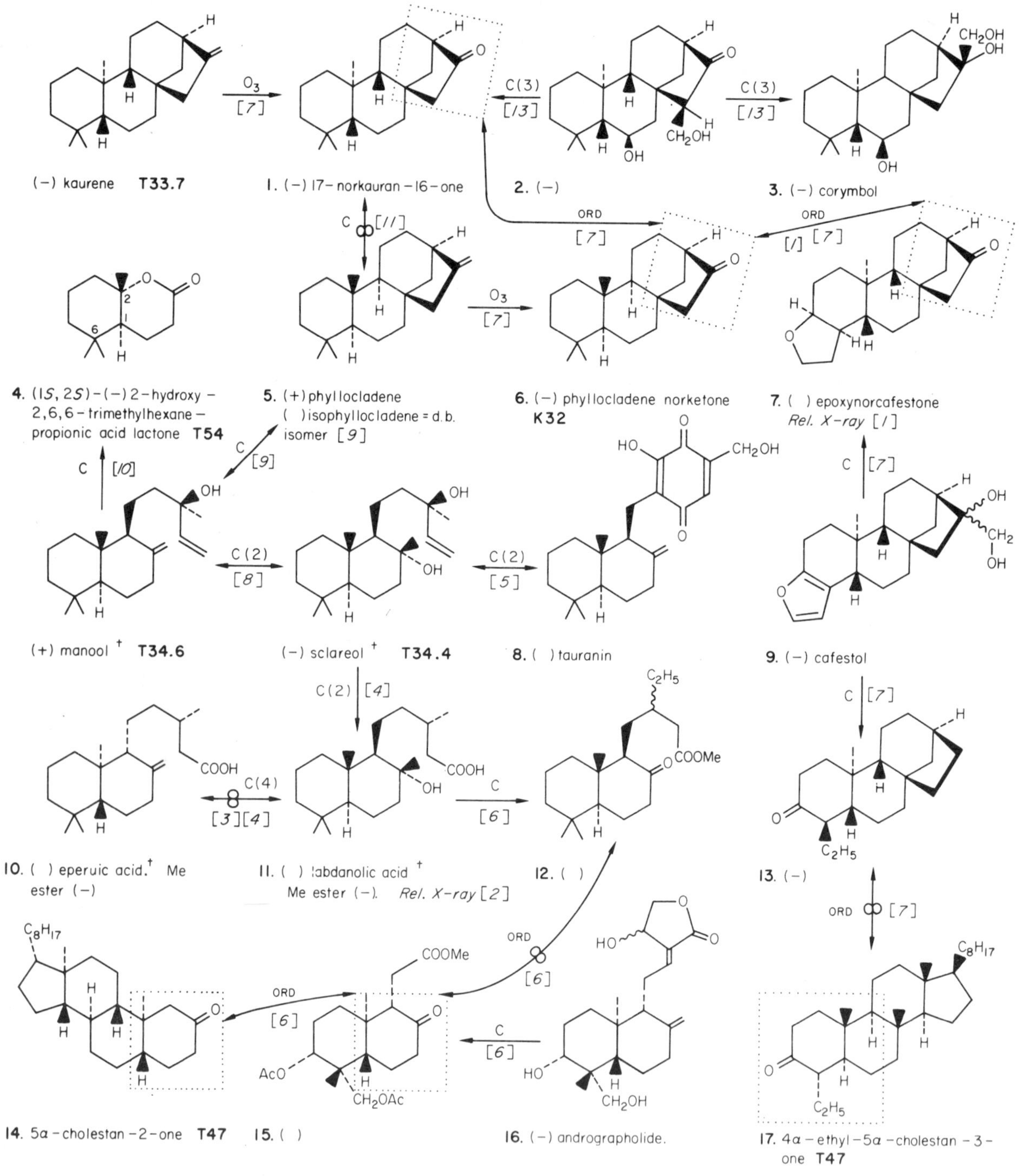

1. (−) 17-norkauran-16-one
2. (−)
3. (−) corymbol
4. (1S,2S)-(−)-2-hydroxy-2,6,6-trimethylhexane-propionic acid lactone **T54**
5. (+) phyllocladene ( ) isophyllocladene = d.b. isomer [9]
6. (−) phyllocladene norketone **K32**
7. ( ) epoxynorcafestone Rel. X-ray [1]
8. ( ) tauranin
9. (−) cafestol
10. ( ) eperuic acid.† Me ester (−)
11. ( ) labdanolic acid † Me ester (−). Rel. X-ray [2]
12. ( )
13. (−)
14. 5α-cholestan-2-one **T47**
15. ( )
16. (−) andrographolide.
17. 4α-ethyl-5α-cholestan-3-one **T47**

(−) kaurene **T33.7**

(+) manool † **T34.6**

(−) sclareol † **T34.4**

---

1. A. I. Scott, G. A. Sim, G. Ferguson, D. W. Young and F. McCapra, *J. Amer. Chem. Soc.*, 1962, **84**, 3197.
2. K. Bjamer, G. Ferguson and R. D. Melville, *Acta Cryst.*, 1968, **B24**, 855.
3. C. A. Henrick and P. R. Jefferies, *Tetrahedron*, 1965, **21**, 1175.
4. E. M. Graham and K. H. Overton, *J. Chem. Soc.*, 1965, 126.
5. K. Kawashima, K. Nakanishi, M. Tada and H. Nishikawa, *Tetrahedron Letters*, 1964, 1227.
6. M. P. Cava, W. R. Chan, L. J. Haynes, L. F. Johnson and B. Weinstein, *Tetrahedron*, 1962, **18**, 397; W. R. Chan, C. Willis, M. P. Cava and R. P. Stein, *Chem. and Ind.*, 1963, 495.
7. C. Djerassi, M. Cais and L. A. Mitscher, *J. Amer. Chem. Soc.*, 1959, **81**, 2386 and references therein.
8. J. R. Hosking and C. W. Brandt, *Ber.*, 1935, **68**, 1311.
9. P. K. Grant and R. Hodges, *Tetrahedron*, 1960, **8**, 261.
10. C. H. Eugster, R. Buchecker, Ch. Tscharner, G. Uhde and G. Ohloff, *Helv. Chim. Acta*, 1969, **52**, 1729.
11. J. R. Hanson, *J. Chem. Soc.*, 1963, 5061.
12. K. Bruns, *Tetrahedron Letters*, 1970, 3263.
13. M. C. Pérezamador and F. G. Jiménez, *Tetrahedron*, 1966, **22**, 1937; F. G. Jiménez, M. C. Pérezamador, S. E. Flores and J. Herrán, *Tetrahedron Letters*, 1965, 621.

†*Note on the stereochemistry of the labdane group of diterpenoids* (Examples: manool, **T34.6**, sclareol **T34.4**, 13-epimanool **T34.7**, eperuic acid **T36.10**, labdanolic acid **T36.11**).

Diterpenes of this series, of which many are known, are derived from both antipodal series, i.e. from (5*S*, 9*S*, 10*R*)labdane and from (5*R*, 9*R*, 10*S*)labdane (eperuane, *ent.*-labdane).

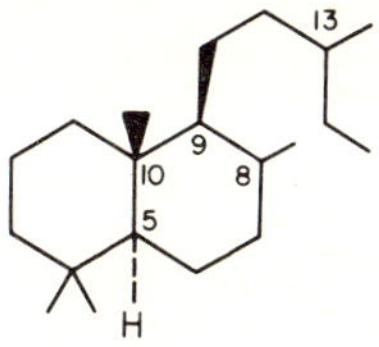

(5*S*, 9*S*, 10*R*) 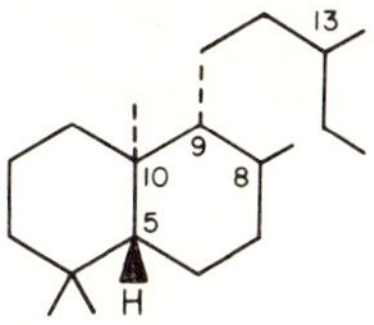   (5*R*, 9*R*, 10*S*)

Since all known natural products have the *trans-anti* stereochemistry at C_5, C_{10} and C_9, specification of the configuration at C_{10} fixes those at C_5 and C_9, and the two series are frequently referred to as 10β and 10α respectively. The configurations at C_8 and C_{13} are, however, variable. (Cf. manool and 13-epimanool.)

For a note on the configuration at C_{13}, see [*12*].

T37 Marrubiin and clerodin types.

1. (−) nepetaefolin
2. (+) marrubiin [1] Related compd; (−) leonotin [6]
   (+) ambrienolide T34.8
3. (−) clerodin. Abs. X-ray [2]
4. (−) solidagenone
5. (−) gutierolide. Abs. X-ray [3]
   (R)-(−)-2-acetoxy-2-methylbutyric acid A33.8
6. ( ) clerodendrin A. Abs. X-ray [4]
7. (−) tinophyllone. Rel. X-ray [5]
8. ( ) 3,4-dihydrotinophyllone. By ORD [5]
9. (−) hardwickiic acid*
10. (−) kolavic acid (related compds; (−) kolavenic acid, (−) kolavenol)
11. (−)

5α-androstan-17-one T27.13

1. I. Mangoni and M. Adinolfi, *Tetrahedron Letters*, 1968, 269.
2. I. C. Paul, G. A. Sim, T. A. Hamor and J. M. Robertson, *J. Chem. Soc.*, 1962, 4133.
3. W. B. T. Cruse, M. N. G. James, A. A. Al-Shamma, J. K. Beal and R. W. Doskotch, *Chem. Comm.*, 1971, 1278.
4. N. Kato, S. Shibayama, K. Munakata and C. Katayama, *Chem. Comm.*, 1971, 1632.
5. L. Brehm, O. J. R. Hodder and T. G. Halsall, *J. Chem. Soc. (C)*, 1971, 2529.
6. J. D. White, P. S. Manchand and W. B. Whalley, *Chem. Comm.*, 1969, 1315.
7. J. D. White and P. S. Manchand, *J. Amer. Chem. Soc.*, 1970, 92, 5527.
8. T. Anthonsen, P. H. McCabe, R. McCrindle and R. D. H. Murray, *Tetrahedron*, 1969, 25, 2233.
9. R. Misra, R. C. Pandey and S. Dev, *Tetrahedron Letters*, 1968, 2681.
10. R. A. Appleton, J. W. B. Fulke, M. S. Henderson and R. McCrindle, *J. Chem. Soc. (C)*, 1967, 1943.

Columbin group; grayanotoxin and nagilactone groups.

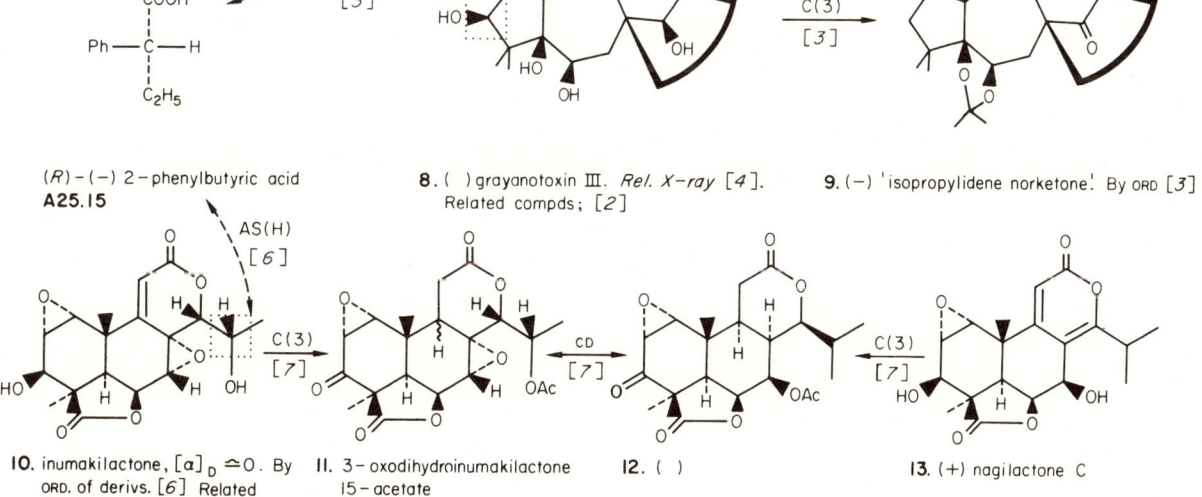

1. (+) isocolumbin (8-*epi*-columbin). *Rel. X-ray* [*1*]
2. (+) columbin. ( ) jateorin = columbin-2,3-epoxide.
3. chasmanthin (12-*epi*-jateorin). $[\alpha]_D \simeq 0$
4. (+) palmarin (8-*epi*-chasmanthin)
5. 5β-cholestan-1-one **T47**
6. ( )
7. ( )
   (9R,10R)-(+) 9-methyl-1-decalone **A53.16**

(R)-(−) 2-phenylbutyric acid **A25.15**

8. ( ) grayanotoxin III. *Rel. X-ray* [*4*]. Related compds; [*2*]
9. (−) 'isopropylidene norketone'. By ORD [*3*]
10. inumakilactone, $[\alpha]_D \simeq 0$. By ORD. of derivs. [*6*] Related compds; podolactones [*5*]
11. 3-oxodihydroinumakilactone 15-acetate
12. ( )
13. (+) nagilactone C

1. K. K. Cheung, D. Melville, K. H. Overton, J. M. Robertson and G. A. Sim, *J. Chem. Soc. (B)*, 1966, 853.
2. H. Hiniko, M. Ogura, T. Ohta and T. Takemoto, *Chem. Pharm. Bull. Japan*, 1970, **18**, 1071 and references therein.
3. H. Kakisawa, T. Kozima, M. Yanai and K. Nakanishi, *Tetrahedron*, 1965, **21**, 3091.
4. P. Narayanan, M. Röhrl, K. Zechmeister and W. Hoppe, *Tetrahedron Letters*, 1970, 3943.
5. M. N. Galbraith, D. H. S. Horn and J. M. Sasse, *Chem. Comm.*, 1971, 1362.
6. S. Itô, M. Kodama, M. Sunagawa, T. Takahashi, H. Imamura and O. Honda, *Tetrahedron Letters*, 1968, 2065.
7. S. Itô, M. Kodama, M. Sunagawa, H. Honma, Y. Hayashi, S. Takahashi, H. Ona and T. Sakan, *Tetrahedron Letters*, 1969, 2951.
8. K. H. Overton, N. G. Weir and A. Wylie, *J. Chem. Soc. (C)*, 1966, 1482 and references therein.

T 39 Phorbol group; miscellaneous diterpene types.

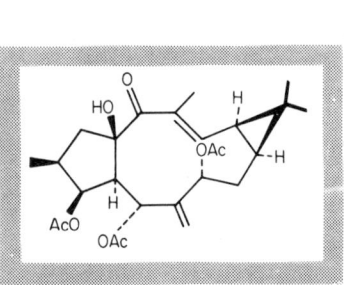

1. ( ) 7-hydroxylathyrol-3,5,7-triacetate. Abs. X-ray [4]

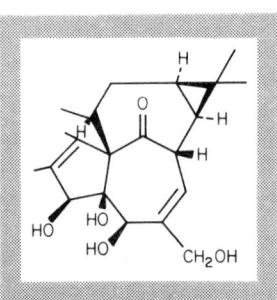

2. ( ) ingenol. Abs. X-ray [1]
Related compounds; milliamines [13]

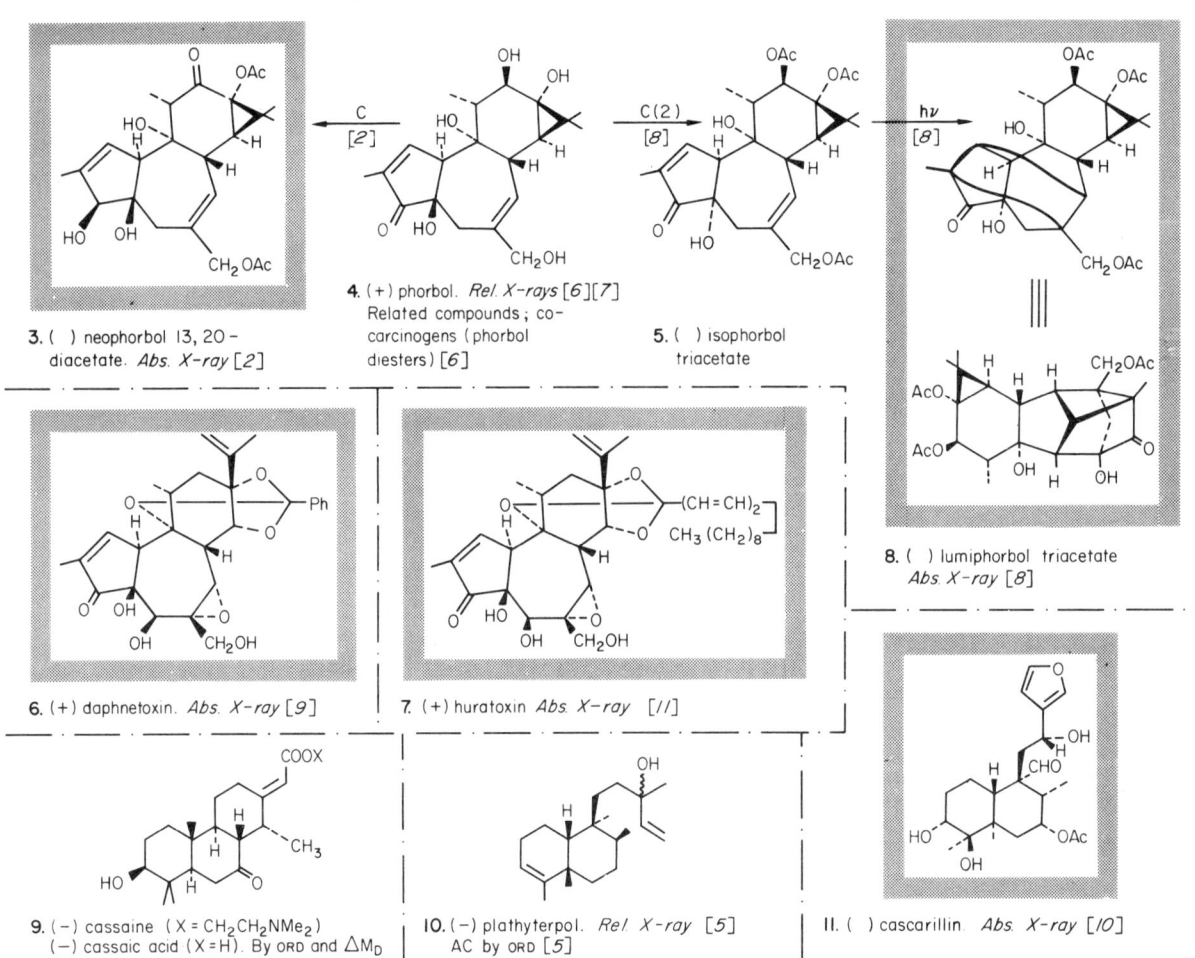

3. ( ) neophorbol 13,20-diacetate. Abs. X-ray [2]

4. (+) phorbol. Rel. X-rays [6][7]
Related compounds; co-carcinogens (phorbol diesters) [6]

5. ( ) isophorbol triacetate

8. ( ) lumiphorbol triacetate Abs. X-ray [8]

6. (+) daphnetoxin. Abs. X-ray [9]

7. (+) huratoxin Abs. X-ray [11]

9. (−) cassaine (X = CH$_2$CH$_2$NMe$_2$)
(−) cassaic acid (X = H). By ORD and ΔM_D [3]   See also [12]

10. (−) plathyterpol. Rel. X-ray [5]
AC by ORD [5]

11. ( ) cascarillin. Abs. X-ray [10]

1. K. Zechmeister, F. Brandl, W. Hoppe, E. Hecker, H. J. Opferkuch & W. Adolf, *Tetrahedron Letters*, 1970, 4075
2. W. Hoppe, F. Brandl, I. Strell, M. Röhrl, I. Gassman, E. Hecker, H. Bartsch, G. Krebich & C. v. Szczepanski, *Angew. Chem. (Internat. Edn.)*, 1967, 6, 809.
3. R. B. Turner, O. Buchardt, E. Herzog, R. B. Morin, A. Riebel & J. M. Sanders, *J. Amer. Chem. Soc.*, 1966, 88, 1766.
4. P. Narayanan, M. Röhrl, K. Zechmeister, D. W. Engel, W. Hoppe, E. Hecker & W. Adolf, *Tetrahedron Letters*, 1971, 1325.
5. T. J. King, S. Rodrigo & S. C. Wallwork, *Chem. Comm.*, 1969, 683.
6. L. Crombie, M. L. Games & D. J. Pointer, *J. Chem. Soc. (C)*, 1968, 1347.
7. W. Hoppe, K. Zechmeister, M. Röhrl, F. Brandl, E. Hecker, G. Kreibich & H. Bartsch, *Tetrahedron Letters*, 1969, 667.
8. E. Hecker, E. Härle, H. U. Schairer, P. Jacobi, W. Hoppe, J. Gassmann, M. Röhrl & H. Abel, *Angew. Chem. (Internat. Edn.)*, 1968, 7, 890.
9. G. H. Stout, W. G. Balkenhol, M. Poling & G. L. Hickernell, *J. Amer. Chem. Soc.*, 1970, 92, 1070.
10. C. E. McEachan, A. T. McPhail & G. A. Sim, *J. Chem. Soc. (B.)*, 1966, 633.
11. K. Sakata, K. Kawazu, T. Mitsui & N. Masaki, *Tetrahedron Letters*, 1971, 1141.
12. R. L. Clarke, S. J. Daum, P. E. Shaw & R. K. Kullnig, *J. Amer. Chem. Soc.*, 1966, 88, 5865.
13. D. Uemura & Y. Hirata, *Tetrahedron Letters*, 1971, 3673.

Taxane group; ginkolides.

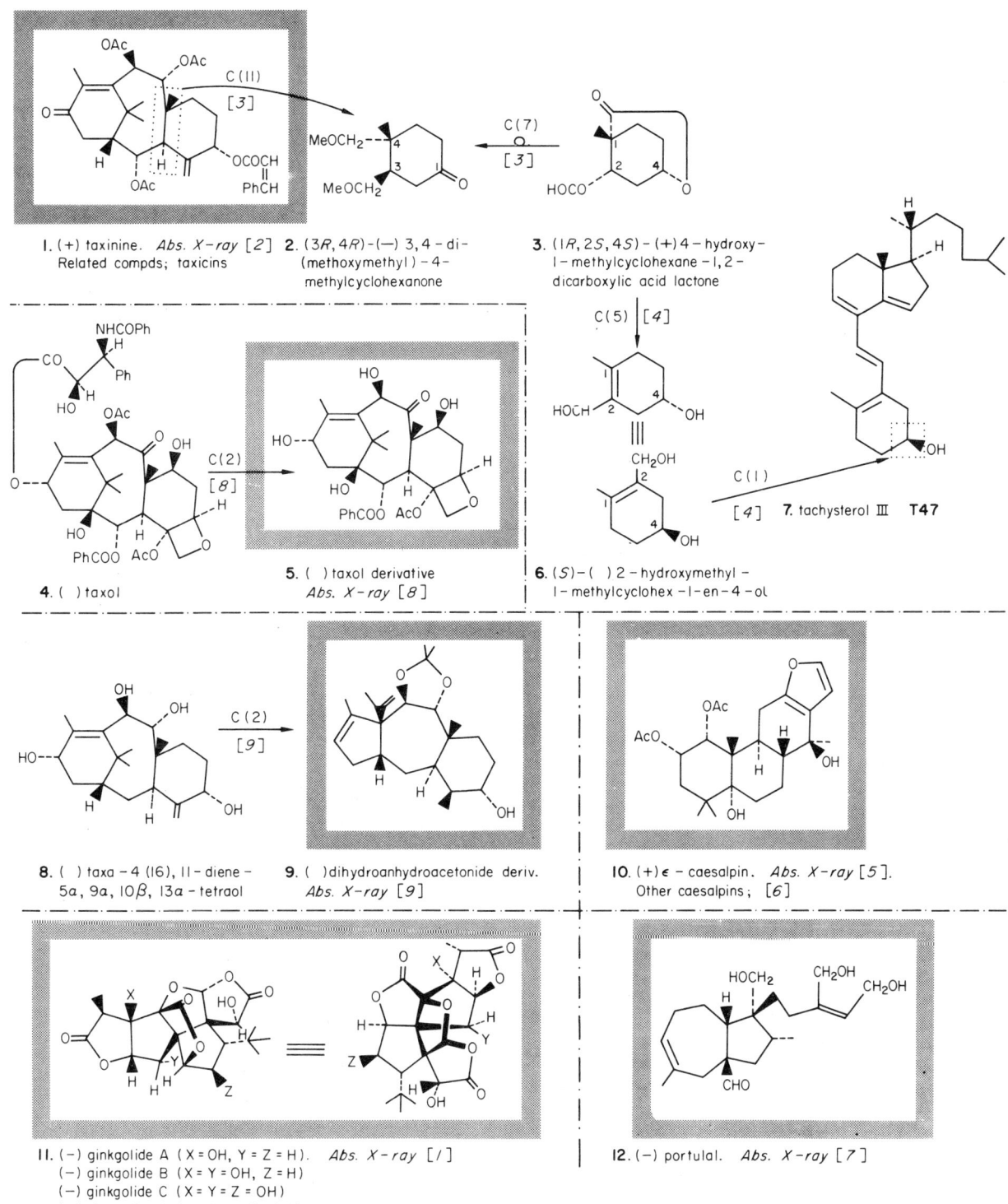

1. (+) taxinine. Abs. X-ray [2]
   Related compds; taxicins
2. (3R,4R)-(−) 3,4-di-(methoxymethyl)-4-methylcyclohexanone
3. (1R,2S,4S)-(+) 4-hydroxy-1-methylcyclohexane-1,2-dicarboxylic acid lactone
4. ( ) taxol
5. ( ) taxol derivative Abs. X-ray [8]
6. (S)-( ) 2-hydroxymethyl-1-methylcyclohex-1-en-4-ol
7. tachysterol III T47
8. ( ) taxa-4(16),11-diene-5α,9α,10β,13α-tetraol
9. ( ) dihydroanhydroacetonide deriv. Abs. X-ray [9]
10. (+) ε-caesalpin. Abs. X-ray [5]. Other caesalpins; [6]
11. (−) ginkgolide A (X = OH, Y = Z = H). Abs. X-ray [1]
    (−) ginkgolide B (X = Y = OH, Z = H)
    (−) ginkgolide C (X = Y = Z = OH)
    (−) ginkgolide D (X = H, Y = Z = OH)
12. (−) portulal. Abs. X-ray [7]

1. N. Sakabe, S. Takada and K. Okabe, *Chem. Comm.*, 1967, 259.
2. M. Shiro and H. Koyama, *J. Chem. Soc. (B)*, 1971, 1342.
3. M. Dukes, H. Eyre, J. W. Harrison and B. Lythgoe, *Tetrahedron Letters*, 1965, 4765.
4. R. S. Davidson, P. S. Littlewood, T. Medcalfe, S. M. Waddington-Feather, D. H. Williams and B. Lythgoe, *Tetrahedron Letters*, 1963, 1413.
5. A. Balmain, K. Bjamer, J. D. Connolly and G. Ferguson, *Tetrahedron Letters*, 1967, 5027.
6. A. Balmain, J. D. Connolly, M. Ferrari, E. L. Ghisalberti, U. M. Pagnoni and F. Pelizzoni, *Chem. Comm.*, 1970, 1244.
7. S. Yamazaki, S. Tamura, F. Marumo and Y. Saito, *Tetrahedron Letters*, 1969, 358.
8. M. C. Wani, H. L. Taylor, M. E. Wall, P. Coggon and A. T. McPhail, *J. Amer. Chem. Soc.*, 1971, 93, 2325.
9. K. Bjamer, G. Ferguson and J. M. Robertson, *J. Chem. Soc. (B)*, 1967, 1272.

T41 Macrocyclic diterpenes; Sesterterpenes

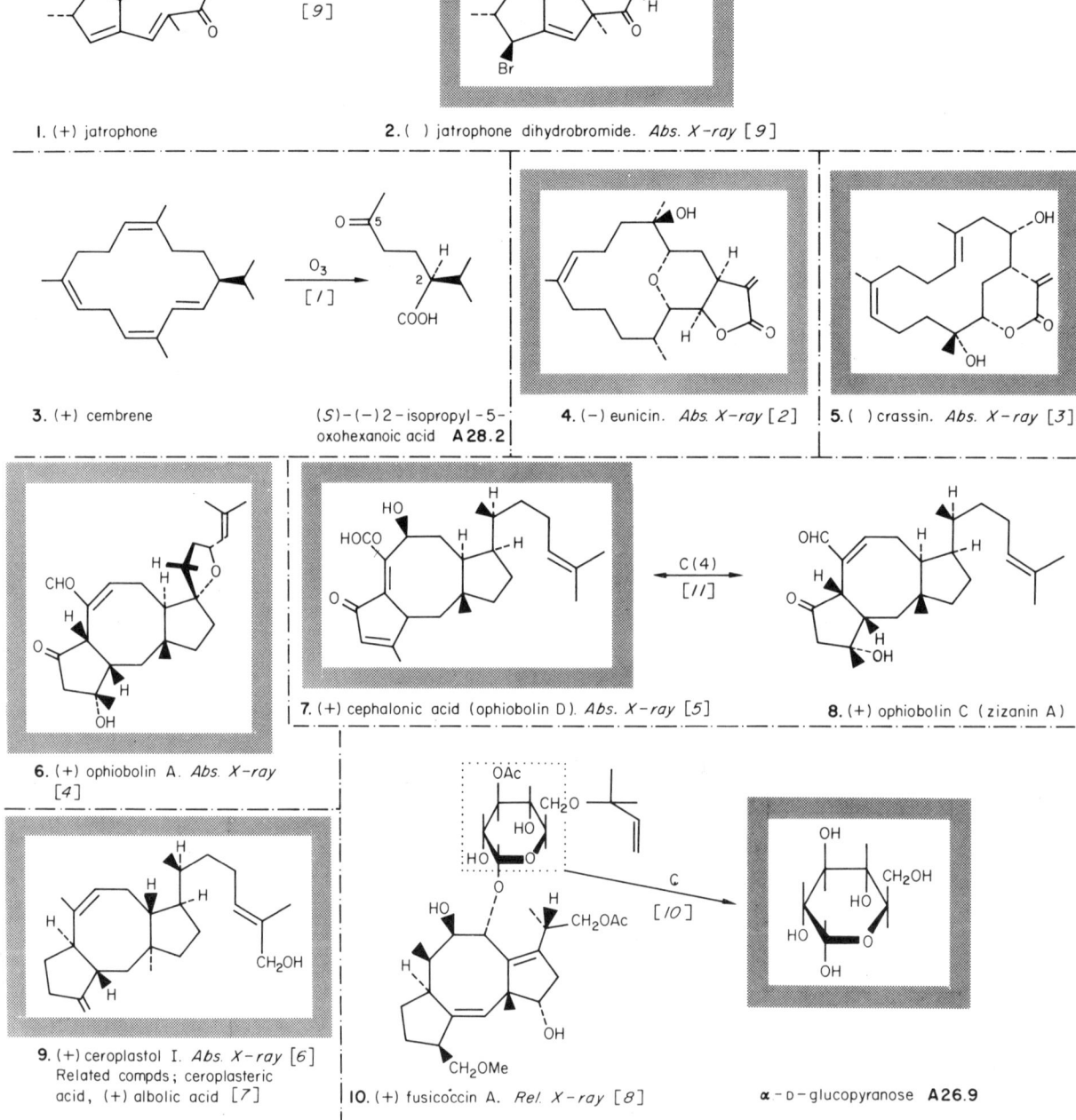

1. (+) jatrophone
2. ( ) jatrophone dihydrobromide. *Abs. X-ray* [9]
3. (+) cembrene
   (S)-(−)2-isopropyl-5-oxohexanoic acid **A28.2**
4. (−) eunicin. *Abs. X-ray* [2]
5. ( ) crassin. *Abs. X-ray* [3]
6. (+) ophiobolin A. *Abs. X-ray* [4]
7. (+) cephalonic acid (ophiobolin D). *Abs. X-ray* [5]
8. (+) ophiobolin C (zizanin A)
9. (+) ceroplastol I. *Abs. X-ray* [6]
   Related compds; ceroplasteric acid, (+) albolic acid [7]
10. (+) fusicoccin A. *Rel. X-ray* [8]
    α-D-glucopyranose **A26.9**

1. W. G. Dauben, W. E. Thiessen and P. R. Resnick, *J. Org. Chem.*, 1965, **30**, 1693.
2. M. B. Hossain, A. F. Nicholas and D. van der Helm, *Chem. Comm.*, 1968, 385.
3. M. B. Hossain and D. van der Helm, *Rec. Trav. chim.*, 1969, **88**, 1413.
4. S. Nozoe, M. Morisaki, K. Tsuda, Y. Iitaka, N. Takahashi, S. Tamura, K. Ishibashi and M. Shirasaka, *J. Amer. Chem. Soc.*, 1965, **87**, 4968.
5. A. Itai, S. Nozoe, K. Tsuda, S. Okuda, Y. Iitaka and Y. Nakayama, *Tetrahedron Letters*, 1967, 4111.
6. Y. Iitaka, I. Watanabe, I. T. Harrison and S. Harrison, *J. Amer. Chem. Soc.*, 1968, **90**, 1092.
7. T. Rios and F. Gomez, *Tetrahedron Letters*, 1969, 2929.
8. M. Brufani, S. Cerrini, W. Fedeli and A. Vaciago, *J. Chem. Soc. (B)*, 1971, 2021.
9. S. M. Kupchan, C. W. Sigel, M. J. Matz, J. A. S. Renauld, R. C. Haltiwanger and R. F. Bryan, *J. Amer. Chem. Soc.*, 1970, **92**, 4476.
10. K. D. Barrow, D. H. R. Barton, E. B. Chain, C. Conlay, T. V. Smale, R. Thomas and E. S. Waight, *Chem. Comm.*, 1968, 1195.
11. S. Nozoe, A. Itai, K. Tsuda and S. Okuda, *Tetrahedron Letters*, 1967, 4113.

17. W. Laird, F. S. Spring and R. Stevenson, *J. Chem. Soc.*, 1961, 2638. ▶
18. D. Arigoni, H. Bosshard, J. Dreiding and O. Jeger, *Helv. Chim. Acta*, 1954, **37**, 2173.
19. L. Ruzicka, H. Gutmann, O. Jeger and E. Lederer, *Helv. Chim. Acta*, 1948, **31**, 1746.
20. D. H. R. Barton and N. J. Holness, *J. Chem. Soc.*, 1952, 78 and references therein.

T42

Triterpenes; ambrein, oleanane and related compounds.

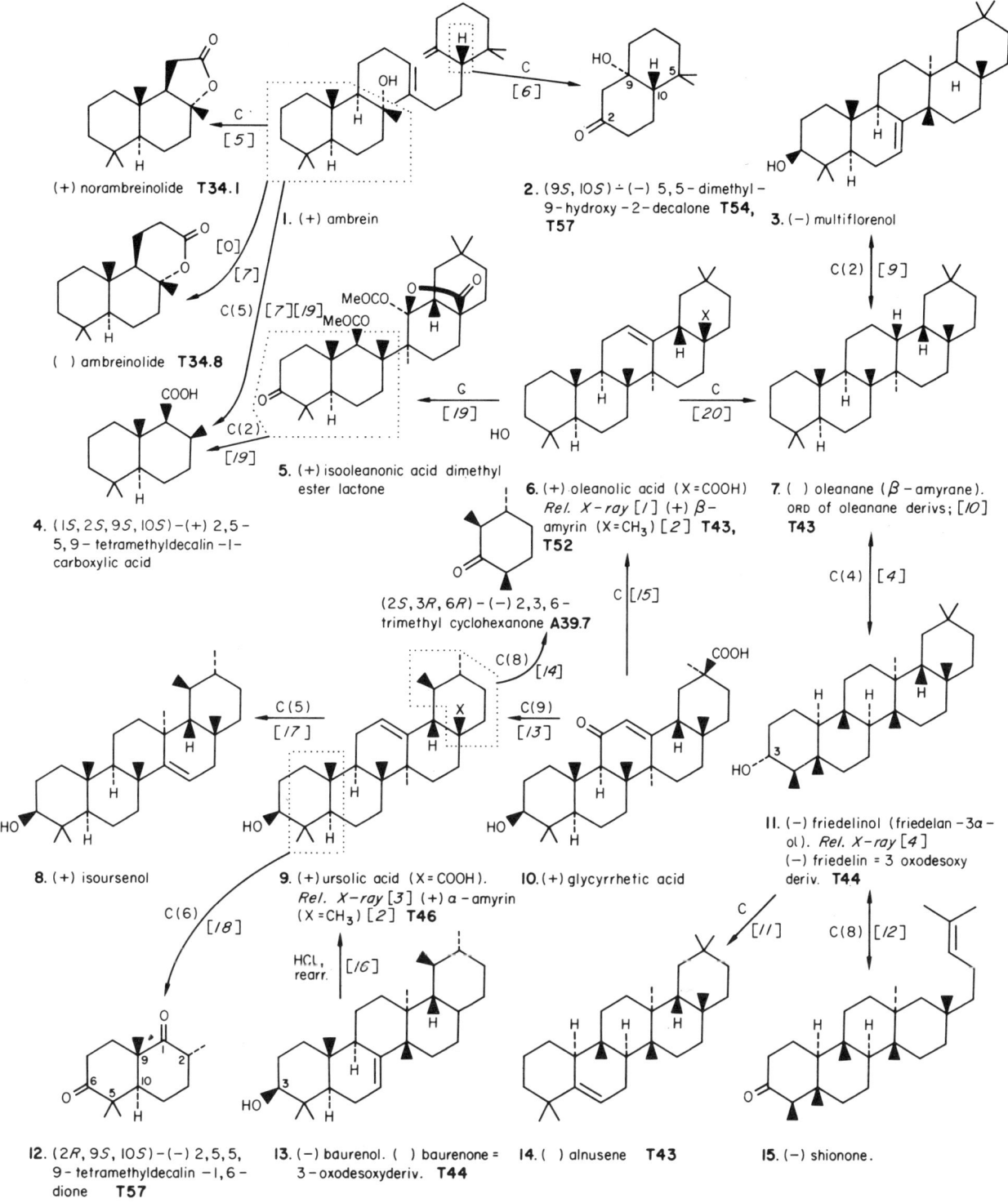

1. (+) norambreinolide **T34.1**
2. (9S,10S)-(−)-5,5-dimethyl-9-hydroxy-2-decalone **T54, T57**
3. (−) multiflorenol
4. (1S,2S,9S,10S)-(+) 2,5-5,9-tetramethyldecalin-1-carboxylic acid
5. (+) isooleanonic acid dimethyl ester lactone
6. (+) oleanolic acid (X=COOH) Rel. X-ray [1] (+) β-amyrin (X=CH₃) [2] **T43, T52**
7. ( ) oleanane (β-amyrane). ORD of oleanane derivs; [10] **T43**
8. (+) isoursenol
9. (+) ursolic acid (X=COOH). Rel. X-ray [3] (+) α-amyrin (X=CH₃) [2] **T46**
10. (+) glycyrrhetic acid
11. (−) friedelinol (friedelan-3α-ol). Rel. X-ray [4] (−) friedelin ≡ 3 oxodesoxy deriv. **T44**
12. (2R,9S,10S)-(−) 2,5,5,9-tetramethyldecalin-1,6-dione **T57**
13. (−) baurenol. ( ) baurenone ≡ 3-oxodesoxyderiv. **T44**
14. ( ) alnusene **T43**
15. (−) shionone.

( ) ambreinolide **T34.8**

(2S,3R,6R)-(−) 2,3,6-trimethyl cyclohexanone **A39.7**

1. A. M. Abd El Rahim and C. H. Carlisle, *Chem. and Ind.*, 1954, 279.
2. R. B. Boar, D. C. Knight, J. F. McGhie and D. H. R. Barton, *J. Chem. Soc. (C)*, 1970, 678.
3. G. H. Stout and K. L. Stevens, *J. Org. Chem.*, 1963, 28, 1259.
4. E. J. Corey and J. J. Ursprung, *J. Amer. Chem. Soc.*, 1952, 78, 5041.
5. E. Lederer and D. Mercier, *Experientia*, 1947, 3, 188.
6. G. Büchi, O. Jeger and L. Ruzicka, *Helv. Chim. Acta*, 1948, 31, 241.
7. L. Ruzicka, O. Dürst and O. Jeger, *Helv. Chim. Acta*, 1947, 30, 353.
8. L. Ruzicka and K. Hofmann, *Helv. Chim. Acta*, 1936, 19, 114.
9. P. Sengupta and H. N. Khastgir, *Tetrahedron*, 1963, 19, 123.
10. C. Djerassi, J. Osiecki and W. Closson, *J. Amer. Chem. Soc.*, 1959, 81, 4587.
11. J. M. Beaton, F. S. Spring, R. Stevenson and J. L. Stewart, *Tetrahedron*, 1958, 2, 246.
12. T. Takahashi, T. Tsuyuki, T. Hoshino and M. Ito, *Tetrahedron Letters*, 1967, 2997.
13. E. J. Corey and W. E. Cantrall, *J. Amer. Chem. Soc.*, 1959, 81, 1745.
14. A. Melera, D. Arigoni, A. Eschenmoser, O. Jeger and L. Ruzicka, *Helv. Chim. Acta*, 1956, 39, 441.
15. L. Ruzicka and A. Marxer, *Helv. Chim. Acta*, 1939, 22, 195.
16. F. N. Lahey and M. V. Leeding, *Proc. Chem. Soc.*, 1958, 342.

References continued on page 118

T43 Oleanane, lupane groups.

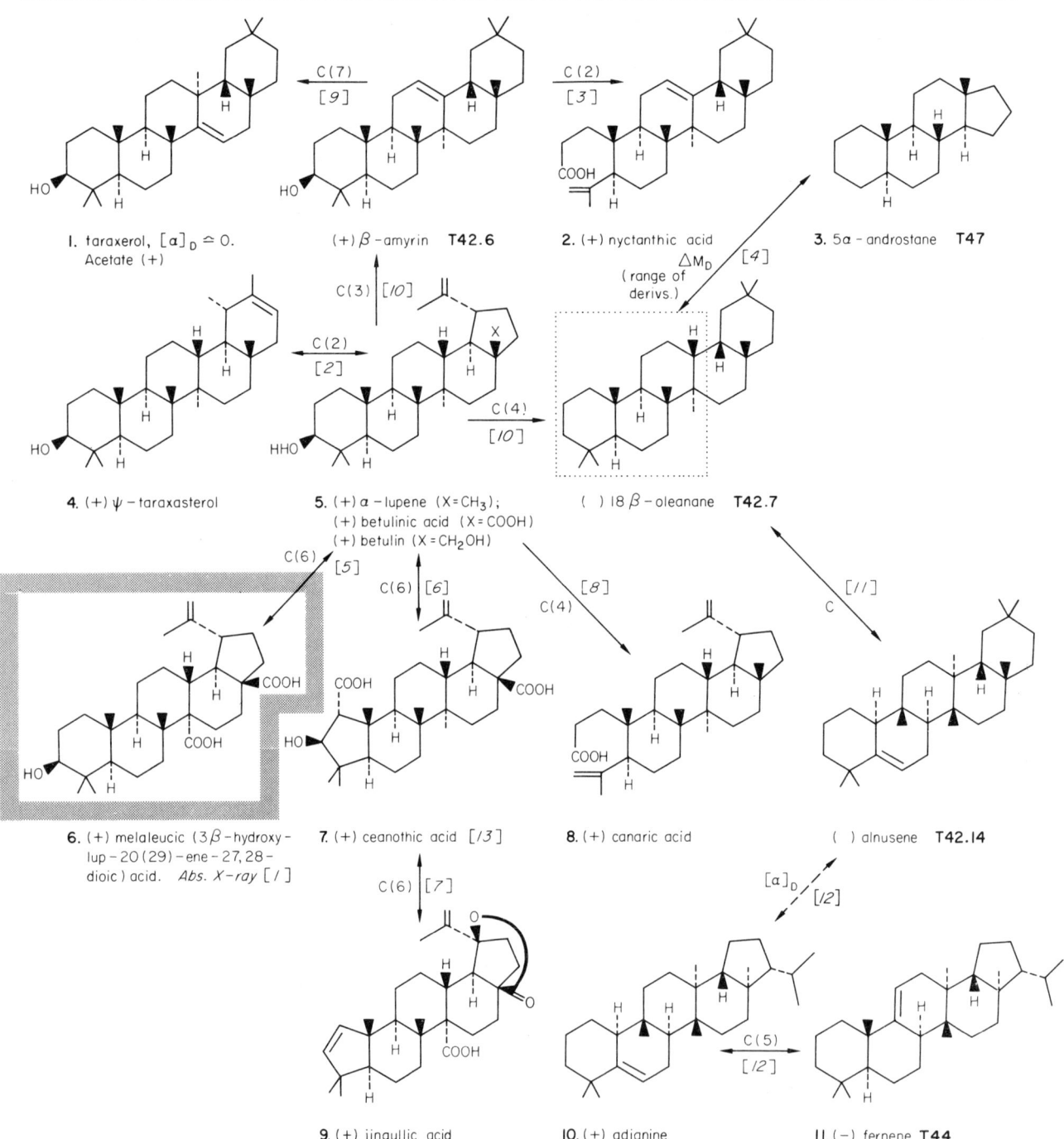

1. taraxerol, $[\alpha]_D \simeq 0$. Acetate (+)
2. (+) nyctanthic acid
3. 5α-androstane T47
4. (+) ψ-taraxasterol
5. (+) α-lupene (X=CH$_3$); (+) betulinic acid (X=COOH); (+) betulin (X=CH$_2$OH)
6. (+) melaleucic (3β-hydroxy-lup-20(29)-ene-27,28-dioic) acid. Abs. X-ray [1]
7. (+) ceanothic acid [13]
8. (+) canaric acid
9. (+) jingullic acid
10. (+) adianine
11. (−) fernene T44

( ) 18β-oleanane T42.7
(+)β-amyrin T42.6
( ) alnusene T42.14

1. S. R. Hall and E. N. Maslen, *Acta Cryst.*, 1965, **18**, 265.
2. T. G. Halsall, E. R. H. Jones and R. E. H. Swayne, *J. Chem. Soc.*, 1954, 1902.
3. G. H. Whitham, *J. Chem. Soc.*, 1960, 2016.
4. W. Klyne, *J. Chem. Soc.*, 1952, 2916.
5. C. S. Chopra, A. R. H. Cole, K. J. L. Theiberg, D. E. White and H. R. Arthur, *Tetrahedron*, 1965, **21**, 1529.
6. P. de Mayo and A. N. Starratt, *Canad. J. Chem.*, 1962, **40**, 788.
7. R. A. Eade, J. Ellis, P. Harper and J. J. H. Simes, *Chem. Comm.*, 1969, 579.
8. R. M. Carman and D. E. Cowley, *Tetrahedron Letters*, 1964, 627.
9. J. M. Beaton, F. S. Spring, R. Stevenson and J. L. Stewart, *Chem. and Ind.*, 1955, 35.
10. T. R. Ames, T. G. Halsall and E. R. H. Jones, *J. Chem. Soc.*, 1951, 450.
11. F. S. Spring, J. M. Beaton, R. Stevenson and J. L. Stewart, *Chem. and Ind.*, 1956, 1054.
12. H. Ageta, K. Iwata and S. Natori, *Tetrahedron Letters*, 1964, 3413.
13. R. A. Eade, P. K. Grant, M. J. A. McGrath, J. J. H. Simes and M. Wootton, *Chem. Comm.*, 1967, 1204.

Serratane and onocerane and related groups.

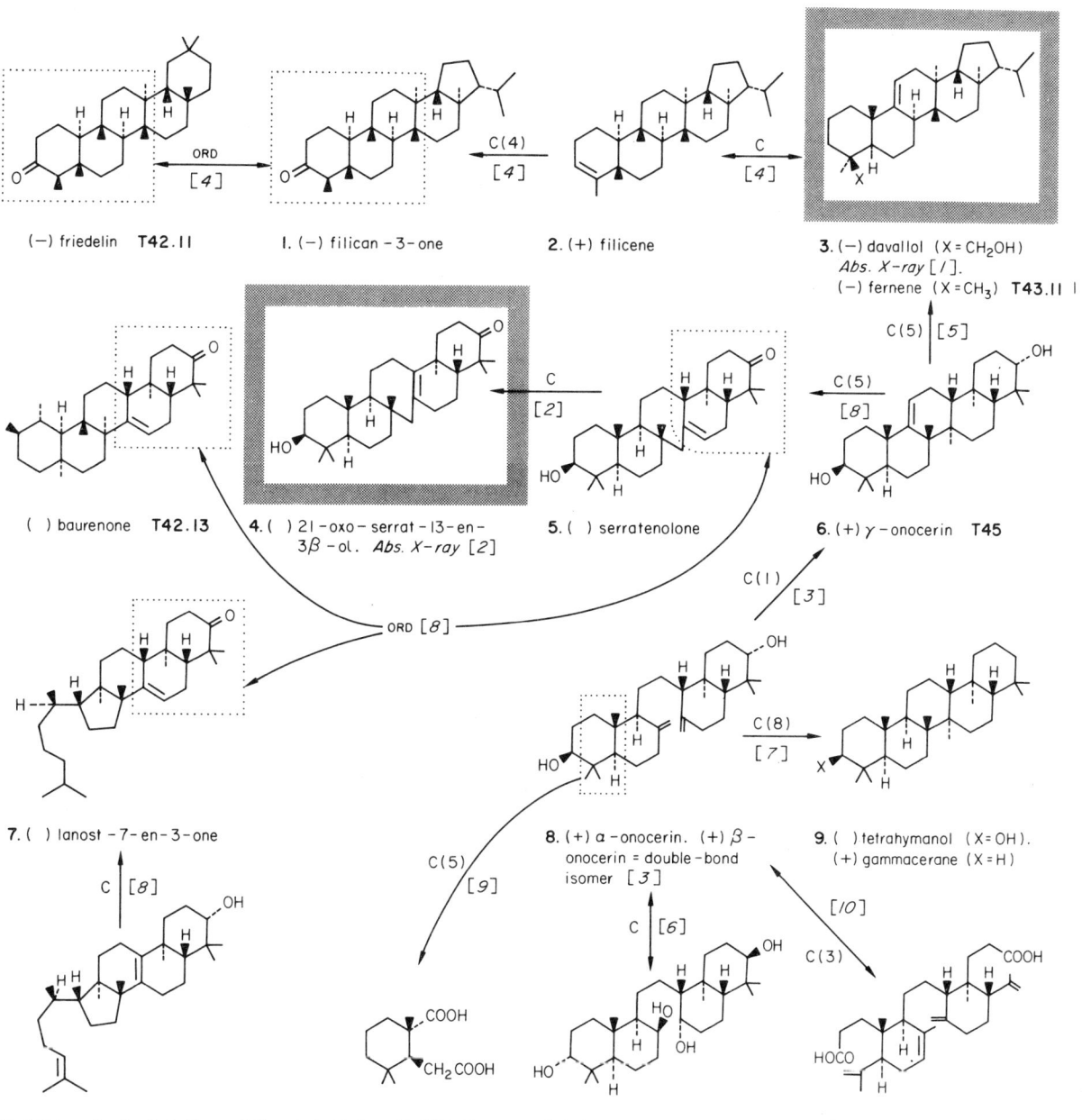

1. (−) filican-3-one
2. (+) filicene
3. (−) davallol (X=CH$_2$OH) Abs. X-ray [1].
   (−) fernene (X=CH$_3$) T43.11
4. ( ) 21-oxo-serrat-13-en-3β-ol. Abs. X-ray [2]
5. ( ) serratenolone
6. (+) γ-onocerin T45
7. ( ) lanost-7-en-3-one
8. (+) α-onocerin. (+) β-onocerin = double-bond isomer [3]
9. ( ) tetrahymanol (X=OH). (+) gammacerane (X=H)
10. (+) lanosterol. Rel. X-ray [1] T46, T50
(−) drimic acid T32.7
11. ( ) clavatol (bisnortriterpene)
12. (−) lansic acid

(−) friedelin T42.11
( ) baurenone T42.13

1. Y-L. Oh and E. N. Maslen, *Acta Cryst.*, 1966, **20**, 852.
2. F. H. Allen and J. Trotter, *Acta Cryst.*, 1969 (suppl.), S137.
3. D. H. R. Barton and K. H. Overton, *J. Chem. Soc.*, 1955, 2639.
4. H. Ageta, K. Iwata and S. Natori, *Tetrahedron Letters*, 1964, 3413.
5. K. Iguchi and H. Kakisawa, *Chem. Comm.*, 1970, 1486.
6. T. Sano, T. Fujimoto and Y. Tsuda, *Chem. Comm.*, 1970, 1274.
7. Y. Tsuda, A. Morimoto, T. Sano, Y. Inubushi, F. B. Mallory and J. T. Gordon, *Tetrahedron Letters*, 1965, 1427.
8. Y. Tsuda, T. Sano, K. Kawaguchi and Y. Inubushi, *Tetrahedron Letters*, 1964, 1279.
9. K. Schaffner, R. Viterbo, D. Arigoni and O. Jeger, *Helv. Chim. Acta*, 1956, **39**, 174.
10. K. Habaguchi, M. Watanabe, Y. Nakadaira, K. Nakanishi, A. K. Kiang and F. Y. Lim, *Tetrahedron Letters*, 1968, 3731.
11. J. Fridrichsons and A. McL. Mathieson, *J. Chem. Soc.*, 1953, 2159.

T45 Onocerane group; miscellaneous pentacyclic triterpenes.

(+) γ-onocerin **T44.6**

1. (+) hopenone I (hop-17(21)-ene-3-one)
2. ( ) hopane† ( ) isohopane = 21-epimer†
3. (+) zeorin. *Rel. X-ray* [1]
4. (+) adiantol (19a-nor-hopanol). *Abs. X-ray* [3]
5. (+) adipedatol
6. ( ) 6-oxoleucotylin. *Rel. X-ray* [4]
7. (+) arborinol. *Rel. X-ray* [8] AC by ORD [9]
8. (+) ketohakonanol (nor-triterpene)
9. (+) platicodigenin. *Abs. X-ray* [5]
10. (+) eupteleogenin. *Rel. X-ray* [6] AC by ORD of derivs. [7]
11. ( ) pristimerin.
12. ( ) pristimerol. *Abs. X-ray* [10]

1. T. Nakanishi, H. Yamauchi, T. Fujiwara and K. Tomita, *Tetrahedron Letters*, 1971, 1157.
2. I. Yosioka, T. Nakanishi, H. Yamauchi and I. Kitagawa, *Tetrahedron Letters*, 1971, 1161.
3. H. Koyama and H. Nakai, *J. Chem. Soc. (B)*, 1970, 546.
4. T. Nakanishi, T. Fujiwara and K. Tomita, *Tetrahedron Letters*, 1968, 1491.
5. T. Akiyama, Y. Iitaka and O. Tanaka, *Tetrahedron Letters*, 1968, 5577.
6. M. Nishikawa, K. Kamiya, T. Murata, Y. Tomiie and I. Nitta, *Tetrahedron Letters*, 1965, 3223.
7. T. Murata, S. Imai, M. Imanishi, M. Goto and K. Morita, *Tetrahedron Letters*, 1965, 3215.
8. O. Kennard, L. R. di Sanseverino and J. S. Rollett, *Tetrahedron*, 1967, **23**, 131.
9. H. Vorbrüggen, S. C. Pakrashi and C. Djerassi, *Annalen*, 1963, **668**, 57.
10. P. J. Ham and D. A. Whiting, *Chem. and Ind.*, 1970, 1379.
11. K. Schaffner, L. Caglioti, D. Arigoni and O. Jeger, *Helv. Chim. Acta*, 1958, **41**, 152.
12. H. Fazakerley, T. G. Halsall and E. R. H. Jones, *J. Chem. Soc.*, 1959, 1877.
13. H. Ageta, K. Iwata, Y. Irai, Y. Tsuda, K. Isobe and S. Fukushima, *Tetrahedron Letters*, 1966, 5679.
14. H. Ageta and K. Shiojima, *Chem. Comm.*, 1968, 1372.
15. I. Yosioka, T. Nakanishi and I. Kitagawa, *Chem. Pharm. Bull. (Japan)*, 1969, **17**, 291 and references therein.

† Derivatives of both hopane and 21-epihopane (isohopane) occur naturally, frequently in the same plant. The situation regarding C_{21} stereochemistry in these compounds has been confused in the past. The latest work [2] indicates that hopane has the (21R) or 21β-H configuration shown, and this is corroborated by the X-ray structure determination of adipedatol [3].

T46

Steroids and trimethylsteroids.

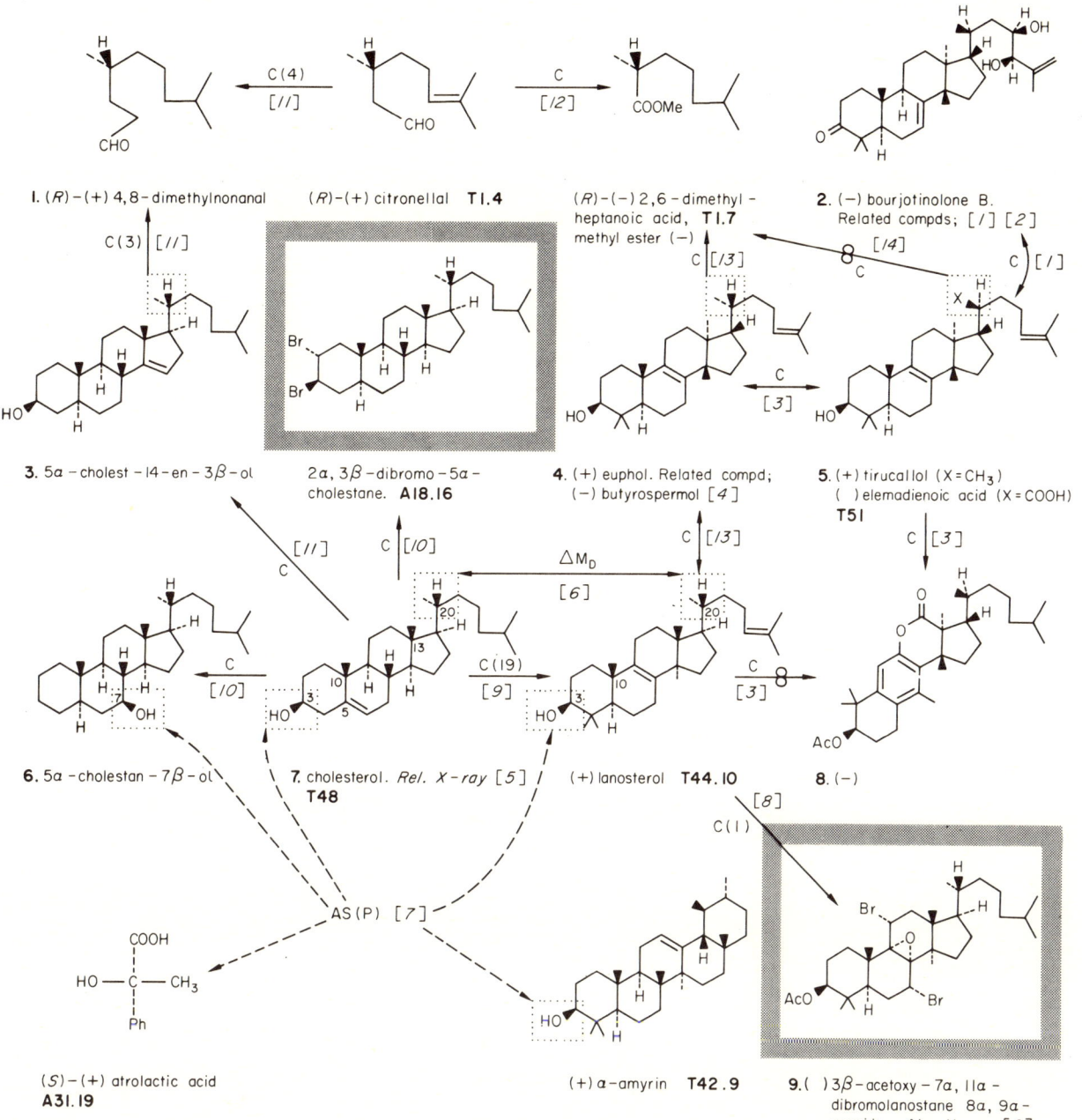

1. G. J. W. Breen, E. Ritchie, W. T. L. Sidwell and W. C. Taylor, *Austral. J. Chem.*, 1966, **19**, 455.
2. W. R. Chan, D. R. Taylor and T. Yee, *J. Chem. Soc. (C)*, 1970, 311.
3. E. Ménard, H. Wyler, A. Hiestand, D. Arigoni, O. Jeger and L. Ruzicka, *Helv. Chim. Acta*, 1955, **38**, 1517.
4. D. S. Irvine, W. Lawrie, A. S. McNab and F. S. Spring, *Chem. and Ind.*, 1955, 626.
5. C. H. Carlisle and D. Crowfoot, *Proc. Roy. Soc. (A)*, 1945, **184**, 64.
6. C. S. Barnes, D. H. R. Barton, J. S. Fawcett and B. R. Thomas, *J. Chem. Soc.*, 1953, 576.
7. W. G. Dauben, D. F. Dickel, O. Jeger and V. Prelog, *Helv. Chim. Acta*, 1953, **36**, 325.
8. J. K. Fawcett and J. Trotter, *J. Chem. Soc. (B)*, 1966, 174.
9. R. B. Woodward, A. A. Patchett, D. H. R. Barton, D. A. J. Ives and R. B. Kelly, *J. Chem. Soc.*, 1957, 1131.
10. C. J. W. Brooks in Rodd's 'Chemistry of Carbon Compounds', 2nd edn., vol. IID.
11. B. Riniker, D. Arigoni and O. Jeger, *Helv. Chim. Acta*, 1954, **37**, 546.
12. J. v. Braun and W. Teuffert, *Ber.*, 1929, **62**, 235.
13. D. Arigoni, R. Viterbo, M. Dünnenberger, O. Jeger and L. Ruzicka, *Helv. Chim. Acta*, 1954, **37**, 2306.
14. D. Arigoni, O. Jeger and L. Ruzicka, *Helv. Chim. Acta*, 1955, **38**, 222.

T⁴⁷

In addition to **A18.16**, **T.32.13** and **A46.12**, the following Bijvoet X-ray determinations have been carried out on steroids:

17β-hydroxy-9β,10a-androst-4-en-3-one [*1*]; 17aβ-hydroxy-17a-methyl-19-nor-9β,10a-D-homo-androst-4-en-3-one [*2*]; 4a-bromo-5a-androst-2-ene-1,17-dione [*3*]; 5a-bromo-6β,19-oxidopregnan-3β-ol-20-one [*4*]; 6β-bromoprogesterone [*5*]; 12a-bromo-11β-hydroxyprogesterone [*6*]; pancuronium bromide (3a,17β-diacetoxy-2β,16β-dipiperidino-5a-androstane dimethobromide) [*7*]; 3β-bromoacetoxy-16a-ethyl-16^2-cyano-16^2,21-cyclo-5a-pregna-17,21-diene [*8*].

A comprehensive compilation of steroid X-ray determinations (absolute and relative) is to appear in *Atlas of Steroid Structure* (D. A. Norton, Editor-in-Chief), in preparation (1971).

Unequivocal chemical correlations exist [*9*] between cholesterol and related steroids on page T46 and the following steroids which appear in other parts of the 'Atlas' **A53.5, T20.18, T23.17, T26.19, T27.13, T32.9, T32.10, T32.19, T35.3, T35.14, T36.14, T36.17, T38.5, T40.7, T43.3, T49.5, T53.11, T57.7, K34.2, K34.3, K34.4, Y21.7**.

1. W. E. Oberhansli and J. M. Robertson, *Helv. Chim. Acta*, 1967, **50**, 53.
2. R. T. Puckett, G. A. Sim, A. D. Cross and J. B. Siddall, *J. Chem. Soc. (B)*, 1967, 783.
3. J. R. Hanson, T. D. Organ, G. A. Sim and D. N. J. White, *J. Chem. Soc. (C)*, 1970, 2111.
4. E. M. Gopalakrishna, A. Cooper and D. A. Norton, *Acta Cryst.*, 1969, **B25**, 2473.
5. E. M. Gopalakrishna, A. Cooper and D. A. Norton, *Acta Cryst.*, 1969, **B25**, 639.
6. A. Cooper and D. A. Norton, *Acta Cryst.*, 1968, **B24**, 811.
7. D. S. Savage, A. F. Cameron, G. Ferguson, C. Hannaway and I. R. Mackay, *J. Chem. Soc. (B)*, 1971, 410.
8. D. R. Pollard and F. R. Ahmed, *Acta Cryst.*, 1971, **B27**, 1976.
9. C. J. W. Brooks in Rodd's 'Chemistry of Carbon Compounds', 2nd edn., vol. IID.

---

14. W. G. Dauben and P. Baumann, *Tetrahedron Letters*, 1961, 565. ▶
15. S. Bergström, A. Lardon and T. Reichstein, *Helv. Chim. Acta*, 1949, **32**, 1617.
16. W. Bergmann and H. A. Stansbury, *J. Org. Chem.*, 1944, **9**, 281.

† For other D vitamins and their precursors and for further interrelationships between these and mono- and bicyclic intermediates, see T. M. Dawson, P. S. Littlewood, B. Lythgoe and A. K. Saksena, *J. Chem. Soc. (C)*, 1971, 2960 and references therein.

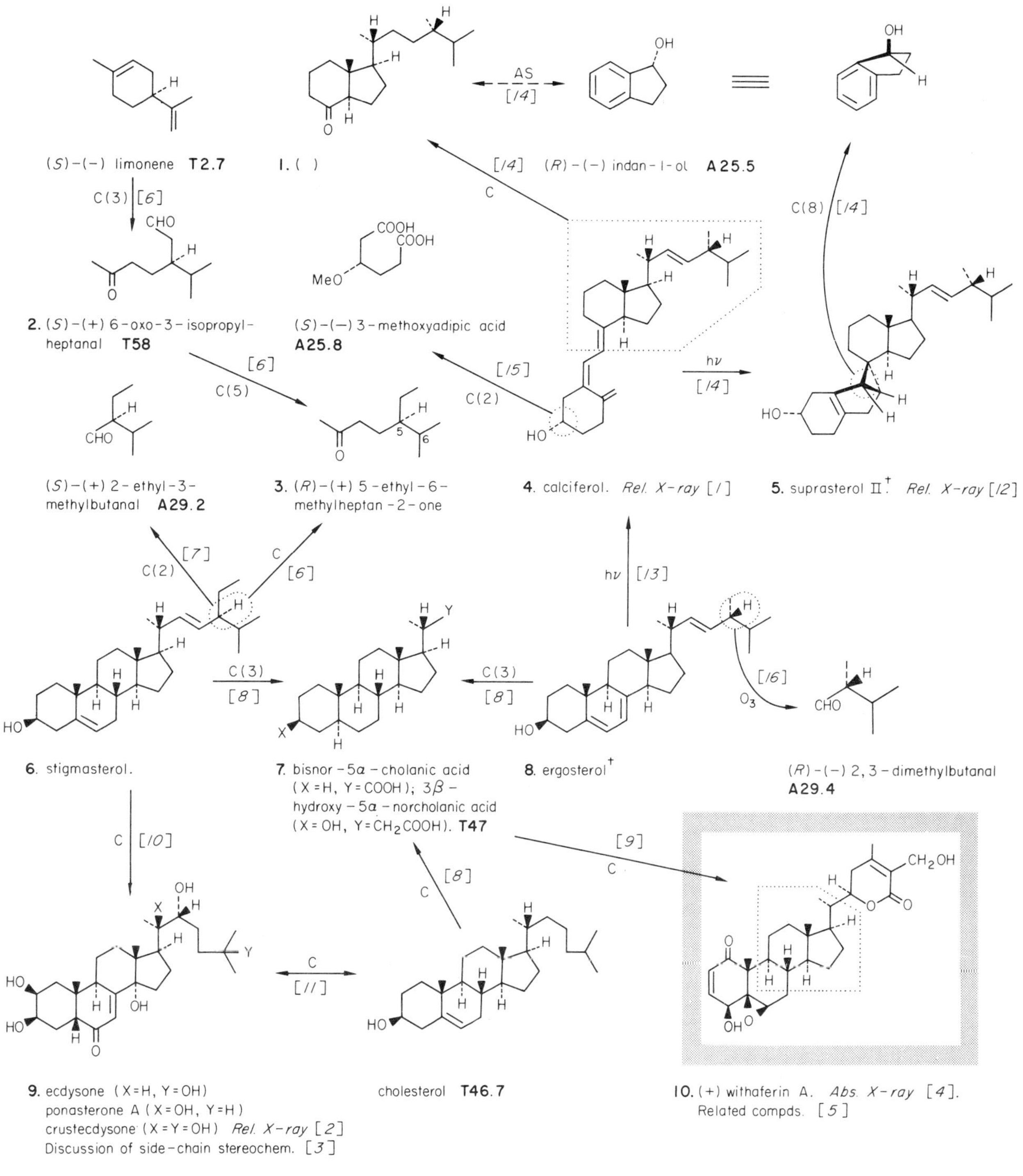

Steroids (contd.).

(S)-(−) limonene **T2.7**

1. ( )

[14] (R)-(−) indan-1-ol **A25.5**

2. (S)-(+) 6-oxo-3-isopropyl-heptanal **T58**

(S)-(−) 3-methoxyadipic acid **A25.8**

(S)-(+) 2-ethyl-3-methylbutanal **A29.2**

3. (R)-(+) 5-ethyl-6-methylheptan-2-one

4. calciferol. *Rel. X-ray* [1]

5. suprasterol II.† *Rel. X-ray* [12]

6. stigmasterol.

7. bisnor-5α-cholanic acid (X=H, Y=COOH); 3β-hydroxy-5α-norcholanic acid (X=OH, Y=CH₂COOH). **T47**

8. ergosterol†

(R)-(−) 2,3-dimethylbutanal **A29.4**

9. ecdysone (X=H, Y=OH)
ponasterone A (X=OH, Y=H)
crustecdysone (X=Y=OH) *Rel. X-ray* [2]
Discussion of side-chain stereochem. [3]

cholesterol **T46.7**

10. (+) withaferin A. *Abs. X-ray* [4].
Related compds. [5]

---

1. D. C. Hodgkin, B. M. Rimmer, J. D. Dunitz and K. N. Trueblood, *J. Chem. Soc.*, 1963, 4945.
2. R. Huber and W. Hoppe, *Chem. Ber.*, 1965, **98**, 2403.
3. M. Koreeda, D. A. Schooley, K. Nakanishi and H. Hagiwara, *J. Amer. Chem. Soc.*, 1971, **93**, 4084.
4. A. T. McPhail and G. A. Sim, *J. Chem. Soc. (B)*, 1968, 962.
5. I. Kirson, E. Glotter, D. Lavie and A. Abraham, *J. Chem. Soc. (C)*, 1971, 2032 and references therein: R. Tschesche, K. Annen and P. Welzel, *Chem. Ber.*, 1971, **104**, 3556.
6. Y. Kishida, *Chem. and Ind.*, 1960, 465.
7. G. Slomp and J. L. Johnson, *J. Amer. Chem. Soc.*, 1958, **80**, 915.
8. E. Fernholz and P. N. Chakravorty, *Ber.*, 1934, **67**, 2021.
9. D. Lavie, S. Greenfield and E. Glotter, *J. Chem. Soc. (C)*, 1966, 1753.
10. H. Mori, K. Shibata, K. Tsuneda and M. Sawai, *Tetrahedron*, 1971, **27**, 1157.
11. J. B. Siddall, A. D. Cross and J. H. Fried, *J. Amer. Chem. Soc.*, 1966, **88**, 862.
12. C. P. Sanderson and D. C. Hodgkin, *Tetrahedron Letters*, 1961, 573.
13. L. Velluz, G. Amiard and B. Goffinet, *Bull. Soc. chim. France*, 1955, 1341.

References continued on page 124

T 49
Steroid aglycones, sapogenins and alkaloids.

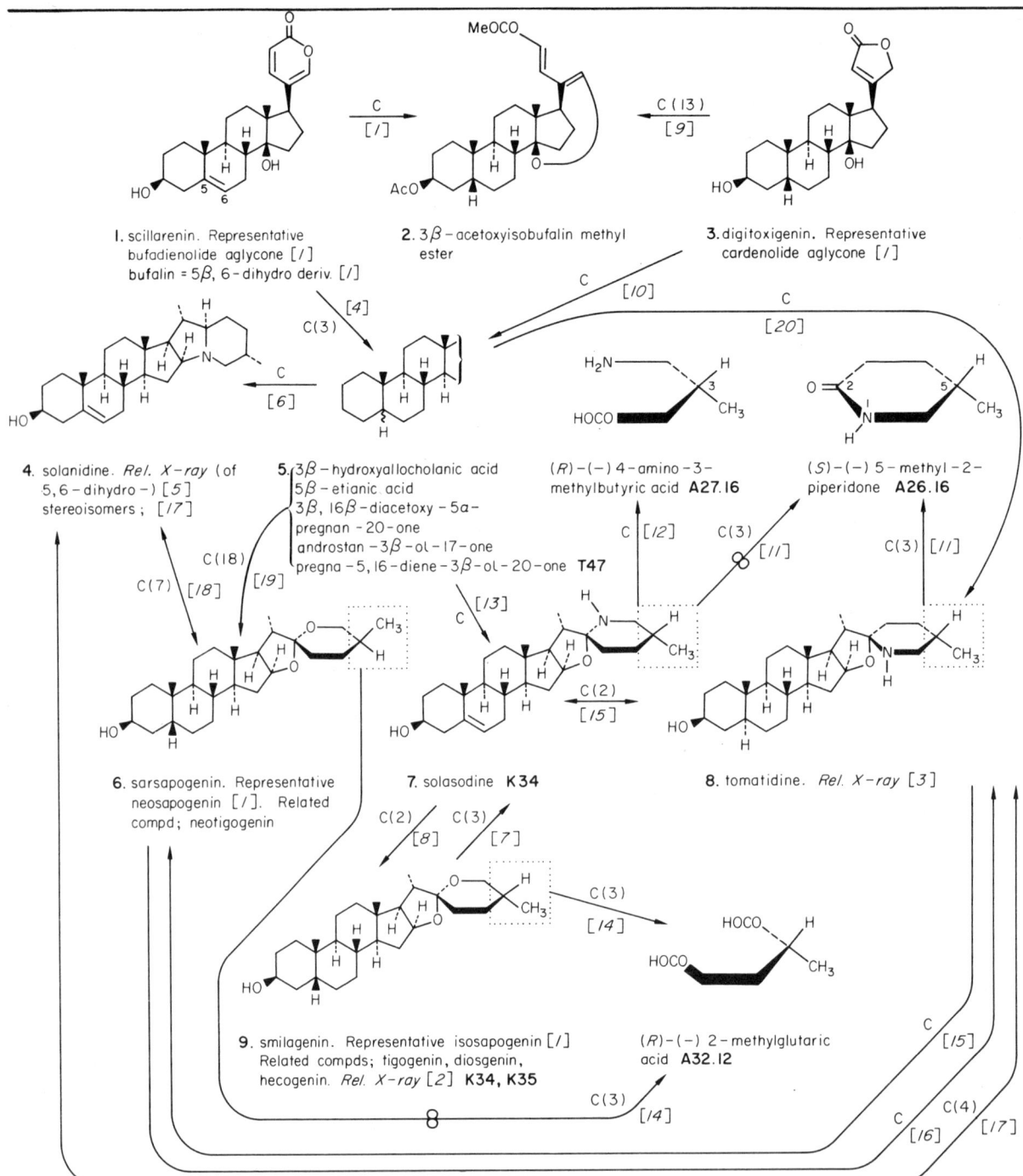

1. scillarenin. Representative bufadienolide aglycone [/] bufalin = 5β, 6-dihydro deriv. [/]
2. 3β-acetoxyisobufalin methyl ester
3. digitoxigenin. Representative cardenolide aglycone [/]
4. solanidine. Rel. X-ray (of 5,6-dihydro-) [5] stereoisomers; [17]
5. 3β-hydroxyallocholanic acid
   5β-etianic acid
   3β,16β-diacetoxy-5α-pregnan-20-one
   androstan-3β-ol-17-one
   pregna-5,16-diene-3β-ol-20-one T47
6. sarsapogenin. Representative neosapogenin [/]. Related compd; neotigogenin
7. solasodine K34
8. tomatidine. Rel. X-ray [3]
9. smilagenin. Representative isosapogenin [/] Related compds; tigogenin, diosgenin, hecogenin. Rel. X-ray [2] K34, K35

(R)-(−) 4-amino-3-methylbutyric acid A27.16
(S)-(−) 5-methyl-2-piperidone A26.16
(R)-(−) 2-methylglutaric acid A32.12

1. K. Meyer, *Helv. Chim. Acta*, 1949, **37**, 1238; C. J. W. Brooks in Rodd's 'Chemistry of Carbon Compounds', 2nd edn., vol. IID.
2. E. A. O'Donnell and M. F. C. Ladd, *Chem. and Ind.*, 1963, 1984.
3. O. Kennard, L. R. di Sanseverino and J. S. Rollett, *J. Chem. Soc. (C)*, 1967, 956.
4. A. Stoll, A. Hofmann and J. Peyer, *Helv. Chim. Acta*, 1935, **18**, 1247.
5. E. Höhne, K. Schreiber, H. Ripperger and H. H. Worch, *Tetrahedron*, 1966, **22**, 673.
6. S. V. Kessar, A. L. Rampal, S. S. Gandhi and R. K. Mahajan, *Tetrahedron*, 1971, **27**, 2153.
7. F. C. Uhle, *J. Org. Chem.*, 1962, **27**, 656.
8. L. H. Briggs and T. O'Shea, *J. Chem. Soc.*, 1952, 1654.
9. T. R. Kasturi, G. R. Pettit and K. A. Jaeggi, *Chem. Comm.*, 1967, 644.
10. W. A. Jacobs and R. C. Elderfield, *J. Biol. Chem.*, 1935, **108**, 497.
11. K. Schreiber, *Annalen*, 1965, **682**, 219.
12. K. Schreiber, *Chem. Ber.*, 1965, **98**, 323.
13. K. Schreiber, A. Walther and H. Rönsch, *Tetrahedron*, 1964, **20**, 1939.
14. I. Scheer, R. B. Kostic and E. Mosettig, *J. Amer. Chem. Soc.*, 1953, **75**, 4871.
15. Y. Sato, H. G. Latham, L. H. Briggs and R. N. Seelye, *J. Amer. Chem. Soc.*, 1957, **79**, 6089.
16. F. C. Uhle and J. A. Moore, *J. Amer. Chem. Soc.*, 1954, **76**, 6412.
17. Y. Sato and H. G. Latham, *J. Amer. Chem. Soc.*, 1956, **78**, 3146.
18. F. C. Uhle and W. A. Jacobs, *J. Biol. Chem.*, 1945, **160**, 243.
19. F. Sondheimer, Y. Mazur and N. Danieli, *J. Amer. Chem. Soc.*, 1960, **82**, 5889.
20. K. Schreiber and G. Adam, *Annalen*, 1963, **666**, 155.

T 50
9,19-cyclosteroids and trimethylsteroids; protostanes.

1. (+) cycloeucalenol **K36**
2. (+) cycloartenol **K36**
3. ( ) ambonic acid
   (S)-(+) propane-1,2-diol **A14.8**
4. (+) cycloneolitisin
   (+) lanosterol **T44.10**
5. (+) eburicoic acid
6. (+) cucurbitatin.A. CD of derivs; [3]
7. (+) polyporenic acid A
8. (+) echinodol
9. 3β-hydroxyprotost-13(17)-ene
10. (−) fusidic acid. Abs. X-ray [4]
    Related compd;
    (−) helvolic acid [5]
11. (+) cephalosporin P₁. By ORD of derivs. [6]
12. ( ) cyclograndisolide. Abs. X-ray [7]
13. ( ) gorgosterol. Rel. X-ray [1]
    AC by ORD of 3-ketone (octant rule) [1]

1. N. C. Ling, R. L. Hale and C. Djerassi, *J. Amer. Chem. Soc.*, 1970, **92**, 5281.
2. S. Corsano and E. Mincione, *Chem. Comm.*, 1968, 738.
3. G. Snatzke, P. R. Enslin, C. W. Holzapfel and K. B. Norton, *J. Chem. Soc. (C)*, 1967, 972.
4. A. Cooper and D. C. Hodgkin, *Tetrahedron*, 1968, **24**, 909.
5. S. Iwasaki, M. I. Sair, H. Igarashi and S. Okuda, *Chem. Comm.*, 1970, 1119.
6. T. S. Chou, E. J. Eisenbraun and R. T. Rapala, *Tetrahedron Letters*, 1967, 409.
7. F. H. Allen, J. P. Kutney, J. Trotter and N. D. Westcott, *Tetrahedron Letters*, 1971, 283.
8. J. S. G. Cox, F. E. King and T. J. King, *J. Chem. Soc.*, 1959, 514.
9. D. H. R. Barton, R. P. Budhiraja and J. F. McGhie, *Proc. Chem. Soc.*, 1963, 170.
10. T. G. Halsall and R. Hodges, *J. Chem. Soc.*, 1954, 2385.
11. J. S. E. Holker, A. D. G. Powell, A. Robertson, J. J. H. Simes, R. S. Wright and R. M. Gascoigne, *J. Chem. Soc.*, 1953, 2422.
12. D. H. R. Barton, C. F. Garbers, D. Giacopello, R. G. Harvey, J. Lessard and D. R. Taylor, *J. Chem. Soc. (C)*, 1969, 1050.
13. S. Corsano and E. Mincione, *Ann. Chim. (Rome)*, 1967, **57**, 522.
14. E. Ritchie, R. G. Senior and W. C. Taylor, *Austral. J. Chem.*, 1969, **22**, 2371.
15. F. T. Bond, D. S. Fullerton, L. A. Sciuchetti and P. Catalfomo, *J. Amer. Chem. Soc.*, 1966, **88**, 3882.
16. T. Hattori, H. Igarashi, S. Iwasaki and S. Okuda, *Tetrahedron Letters*, 1969, 1023.

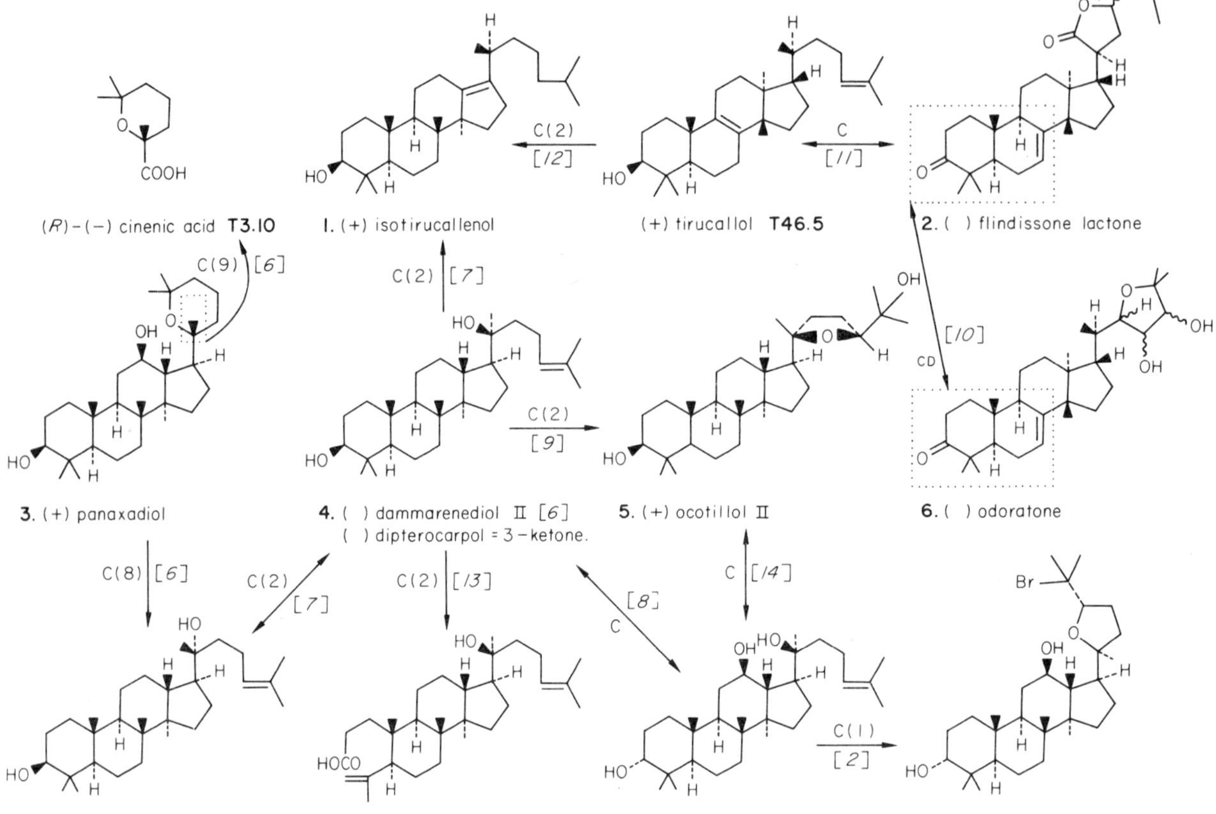

Steroids (contd.).

(R)-(−) cinenic acid T3.10

1. (+) isotirucallenol

(+) tirucallol T46.5

2. (−) flindissone lactone

3. (+) panaxadiol

4. (−) dammarenediol II [6]
(−) dipterocarpol = 3-ketone.

5. (+) ocotillol II

6. (−) odoratone

7. (−) dammarenediol I [6]

8. (+) dammarenolic acid

9. (−) betulafolienetriol

10. (−) betulafolienetriol deriv. Rel. X-ray [2]

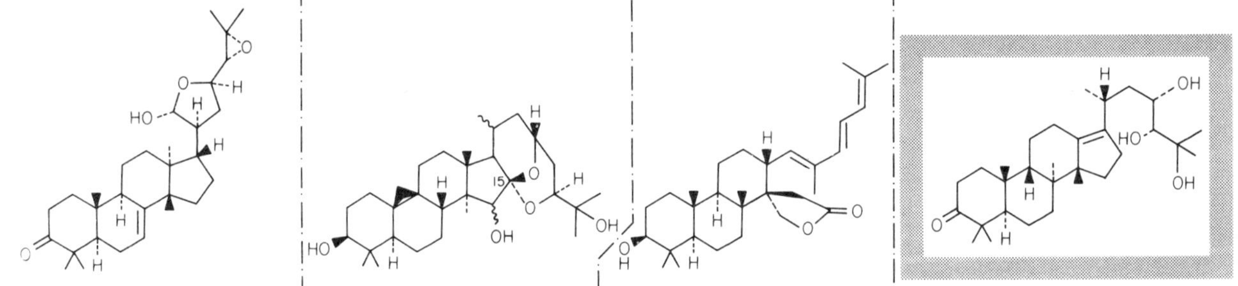

11. (−) melianone. By ORD (octant rule, comparison with steroids) [1]

12. (+) cimigenol. By CD of 15-ketones [3]

13. (−) ebelin lactone. Rel. X-ray [4]. AC by ORD of derivs. [4]

14. (+) alisol A. Abs. X-ray [5]

---

1. D. Lavie, M. K. Jain and I. Kirson, *J. Chem. Soc. (C)*, 1967, 1347.
2. O. Tanaka, N. Tanaka, T. Ohsawa, Y. Iitaka and S. Shibata, *Tetrahedron Letters*, 1968, 4235.
3. S. Corsano, J. M. Mellor and G. Ourisson, *Chem. Comm.*, 1965, 185.
4. G. A. Barclay, R. A. Eade, H. V. Simes, J. J. H. Simes and J. C. Taylor, *Chem. and Ind.*, 1963, 1206.
5. T. Murata, M. Shinohara, T. Hirata, K. Kamiya, M. Nishikawa and M. Miyamoto, *Tetrahedron Letters*, 1968, 103.
6. M. Nagai, O. Tanaka and S. Shibata, *Tetrahedron Letters*, 1966, 4797.
7. J. S. Mills, *J. Chem. Soc.*, 1956, 2196.
8. F. G. Fischer and N. Seiler, *Annalen*, 1959, 626, 185.
9. E. W. Warnhoff and C. M. M. Halls, *Canad. J. Chem.*, 1965, 43, 3311.
10. W. R. Chan, D. R. Taylor, G. Snatzke and H-W. Fehlhaber, *Chem. Comm.*, 1967, 548.
11. A. J. Birch, D. J. Collins, S. Muhammad and J. P. Turnbull, *J. Chem. Soc.*, 1963, 2762.
12. D. Arigoni, O. Jeger and L. Ruzicka, *Helv. Chim. Acta*, 1955, 38, 222.
13. D. Arigoni, D. H. R. Barton, R. Bernasconi, C. Djerassi, J. S. Mills and R. E. Wolff, *J. Chem. Soc.*, 1960, 1900.
14. M. Nagai, N. Tanaka, S. Ichikawa and O. Tanaka, *Tetrahedron Letters*, 1968, 4239.

Limonoid bitter principles. Review: [14]

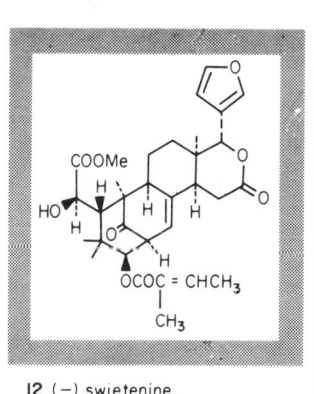
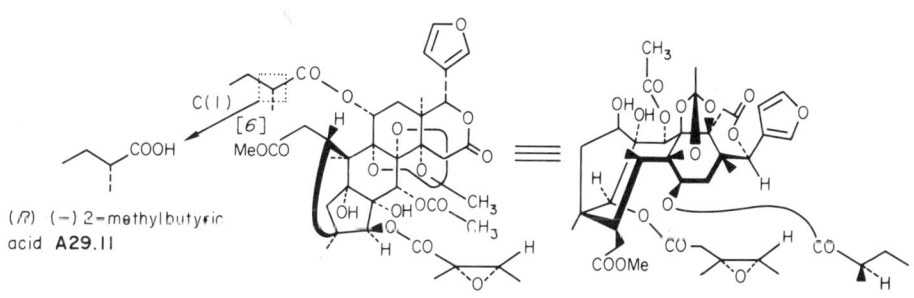

1. (+) β-amyrin T42.6
2. (−) deoxyandirobin
3. (−) methyl angolensate
4. (−) methyl ivorensate
5. (−) ichangin
6. (+) gedunin. Rel. X-ray [1]
7. (−) khivorin
8. (−) mexicanolide. Rel. X-ray [15]
9. (−) limonin. Rel. X-ray [2]. AC by ORD; [3]
10. (−) obacunone
11. (−) veprisone
12. (−) swietenolide. AC by CD of derivs; [17]
13. (−) swietenine. Abs. X-ray [4]
14. (−) utilin. Rel. X-ray [5]

(R) (−) 2-methylbutyric acid A29.11

1. S. A. Sutherland, G. A. Sim and J. M. Robertson, *Proc. Chem. Soc.*, 1962, 222.
2. S. Arnott, A. W. Davie, J. M. Robertson, G. A. Sim and D. G. Watson, *J. Chem. Soc.*, 1961, 4183.
3. D. Arigoni, D. H. R. Barton, E. J. Corey and O. Jeger, *Experientia*, 1960, 16, 41.
4. A. T. McPhail and G. A. Sim, *J. Chem. Soc. (B)*, 1966, 318.
5. H. R. Harrison, O. J. R. Hodder, C. W. L. Bevan, D. A. H. Taylor and T. G. Halsall, *Chem. Comm.*, 1970, 1388.
6. D. H. Calam and D. A. H. Taylor, *J. Chem. Soc. (C)*, 1966, 949.
7. A. Akisanya, E. O. Arene, C. W. L. Bevan, D. E. U. Ekong, N. M. Nwaji, J. I. Okugun, J. W. Powell and D. A. H. Taylor, *J. Chem. Soc. (C)*, 1966, 506.
8. D. E. U. Ekong and E. O. Olagbemi, *J. Chem. Soc. (C)*, 1966, 944.
9. E. K. Adesogan and D. A. H. Taylor, *Chem. Comm.*, 1969, 89.
10. A. Akisanya, C. W. L. Bevan, T. G. Halsall, J. W. Powell and D. A. H. Taylor, *J. Chem. Soc.*, 1961, 3705.
11. D. L. Dreyer, *J. Org. Chem.*, 1966, 31, 2279.
12. T. Kubota, T. Matsuura, T. Tokoroyama, T. Kamikawa and T. Matsumoto, *Tetrahedron Letters*, 1961, 325.
13. T. R. Govindachari, B. S. Joshi and V. N. Sundrarajan, *Tetrahedron*, 1964, 20, 2985.
14. D. L. Dreyer, *Fortschritte Chem. Org. Naturstoffe*, 1968, 26, 190.
15. S. A. Adeoye and D. A. Bekoe, *Chem. Comm.*, 1965, 30.
16. J. D. Connolly, I. M. S. Thornton and D. A. H. Taylor, *Chem. Comm.*, 1971, 17.
17. J. D. Connolly, R. McCrindle, K. H. Overton and W. D. C. Warnock, *Tetrahedron*, 1968, 24, 1507.

T 53
Further nortriterpenes; quassinoids.

1. ( ) havanensin. By CD of derivs. [1]
2. (−) mahoganin. By ORD [2] (octant rule) (?)
3. ( ) nimbin
4. ( ) pyronimbic acid. By ORD (cisoid diene) (?) [3]
5. ( ) physalin A. Abs. X-ray [4]
6. ( ) physalin B.
7. (+) odoratin. By CD [10]
8. (+) glaucanol
9. (+) glaucarubin
   (S)-(+) 2-hydroxy-2-methylbutyric acid A33.8
10. (+) simarolide. Abs. X-ray [9] Related compd; picrasin A [7]
11. 5α-cholestan-1-one T47
12. (+)
13. (−) chapparin. Also by ORD [8]
14. (S)-(+) 9,10-dihydro-3-hydroxy-4-methoxy-2,5,8-trimethylphenanthrene-1-acetic acid X5

---

1. W. R. Chan, J. A. Gibbs and D. R. Taylor, *Chem. Comm.*, 1967, 720.
2. D. P. Chakraborty, K. C. Das and C. F. Hammer, *Tetrahedron Letters*, 1968, 5015.
3. H. Ziffer, U. Weiss, G. R. Narayanan and R. V. Pachapurkar, *J. Org. Chem.*, 1966, **31**, 2691.
4. M. Kawai, T. Taga, K. Osaki and T. Matsuura, *Tetrahedron Letters*, 1969, 1087.
5. T. Matsuura and M. Kawai, *Tetrahedron Letters*, 1969, 1765.
6. J. Polonsky, C. Fouquey and M. A. Gaudemer, *Bull. Soc. chim. France*, 1964, 1827.
7. H. Hiniko, T. Ohta and T. Takemoto, *Chem. Pharm. Bull. Japan*, 1970, **18**, 1082.
8. T. R. Hollands, P. de Mayo, M. Nisbet and P. Crabbé, *Canad. J. Chem.*, 1965, **43**, 3008.
9. W. A. C. Brown and G. A. Sim, *Proc. Chem. Soc.*, 1964, 293.
10. W. R. Chan, D. R. Taylor and R. T. Aplin, *Chem. Comm.*, 1966, 576.

Carotenoids; ionone C$_{13}$ group.

T54

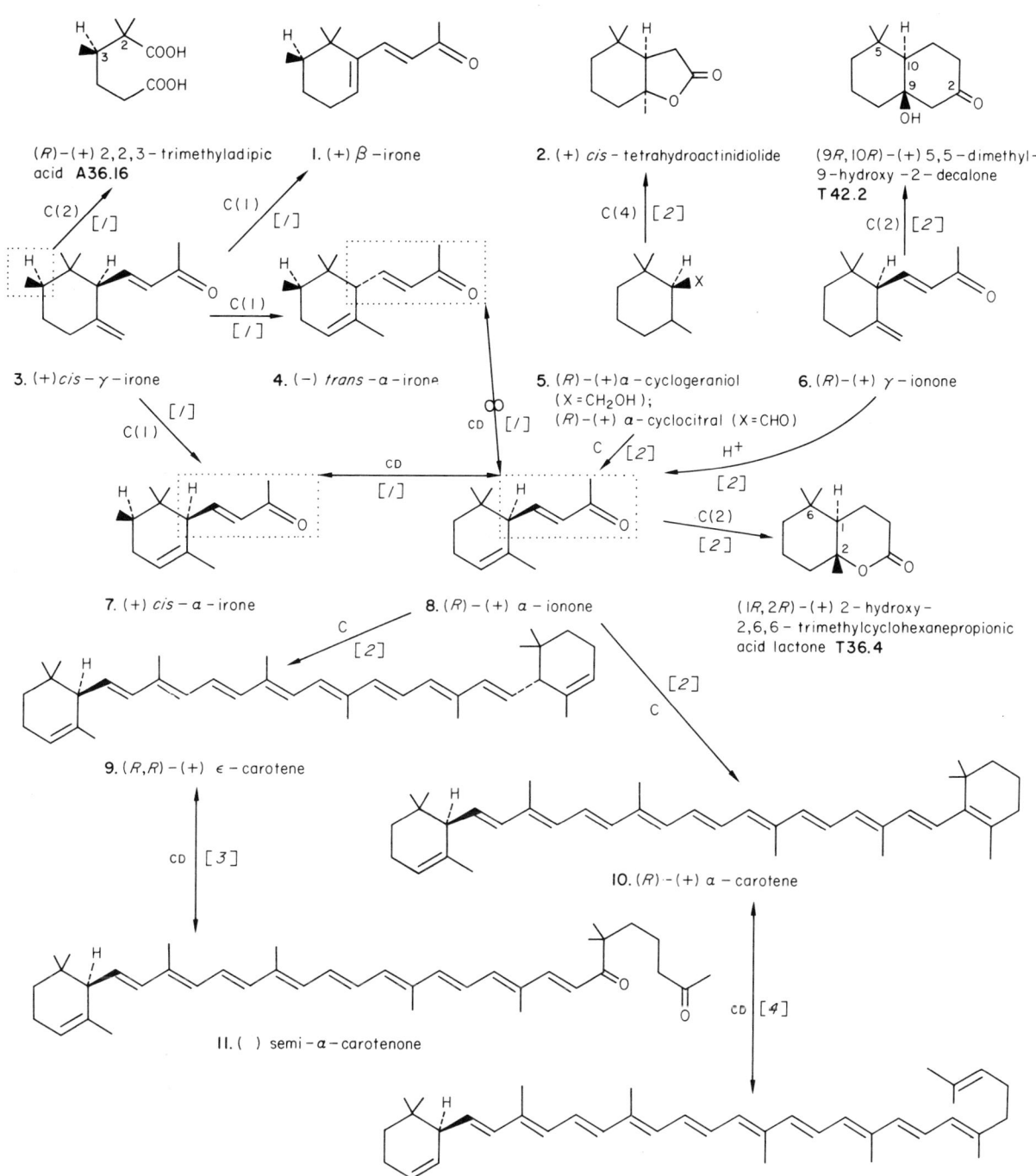

1. V. Rautenstrauch and G. Ohloff, *Helv. Chim. Acta*, 1971, **54**, 1776.
2. C. H. Eugster, R. Buchecher, Ch. Tscharner, G. Uhde and H. Ohloff, *Helv. Chim. Acta*, 1969, **52**, 1729; M. Ribi and C. H. Eugster, ibid., 1732.
3. R. Buchecher, H. Yokoyama and C. H. Eugster, *Helv. Chim. Acta*, 1970, **53**, 1210.
4. R. Buchecher and C. H. Eugster, *Helv. Chim. Acta*, 1971, **54**, 327.

T55 Further carotenoids.

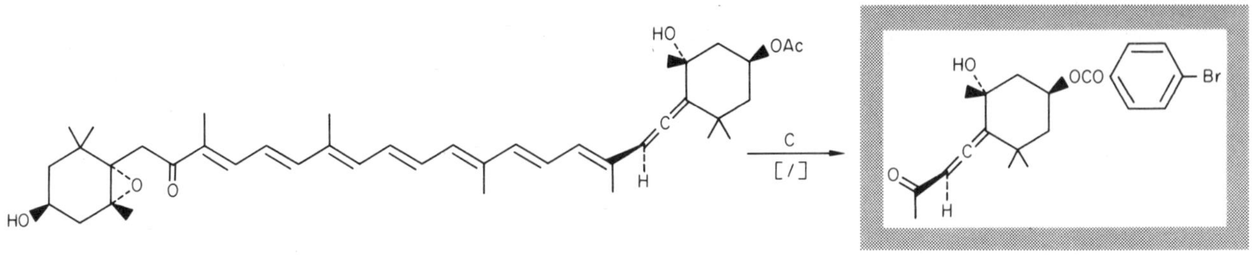

1. (−) fucoxanthin

2. (−) fucoxanthin deriv. *Abs. X-ray* [*1*]

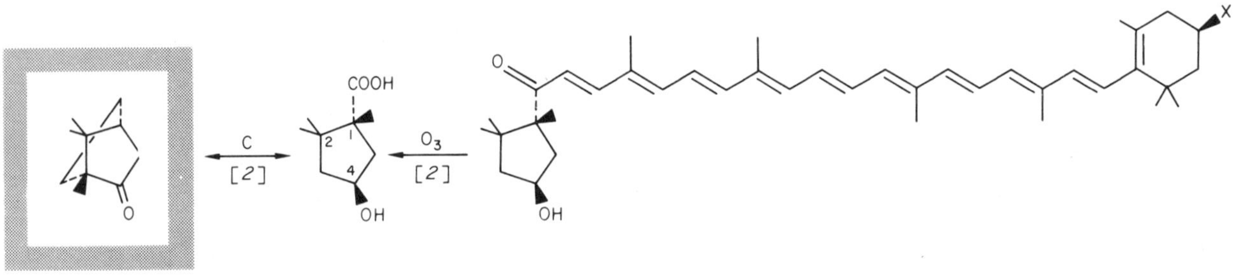

(+) camphor
TII.11

3. (1*R*,4*S*)-(−)-4-hydroxy-1,2,2-trimethylcyclopentane-1-carboxylic acid

4. (+) capsanthin (X = OH)
(−) kryptocapsin (X = H)

The following natural carotenoids have also been correlated with fucoxanthin or capsanthin by chemical or chiroptical means; neoxanthin, violaxanthin, zeaxanthin, β-cryptoxanthin, α-cryptoxanthin, rubixanthin, gazaniaxanthin, β-citraurin, reticulatoxanthin, alloxanthin, diatoxanthin [*1*]; lutein [*3*] and many others [*4*].

1. T. E. De Ville, M. B. Hursthouse, S. W. Russell and B. C. L. Weedon, *Chem. Comm.*, 1969, 1311.
2. B. C. L. Weedon, *Chem. Brit.*, 1967, 3, 424.
3. D. Goodfellow, G. P. Moss and B. C. L. Weedon, *Chem. Comm.*, 1970, 1578.
4. L. Bartlett, W. Klyne, W. P. Mose, P. M. Scopes, G. Galasko, A. K. Mallams, B. C. L. Weedon, J. Szabolcs and G. Tóth, *J. Chem. Soc. (C)*, 1969, 2527.

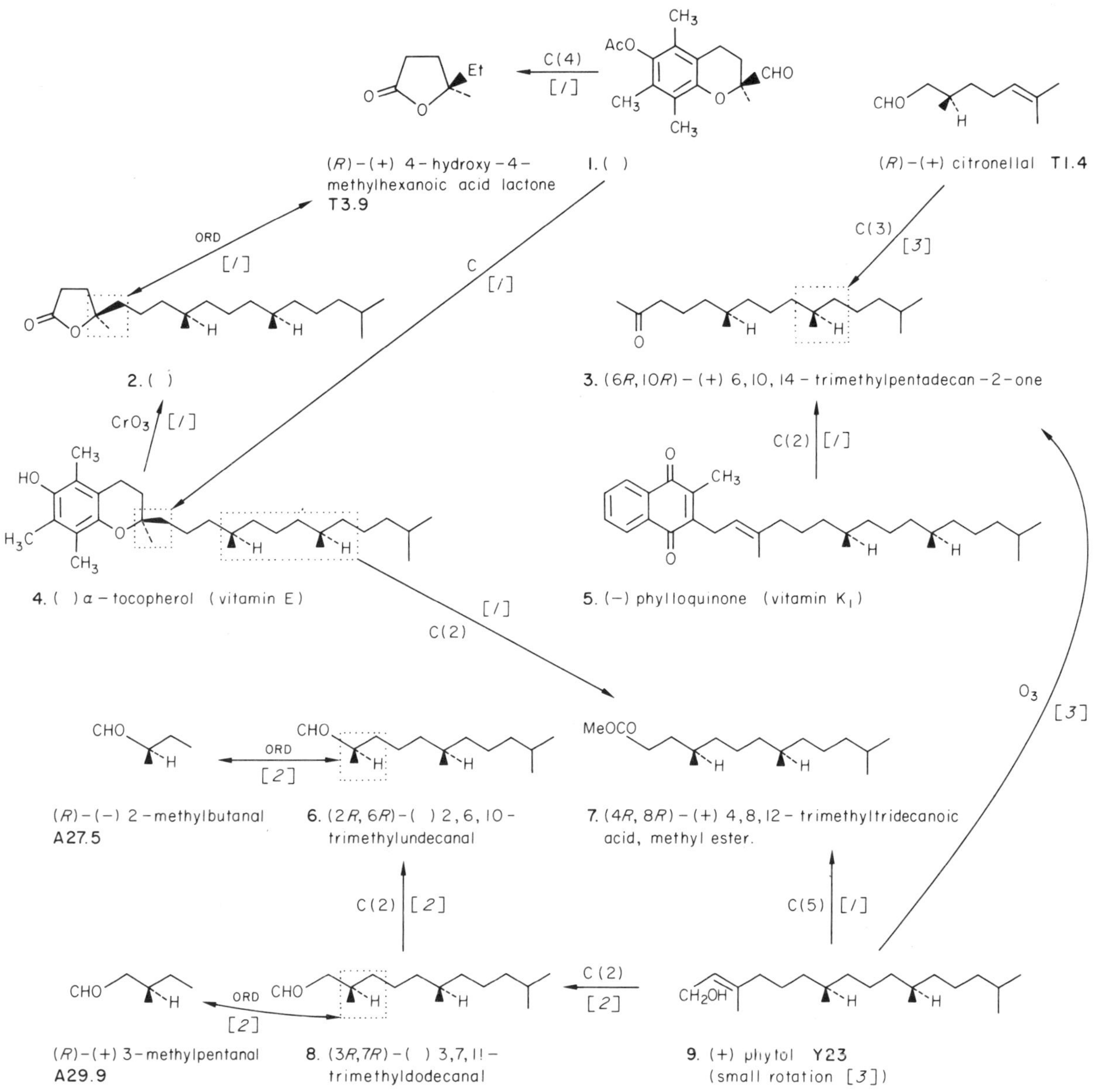

Phytol, tocopherol and phylloquinone.

1. ( )

2. ( )

3. (6R,10R)-(+) 6,10,14-trimethylpentadecan-2-one

4. ( ) α-tocopherol (vitamin E)

5. (−) phylloquinone (vitamin K₁)

6. (2R,6R)-( ) 2,6,10-trimethylundecanal

7. (4R,8R)-(+) 4,8,12-trimethyltridecanoic acid, methyl ester.

8. (3R,7R)-( ) 3,7,11-trimethyldodecanal

9. (+) phytol Y23 (small rotation [3])

(R)-(+) 4-hydroxy-4-methylhexanoic acid lactone T3.9

(R)-(+) citronellal T1.4

(R)-(−) 2-methylbutanal A27.5

(R)-(+) 3-methylpentanal A29.9

1. H. Mayer, U. Gloor, O. Isler, R. Rüegg and O. Wiss, *Helv. Chim. Acta*, 1964, **47**, 221 and references therein; H. Mayer, P. Schudel, R. Rüegg and O. Isler, ibid, 1963, **46**, 963.
2. P. Crabbé, C. Djerassi, E. J. Eisenbraun and S. Liu, *Proc. Chem. Soc.*, 1959, 264.
3. J. W. K. Burrell, L. M. Jackman and B. C. L. Weedon, *Proc. Chem. Soc.*, 1959, 263.

T 57

Miscellaneous terpenoids and norterpenoid derivatives.

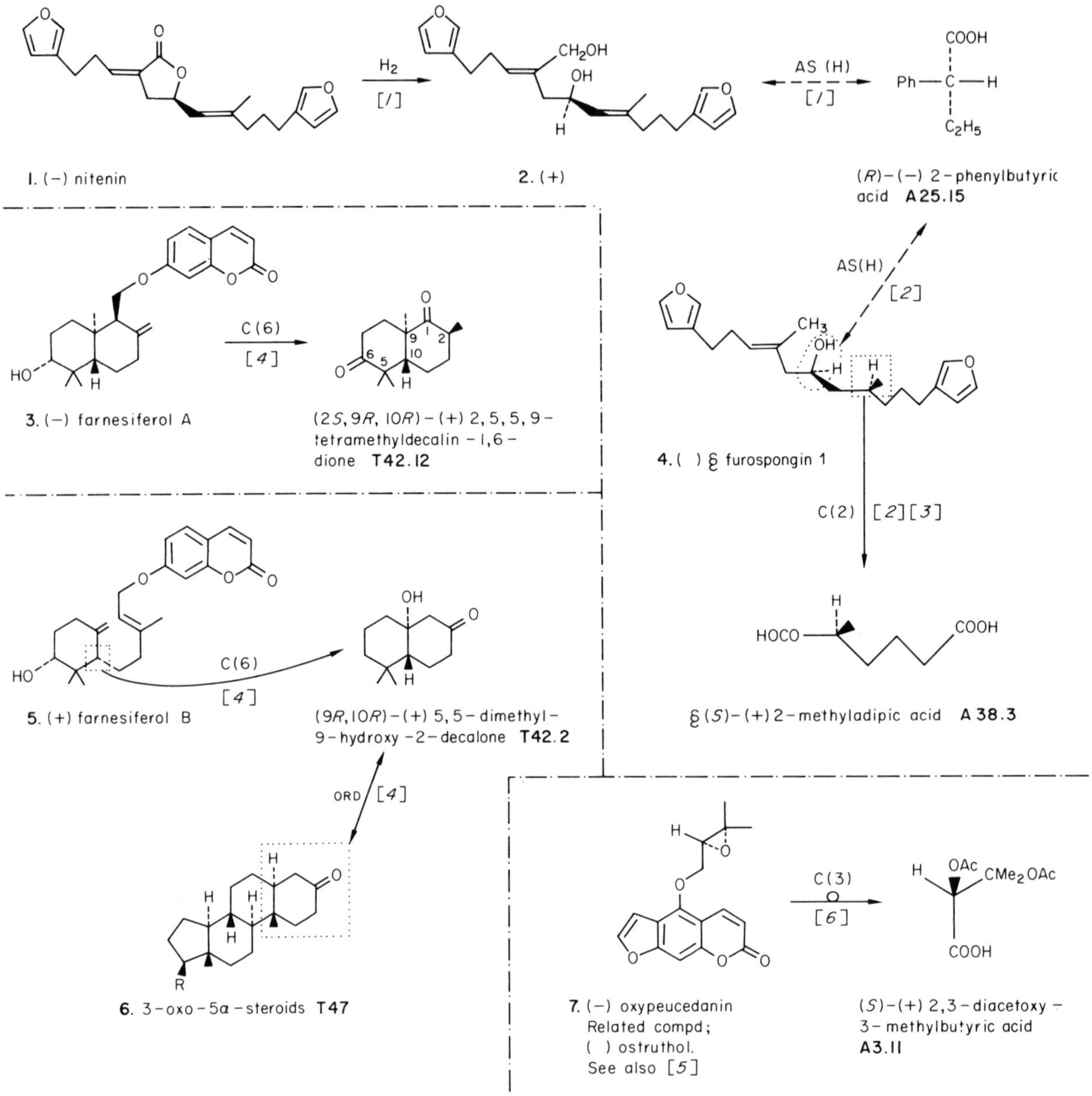
H2
[1]
AS (H)
[1]
COOH
Ph—C—H
C2H5
CH2OH
OH
1. (−) nitenin
2. (+)
(R)-(−) 2-phenylbutyric acid A25.15
C(6)
[4]
3. (−) farnesiferol A
(2S, 9R, 10R)-(+) 2,5,5,9-tetramethyldecalin-1,6-dione T42.12
AS(H)
[2]
CH3
OH
4. ( ) § furospongin 1
C(2)
[2][3]
HOCO
COOH
§ (S)-(+) 2-methyladipic acid A 38.3
C(6)
[4]
5. (+) farnesiferol B
(9R, 10R)-(+) 5,5-dimethyl-9-hydroxy-2-decalone T42.2
ORD
[4]
6. 3-oxo-5α-steroids T47
C(3)
[6]
OAc
CMe2OAc
COOH
7. (−) oxypeucedanin
Related compd;
( ) ostruthol.
See also [5]
(S)-(+) 2,3-diacetoxy-3-methylbutyric acid A3.11

1. E. Fattorusso, L. Minale, G. Sodano and E. Trivellone, *Tetrahedron*, 1971, **27**, 3909.
2. G. Cimino, S. De Stefano, L. Minale and E. Fattorusso, *Tetrahedron*, 1971, **27**, 4673.
3. E. Fattorusso, personal communication.
4. L. Caglioti, H. Naef, D. Arigoni and O. Jeger, *Helv. Chim. Acta*, 1958, **41**, 2278; 1959, **42**, 2557.
5. D. L. Dreyer, *J. Org. Chem.*, 1970, **35**, 2294.
6. B. E. Neilsen and J. Lemmich, *Acta Chem. Scand.*, 1969, **23**, 962.

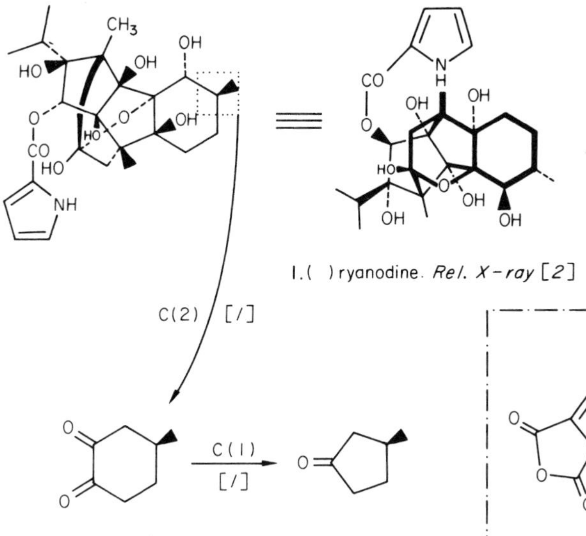

Miscellaneous terpenoids; ryanodine, solanone, nonadrides.

T 58

1. ( ) ryanodine. *Rel. X-ray* [2]
2. (R)-(+) solanone
3. (R)-(−) 8-methyl-5-isopropylnonan-2-one (tetrahydrosolanone)
4. (S)-( ) 4-methyl-cyclohexane-1,2-dione
   (S)-(−) 3-methylcyclopentanone **A36.6**
5. ( ) glauconic acid (X = OH) *Rel. X-ray* [4]
   ( ) glauconic acid (X = H)
   (2S,3S)-(+) 2,3-diethylglutaric acid **A45.11**
   (S)-(+) 6-oxo-3-isopropylheptanal **T48.2**
6. ( ) byssochlamic acid. *Rel. X-ray* [5] 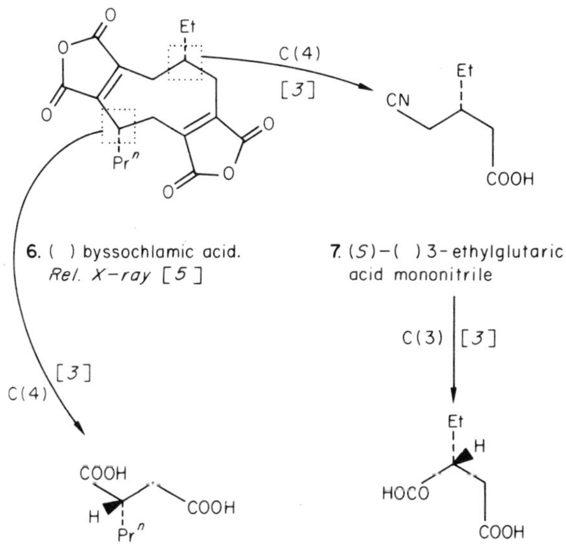
7. (S)-( ) 3-ethylglutaric acid mononitrile
   (S)-(−) n-propylsuccinic acid **A28.9**
   (R)-(+) ethylsuccinic acid **A28.9**
   (S)-(+) 3-methylpent-1-yn-3-ol **A33.4** 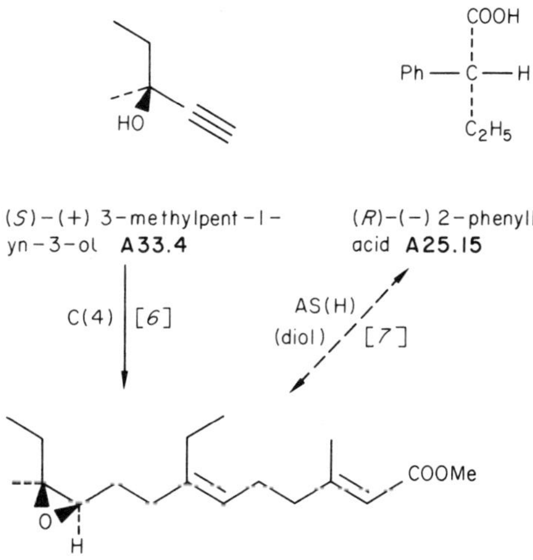
   (R)-(−) 2-phenylbutyric acid **A25.15**
8. (+) C_{18} juvenile hormone. Also by CD [9]

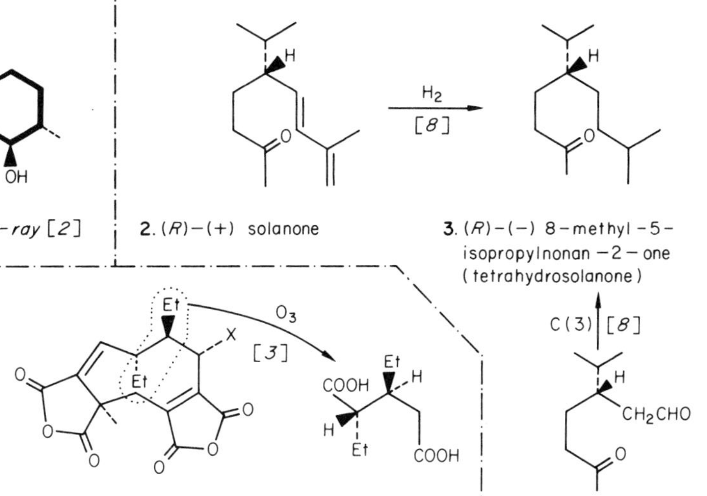

---

1. K. Wiesner, *Pure Appl. Chem.*, 1963, **7**, 285.
2. S. N. Srivastava and M. Przybylska, *Canad. J. Chem.*, 1968, **46**, 795.
3. D. H. R. Barton, L. D. S. Godhino and J. K. Sutherland, *J. Chem. Soc.*, 1965, 1779; J. E. Baldwin, D. H. R. Barton, and J. K. Sutherland, ibid., 1787.
4. G. Ferguson, G. A. Sim and J. M. Robertson, *Proc. Chem. Soc.*, 1962, 385.
5. I. C. Paul, G. A. Sim, T. A. Hamor and J. M. Robertson, *J. Chem. Soc.*, 1963, 5502.
6. D. J. Faulkner and M. R. Petersen, *J. Amer. Chem. Soc.*, 1971, **93**, 3766.
7. A. S. Meyer, E. Hanzmann and R. C. Murphy, *Proc. Nat. Acad. Sci. USA*, 1971, **68**, 2312.
8. R. R. Johnson and J. A. Nicholson, *J. Org. Chem.*, 1965, **30**, 2918.
9. K. Nakanishi, D. A. Schooley, M. Koreeda and J. Dillon, *Chem. Comm.*, 1971, 1235.

-K-

Alkaloids

Introductory Notes to Chapter K

Sources

The major reference work dealing with alkaloid chemistry is *The Alkaloids* (R. H. F. Manske and H. L. Holmes, Eds., Academic Press, 1950-1970), in 13 volumes. A recent one-volume text (*Chemistry of the Alkaloids*, S. W. Pelletier, Ed., Van Nostrand, 1970), was also consulted extensively in preparing this chapter, and the arrangement of material largely follows the latter book.

Alkaloids found in other chapters

Some alkaloids, for reasons of convenience, appear elsewhere in the 'Atlas'. In addition, the term 'Alkaloid' is subject to varying interpretation, and there are a few nitrogenous substances listed elsewhere which may be considered as alkaloids in the widest sense but not according to the narrower definitions. The following is a list, in alphabetical order, of compounds falling into these two categories.

Actinidine **T13.17**
Adrenaline **A22.14**
Amphetamine **A5.10**
Boschniakine **T15.7**
Cassaine **T39.9**
Ephedrine **A21.10**
Epiguiapyridine **T25.4**
Evoninic acid **A31.9**
Julocrotine **Y29.9**
2-Methylpiperidine **A10.5**
Milliamines **T39.2**
Muscarine **Y20.8**
Patchoulipyridine **T27.4**
Ryanodine **T58.1**
Skytanthines **T14.15-T14.18**
Solanidine **T49.4**
Solasodine **T49.7**
Tomatidine **T49.8**

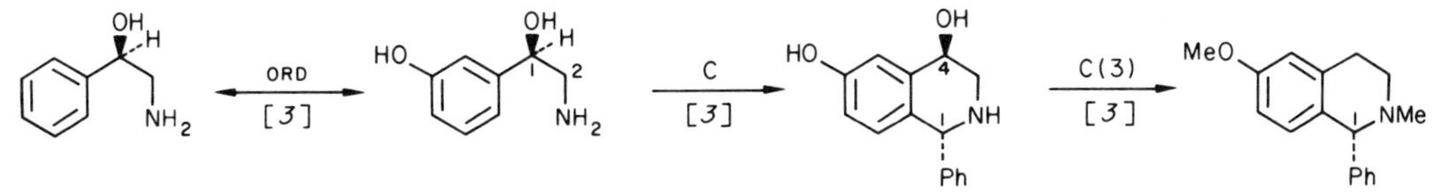

4- and 1-phenyltetrahydroisoquinoline, pavine and isopavine groups. K[1]

1. (R)-(-) adrenaline A22.14
2. (R)-( ) 2-amino-1-(3-hydroxyphenyl)-ethanol. Hydrochloride (+)
3. (1R,4R)-( ) 1,2,3,4-tetrahydro-4,6-dihydroxy-1-phenylisoquinoline. ORD/CD; [3]
4. (R)-(+) 1,2,3,4-tetrahydro-6-methoxy-2-methyl-1-phenylisoquinoline ORD/CD; [3]
5. (-) amurensine. By CD (coupled oscillators method)[7]
6. (1S,4R)-( ) 1,2,3,4-tetrahydro-4,6-dihydroxy-1-phenylisoquinoline
7. (R)-(-) cryptostyline II (non-natural enantiomer)(X=H) Abs. X-ray [4]
   (R)-(-) cryptostyline III* (non-natural enantiomer) (X = OMe)[2]
8. (S)-(-) 6,7-dimethoxy-4-(p-methoxyphenyl)-1,2,3,4-tetrahydroisoquinoline. Abs. X-ray [1] ORD; [1]
9. (+) epiglaudine. By CD (coupled oscillators) [8]
10. (+) oreodine. (+) glaudine = 14-epimer. Both by CD (coupled oscillators) [8]
11. (+) rhoeagenine. Abs. X-ray [9]
12. (-) argemonine. Also by CD [6]
13. (S)-(-) N-formyl-dihydropavine methine
14. (S)-(+) aspartic acid A4.6

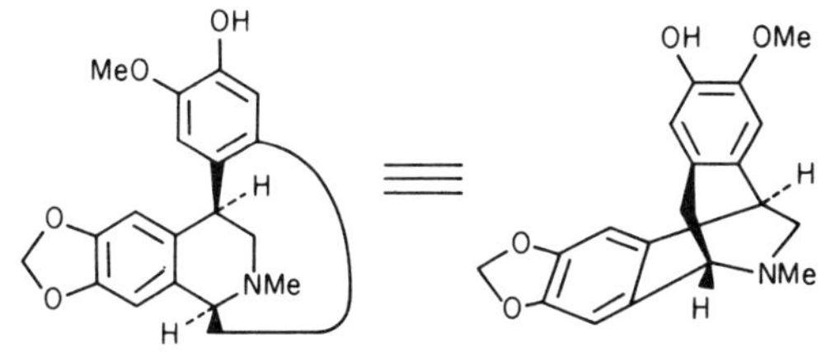

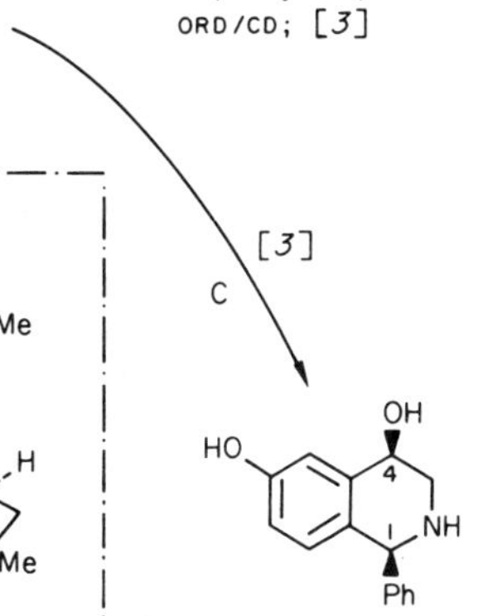

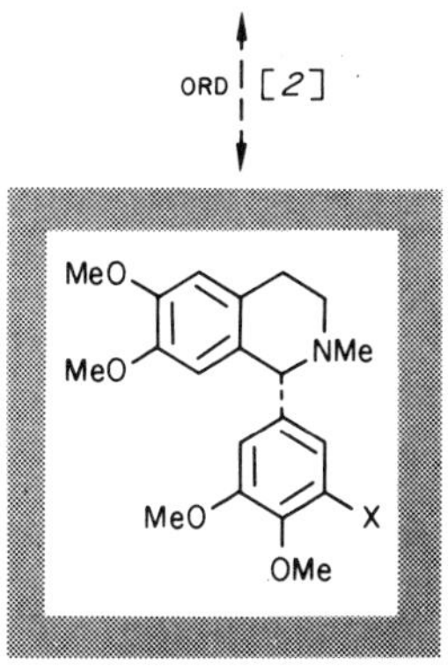

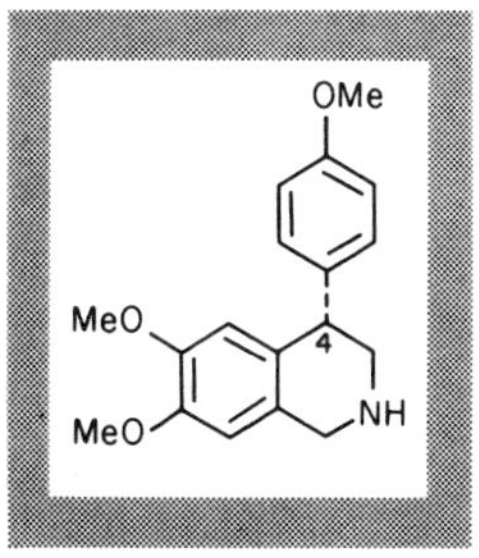

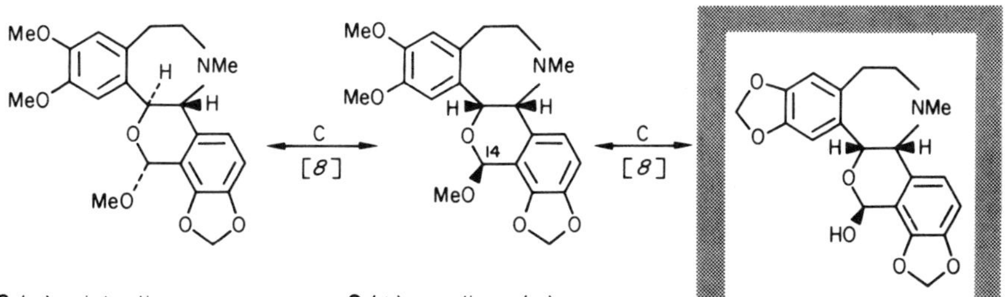

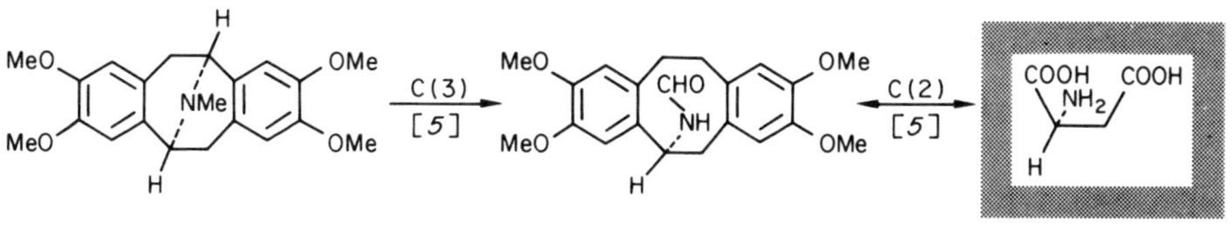

1. V. Toome, J. F. Blount, G. Grethe, and M. Uskoković, *Tetrahedron Letters*, 1970, 49.
2. T. Kametani, H. Sugi, and S. Shibuya, *Tetrahedron*, 1971, 27, 2409.
3. T. Kametani, H. Sugi, H. Yagi, K. Fukumoto, and S. Shibuya, *J. Chem. Soc. (C)*, 1970, 2213.
4. A. Brossi and S. Teitel, *Helv. Chim. Acta*, 1971, 54, 1564.
5. A. C. Barker and A. R. Battersby, *J. Chem. Soc. (C)*, 1967, 1317.
6. S. F. Mason, G. W. Vane, and J. S. Whitehurst, *Tetrahedron*, 1967, 23, 4087.
7. M. Shamma, J. L. Moniot, W. K. Chan, and K. Nakanishi, *Tetrahedron Letters*, 1971, 3425.
8. M. Shamma, J. L. Moniot, W. K. Chan, and K. Nakanishi, *Tetrahedron Letters*, 1971, 4207, and references therein.
9. C. S. Huber, quoted by W. Klötzer, S. Teitel, and A. Brossi, *Helv. Chim. Acta*, 1971, 54, 2057.

K² Tetrahydroisoquinoline, Ipecacuanha & Alangium groups.

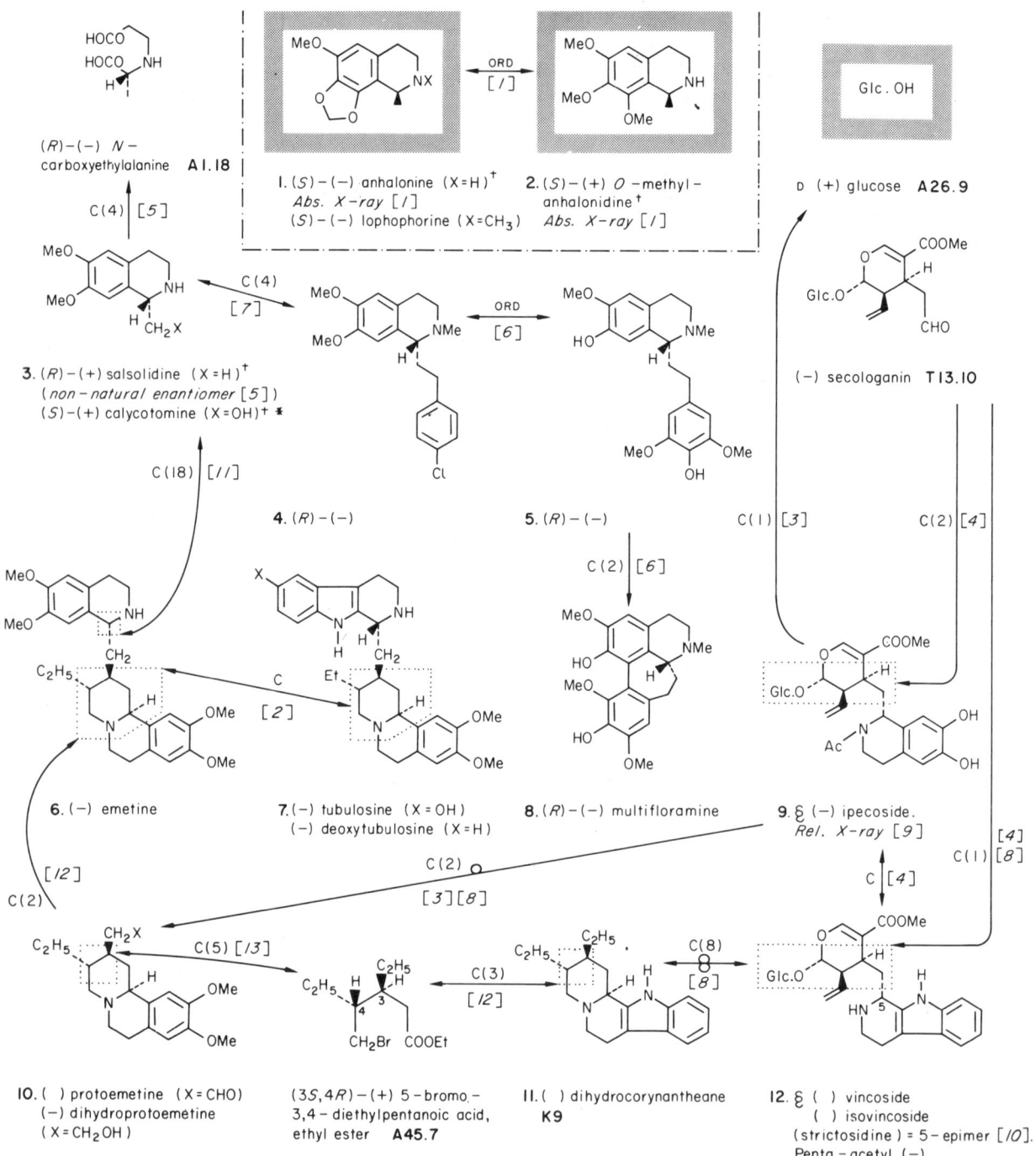

1. (S)-(−) anhalonine (X=H)† Abs. X-ray [1]
   (S)-(−) lophophorine (X=CH₃)
2. (S)-(+) O-methyl-anhalonidine† Abs. X-ray [1]
3. (R)-(+) salsolidine (X=H)† (non-natural enantiomer [5])
   (S)-(+) calycotomine (X=OH)† *
4. (R)-(−)
5. (R)-(−)
6. (−) emetine
7. (−) tubulosine (X=OH)
   (−) deoxytubulosine (X=H)
8. (R)-(−) multifloramine
9. § (−) ipecoside. Rel. X-ray [9]
10. ( ) protoemetine (X=CHO)
    (−) dihydroprotoemetine (X=CH₂OH)
11. ( ) dihydrocorynantheane K9
12. § ( ) vincoside
    ( ) isovincoside
    (strictosidine) = 5−epimer [10].
    Penta−acetyl (−)

(R)-(−) N-carboxyethylalanine A1.18
(3S,4R)-(+) 5-bromo-3,4-diethylpentanoic acid, ethyl ester A45.7
D (+) glucose A26.9
(−) secologanin T13.10

1. A. Brossi, J. F. Blount, J. O'Brien, and S. Teitel, *J. Amer. Chem. Soc.*, 1971, 93, 6248.
2. A. R. Battersby, J. E. Merchant, E. A. Ruveda, and S. S. Salgar, *Chem. Comm.*, 1965, 315; H. T. Openshaw and N. Whittaker, *Chem. Comm.*, 1966, 131.
3. A. R. Battersby, B. Gregory, H. Spencer, J. C. Turner, M. M. Janot, P. Potier, P. Francois, and J. Levisalles, *Chem. Comm.*, 1967, 219.
4. A. R. Battersby, A. R. Burnett, and P. G. Parsons, *J. Chem. Soc. (C)*, 1969, 1187, 1193.
5. A. R. Battersby and T. P. Edwards, *J. Chem. Soc.*, 1960, 1214.
6. A. Brossi, J. O'Brien, and S. Teitel, *Helv. Chim. Acta*, 1969, 52, 678; A. Rheiner and A. Brossi, *Experientia*, 1964, 20, 488.
7. A. Brossi and F. Burkhardt, *Helv. Chim. Acta*, 1961, 44, 1558.
8. W. P. Blackstock, R. T. Brown, and G. K. Lee, *Chem. Comm.*, 1971, 910.
9. O. Kennard, P. J. Roberts, N. W. Isaacs, F. H. Allen, W. D. S. Motherwell, K. H. Gibson, and A. R. Battersby, *Chem. Comm.*, 1971, 899.
10. K. T. D. De Silva, G. N. Smith, and K. E. H. Warren, *Chem. Comm.*, 1971, 905.
11. A. R. Battersby, R. Binks, and T. P. Edwards, *J. Chem. Soc.*, 1960, 3474.
12. C. Szántay, L. Töke, and P. Kolonits, *J. Org. Chem.*, 1966, 31, 1447.
13. A. R. Battersby, S. W. Breuer, and S. Garratt, *J. Chem. Soc. (C)*, 1968, 2467.

†Naturally occurring tetrahydroisoquinoline alkaloids may be of either enantiomeric type or may be racemic [5] [6].

Aporphine, proaporphine, tetrahydroprotoberberine and phthalideisoquinoline groups. K3

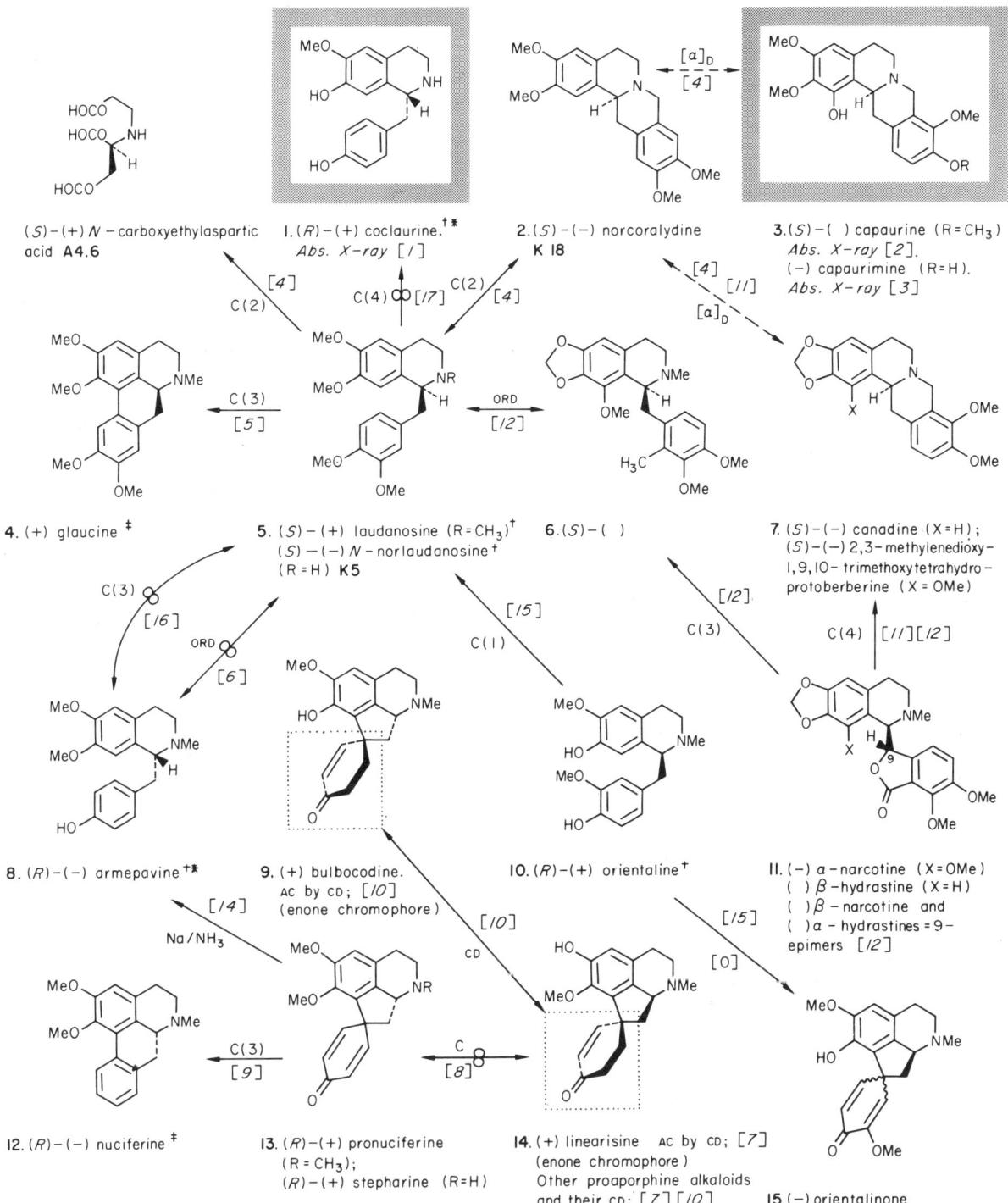

†Naturally occurring benzylisoquinoline alkaloids may be of either enantiomeric series or may be racemic. Alkaloids homochirally analogous with (R)-(+) coclaurine, and their quaternary derivatives, have +veORD and vice-versa [6].
Bisbenzylisoquinoline alkaloids. At least 6 types of alkaloid are known in which one benzylisoquinoline moiety is joined to another by one or more ether linkages. Structural determination of these compounds is usually by fission followed by identification of the benzylisoquinoline components, and they are not therefore specifically treated here.
ORD of bisbenzylisoquinoline types; [13].

‡ see footnote on page 142

References at top of page 142

1. J. Fridrichsons and A. McL. Mathieson, *Tetrahedron*, 1968, **24**, 5785.
2. H. Shimanouchi, Y. Sasada, M. Ihara, and T. Kametani, *Acta Cryst.*, 1969, **B25**, 1310.
3. T. Kametani, M. Ihara, T. Honda, H. Shimanouchi, and Y. Sasada, *J. Chem. Soc. (C)*, 1971, 2541.
4. H. Corrodi and E. Hardegger, *Helv. Chim. Acta*, 1956, **39**, 889.
5. F. Faltis and E. Adler, *Arch. Pharm.*, 1951, **284**, 281.
6. S. M. Albonico, J. Comin, A. M. Kuck, E. Sanchez, P. M. Scopes, R. J. Swan, and J. Vernengo, *J. Chem. Soc. (C)*, 1966, 1340, and references therein.
7. G. Snatzke and G. Wollenberg, *J. Chem. Soc. (C)*, 1966, 1681.
8. L. J. Haynes, K. L. Stuart, D. H. R. Barton, and G. W. Kirby, *Proc. Chem. Soc.*, 1964, 261.
9. K. Bernauer, *Helv. Chim. Acta*, 1963, **46**, 1783.
10. F. Santavý, P. Sedmera, G. Snatzke, and T. Reichstein, *Helv. Chim. Acta*, 1971, **54**, 1084.
11. A. R. Battersby and H. Spencer, *J. Chem. Soc.*, 1965, 1087.
12. M. Ohta, H. Tani, S. Morozumi, and S. Kodaira, *Tetrahedron Letters*, 1963, 1857.
13. A. R. Battersby, I. R. C. Bick, W. Klyne, J. P. Jennings, P. M. Scopes, and M. J. Vernengo, *J. Chem. Soc.*, 1965, 2239.
14. M. P. Cava, K. Nomura, S. K. Talapatra, M. J. Mitchell, R. H. Schlessinger, K. T. Buck, J. L. Beal, B. Douglas, R. F. Raffauf, and J. A. Weisbach, *J. Org. Chem.*, 1968, **33**, 2785.
15. A. R. Battersby, T. H. Brown, and J. H. Clements, *J. Chem. Soc.*, 1965, 4550.
16. C. Ferrari and V. Deulofeu, *Tetrahedron*, 1962, **18**, 419.
17. M. Tomita and J. Kunimoto, *J. Pharm. Soc. Japan*, 1962, **82**, 734; L. J. Haynes, K. L. Stuart, D. H. R. Barton, D. S. Bhakuni, and G. W. Kirby, *Chem. Comm.*, 1964, 141.

---

1. T. Ashida, R. Pepinsky, and Y. Okaya, *Acta Cryst.*, 1963, **16** (suppl.), A48, abstr.5.8.
2. J. H. van den Hende and N. R. Nelson, *J. Amer. Chem. Soc.*, 1967, **89**, 2901.
3. J. Fridrichsons, M. F. Mackay, and A. McL. Mathieson, *Tetrahedron Letters*, 1968, 2887.
4. G. Kartha, F. R. Ahmed, and W. H. Barnes, *Acta Cryst.*, 1962, **15**, 326.
5. M. G. Waite, G. A. Sim, C. R. Olander, R. J. Warnet, and D. M. S. Wheeler, *J. Amer. Chem. Soc.*, 1969, **91**, 7765.
6. D. H. R. Barton, D. S. Bhakuni, R. James, and G. W. Kirby, *J. Chem. Soc. (C)*, 1967, 128.
7. D. H. R. Barton, G. W. Kirby, W. Steglich, G. M. Thomas, A. R. Battersby, T. A. Dobson, and M. Ramuz, *J. Chem. Soc.*, 1965, 2423.
8. J. Kalvoda, P. Buchsacher, and O. Jeger, *Helv. Chim. Acta*, 1955, **38**, 1847.
9. K. W. Bentley, D. G. Hardy, and B. Meek, *J. Amer. Chem. Soc.*, 1967, **89**, 3273.
10. A. R. Battersby, R. B. Herbert, L. Pijewska, and F. Santavý, *Chem. Comm.*, 1965, 228.
11. K. Goto and S. Mitsui, *Bull. Chem. Soc. Japan*, 1931, **6**, 33.
12. C. Djerassi, K. Mislow, and M. Shamma, *Experientia*, 1962, **18**, 53.
13. J. C. Craig and S. K. Roy, *Tetrahedron*, 1965, **21**, 395; S. M. Albonico, J. Comin, A. M. Kuck, E. Sanchez, P. M. Scopes, R. J. Swan, and M. J. Vernengo, *J. Chem. Soc. (C)*, 1966, 1340.
14. Z. J. Barneis, D. M. S. Wheeler, and T. H. Kinstle, *Tetrahedron Letters*, 1965, 275.
15. C. Schöpf and F. Borkowsky, *Annalen*, 1927, **458**, 148.
16. C. Schöpf and H. Hirsch, *Annalen*, 1931, **489**, 224.

†Morphine bases of both enantiomeric types, derived from morphane (e.g. (-) morphine, K4.5, (-) thebaine K4.6) and from *enantio*-morphane (e.g. sinomenine K4.2) occur naturally. One or two alkaloids are known to occur naturally as both enantiomers, e.g. (+) salutaridine K4.4 (morphane type) and (-) sinoacutine (*enantio*-salutaridine) (*enantio*-morphane type).

‡The ORD of aporphine alkaloids has been discussed [12][13]. The overriding factor in determining the sign is the chirality of the biphenyl chromophore, which is in turn uniquely determined by the configuration at the asymmetric carbon atom.

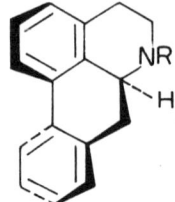

biphenyl system (*S*)
asymmetric C atom (*S*)

The exact form of the ORD curve is, however, strongly determined by the 1,2,10,11-substitution pattern and it is therefore necessary to choose as a model compound for an alkaloid of unknown configuration one having a similar substitution pattern [12].

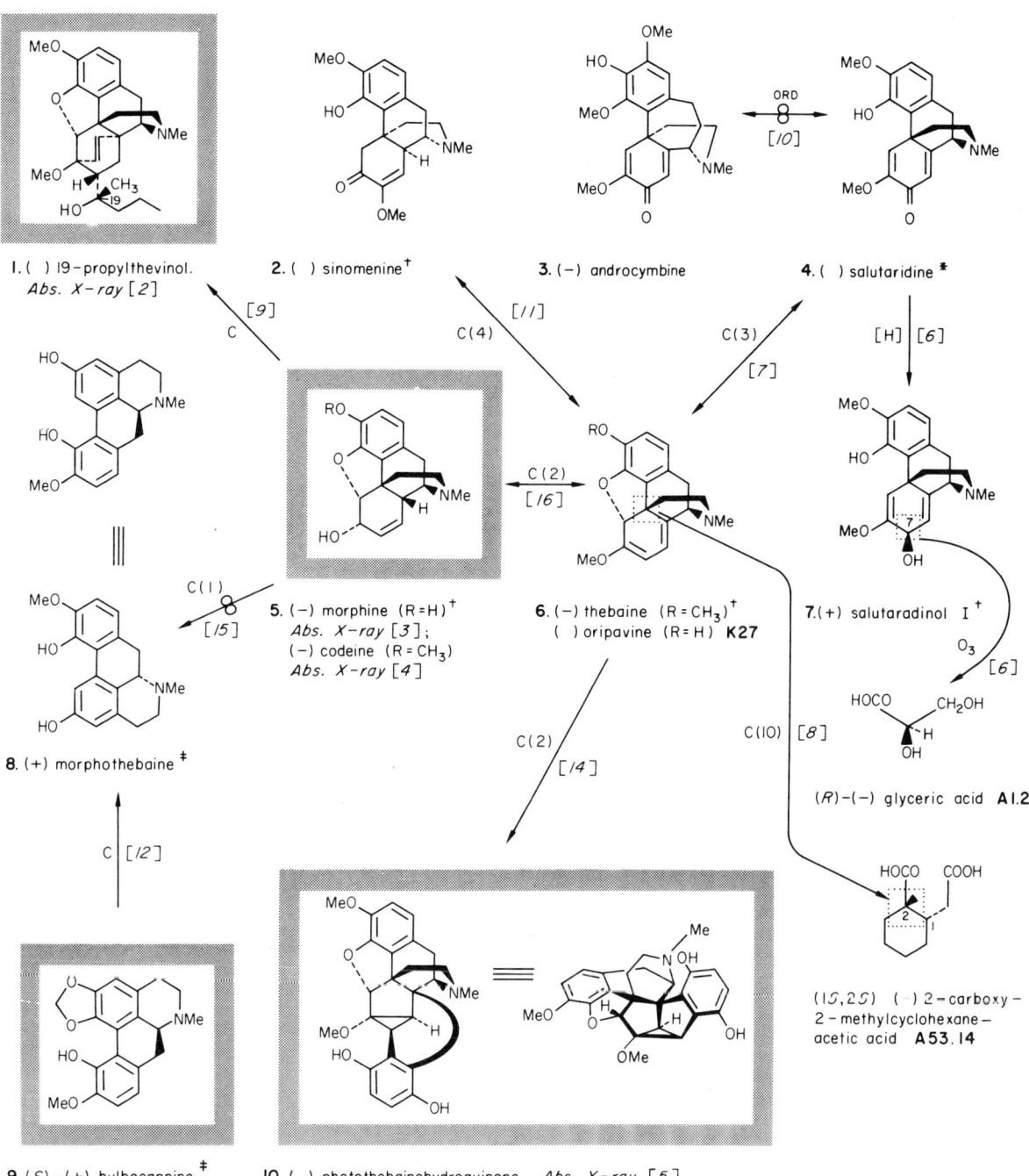

K 5
Morphine and related groups (contd.); Cularine, chelidonine and dibenzopyrrocoline groups.

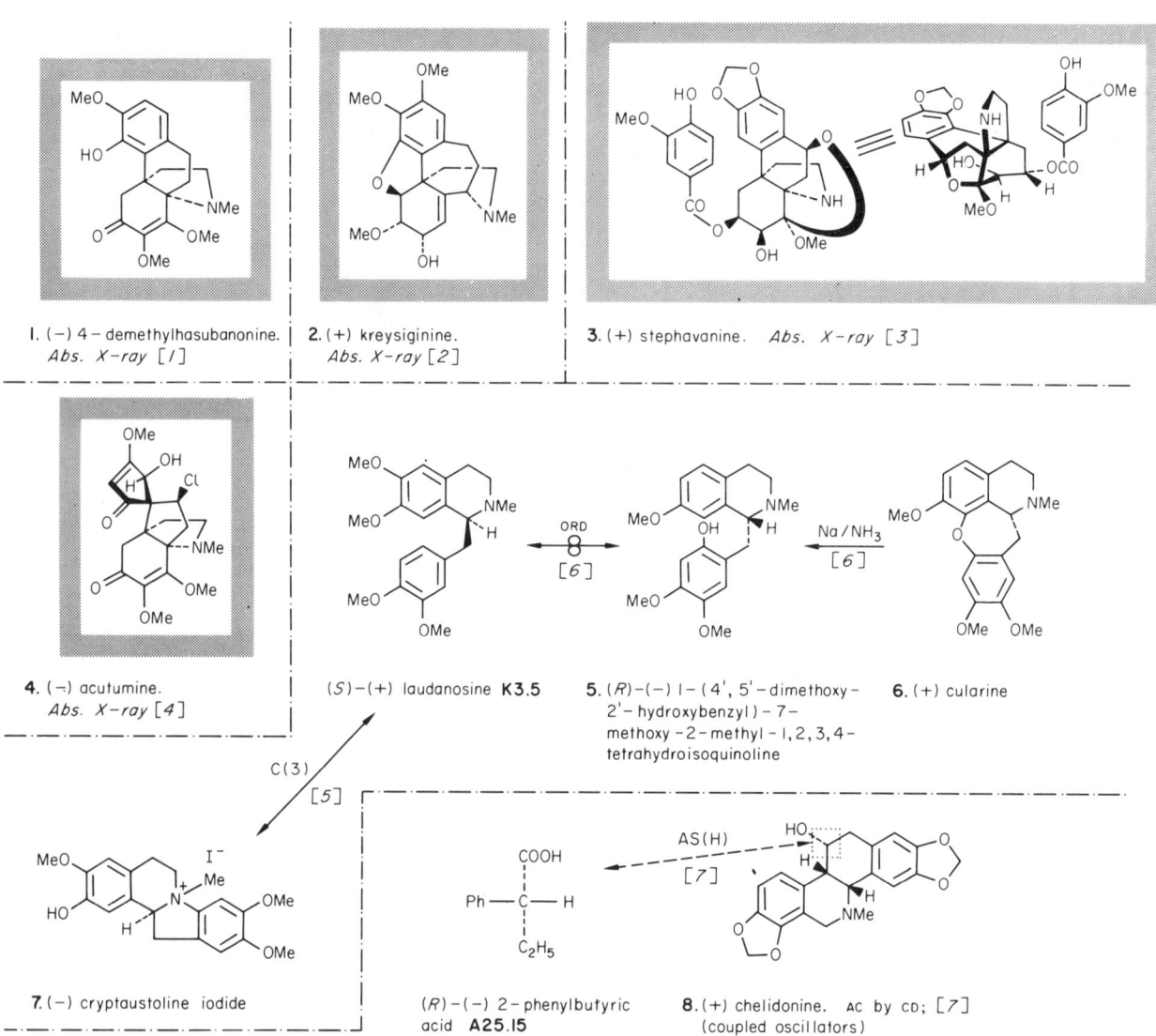

1. (−) 4−demethylhasubanonine. Abs. X−ray [1]
2. (+) kreysiginine. Abs. X−ray [2]
3. (+) stephavanine. Abs. X−ray [3]
4. (−) acutumine. Abs. X−ray [4]
(S)−(+) laudanosine K3.5
5. (R)−(−) 1−(4′,5′−dimethoxy− 2′−hydroxybenzyl)−7− methoxy−2−methyl−1,2,3,4− tetrahydroisoquinoline
6. (+) cularine
7. (−) cryptaustoline iodide
(R)−(−) 2−phenylbutyric acid A25.15
8. (+) chelidonine. AC by CD; [7] (coupled oscillators)

1. S. M. Kupchan, M. I. Suffness, D. N. J. White, A. T. McPhail, and G. A. Sim, *J. Org. Chem.*, 1968, 33, 4529.
2. J. Fridrichsons, M. F. Mackay, and A. McL. Mathieson, *Tetrahedron Letters*, 1968, 2887.
3. S. M. Kupchan, M. I. Suffness, R. J. McClure, and G. A. Sim, *J. Amer. Chem. Soc.*, 1970, 92, 5756.
4. M. Nishikawa, K. Kamiya, M. Tomita, Y. Okamoto, T. Kikuchi, K. Osaki, Y. Tomiie, I. Nitta, and K. Goto, *J. Chem. Soc. (B)*, 1968, 652.
5. G. K. Hughes, E. Ritchie, and W. C. Taylor, *Austral. J. Chem.*, 1953, 6, 315.
6. N. S. Bhacca, J. C. Craig, R. H. F. Manske, S. K. Roy, M. Shamma, and W. A. Slusarchyk, *Tetrahedron*, 1966, 22, 1467.
7. G. Snatzke, J. Hrbek, L. Hruban, A. Horeau, and F. Santavý, *Tetrahedron*, 1970, 26, 5013.

---

16. D. J. Williams and D. Rogers, *Proc. Chem. Soc.*, 1964, 357.
17. D. H. R. Barton and G. W. Kirby, *J. Chem. Soc.*, 1962, 806.
18. K. Kotera, Y. Hamada, K. Tori, K. Aono, and K. Kuriyama, *Tetrahedron Letters*, 1966, 2009.
19. H. M. Fales and W. C. Wildman, *J. Amer. Chem. Soc.*, 1960, 82, 3368.

Amaryllidaceae alkaloids.

ORD/CD; [7] NOMENCLATURE; [4]

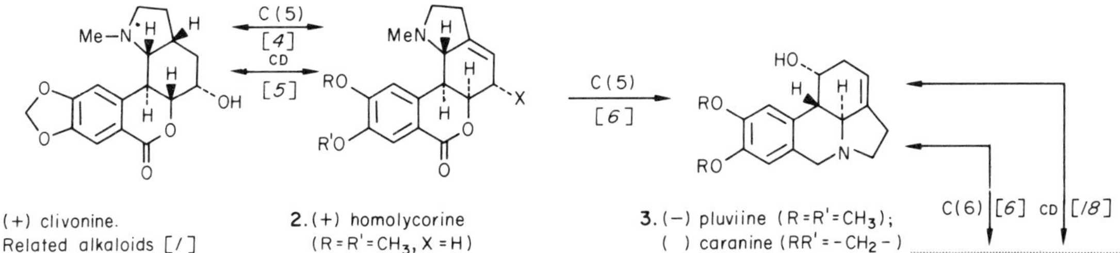

1. (+) clivonine.
Related alkaloids [1]

2. (+) homolycorine
(R = R' = CH₃, X = H)
(+) hippeastrine (RR' = CH₂, X = OH)

3. (−) pluviine (R = R' = CH₃);
( ) caranine (RR' = -CH₂-)
CD; [18] X7

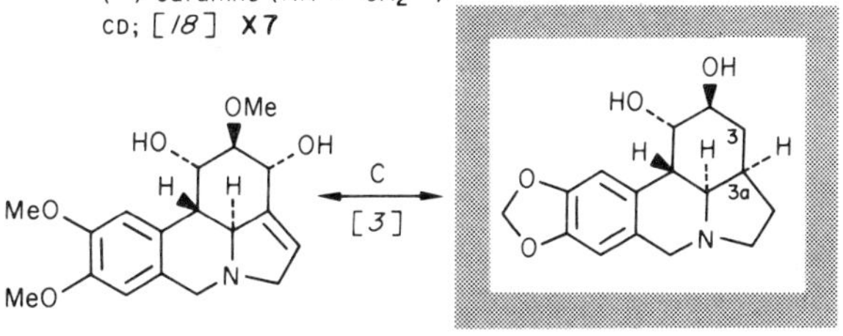

4. (−) narcissidine.
Rel. X-ray [3]

5. ( ) dihydrolycorine.
Abs. X-ray [2] CD; [18]
( ) lycorine = 3, 3a - didehydro - deriv.

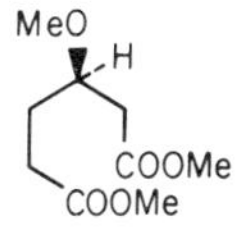

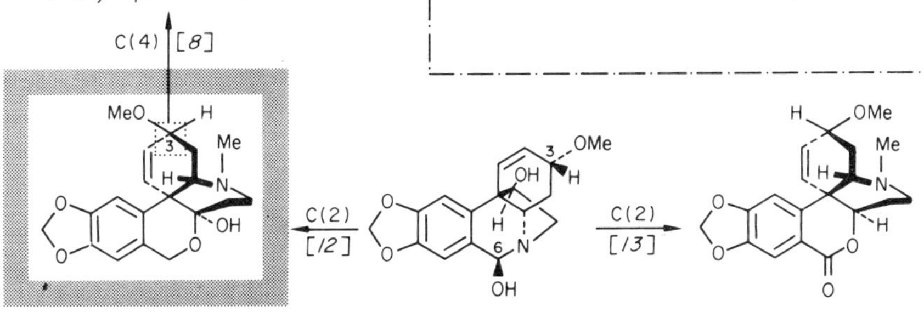

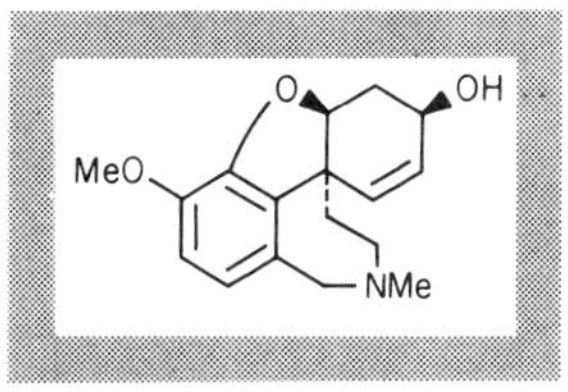

(R)-(+) dimethyl 3-methoxyadipate **A25.8**

6. (+) tazettine.
Abs. X-ray [9].
(+) criwelline = 3-epimer.

7. (+) 6-hydroxycrinamine.
Rel. X-ray [10].
( ) haemanthidine = 3-epimer. These compounds exist as mixtures of 6-epimers in solution [11]

8. (+) macronine

9. (−) galanthamine.
Abs. X-ray [16].
See also [7]
(−) lycoramine = dihydroderiv.

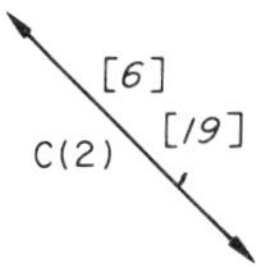

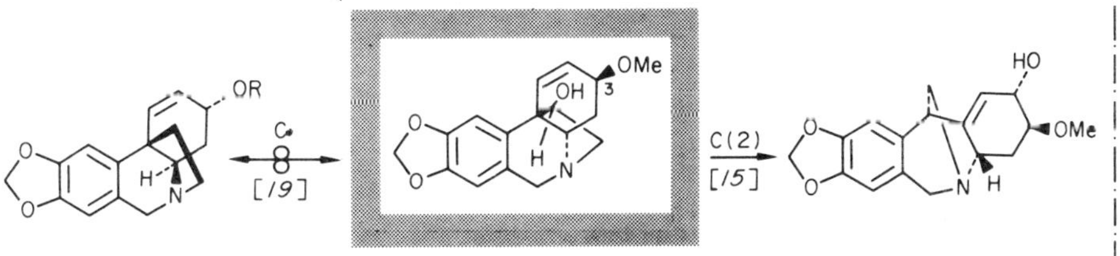

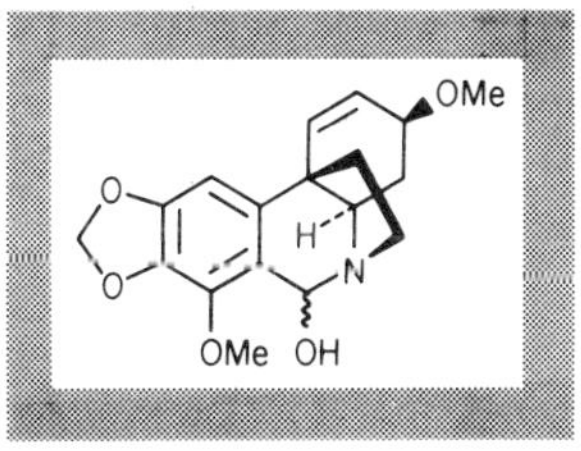

10. (−) crinine (R = H) *
(−) buphanisine (R = CH₃)
Other related alkaloids; [19]

11. ( ) haemanthamine.
Abs. X-ray [14]
( ) crinamine = 3-epimer.
Other related alkaloids; [19]

12. (−) isohaemanthamine.
Related compds;
(−)montanine, (−) coccinine, (−)manthine [15]

13. ( ) 6-hydroxybuphanidrine
Abs. X-ray [14]

---

1. W. Dopke and M. Bienert, *Tetrahedron Letters*, 1970, 3245 and references therein.
2. M. Shiro, T. Sato, and H. Koyama, *Chem. and Ind.*, 1966, 1229.
3. J. C. Clardy, W. C. Wildman, and F. M. Hauser, *J. Amer. Chem. Soc.*, 1970, 92, 1781, and references therein.
4. P. W. Jeffs, J. F. Hansen, W. Döpke, and M. Bienert, *Tetrahedron*, 1971, 27, 5065.
5. W. Döpke, M. Bienert, P. W. Jeffs, and D. S. Farrier, *Tetrahedron Letters*, 1967, 451.
6. T. Kitagawa, S. Uyeo, and N. Yokoyama, *J. Chem. Soc.*, 1959, 3741.
7. K. Kuriyama, T. Iwata, M. Moriyama, K. Kotera, Y. Hamada, R. Mitsui, and K. Takeda, *J. Chem. Soc. (B)*, 1967, 46.
8. R. J. Highet and P. F. Highet, *Tetrahedron Letters*, 1966, 4099.
9. T. Sato and H. Koyama, *J. Chem. Soc. (B)*, 1971, 1070.
10. J. Karle, J. A. Estlin, and I. L. Karle, *J. Amer. Chem. Soc.*, 1967, 89, 6510.
11. R. W. King, C. F. Murphy, and W. C. Wildman, *J. Amer. Chem. Soc.*, 1965, 87, 4912.
12. H. M. Fales, D. H. S. Horn, and W. C. Wildman, *Chem. and Ind.*, 1959, 1415.
13. C. F. Murphy and W. C. Wildman, *Tetrahedron Letters*, 1964, 3857.
14. J. Clardy, F. M. Hauser, D. Dahm, R. A. Jacobson, and W. C. Wildman, *J. Amer. Chem. Soc.*, 1970, 92, 6337.
15. Y. Inubushi, H. M. Fales, E. W. Warnhoff, and W. C. Wildman, *J. Org. Chem.*, 1960, 25, 2153.

K7 Erythrina and homoerythrina groups.

1. (+) erythraline.
   Rel. X-ray [1]
2. (−) 15,16-dimethoxy-cis-5,6-erythrinan
3. ( ) β-tetrahydro-β-erythroidine
4. (+) erythristemine.
   Abs. X-ray [3]
5. ( ) dihydro-β-erythroidine.
   Abs. X-ray [2]
6. (+) α-erythroidine
7. (+) schellhammerine (X = OH)
   Abs. X-ray [8].
   (+) schellhammericine (X = H)
   X7
8. (+) schellhammeridine

(S)-(−) 3-methoxyadipic acid A25.8

(R)-(+) 3-(2-ethylbenzoyl)-tetrahydrofuran A30.4

1. W. Nowacki and G. F. Bonsma, *Z. Krist.*, 1958, **110**, 89.
2. A. W. Hanson, *Proc. Chem. Soc.*, 1963, 52.
3. D. H. R. Barton, P. N. Jenkins, R. Letcher, D. A. Widdowson, E. Hough, and D. Rogers, *Chem. Comm.*, 1970, 391.
4. J. C. Godfrey, D. S. Tarbell, and V. Boekelheide, *J. Amer. Chem. Soc.*, 1955, **77**, 3342.
5. V. Boekelheide and M. Y. Chang, *J. Org. Chem.*, 1964, **29**, 1303 and references therein.
6. G. R. Wenziger and V. Boekelheide, *Proc. Chem. Soc.*, 1963, 53.
7. V. Boekelheide and G. C. Morrison, *J. Amer. Chem. Soc.*, 1958, **80**, 3905.
8. S. R. Johns, C. Kowala, J. A. Lamberton, A. A. Sioumis, and J. A. Wunderlich, *Chem. Comm.*, 1968, 1102.

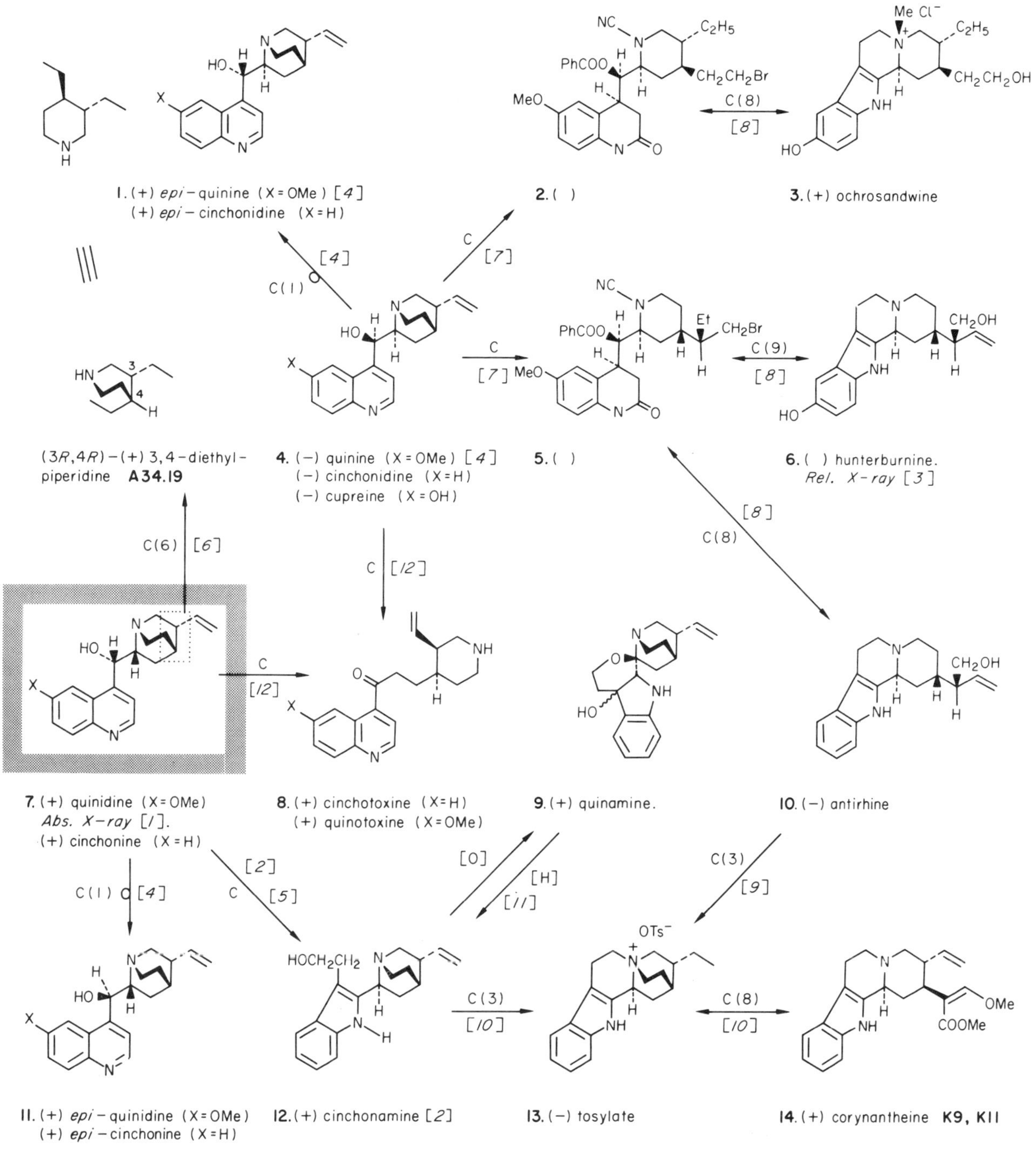

Cinchona and corynantheine-type alkaloids. K⁸

1. (+) epi-quinine (X = OMe) [4]
   (+) epi-cinchonidine (X = H)

2. ( )

3. (+) ochrosandwine

(3R,4R)-(+) 3,4-diethyl-piperidine A34.19

4. (−) quinine (X = OMe) [4]
   (−) cinchonidine (X = H)
   (−) cupreine (X = OH)

5. ( )

6. ( ) hunterburnine.
   Rel. X-ray [3]

7. (+) quinidine (X = OMe)
   Abs. X-ray [1].
   (+) cinchonine (X = H)

8. (+) cinchotoxine (X = H)
   (+) quinotoxine (X = OMe)

9. (+) quinamine.

10. (−) antirhine

11. (+) epi-quinidine (X = OMe)
    (+) epi-cinchonine (X = H)

12. (+) cinchonamine [2]

13. (−) tosylate

14. (+) corynantheine K9, K11

---

1. O. L. Carter, A. T. McPhail, and G. A. Sim, *J. Chem. Soc. (A)*, 1967, 365.
2. Y. K. Sawa and H. Matsumura, *Tetrahedron*, 1970, **26**, 2923.
3. J. D. M. Asher, J. M. Robertson, and G. A. Sim, *J. Chem. Soc.* 1965, 6355.
4. G. G. Lyle and L. K. Keefer, *Tetrahedron*, 1967, **23**, 3253 and references therein.
5. E. Ochiai and M. Ishikawa, *Chem. Pharm. Bull. Japan*, 1958, **6**, 208; E. Ochiai, M. Ishikawa, and Y. Oka, *ibid.*, 1959, **7**, 744.
6. V. Prelog and E. Zalán, *Helv. Chim. Acta*, 1944, **27**, 535.
7. E. Ochiai and M. Ishikawa, *Chem. Pharm. Bull. Japan*, 1959, **7**, 559.
8. Y. K. Sawa and H. Matsumura, *Chem. Comm.*, 1968, 679.
9. S. R. Johns, J. A. Lamberton, and J. L. Occolowitz, *Austral. J. Chem.*, 1967, **20**, 1463.
10. E. Wenkert and N. V. Bringi, *J. Amer. Chem. Soc.*, 1958, **80**, 3484.
11. R. Goutarel, M. M. Janot, V. Prelog, and W. I. Taylor, *Helv. Chim. Acta*, 1950, **33**, 150; B. Witkop, *J. Amer. Chem. Soc.*, 1950, **72**, 2311.
12. R. B. Turner and R. B. Woodward in 'The Alkaloids', 1953, **3**, 25; P. Rabe, *Ber.*, 1922, **55**, 522.

K 9 Indole Alkaloids; Yohimbane, heteroyohimbane and oxindole groups.

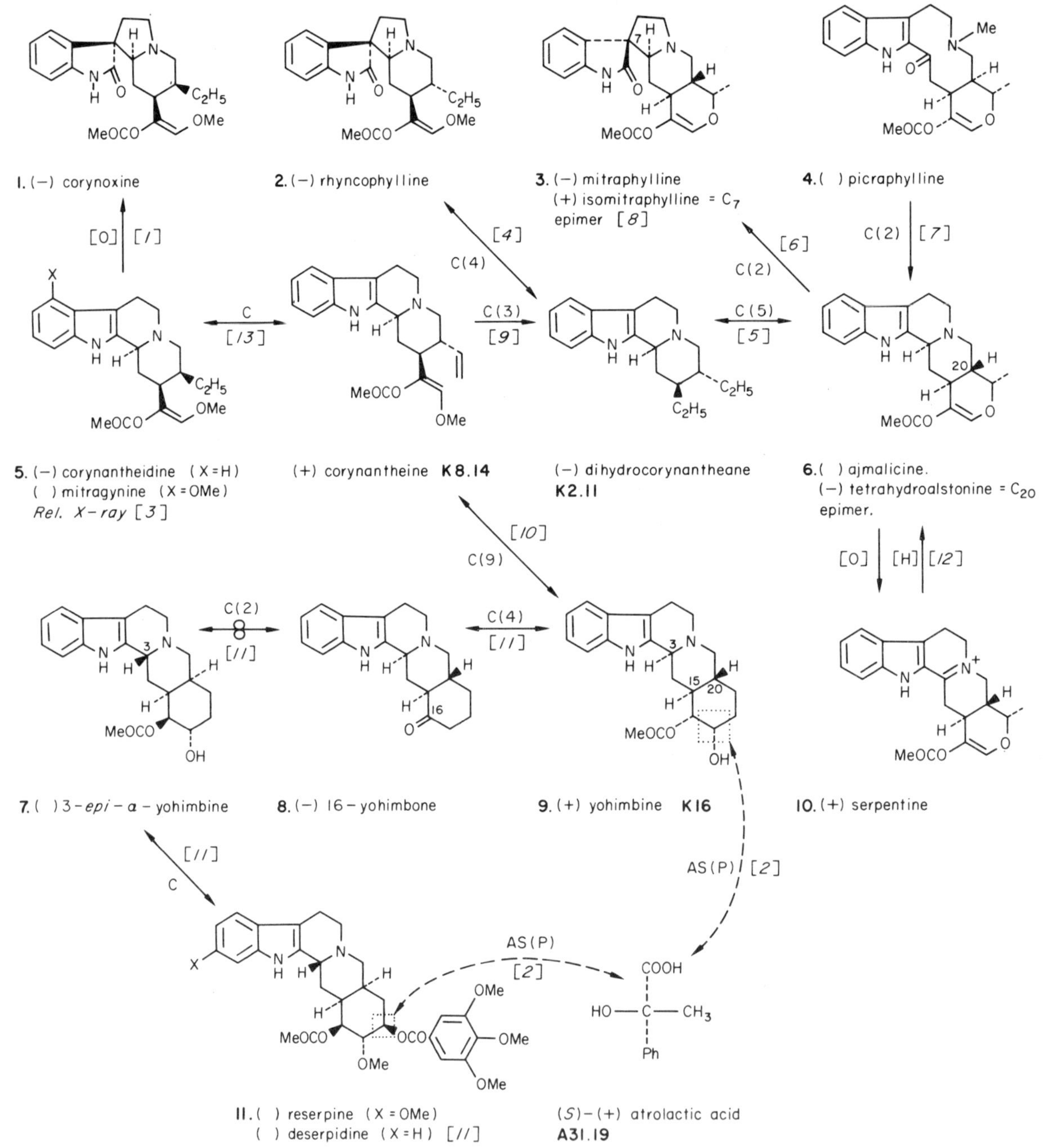
1. (−) corynoxine
2. (−) rhyncophylline
3. (−) mitraphylline
(+) isomitraphylline = C_7 epimer [8]
4. ( ) picraphylline
[O] [1]
[4]
C(4)
[6]
C(2)
C(2) [7]
C
[13]
C(3)
[9]
C(5)
[5]
5. (−) corynantheidine (X = H)
( ) mitragynine (X = OMe)
Rel. X-ray [3]
(+) corynantheine K8.14
(−) dihydrocorynantheane K2.11
6. ( ) ajmalicine.
(−) tetrahydroalstonine = C_{20} epimer.
[10]
C(9)
[O] [H] [12]
C(2)
[11]
C(4)
[11]
7. ( ) 3-epi-α-yohimbine
8. (−) 16-yohimbone
9. (+) yohimbine K16
10. (+) serpentine
[11]
C
AS(P) [2]
AS(P)
[2]
11. ( ) reserpine (X = OMe)
( ) deserpidine (X = H) [11]
(S)-(+) atrolactic acid A31.19

1. J. L. Pousset, J. Poisson, and M. Legrand, *Tetrahedron Letters*, 1966, 6283.
2. Y. Ban and O. Yonemitsu, *Tetrahedron*, 1964, **20**, 2877.
3. D. E. Zacharias, R. D. Rosenstein, and G. A. Jeffrey, *Acta Cryst.*, 1965, **18**, 1039.
4. N. Finch and W. I. Taylor, *J. Amer. Chem. Soc.*, 1962, **84**, 1318.
5. E. Wenkert and N. V. Bringi, *J. Amer. Chem. Soc.*, 1958, **80**, 3484.
6. J. Shavel and H. Zinnes, *J. Amer. Chem. Soc.*, 1962, **84**, 1320.
7. J. Lévy, G. Ledouble, J. Le Men, and M.-M. Janot, *Bull. Soc. chim. France*, 1964, 1917.
8. N. Finch and W. I. Taylor, *J. Amer. Chem. Soc.*, 1962, **84**, 3871; E. J. Shellard and J. D. Phillipson, *Tetrahedron Letters*, 1966, 1113.
9. M.-M. Janot, R. Goutarel, and V. Prelog, *Helv. Chim. Acta*, 1951, **34**, 1207.
10. R. L. Autrey and P. W. Scullard, *Chem. Comm.*, 1966, 841.
11. E. Wenkert, E. W. Robb, and N. V. Bringi, *J. Amer. Chem. Soc.*, 1957, **79**, 6570.
12. F. L. Weisenborn, M. Moore, and P. A. Diassi, *Chem. and Ind.*, 1954, 375.
13. W. F. Trager, C. M. Lee, and A. H. Beckett, *Tetrahedron*, 1967, **23**, 375, and references therein.

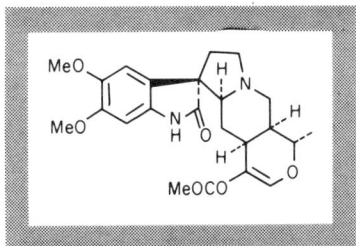

1. (+) rauvoxinine. *Abs. X-ray* [5]

Configurations of yohimbane, corynantheane and heteroyohimbane derivatives

Only a few of the known configurational types of these alkaloids are represented on pp. **K8-K9**. The yohimbane skeleton contains 3 asymmetric centres. All known yohimbane derivatives are homochirally analogous at $C_{(15)}$, having the $C_{(15)}$ configuration corresponding to (15S) in yohimbane. There are thus 4 stereoisomeric types derived from the following 4 parent compounds:

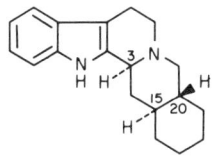 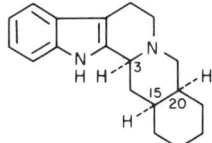 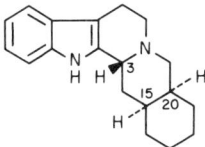

(−) yohimbane (3S, 15S, 20S)   ( ) pseudoyohimbane   (−) alloyohimbane   ( ) epialloyohimbane
Example;   (3R, 15S, 20S)   (3S, 15S, 20R)   (3R, 15S, 20R)
(+) yohimbine **K9.10**   Example;
                                                                                                                                                                              (−) reserpine **K9.12**

The known yohimbane alkaloids have been correlated with the key compounds on p. **K9** by chemical and chiroptical means [1]. ORD/CD of the four yohimbane types; [3].

The situation with alkaloids containing the dihydro-corynantheane skeleton is similar; there are four stereoisomeric types, normal (e.g. corynantheine **K8.14**, *pseudo, allo* (e.g. corynantheidine **K9.5**) and *epiallo*. Their stereochemistry has been fully discussed [2]. On the basis of this work a corynantheine type alkaloid can be allotted to one of the four classes by spectroscopic measurement, making the same assumption about $C_{(15)}$ configuration [2].

The stereochemistry of the heteroyohimbine types resembles that of the yohimbines but with an additional chiral centre at C_{19}. It is again probable that only (15S) type compounds occur naturally and there are thus again 4 basic skeletal types, normal (e.g. ajmalicine **K9.7**, *pseudo, allo* and *epiallo*, each of which may in addition have the (19R) or (19S) configuration [4]. ORD of heteroyohimbine alkaloids; [8].

Oxindole group

There are 16 possible stereochemical types arising from the 4 asymmetric centres at C_3, C_7, C_{19} and C_{20} (cf. mitraphylline **K9.3**) although some of the isomers are sterically unfavourable [7]. As with the previous types of indole alkaloid, the C_{15} configuration is assumed to be always as shown for mitraphylline.

Other correlations by rearrangement of yohimbane-type alkaloids to oxidinole alkaloids in addition to these shown on p. **K9** have been carried out.

CD of oxindole alkaloids; [6].

1. J. E. Saxton, *Quart. Rev.*, 1956, **10**, 108.
2. W. F. Trager, C. M. Lee, and A. H. Beckett, *Tetrahedron*, 1967, **23**, 365, 375.
3. L. Bartlett, N. J. Dastoor, J. Hrbek, W. Klyne, H. Schmid, and G. Snatzke, *Helv. Chim. Acta*, 1971, **54**, 1238.
4. M. Shamma and J. M. Richey, *J. Amer. Chem. Soc.*, 1953, **85**, 2507.
5. C. Pascard-Billy, *Bull. Soc. chim. France*, 1968, 3289.
6. A. F. Beecham, N. K. Hart, S. R. Johns, and J. A. Lamberton, *Tetrahedron Letters*, 1967, 991 and references therein.
7. M. Shamma, R. J. Shine, I. Kompis, T. Sticzay, F. Morsingh, J. Poisson, and J.-L. Pousset, *J. Amer. Chem. Soc.*, 1967, **89**, 1739.
8. N. Finch, W. I. Taylor, T. R. Emerson, W. Klyne, and R. J. Swan, *Tetrahedron*, 1966, **22**, 1327.

K11
Ajmaline, sarpagine and 2-acylindole groups.

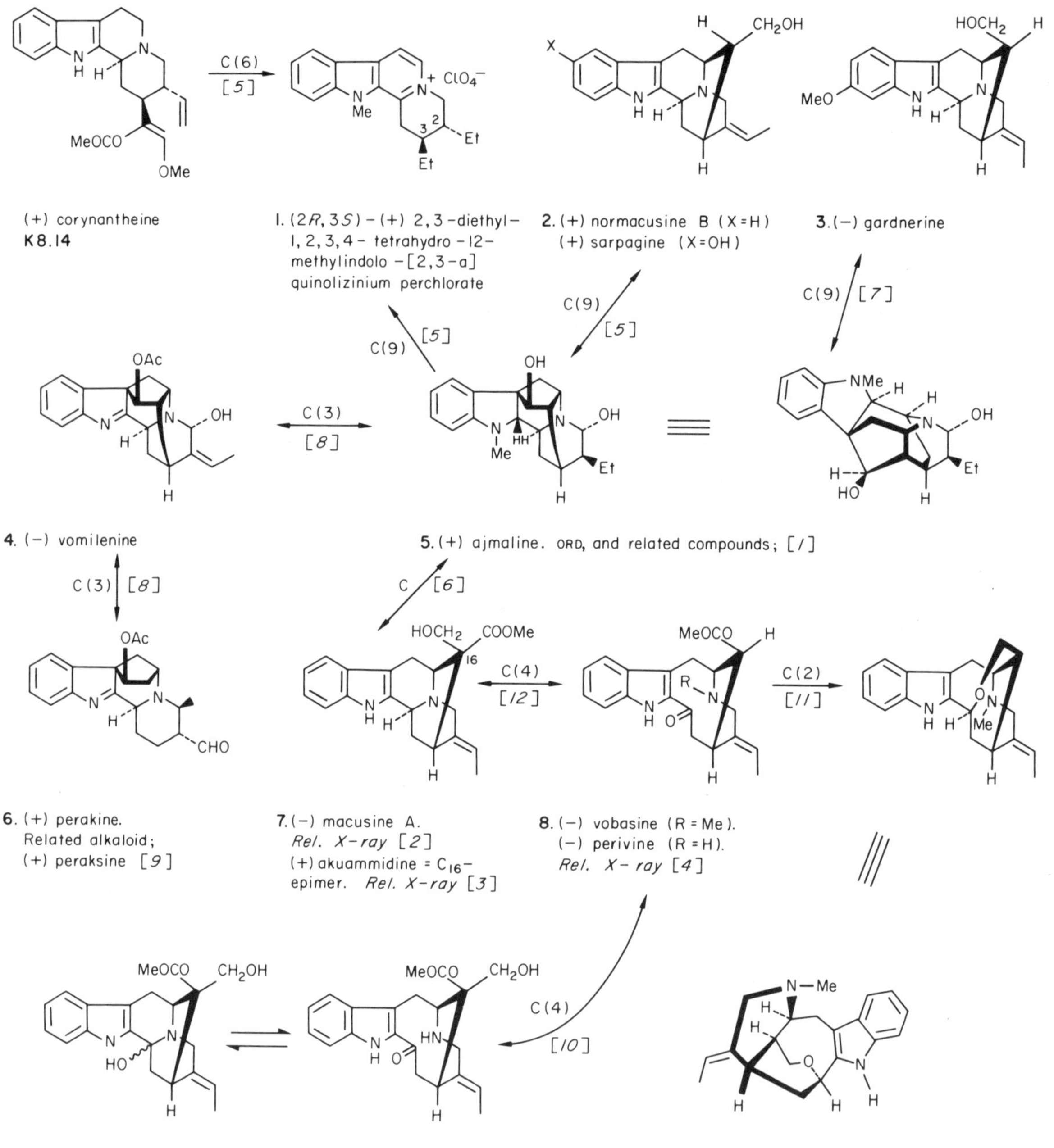

(+) corynantheine
K 8.14

1. (2R,3S)-(+) 2,3-diethyl-1,2,3,4-tetrahydro-12-methylindolo-[2,3-a]quinolizinium perchlorate

2. (+) normacusine B (X=H)
   (+) sarpagine (X=OH)

3. (−) gardnerine

4. (−) vomilenine

5. (+) ajmaline. ORD, and related compounds; [1]

6. (+) perakine.
   Related alkaloid;
   (+) peraksine [9]

7. (−) macusine A.
   Rel. X-ray [2]
   (+) akuammidine = C_{16}-epimer. Rel. X-ray [3]

8. (−) vobasine (R = Me).
   (−) perivine (R = H).
   Rel. X-ray [4]

9. (+) voacarpine.

10. (−) anhydrovobasindiol

1. M. Hanaoka, M. Hesse, and H. Schmid, *Helv. Chim. Acta*, 1970, **53**, 1723.
2. A. T. McPhail, J. M. Robertson, G. A. Sim, A. R. Battersby, H. F. Hodson, and D. A. Yeowell, *Proc. Chem. Soc.*, 1961, 223.
3. S. Silvers and A. Tulinsky, *Tetrahedron Letters*, 1962, 339.
4. H. Jaggi and U. Renner, *Chimia (Switz.)*, 1964, **18**, 173.
5. M. F. Bartlett, R. Sklar, W. I. Taylor, E. Schlittler, R. L. S. Amai, P. Beak, N. V. Bringi, and E. Wenkert, *J. Amer. Chem. Soc.*, 1962, **84**, 622.
6. A. R. Battersby and D. A. Yeowell, *J. Chem. Soc.*, 1964, 4419.
7. S. Sakai, A. Kubo, T. Hamamoto, M. Wakabayashi, K. Takahashi, Y. Ohtani, and J. Haginawa, *Tetrahedron Letters*, 1969, 1489.
8. W. I. Taylor, A. J. Frey, and A. Hofmann, *Helv. Chim. Acta*, 1962, **45**, 611.
9. A. N. Kiang, S. K. Loh, M. Demanczyk, C. W. Gemenden, G. J. Papariello, and W. I. Taylor, *Tetrahedron*, 1966, **22**, 3293.
10. J. C. Braekman, M. Kaisin, J. Pecher, and R. H. Martin, *Bull. Soc. chim. belges*, 1966, **75**, 465.
11. J. J. Dugan, M. Hesse, U. Renner, and H. Schmid, *Helv. Chim. Acta*, 1969, **52**, 701.
12. M. Gorman and J. Sweeny, *Tetrahedron Letters*, 1964, 3105.

ORD/CD; [7]

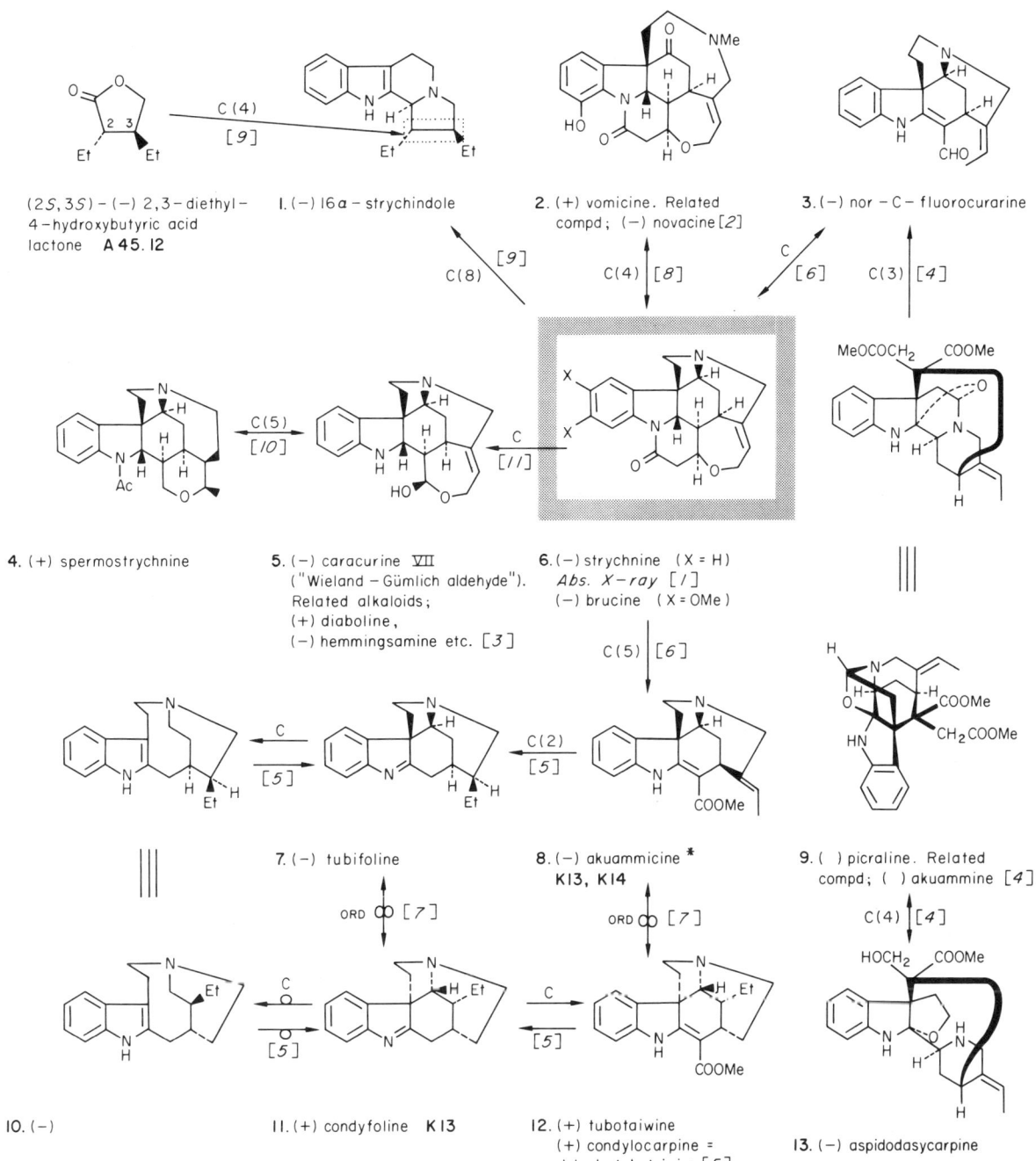

1. A. F. Peerdeman, *Acta Cryst.*, 1956, **9**, 824.
2. W. F. Martin, H. R. Bentley, J. A. Henry, and F. S. Spring, *J. Chem. Soc.*, 1952, 3603.
3. K. Biemann, J. S. Grossert, J. M. Hugo, J. Occolowitz, and F. L. Warren, *J. Chem. Soc.*, 1965, 2815; K. Biemann, J. S. Grossert, J. Occolowitz, and F. L. Warren, *ibid.*, 2818.
4. L. Olivier, J. Lévy, J. Le Men, M.-M. Janot, H. Budzikiewicz, and C. Djerassi, *Bull. Soc. chim. France*, 1965, 868.
5. D. Schumann and H. Schmid, *Helv. Chim. Acta*, 1963, **46**, 1996.
6. Ch. Weissmann, H. Schmid, and P. Karrer, *Helv. Chim. Acta*, 1961, **44**, 1877.
7. W. Klyne, R. J. Swan, B. W. Bycroft, D. Schumann, and H. Schmid, *Helv. Chim. Acta*, 1965, **48**, 443.
8. A. S. Bailey and R. Robinson, *Nature*, 1948, **161**, 433.
9. K. Nagarajan, Ch. Weissman, H. Schmid, and P. Karrer, *Helv. Chim. Acta*, 1963, **46**, 1212.
10. F. A. L. Anet and R. Robinson, *J. Chem. Soc.*, 1955, 2253.
11. H. Wieland and W. Gumlich, *Annalen*, 1932, **494**, 191.

K13

Aspidospermine group.

1. (−) vallesamidine. *Abs. X-ray* [1]
2. (−) tabersonine. (−) vincadifformine = 6,7-dihydro-deriv.
3. (+) retuline
4. (+) vincaminoreine. (+) vincaminorine = 3-epimer [12]
5. (−) quebrachamine.* Related compd; (+) conoflorin [2]
6. (−) compactinervine
7. (−) neblinine. Related alkaloids; [13]
8. (+) aspidospermidine (X = Y = H); (−) aspidospermine (X = Ac, Y = OMe). *Abs. X-ray* [3]
9. (−) rhazidine
10. (+) aspidodispermine. *Rel. X-ray* [4]
11. (+) haplocine
12. (+) 1,2-dehydroaspidospermidine

(−) akuammicine K12.8
(+) condyfoline K12.11

---

1. S. H. Brown, C. Djerassi, and P. G. Simpson, *J. Amer. Chem. Soc.*, 1969, **90**, 2445.
2. J. J. Dugan, M. Hesse, U. Renner, and H. Schmid, *Helv. Chim. Acta*, 1967, **50**, 60.
3. B. M. Craven and D. E. Zacharias, *Experientia*, 1968, **24**, 770.
4. N. C. Ling and C. Djerassi, *Tetrahedron Letters*, 1970, 3015.
5. J. Lévy, P. Maupérin, M. D. de Maindreville, and J. Le Men, *Tetrahedron Letters*, 1971, 1003.
6. W. Klyne, R. J. Swan, B. W. Bycroft, D. Schumann, and H. Schmid, *Helv. Chim. Acta*, 1965, **48**, 443.
7. E. Wenkert and R. Sklar, *J. Org. Chem.*, 1966, **31**, 2689.
8. M. Plat, J. Le Men, M.-M. Janot, J. M. Wilson, H. Budzikiewitz, L. J. Durham, Y. Nakagawa, and C. Djerassi, *Tetrahedron Letters*, 1962, 271; M.-J. Hoizey, L. Olivier, J. Lévy, and J. Le Men, *Tetrahedron Letters*, 1971, 1011.
9. S. Markey, K. Biemann, and B. Witkop, *Tetrahedron Letters*, 1967, 157.
10. B. Gilbert, A. P. Duarte, Y. Nakagawa, J. A. Joule, S. E. Flores, J. A. Brissolese, J. Campello, E. P. Carrazzoni, R. J. Owellen, E. C. Blossey, K. S. Brown, and C. Djerassi, *Tetrahedron*, 1965, **21**, 1141.
11. M. P. Cava, K. Nomura, and S. K. Talapatra, *Tetrahedron*, 1964, **20**, 581.
12. J. Mokry, I. Kompis, M. Shamma, and R. J. Shine, *Chem. and Ind.*, 1964, 1988.
13. K. S. Brown and C. Djerassi, *J. Amer. Chem. Soc.*, 1964, **86**, 2451.
14. B. W. Bycroft, D. Schumann, M. B. Patel, and H. Schmid, *Helv. Chim. Acta*, 1964, **47**, 1147.

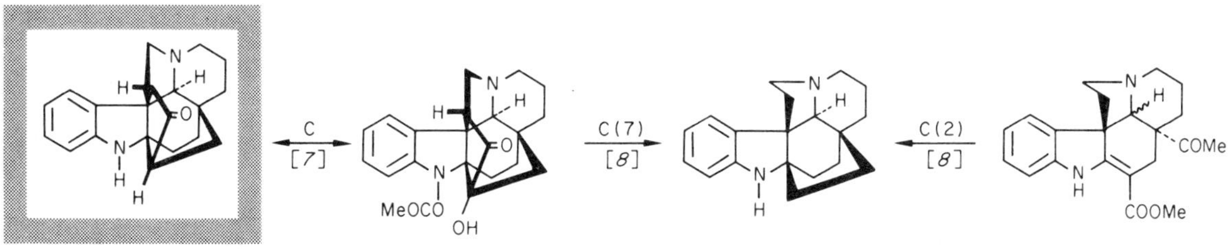

Kopsine and aspidofractinine groups; miscellaneous indole alkaloids

1. (−) kopsanone. Abs. X-ray [1]
2. (−) kopsine
3. (−) aspidofractinine
4. (−) minovincine

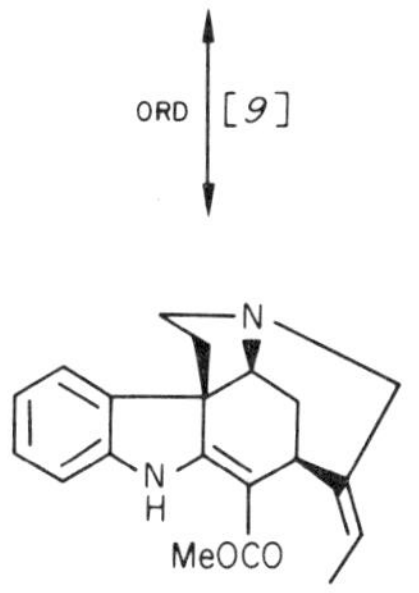

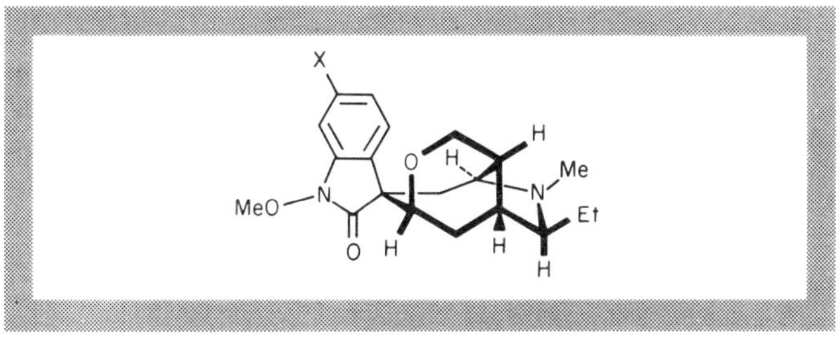

5. (−) gelsemicine (X = OMe). Abs. X-ray [2]   (−) gelsedine (X = H)

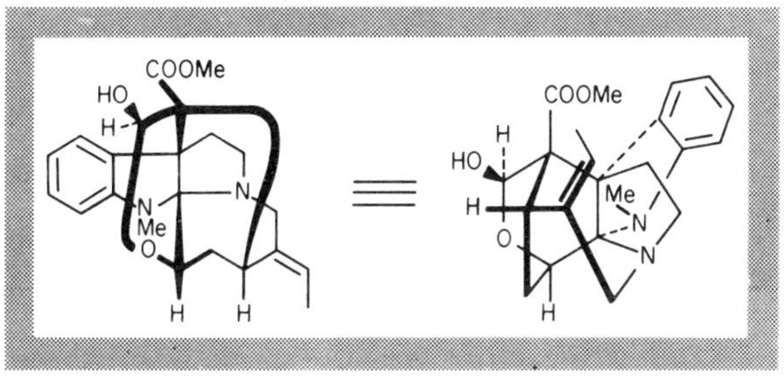

(−) akuammicine K12.8

6. (+) corymine. Abs. X-ray [3] 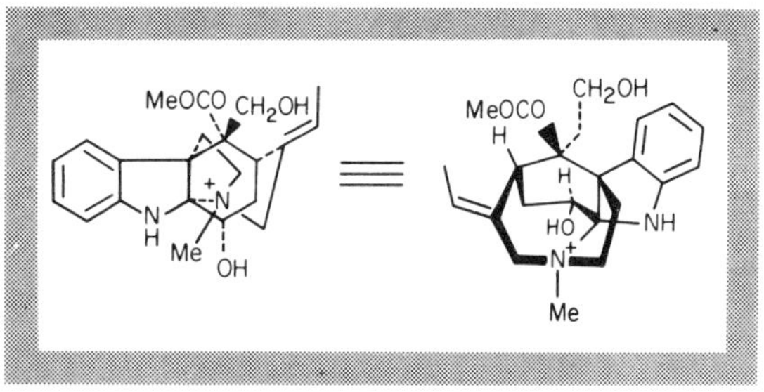
7. ( ) echitamine. Abs. X-ray [4]

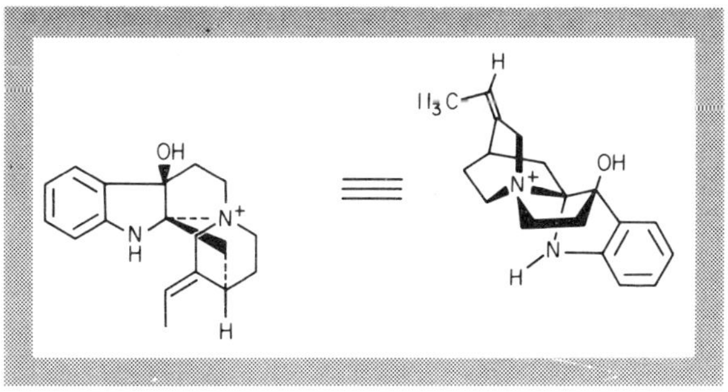

8. (−) hunteracine. Abs. X-ray [5]
9. (−) gardneramine. Abs. X-ray [6] 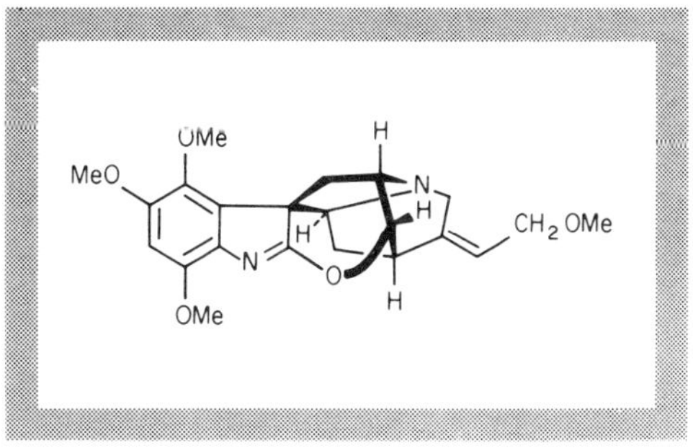

1. B. M. Craven, B. Gilbert, and L. A. P. Leme, *Chem. Comm.*, 1968, 955.
2. M. Przybylska, *Acta Cryst.*, 1962, 15, 301; M. Przybylska and L. Marion, *Canad. J. Chem.*, 1961, 39, 2124.
3. C. W. L. Bevan, M. B. Patel, A. H. Rees, D. R. Harris, M. L. Marshak, and H. H. Mills, *Chem. and Ind.*, 1965, 603.
4. H. Manohar and S. Ramaseshan, *Tetrahedron Letters*, 1961, 814.
5. R. H. Burnell, A. Chapelle, M. F. Khalil, and P. H. Bird, *Chem. Comm.*, 1970, 772.
6. N. Aimi, S. Sakai, Y. Iitaka, and A. Itai, *Tetrahedron Letters*, 1971, 2061.
7. J. M. F. Filho, B. Gilbert, M. Kitagawa, L. A. P. Leme, and L. J. Durham, *J. Chem. Soc. (C)*, 1966, 1260.
8. A. Guggisberg, A. A. Gorman, B. W. Bycroft, and H. Schmid, *Helv. Chim. Acta*, 1969, 52, 76.
9. W. Klyne, R. J. Swan, B. W. Bycroft, D. Schumann, and H. Schmid, *Helv. Chim. Acta*, 1965, 48, 443.

K15 Vinca and iboga alkaloids.

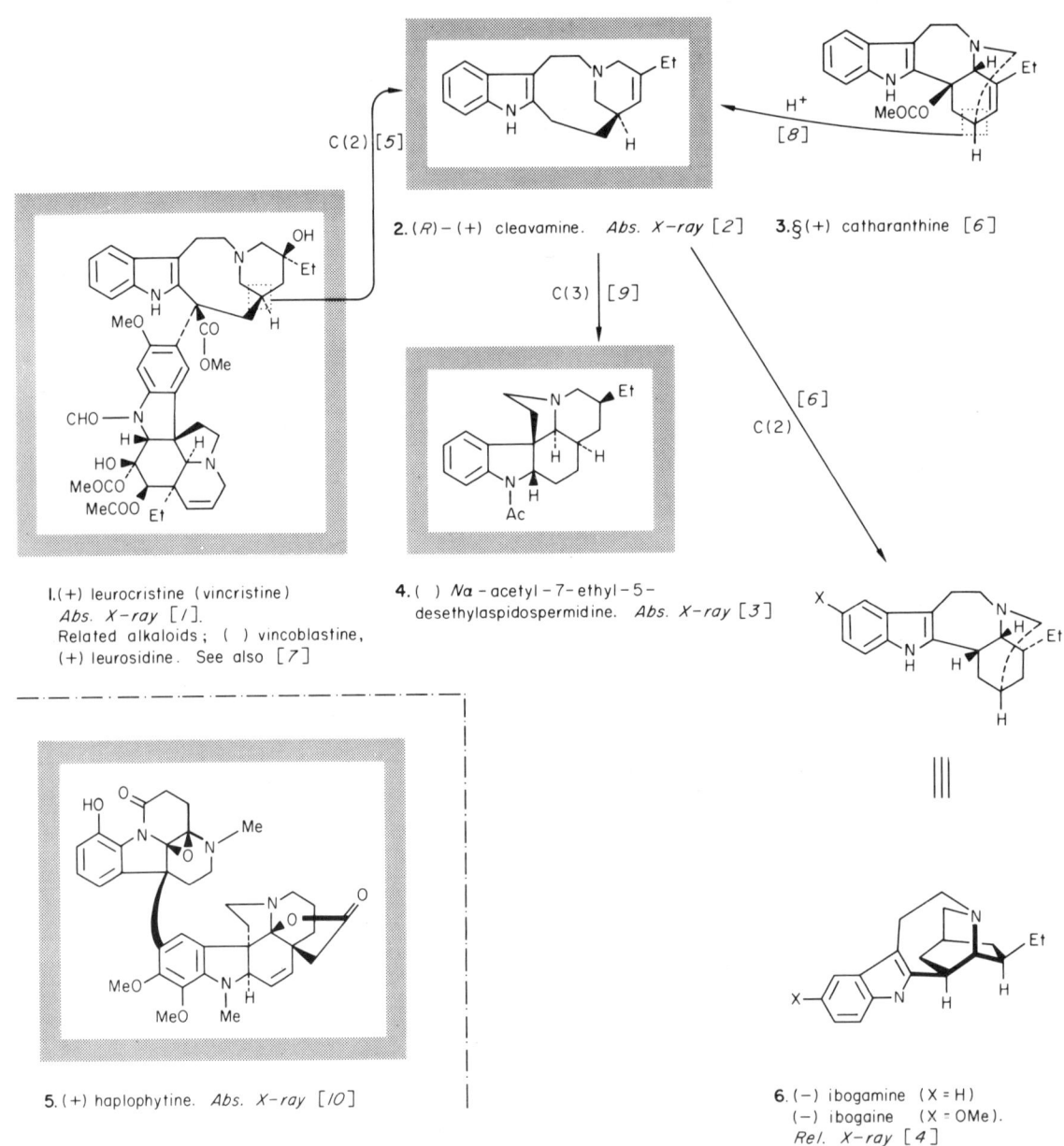

1. (+) leurocristine (vincristine) Abs. X-ray [1]. Related alkaloids; (−) vincoblastine, (+) leurosidine. See also [7]
2. (R)−(+) cleavamine. Abs. X-ray [2]
3. §(+) catharanthine [6]
4. ( ) Na-acetyl-7-ethyl-5-desethylaspidospermidine. Abs. X-ray [3]
5. (+) haplophytine. Abs. X-ray [10]
6. (−) ibogamine (X = H)
   (−) ibogaine (X = OMe).
   Rel. X-ray [4]

Other dimeric indole alkaloids. Some of these have been identified only by mass spectrometry and their absolute configurations are at present therefore unknown; others have been identified by fission followed by correlation of the constituent halves with other known indole alkaloids [5].

1. J. W. Moncrief and W. N. Lipscomb, *J. Amer. Chem. Soc.*, 1965, 87, 4963.
2. N. Camerman and J. Trotter, *Acta Cryst.*, 1964, 17, 384.
3. A. Camerman, N. Camerman, and J. Trotter, *Acta Cryst.*, 1965, 19, 314.
4. G. Arai, J. Coppola, and G. A. Jeffrey, *Acta Cryst.*, 1960, 13, 553.
5. A. A. Gorman, N. J. Dastoor, M. Hesse, W. von Philipsborn, U. Renner, and H. Schmid, *Helv. Chim. Acta,* 1969, 52, 33.
6. J. P. Kutney, R. T. Brown, and E. Piers, *Canad. J. Chem.*, 1966, 44, 637.
7. N. Neuss, M. Gorman, N. J. Cone, and L. L. Huckstep, *Tetrahedron Letters*, 1968, 783.
8. J. P. Kutney, J. Trotter, T. Tabata, A. Kerigan, and N. Camerman, *Chem. and Ind.*, 1963, 648.
9. J. P. Kutney, E. Piers, and R. T. Brown, *J. Amer. Chem. Soc.*, 1970, 92, 1700.
10. I. D. Rae, M. Rosenberger, A. G. Szabo, C. R. Willis, P. Yates, D. E. Zacharias, G. A. Jeffrey, B. Douglas, J. L. Kirkpatrick, and J. A. Weisbach, *J. Amer. Chem. Soc.*, 1967, 89, 3061.

Eburnamine group Talbotine. K¹⁶

Other ebornamine alkaloids and their ORD; [1]

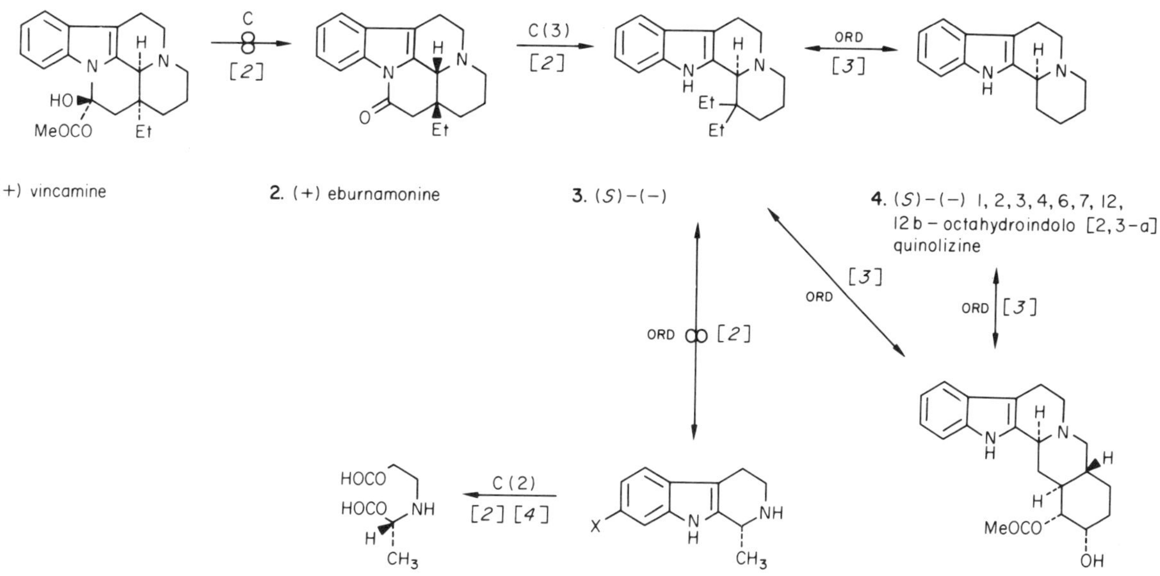

1. (+) vincamine
2. (+) eburnamonine
3. (S)-(−)
4. (S)-(−) 1,2,3,4,6,7,12,12b − octahydroindolo [2,3-a] quinolizine
5. (R)-(+) 1,2,3,4- tetrahydroharman (X=H); (R)-(+) 1,2,3,4- tetrahydroharmine (X = OMe)

(R)-(−) N-carboxyethylalanine **A1.18**

(+) yohimbine **K9.9**

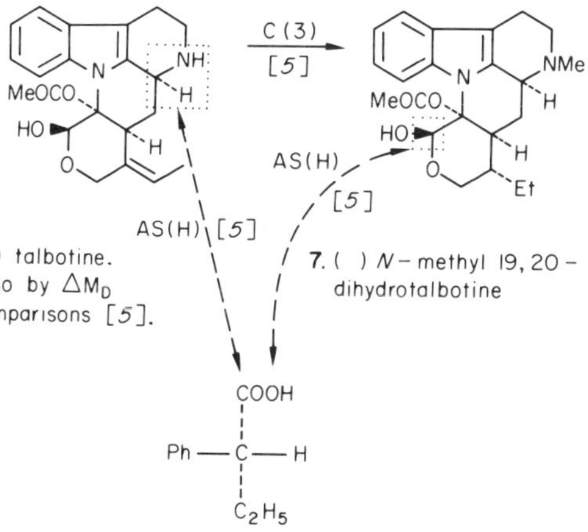

6. (−) talbotine. Also by ΔM_D comparisons [5].
7. ( ) N-methyl 19,20-dihydrotalbotine

(R)-(−) 2-phenylbutyric acid **A25.15**

1. K. Bláha, K. Kavková, Z. Koblicová, and J. Trojánek, *Coll. Czech. Chem. Comm.*, 1968, **33**, 3833.
2. J. Trojánek, Z. Koblicová, and K. Bláha, *Chem. and Ind.*, 1965, 1261.
3. J. Pospisek, Z. Koblicová, J. Trojánek, and K. Bláha, *Chem. and Ind.*, 1969, 25.
4. Z. Koblicová and J. Trojánek, *Chem. and Ind.*, 1966, 1342.
5. M. Pinar, M. Hanaoka, M. Hesse, and H. Schmid, *Helv. Chim. Acta*, 1971, **54**, 15.

K17 Ergot alkaloids; betanin, indicaxanthin. For further ergot alkaloids, see [10]

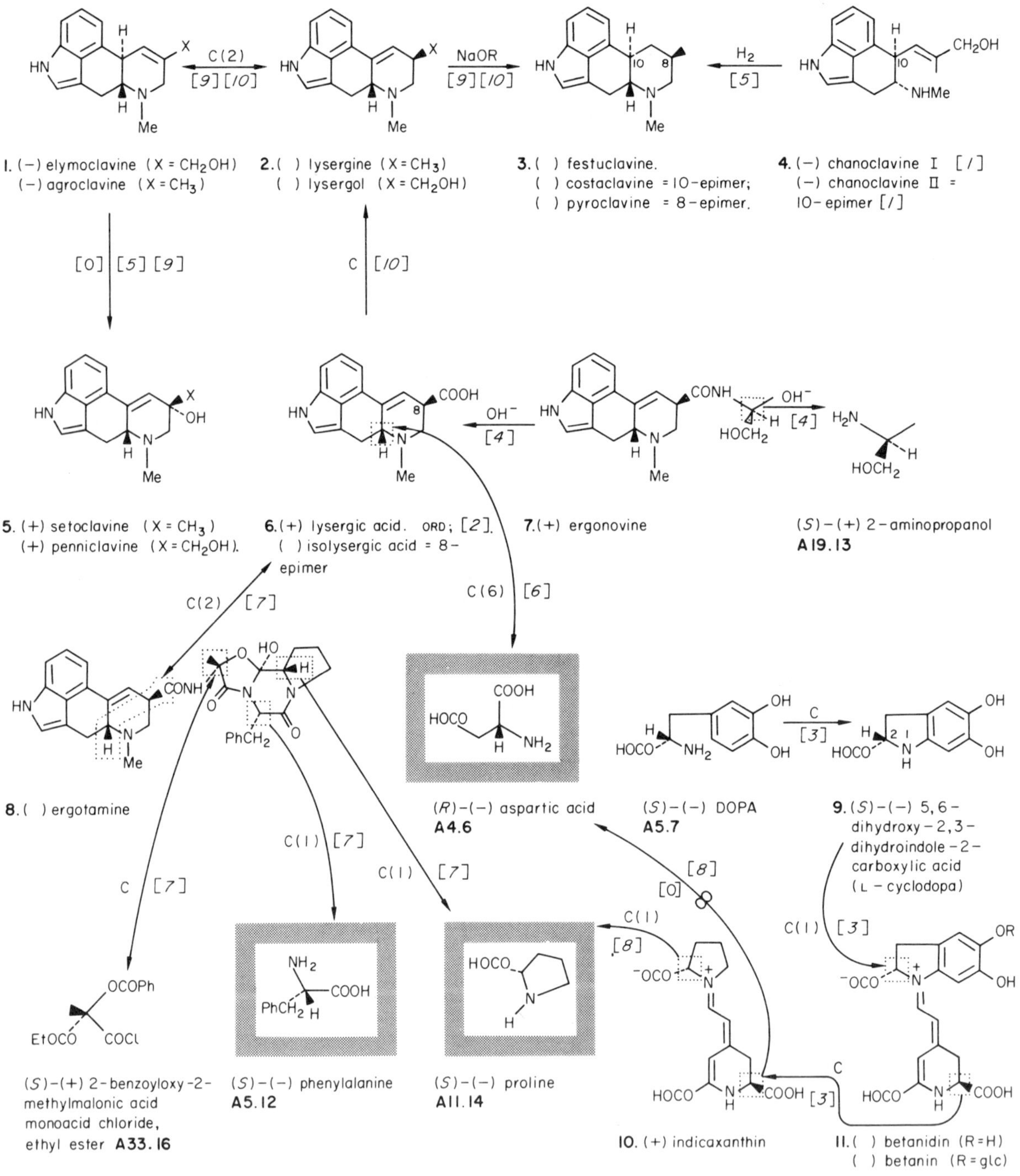

1. (−) elymoclavine (X = CH$_2$OH)
   (−) agroclavine (X = CH$_3$)
2. ( ) lysergine (X = CH$_3$)
   ( ) lysergol (X = CH$_2$OH)
3. ( ) festuclavine.
   ( ) costaclavine = 10-epimer;
   ( ) pyroclavine = 8-epimer.
4. (−) chanoclavine I [1]
   (−) chanoclavine II = 10-epimer [1]
5. (+) setoclavine (X = CH$_3$)
   (+) penniclavine (X = CH$_2$OH).
6. (+) lysergic acid. ORD; [2].
   ( ) isolysergic acid = 8-epimer
7. (+) ergonovine

(S)-(+) 2-aminopropanol A19.13

8. ( ) ergotamine

(S)-(+) 2-benzoyloxy-2-methylmalonic acid monoacid chloride, ethyl ester A33.16

(S)-(−) phenylalanine A5.12

(S)-(−) proline A11.14

(R)-(−) aspartic acid A4.6

(S)-(−) DOPA A5.7

9. (S)-(−) 5,6-dihydroxy-2,3-dihydroindole-2-carboxylic acid (L-cyclodopa)

10. (+) indicaxanthin
11. ( ) betanidin (R = H)
    ( ) betanin (R = glc)

1. W. Acklin, T. Fehr, and D. Arigoni, *Chem. Comm.*, 1966, 799.
2. H. G. Leeman and S. Fabbri, *Helv. Chim. Acta*, 1959, **42**, 2696.
3. H. Wyler, M. E. Wilcox, and A. S. Dreiding, *Helv. Chim. Acta*, 1965, **48**, 361, 1134.
4. W. A. Jacobs and L. C. Craig, *Science*, 1935, **82**, 16.
5. A. Hofmann, R. Brunner, H. Kobel, and A. Brack, *Helv. Chim. Acta*, 1957, **40**, 1358.
6. P. A. Stadler and A. Hofmann, *Helv. Chim. Acta*, 1962, **45**, 2005.
7. A. Hofmann, H. Ott, R. Griot, P. A. Stadler, and A. J. Frey, *Helv. Chim. Acta*, 1963, **46**, 2306, and references therein.
8. M. Piattelli, L. Minale, and G. Prota, *Tetrahedron*, 1964, **20**, 2325.
9. S. Yamatodani and M. Abe, *Bull. Agr. Chem. Soc. Japan*, 1955, **19**, 94, and references therein.
10. A. Stoll and A. Hofmann, *The Alkaloids*, 1965, **8**, 726.

Piperidine alkaloids.

K[18]

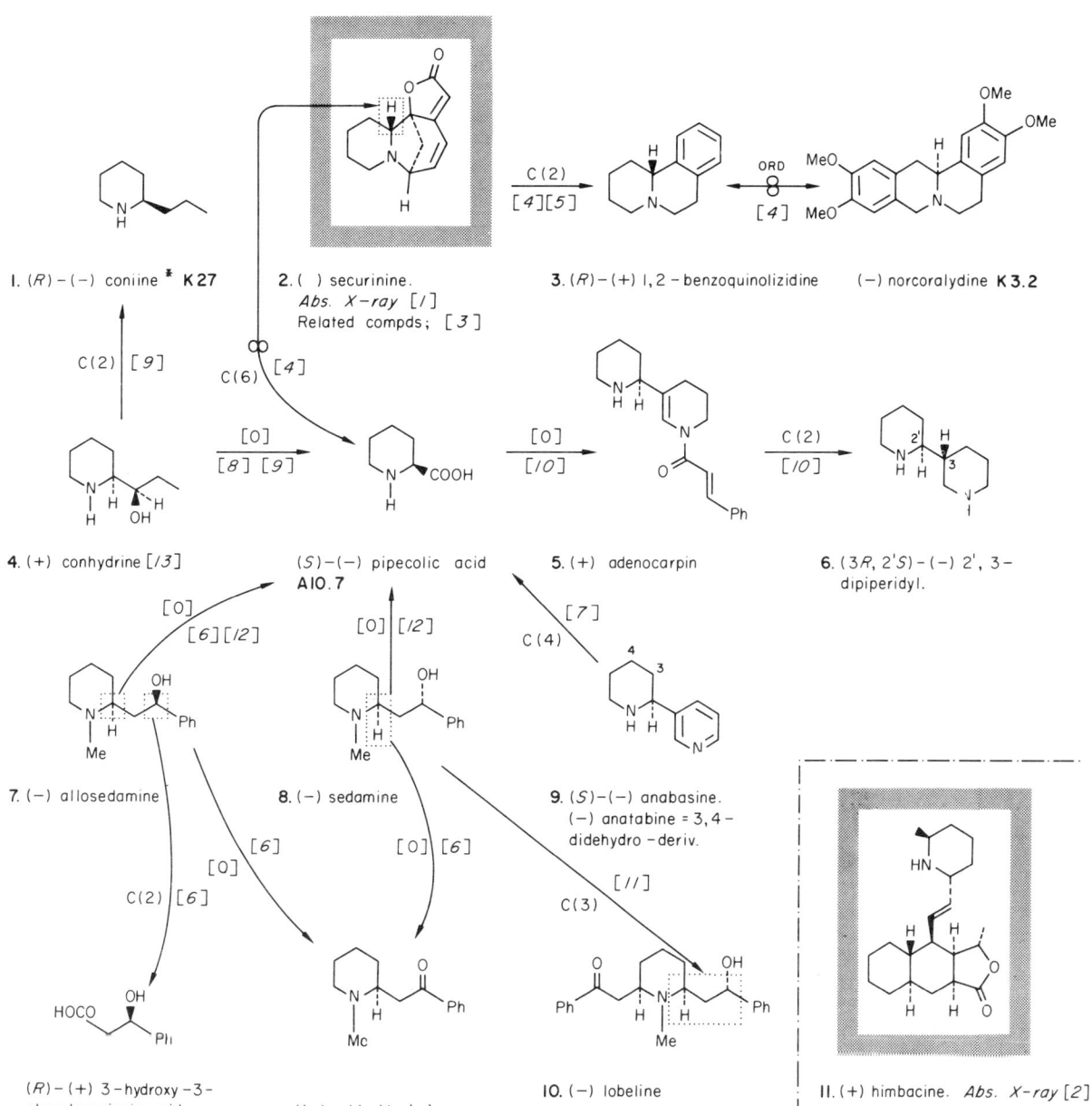

1. (R)-(−) coniine * K27
2. ( ) securinine. Abs. X-ray [1] Related compds; [3]
3. (R)-(+) 1,2-benzoquinolizidine
   (−) norcoralydine K3.2
4. (+) conhydrine [13]
   (S)-(−) pipecolic acid A10.7
5. (+) adenocarpin
6. (3R, 2'S)-(−) 2',3-dipiperidyl.
7. (−) allosedamine
8. (−) sedamine
9. (S)-(−) anabasine. (−) anatabine = 3,4-didehydro-deriv.
10. (−) lobeline
11. (+) himbacine. Abs. X-ray [2]

(R)-(+) 3-hydroxy-3-phenylpropionic acid A23.1

Hydrochloride (−)

1. S. Imado, M. Shiro, and Z. Horii, *Chem. and Ind.*, 1964, 1691.
2. J. Fridrichsons and A. McL. Mathieson, *Acta Cryst.*, 1962, **15**, 119.
3. Z. Horii, M. Yamauchi, M. Ikeda, and T. Momose, *Chem. Pharm. Bull. Japan*, 1970, **18**, 2009, and references therein.
4. Z. Horii, M. Ikeda, Y. Yamawaki, Y. Tamura, S. Saito, and K. Kodera, *Tetrahedron*, 1963, **19**, 2021.
5. J. C. Craig, R. P. K. Chan, and S. K. Roy, *Tetrahedron*, 1967, **23**, 3573.
6. C. Schöpf, G. Dummer, and W. Wüst, *Annalen*, 1959, **626**, 134.
7. R. Lukes, A. A. Arojan, J. Kovár, and K. Bláha, *Coll. Czech. Chem. Comm.*, 1962, **27**, 751.
8. R. Willstätter, *Ber.*, 1901, **34**, 3166.
9. W. Leithe, *Ber.*, 1932, **65**, 927.
10. C. Schöpf, F. Braun, H. Koop, and G. Werner, *Annalen*, 1962, **658**, 156.
11. C. Schöpf and E. Müller, *Annalen*, 1965, **687**, 241.
12. H. C. Beyerman, J. Eenshuistra, and W. Eveleens, *Rec. Trav. chim.*, 1957, **76**, 415.
13. R. K. Hill, *J. Org. Chem.*, 1958, **80**, 1609.

K19 Piperidine and pyrrolidine alkaloids.

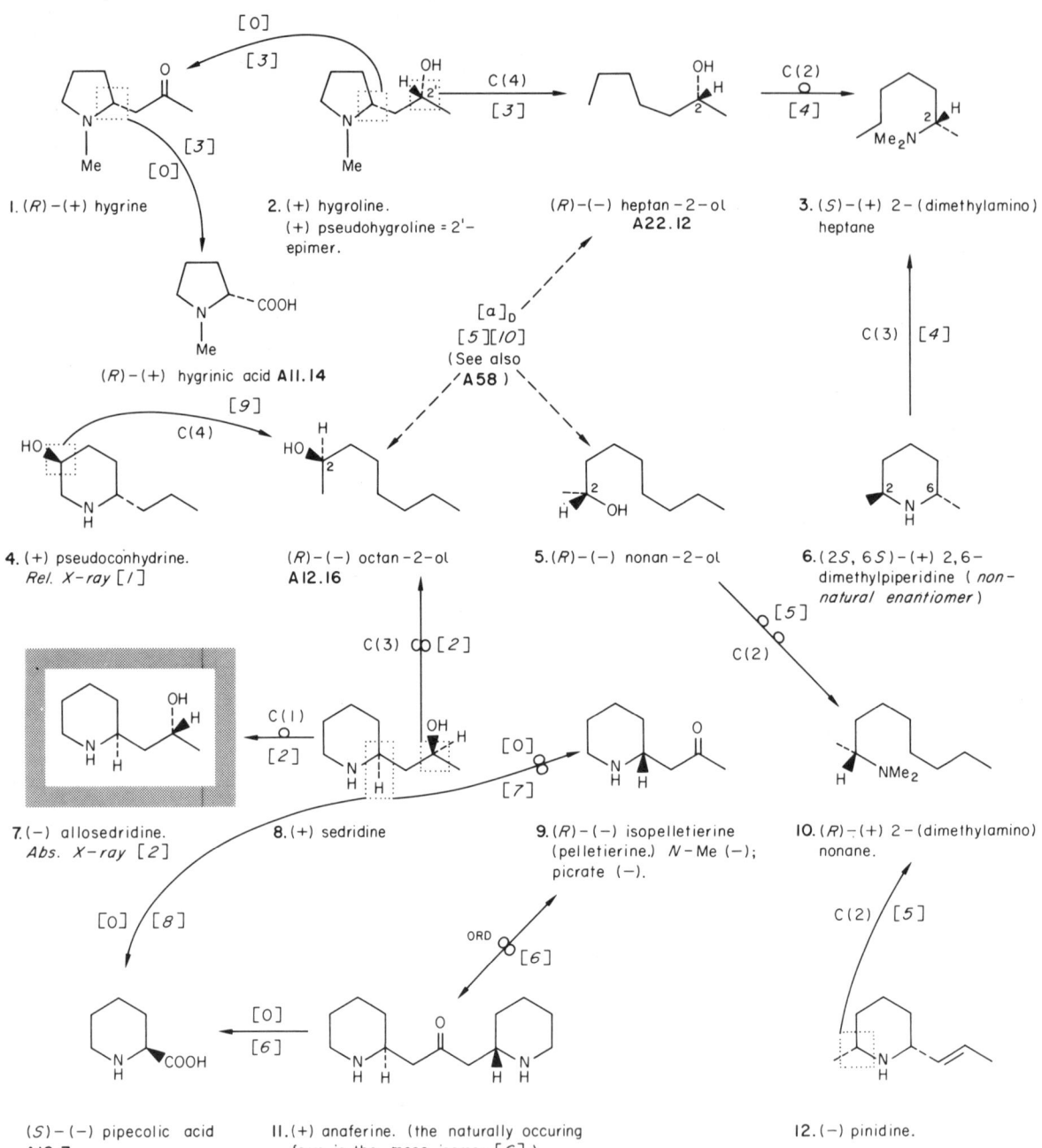

1. (R)-(+) hygrine
2. (+) hygroline. (+) pseudohygroline = 2'-epimer.
   (R)-(-) heptan-2-ol A22.12
3. (S)-(+) 2-(dimethylamino) heptane
   (R)-(+) hygrinic acid A11.14
4. (+) pseudoconhydrine. Rel. X-ray [1]
   (R)-(-) octan-2-ol A12.16
5. (R)-(-) nonan-2-ol
6. (2S,6S)-(+) 2,6-dimethylpiperidine (non-natural enantiomer)
7. (-) allosedridine. Abs. X-ray [2]
8. (+) sedridine
9. (R)-(-) isopelletierine (pelletierine.) N-Me (-); picrate (-).
10. (R)-(+) 2-(dimethylamino) nonane.
11. (+) anaferine. (the naturally occuring form is the meso isomer [6])
12. (-) pinidine.

(S)-(-) pipecolic acid A10.7

1. H. S. Yanai and W. N. Lipscomb, *Tetrahedron*, 1959, **6**, 103.
2. D. Butruille, G. Fodor, C. S. Huber, and F. Letourneau, *Tetrahedron*, 1971, **27**, 2055.
3. R. Lukes, J. Kovár, J. Kloubek, and K. Bláha, *Coll. Czech. Chem. Comm.*, 1960, **25**, 483, and references therein.
4. R. K. Hill and J. W. Morgan, *J. Org. Chem.*, 1966, **31**, 3451.
5. R. K. Hill, T. H. Chan, and J. A. Joule, *Tetrahedron*, 1965, **21**, 147.
6. M. M. El-Olemy and A. E. Schwarting, *J. Org. Chem.*, 1969, **34**, 1352.
7. H. C. Beyerman and L. Maat, *Rec. Trav. chim.*, 1963, **82**, 1033.
8. H. C. Beyerman, L. Maat, A. Van Veen, A. Zweistra, and A. von Phillipsborn, *Rec. Trav. chim.*, 1965, **84**, 1367.
9. R. K. Hill, *J. Amer. Chem. Soc.*, 1958, **80**, 1611.
10. J. A. Mills and W. Klyne, *Progr. Stereochem.*, 1954, **1**, 177.

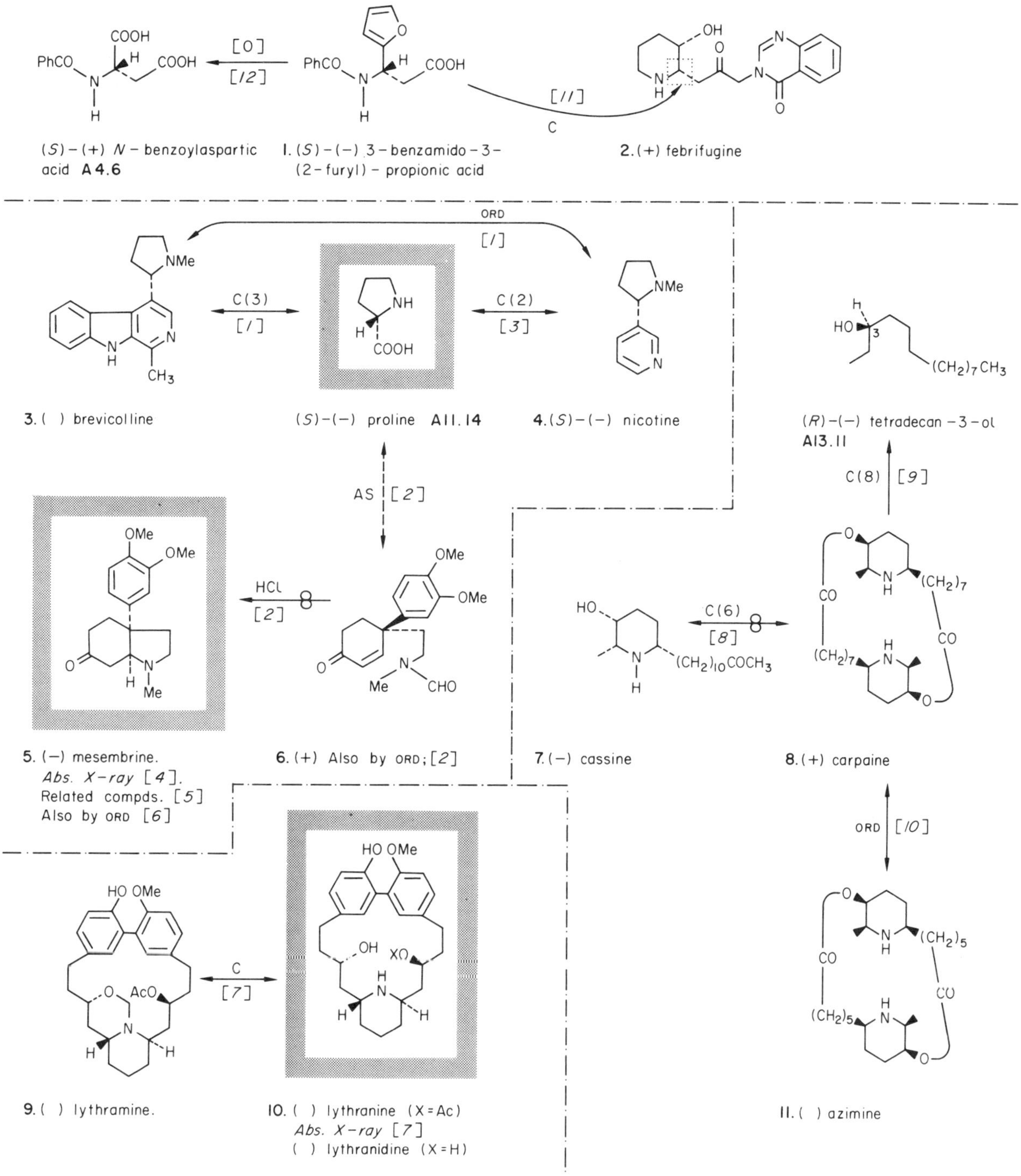

Piperidine alkaloids (contd.); miscellaneous types.

1. (S)-(-) 3-benzamido-3-(2-furyl)-propionic acid
   → [O]/[12] → (S)-(+) N-benzoylaspartic acid **A 4.6**
   → [11] → 2. (+) febrifugine

3. ( ) brevicolline
   ← C(3) [1] ← (S)-(-) proline **A11.14** → C(2) [3] → 4. (S)-(-) nicotine
   ORD [1]

(R)-(-) tetradecan-3-ol **A13.11**
↑ C(8) [9]
8. (+) carpaine
↕ ORD [10]
11. ( ) azimine

5. (-) mesembrine.
   Abs. X-ray [4].
   Related compds. [5]
   Also by ORD [6]
   ← HCl [2] ← 6. (+) Also by ORD; [2]

AS [2]

7. (-) cassine ← C(6) [8] ← 8. (+) carpaine

9. ( ) lythramine → C [7] → 10. ( ) lythranine (X=Ac)
   Abs. X-ray [7]
   ( ) lythranidine (X=H)

---

1. K. Bláha, Z. Koblicová, J. Pospisek, and J. Trojánek, *Coll. Czech. Chem. Comm.*, 1971, **36**, 3448.
2. S. Yamada and G. Otani, *Tetrahedron Letters*, 1971, 1133.
3. P. Karrer and R. Widmer, *Helv. Chim. Acta*, 1925, **8**, 364.
4. P. Coggon, D. S. Farrier, P. W. Jeffs, and A. T. McPhail, *J. Chem. Soc. (B)*, 1970, 1267.
5. P. W. Jeffs, G. Ahmann, H. F. Campbell, D. S. Farrier, G. Ganguli, and R. C. Hawks, *J. Org. Chem.*, 1970, **35**, 3512.
6. P. W. Jeffs, R. L. Hawks, and D. S. Farrier, *J. Amer. Chem. Soc.*, 1969, **91**, 3831.
7. E. Fujita and K. Fuji, *J. Chem. Soc. (C)*, 1971, 1651.
8. W. Y. Rice and J. L. Coke, *J. Org. Chem.*, 1966, **31**, 1010.
9. J. L. Coke and W. Y. Rice, *J. Org. Chem.*, 1965, **30**, 3420.
10. T. M. Smalberger, G. J. H. Rall, H. L. de Waal, and R. R. Arndt, *Tetrahedron*, 1968, **24**, 6417.
11. B. R. Baker, F. J. McEvoy, R. E. Schaub, J. P. Joseph, and J. H. Williams, *J. Org. Chem.*, 1953, **18**, 178.
12. R. K. Hill and A. G. Edwards, *Chem. and Ind.*, 1962, 858.

K²¹ Quinolizidine alkaloids, including sparteine group. For a tabulation of others not shown here, see [9]

1. (R)-(−) 4-methylnonane A 59.12
2. (−) lupinine
3. (1S,10R)-(+) 1-(aminomethyl)-quinolizidine
4. (+) lamprolobine
5. (−) cytisine
6. (−) anagyrine (X = H) (−) baptifoline (X = OH)
7. (−) lupanine *
8. (+) angustifoline *
9. (+) sparteine (X = H) * Rel. X-ray [1]. (+) retamine (X = OH)
10. (+) leontiformine
11. (−) isolupanine *
12. (−) thermopsine *
13. (−) β-isosparteine
14. (+) α-isosparteine [9] Rel. X-ray [8]
15. (+) monspessaulanine

1. S. N. Srivastava and M. Przybylska, *Tetrahedron Letters*, 1968, 2697.
2. R. C. Cookson, *Chem. and Ind.*, 1953, 337.
3. S. Okuda, K. Tsuda, and H. Kataoka, *Chem. and Ind.*, 1961, 1115.
4. N. K. Hart, S. R. Johns, and J. A. Lamberton, *Chem. Comm.*, 1968, 302.
5. F. Bohlmann, E. Winterfeldt, H. Overwien, and H. Pagel, *Chem. Ber.*, 1962, **95**, 944.
6. L. Marion, M. Wiewiorowski, and M. D. Bratek, *Tetrahedron Letters*, 1960, **no. 19**, 1.
7. E. P. White, *J. Chem. Soc.*, 1964, 4613.
8. M. Przybylska and W. H. Barnes, *Acta Cryst.*, 1963, **6**, 377.
9. S. Okuda, H. Kataoka, and K. Tsuda, *Chem. Pharm. Bull. Japan*, 1965, **13**, 487, 491, and references therein.
10. N. M. Mollov and I. C. Ivanov, *Tetrahedron*, 1970, **26**, 3805.
11. L. Marion and N. J. Leonard, *Canad. J. Chem.*, 1951, **29**, 355, and references therein.

---

6. S. Masamune, *J. Amer. Chem. Soc.*, 1960, **82**, 5253, and references therein.
7. R. Adams and D. Fles, *J. Amer. Chem. Soc.*, 1959, **81**, 5803.
8. L. J. Dry, M. J. Koekemoer, and F. L. Warren, *J. Chem. Soc.*, 1955, 59.
9. R. Adams and N. J. Leonard, *J. Amer. Chem. Soc.*, 1944, **66**, 257.
10. N. J. Leonard in *The Alkaloids*, 1950, **1**, 107; 1960, **6**, 35.
11. F. L. Warren and M. E. von Klemperer, *J. Chem. Soc.*, 1958, 4574.
12. A. V. Danilova and L. M. Utkin, *Z. Obshch. Khim.*, 1960, **30**, 345.
13. A. J. Aasen and C. C. J. Culvenor, *Chem. Comm.*, 1969, 34.
14. G. P. Menshikov and A. D. Kuzovkov, *Zh. obshch. Khim.*, 1949, **19**, 1702.
15. A. J. Aasen, C. C. J. Culvenor, and L. W. Smith, *J. Org. Chem.*, 1969, **34**, 4137.

K22

Pyrrolizidine alkaloids

These are usually mono- or diesters of a pyrrolizidine component (the necine) with a variety of carboxylic acids (the necic acids), some of which are achiral. The chiral necic acids appear on pp. K23-K24.

Reviews of necine stereochemistry; [3][4]. ORD/CD and tabulations of pyrrolizidine alkaloids; [5]

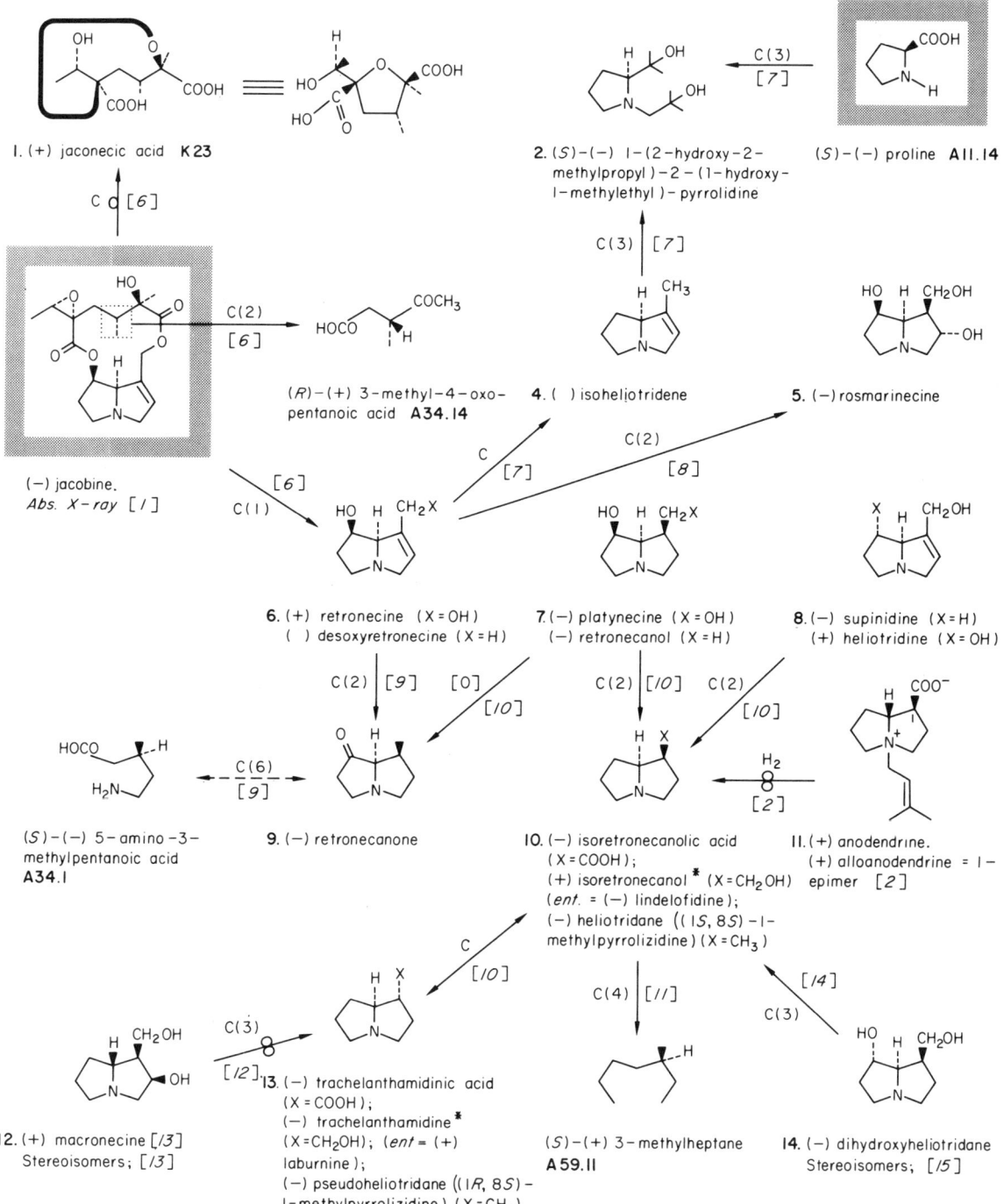

1. J. Fridrichsons, A. McL. Mathieson, and D. J. Sutor, *Acta Cryst.*, 1963, **16**, 1075.
2. K. Sasaki and Y. Hirata, *Tetrahedron Letters*, 1969, 4065.
3. N. K. Kochetov and A. M. Likhosherstov, *Adv. Heterocyclic Chem.*, 1965, **5**, 315.
4. L. B. Bull, C. C. J. Culvenor, and A. T. Dick, *The Pyrrolizidine Alkaloids*, North Holland Publishing Co., 1968.
5. C. C. J. Culvenor, D. H. G. Crout, W. Klyne, W. P. Mose, J. D. Renwick, and P. M. Scopes, *J. Chem. Soc. (C)*, 1971, 3653.

References continued on page 160

Necic acids.
For others not given here, see [3]

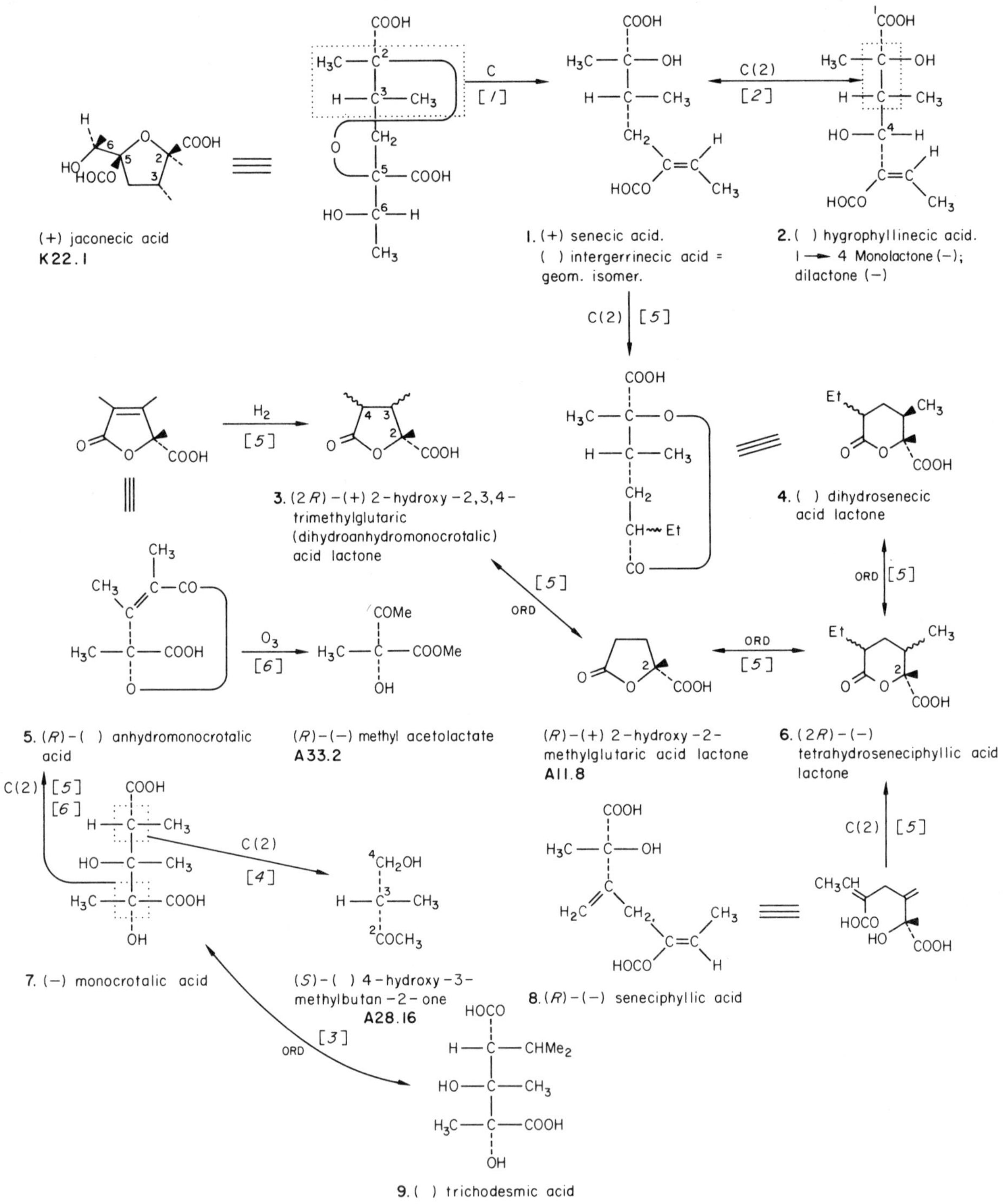
(+) jaconecic acid
K22.1
C
[1]
1. (+) senecic acid.
( ) intergerrinecic acid = geom. isomer.
C(2)
[2]
2. ( ) hygrophyllinecic acid.
1 → 4 Monolactone (−); dilactone (−)
C(2) [5]
H2
[5]
3. (2R)-(+) 2-hydroxy-2,3,4-trimethylglutaric (dihydroanhydromonocrotalic) acid lactone
4. ( ) dihydrosenecic acid lactone
ORD [5]
O3
[6]
5. (R)-( ) anhydromonocrotalic acid
(R)-(−) methyl acetolactate
A33.2
ORD
[5]
(R)-(+) 2-hydroxy-2-methylglutaric acid lactone
A11.8
ORD
[5]
6. (2R)-(−) tetrahydroseneciphyllic acid lactone
C(2) [5]
C(2) [5] [6]
C(2)
[4]
(S)-( ) 4-hydroxy-3-methylbutan-2-one
A28.16
7. (−) monocrotalic acid
8. (R)-(−) seneciphyllic acid
ORD
[3]
9. ( ) trichodesmic acid

1. C. C. J. Culvenor, *Austral. J. Chem.*, 1964, **17**, 233.
2. F. D. Schlosser and F. L. Warren, *J. Chem. Soc.*, 1965, 5707.
3. C. C. J. Culvenor, D. H. G. Crout, W. Klyne, W. P. Mose, J. D. Renwick, and P. M. Scopes, *J. Chem. Soc. (C)*, 1971, 3653.
4. D. J. Robins and D. H. G. Crout, *J. Chem. Soc. (C)*, 1969, 1386.
5. O. Cervinka, L. Hub, A. Klásek and F. Santavý, *Chem. Comm.*, 1968, 261.
6. D. J. Robins and D. H. G. Crout, *J. Chem. Soc. (C)*, 1970, 1334.

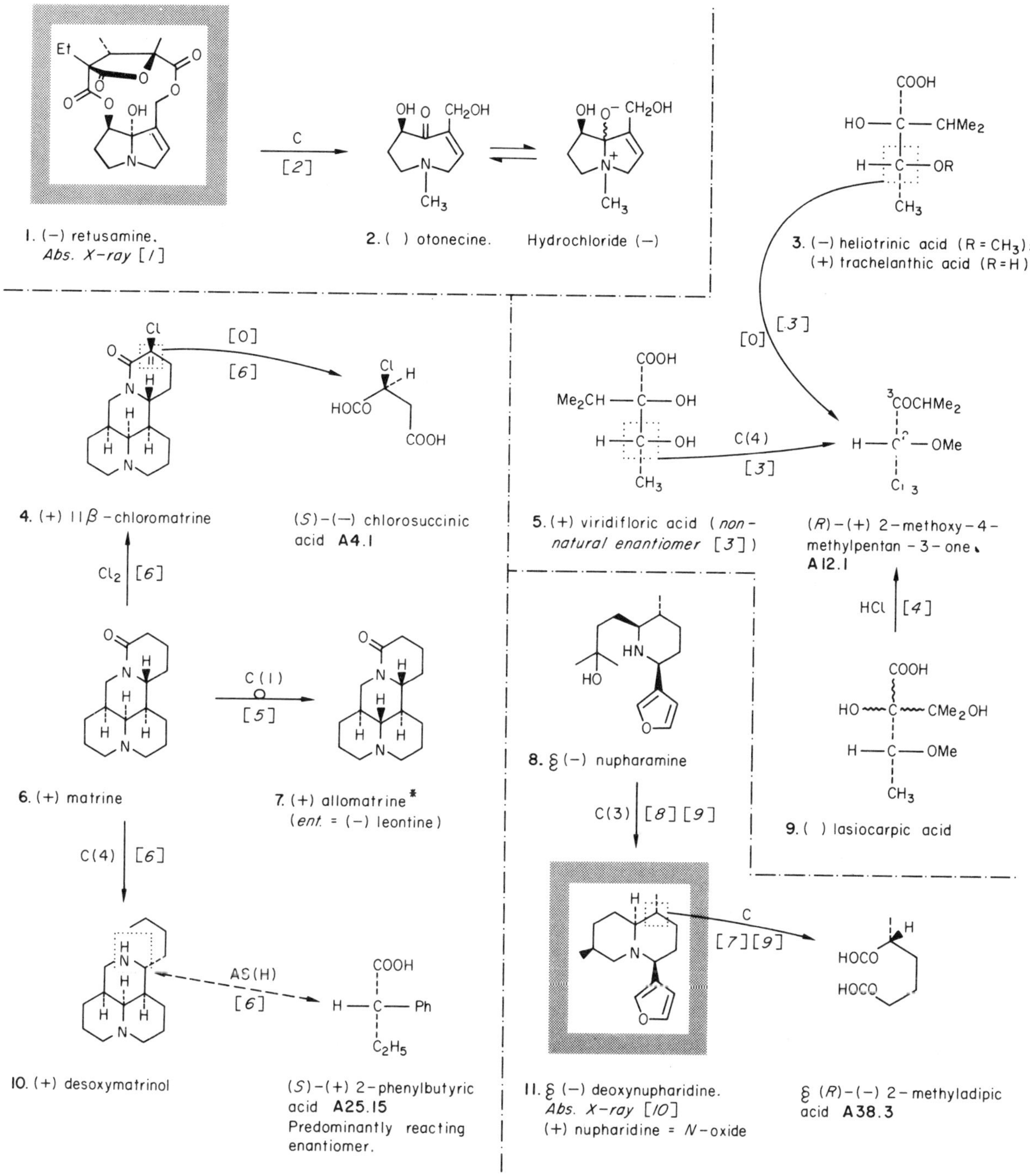

Pyrrolizidine alkaloids and necic acids (contd.); matrine and nupharidine groups. K[24]

1. (−) retusamine. Abs. X-ray [1]
2. ( ) otonecine. Hydrochloride (−)
3. (−) heliotrinic acid (R = CH$_3$); (+) trachelanthic acid (R = H)
4. (+) 11β-chloromatrine
   (S)-(−) chlorosuccinic acid A4.1
5. (+) viridifloric acid (non-natural enantiomer [3])
   (R)-(+) 2-methoxy-4-methylpentan-3-one, A12.1
6. (+) matrine
7. (+) allomatrine* (ent. = (−) leontine)
8. ξ (−) nupharamine
9. ( ) lasiocarpic acid
10. (+) desoxymatrinol
    (S)-(+) 2-phenylbutyric acid A25.15
    Predominantly reacting enantiomer.
11. ξ (−) deoxynupharidine. Abs. X-ray [10]
    (+) nupharidine = N-oxide
    ξ (R)-(−) 2-methyladipic acid A38.3

1. J. A. Wunderlich, *Acta Cryst.*, 1967, **23**, 846.
2. C. C. J. Culvenor, G. M. O'Donovan, and L. W. Smith, *Austral. J. Chem.*, 1967, **20**, 801.
3. N. K. Kochetov, A. M. Likhosherstov, and V. N. Kulakov, *Tetrahedron*, 1967, **25**, 2313.
4. H. C. Crowley and C. C. J. Culvenor, *Austral. J. Chem.*, 1960, **13**, 269.
5. E. Ochiai, S. Okuda, and H. Minato, *J. Pharm. Soc. Japan*, 1952, **72**, 781.
6. S. Okuda, M. Yoshimoto, K. Tsuda, and N. Utzugi, *Chem. Pharm. Bull. Japan*, 1966, **14**, 314.
7. M. Kotake, I. Kawasaki, T. Okamoto, S. Matsutani, and T. Kaneko, *Bull. Chem. Soc. Japan*, 1962, **35**, 1335; M. Kotake, S. Kusumoto, and T. Ohara, *Annalen*, 1957, **606**, 148.
8. I. Kawasaki, S. Matsutani, and T. Kaneko, *Bull. Chem. Soc. Japan*, 1963, **36**, 1414.
9. D. C. Aldridge, J. J. Armstrong, R. N. Speake, and W. B. Turner, *J. Chem. Soc. (C)*, 1967, 1667.
10. K. Oda and H. Koyama, *J. Chem. Soc. (B)*, 1970, 1450.

Lycopodium alkaloids.

1. (−) serratidine
2. ( ) lycopodine. AC by ORD [1][8]
3. (−) lycodine.
   ( ) flabellidine = 9, 10, 11, 12 - tetrahydro deriv. [2]
   Also related; ( ) α- and β- obscurines [3].
4. ( ). By ORD (octant rule) [1]
5. (−) annofoline
6. ( ) annotinine. Rel. X-ray [6]
7. (+) fawcettidine
8. (−) serratinine. Abs. X-ray [4] Related compds; ( ) serratine, ( ) alolycopine, (+) serratinidine.
9. ( )
   (S)-(+) atrolactic acid
   A31.19
10. (+) fawcettimine

1. K. Wiesner, J. E. Francis, J. A. Findlay, and Z. Valenta, *Tetrahedron Letters*, 1961, 187.
2. S. N. Alam, K. A. H. Adams, and D. B. MacLean, *Canad. J. Chem.*, 1964, **42**, 2456.
3. W. A. Ayer and G. G. Iverach, *Canad. J. Chem.*, 1960, **38**, 1823.
4. K. Nishio, T. Fujiwara, K. Tomita, H. Ishii, Y. Inubushi, and T. Harayama, *Tetrahedron Letters*, 1969, 861.
5. F. A. L. Anet and M. V. Rao, *Tetrahedron Letters*, 1960, no. 20, 9.
6. M. Przybylska and F. R. Ahmed, *Acta Cryst.*, 1958, **11**, 718.
7. W. A. Ayer, D. A. Law, and K. Piers, *Tetrahedron Letters*, 1964, 2959.
8. R. H. Burnell and D. R. Taylor, *Tetrahedron*, 1961, **15**, 173.
9. H. Ishii, B. Yasui, T. Harayama, and Y. Inubushi, *Tetrahedron Letters*, 1966, 6215.
10. Y. Inubushi, H. Ishii, T. Harayama, and R. H. Burnell, *Tetrahedron Letters*, 1967, 1069.
11. Y. Inubushi, T. Harayama, M. A. Katsu, and H. Ishii, *Chem. Comm.*, 1968, 1138.

Lycopodium alkaloids (contd.); daphniphyllum alkaloids; calycanthine group. K26

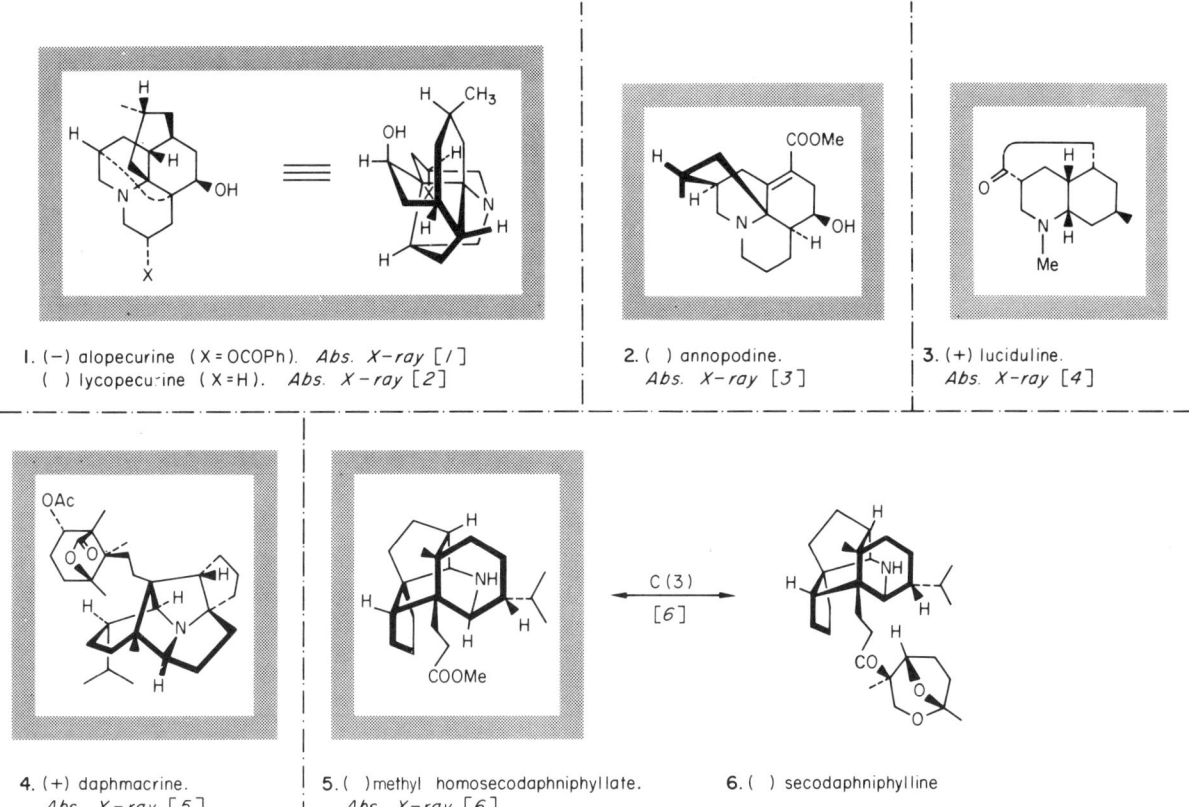

1. (−) alopecurine (X = OCOPh). Abs. X-ray [1]
   ( ) lycopecurine (X = H). Abs. X-ray [2]
2. ( ) annopodine. Abs. X-ray [3]
3. (+) luciduline. Abs. X-ray [4]
4. (+) daphmacrine. Abs. X-ray [5]
5. ( ) methyl homosecodaphniphyllate. Abs. X-ray [6]
6. ( ) secodaphniphylline

Most of the other known daphniphyllum alkaloids do not yet appear to have been rigorously correlated with daphmacrine or methyl homosecodaphniphyllate [7].

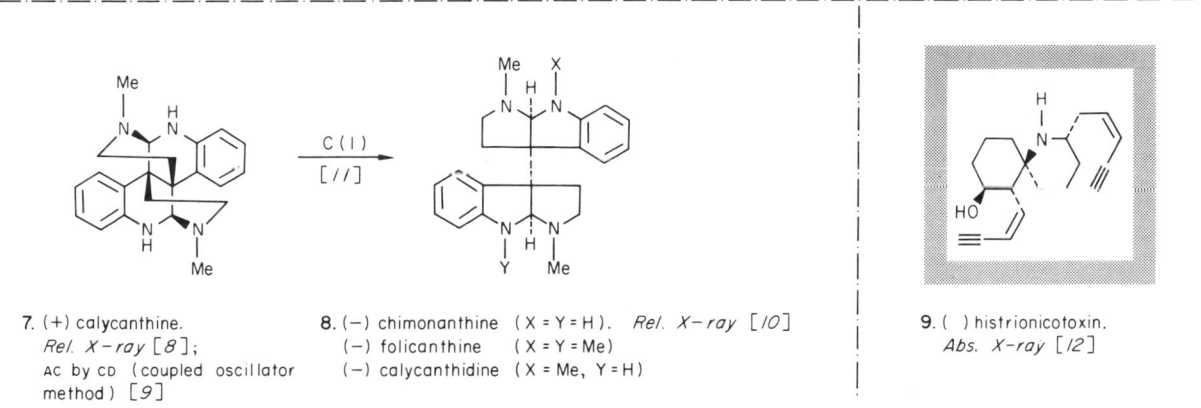

7. (+) calycanthine. Rel. X-ray [8]; AC by CD (coupled oscillator method) [9]
8. (−) chimonanthine (X = Y = H). Rel. X-ray [10]
   (−) folicanthine (X = Y = Me)
   (−) calycanthidine (X = Me, Y = H)
9. ( ) histrionicotoxin. Abs. X-ray [12]

1. W. A. Ayer, B. Altenkirk, N. Masaki, and S. Valverde-Lopez, *Canad. J. Chem.*, 1969, **47**, 2449.
2. W. A. Ayer and N. Masaki, *Canad. J. Chem.*, 1971, **49**, 524.
3. W. A. Ayer, G. G. Iverach, J. A. Jenkins, and N. Masaki, *Tetrahedron Letters*, 1968, 4597.
4. W. A. Ayer, N. Masaki, and D. S. Nkunika, *Canad. J. Chem.*, 1968, **46**, 3631.
5. C. S. Gibbons and J. Trotter, *J. Chem. Soc. (B)*, 1969, 840.
6. K. Sasaki and Y. Hirata, *J. Chem. Soc. (B)*, 1971, 1565.
7. M. Toda, S. Yamamura, and Y. Hirata, *Tetrahedron Letters*, 1969, 2585, and references therein; T. Nakano and B. Nilsson, *Tetrahedron Letters*, 1969, 2883, and references therein.
8. T. A. Hamor, J. M. Robertson, H. N. Shrivastava, and J. V. Silverton, *Proc. Chem. Soc.*, 1960, 78.
9. S. F. Mason and G. W. Vane, *J. Chem. Soc. (B)*, 1966, 370.
10. I. J. Grant, T. A. Hamor, J. M. Robertson, and G. A. Sim, *J. Chem. Soc.*, 1965, 5678.
11. J. B. Hendrickson, R. Gösche, and R. Rees, *Tetrahedron*, 1964, **20**, 565.
12. J. W. Daly, I. Karle, C. W. Myers, T. Tokuyama, J. A. Waters, and B. Witkop, *Proc. Nat. Acad. Sci. USA*, 1971, **68**, 1870.

K27 Lythraceae alkaloids.

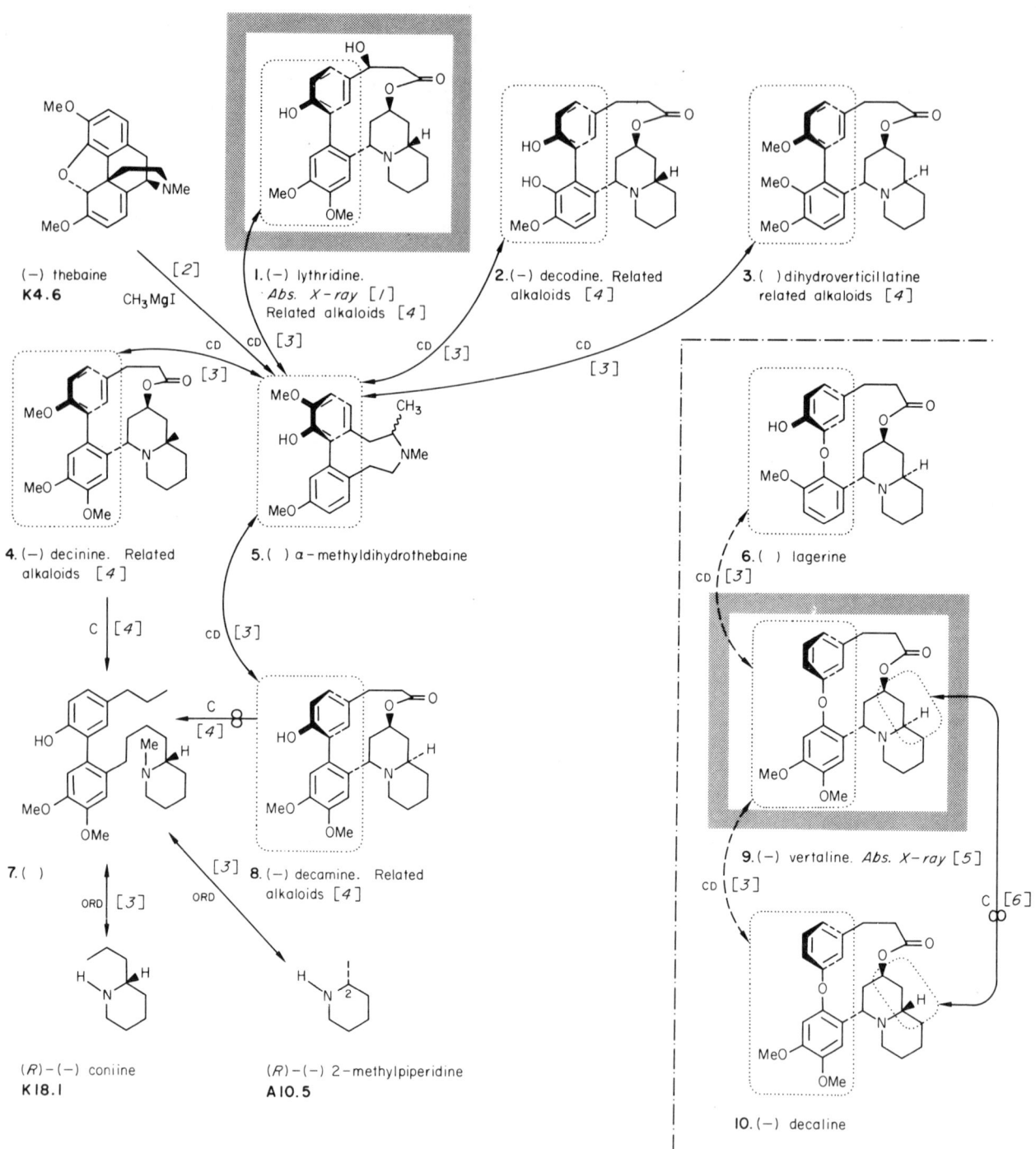

1. S. C. Chu, G. A. Jeffrey, B. Douglas, J. L. Kirkpatrick, and J. A. Weisbach, *Chem. and Ind.*, 1966, 1795.
2. J. A. Berson, *J. Amer. Chem. Soc.*, 1956, **78**, 4170.
3. J. P. Ferris, C. B. Boyce, R. C. Briner, U. Weiss, I. H. Qureshi, and N. E. Sharpless, *J. Amer. Chem. Soc.*, 1971, **93**, 2963.
4. J. P. Ferris, C. B. Boyce, and R. C. Briner, *J. Amer. Chem. Soc.*, 1971, **93**, 2942, and references therein.
5. J. A. Hamilton and L. K. Steinrauf, *J. Amer. Chem. Soc.*, 1971, **93**, 2939.
6. J. P. Ferris, R. C. Briner and C. B. Boyce, *J. Amer. Chem. Soc.*, 1971, **93**, 2953.

Tropane alkaloids.

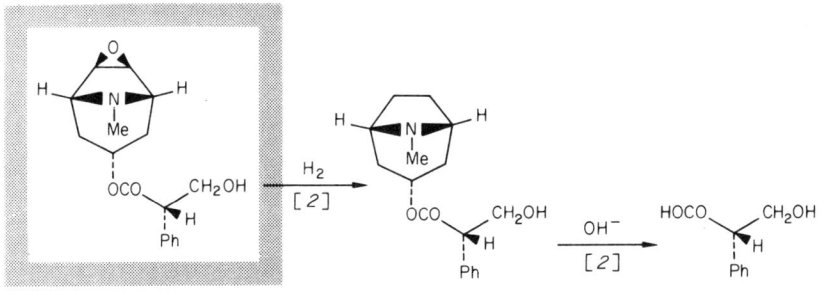

1. (−) hyoscine (scopolamine)
Abs. X-ray [1]
2. (−) hyoscyamine
(S)-(−) tropic acid **A41.8**

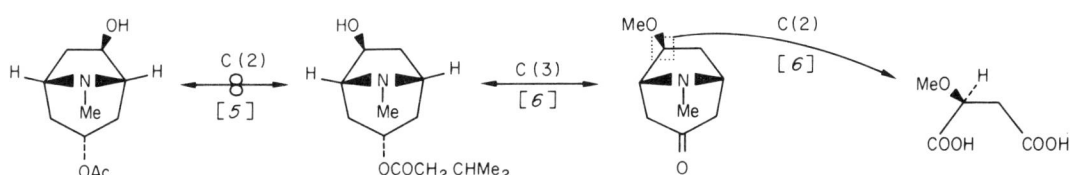

3. (3R, 6R)-(+) 3-acetoxy-6-hydroxytropane
4. (−) valeroidine *
5. (+) 6-methoxytropinone
(S)-(−) methoxysuccinic acid **A2.12**

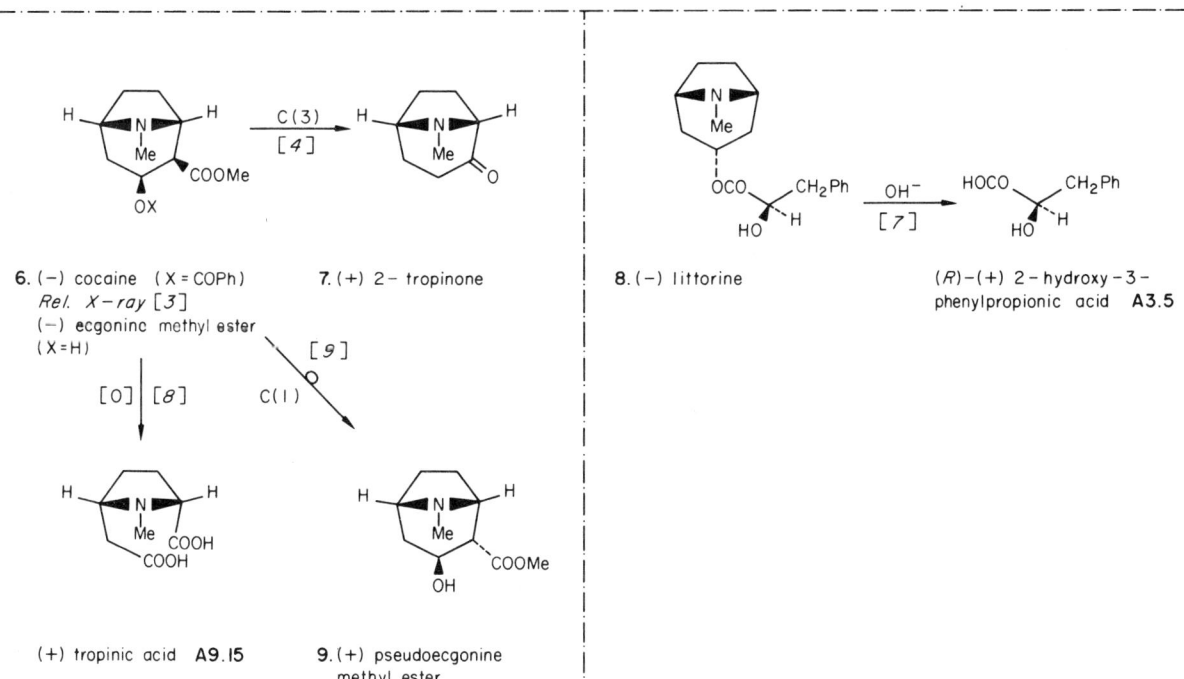

6. (−) cocaine (X = COPh)
Rel. X-ray [3]
(−) ecgonine methyl ester (X = H)
7. (+) 2-tropinone
8. (−) littorine
(R)-(+) 2-hydroxy-3-phenylpropionic acid **A3.5**

(+) tropinic acid **A9.15**

9. (+) pseudoecgonine methyl ester

1. C. S. Huber, G. Fodor, and N. Mandava, *Canad. J. Chem.*, 1971, **49**, 3258.
2. G. Fodor, I. Koczor, and G. Janzsó, *Arch. Pharm.*, 1962, **295**, 91, and references therein.
3. E. J. Gabe and W. H. Barnes, *Acta Cryst.*, 1963, **16**, 796.
4. M. R. Bell and S. Archer, *J. Amer. Chem. Soc.*, 1960, **82**, 4642.
5. S. R. Johns, J. A. Lamberton, and A. A. Sioumis, *Austral. J. Chem.*, 1971, **24**, 2399.
6. G. Fodor and F. Sóti, *J. Chem. Soc.*, 1965, 6830.
7. J. R. Cannon, K. R. Joshi, G. V. Meehan, and J. R. Williams, *Austral. J. Chem.*, 1969, **22**, 221.
8. C. Liebermann, *Ber.*, 1890, **23**, 2518; 1891, **24**, 606.
9. A. Einhorn and A. Marquardt, *Ber.*, 1890, **23**, 468.

K29 Miscellaneous alkaloid types.

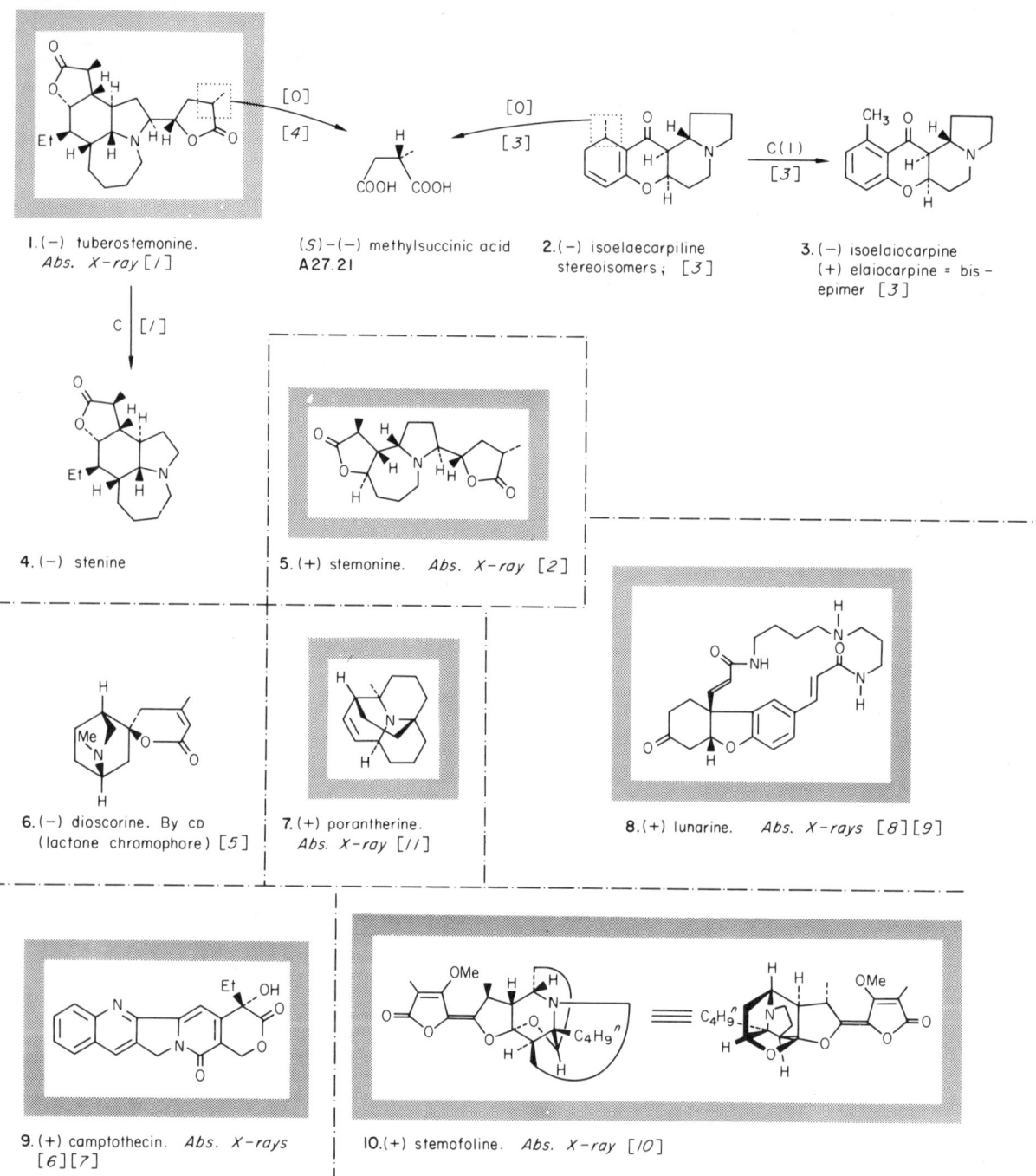

1. H. Harada, H. Irie, N. Masaki, K. Osaki, and S. Uyeo, *Chem. Comm.*, 1967, 460, and references therein.
2. H. Koyama and K. Oda, *J. Chem. Soc. (B)*, 1970, 1330.
3. S. R. Johns, J. A. Lamberton, A. A. Sioumis, H. Suares, and R. I. Willing, *Chem. Comm.*, 1970, 804, and references therein.
4. M. Götz, T. Bögri, and A. H. Gray, *Tetrahedron Letters*, 1961, 707.
5. A. F. Beecham, H. H. Mills, F. B. Wilson, C. B. Page, and A. R. Pinder, *Tetrahedron Letters*, 1969, 3745.
6. M. E. Wall, M. C. Wani, C. E. Cook, K. H. Palmer, A. T. McPhail, and G. A. Sim, *J. Amer. Chem. Soc.*, 1966, 88, 3888.
7. A. T. McPhail and G. A. Sim, *J. Chem. Soc. (B)*, 1968, 923.
8. J. A. D. Jeffreys and G. Ferguson, *J. Chem. Soc. (B)*, 1970, 826.
9. C. Tamura and G. A. Sim, *J. Chem. Soc. (B)*, 1970, 991.
10. H. Irie, N. Masaki, K. Ohno, K. Osaki, T. Taga, and S. Oyeo, *Chem. Comm.*, 1970, 1066.
11. W. A. Denne, S. R. Johns, J. A. Lamberton, and A. McL. Mathieson, *Tetrahedron Letters*, 1971, 3107.

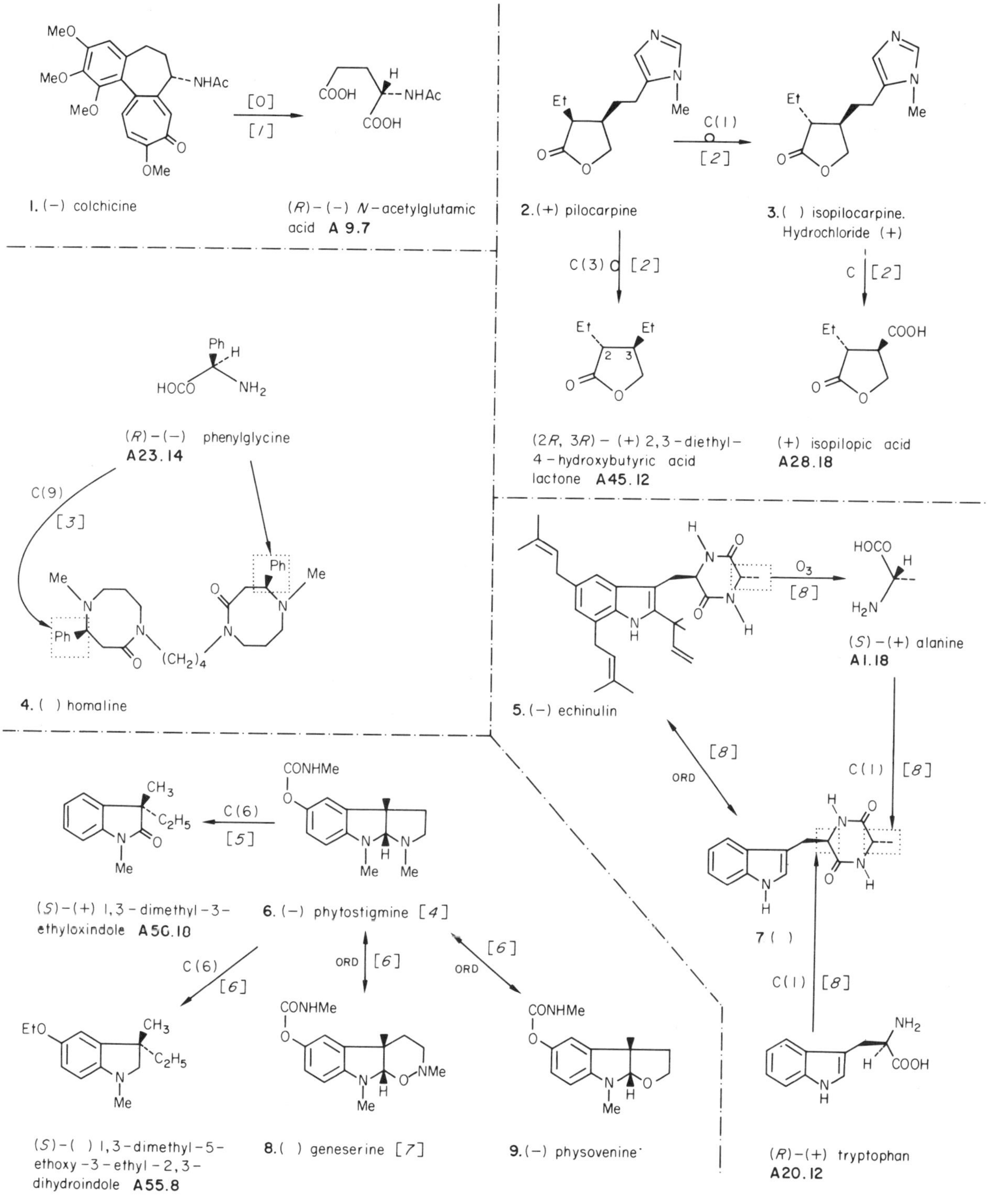

Colchicine, pilocarpine, phytostigmine, echinulin and related alkaloids.

1. (−) colchicine → (R)-(−) N-acetylglutamic acid **A 9.7**

2. (+) pilocarpine → 3. ( ) isopilocarpine. Hydrochloride (+)

(2R, 3R)-(+) 2,3-diethyl-4-hydroxybutyric acid lactone **A 45.12**

(+) isopilopic acid **A 28.18**

(R)-(−) phenylglycine **A 23.14**

4. ( ) homaline

5. (−) echinulin → (S)-(+) alanine **A 1.18**

6. (−) phytostigmine [4]

(S)-(+) 1,3-dimethyl-3-ethyloxindole **A 50.10**

(S)-(−) 1,3-dimethyl-5-ethoxy-3-ethyl-2,3-dihydroindole **A 55.8**

8. ( ) geneserine [7]

9. (−) physovenine

7. ( )

(R)-(+) tryptophan **A 20.12**

---

1. H. Corrodi and E. Hardegger, *Helv. Chim. Acta*, 1955, **38**, 2030.
2. R. K. Hill and S. Barcza, *Tetrahedron*, 1966, **22**, 2889.
3. M. Païs, R. Sarfati, F. Jarreau, and R. Goutarel, *Comptes Rend. (C)*, 1971, **272**, 1728.
4. G. R. Newkome and N. S. Bhacca, *Chem. Comm.*, 1969, 385.
5. R. K. Hill and G. R. Newkome, *Tetrahedron*, 1969, **25**, 1249.
6. R. B. Longmore and B. Robinson, *Chem. and Ind.*, 1969, 622.
7. B. Robinson and D. Moorcroft, *J. Chem. Soc. (C)*, 1970, 2077.
8. R. Nakashima and G. P. Slater, *Tetrahedron Letters*, 1967, 4433; 1971, 2649.

K31 Miscellaneous alkaloid types.

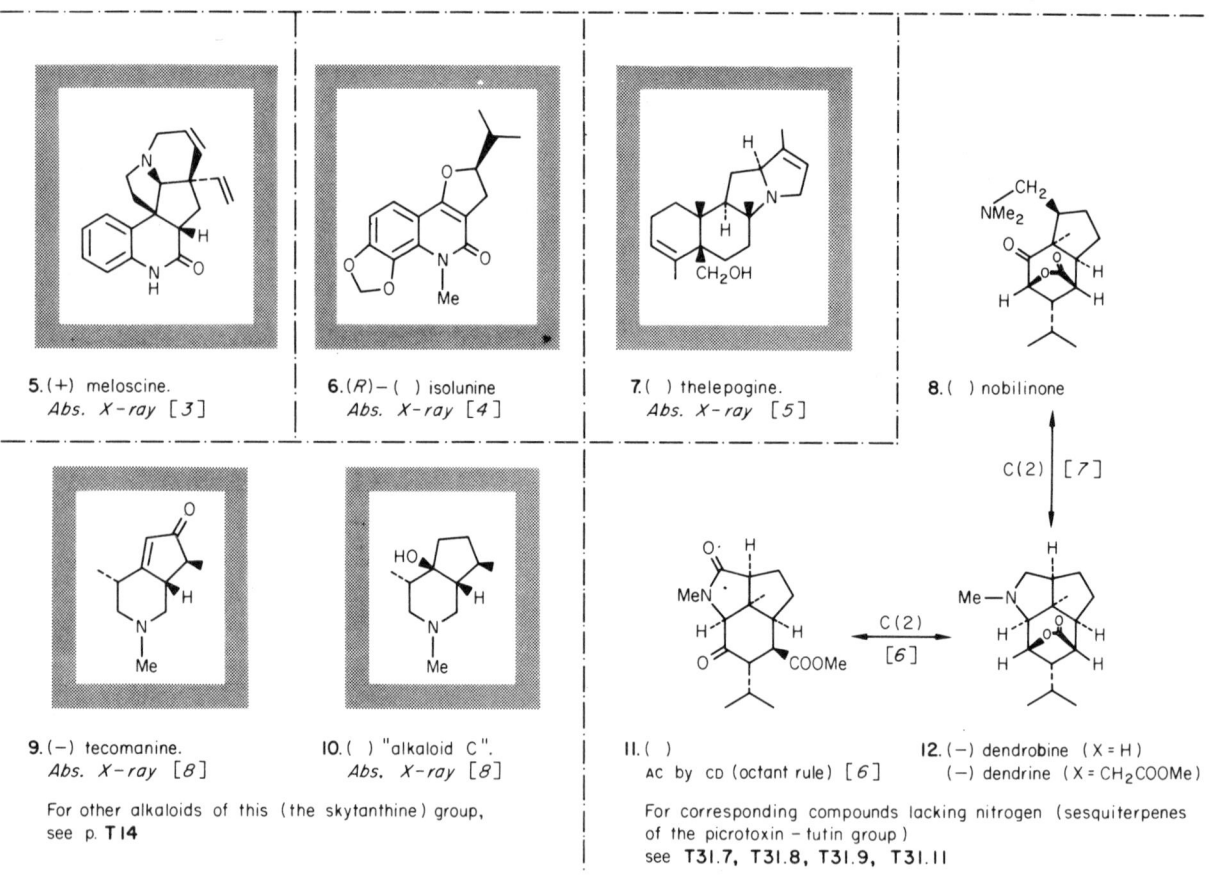

1. (R)-(+) balfourodine *
2. (R)-(−) balfourolone
3. (−) isobalfourodine *
4. (R)-(+) platydesmine

(R)-(+) 4,4-dimethyl-3-hydroxy-γ-butyrolactone **A8.5**

5. (+) meloscine. Abs. X-ray [3]
6. (R)-( ) isolunine. Abs. X-ray [4]
7. ( ) thelepogine. Abs. X-ray [5]
8. ( ) nobilinone
9. (−) tecomanine. Abs. X-ray [8]
10. ( ) "alkaloid C". Abs. X-ray [8]
11. ( ) AC by CD (octant rule) [6]
12. (−) dendrobine (X = H)
     (−) dendrine (X = CH₂COOMe)

For other alkaloids of this (the skytanthine) group, see p. T14

For corresponding compounds lacking nitrogen (sesquiterpenes of the picrotoxin - tutin group) see T31.7, T31.8, T31.9, T31.11

1. J. F. Collins and M. F. Grundon, *Chem. Comm.*, 1969, 1078.
2. R. M. Bowman, J. F. Collins, and M. F. Grundon, *Chem. Comm.*, 1967, 1131.
3. W. E. Oberhansli, *Helv. Chim. Acta*, 1969, **52**, 1905.
4. M. F. Mackay and A. McL. Mathieson, *Acta Cryst.*, 1969, **B25**, 1925.
5. J. Fridrichsons and A. McL. Mathieson, *Acta Cryst.*, 1963, **16**, 206.
6. Y. Inubushi, E. Katarao, Y. Tsuda, and B. Yasui, *Chem. and Ind.*, 1964, 1689.
7. T. Onaka, S. Kamata, T. Maeda, Y. Kawazoe, M. Natsume, T. Okamoto, F. Uchimaru, and M. Shimizu, *Chem. Pharm. Bull. Japan*, 1965, **13**, 745.
8. G. Jones, G. Ferguson, and W. C. Marsh, *Chem. Comm.*, 1971, 994.

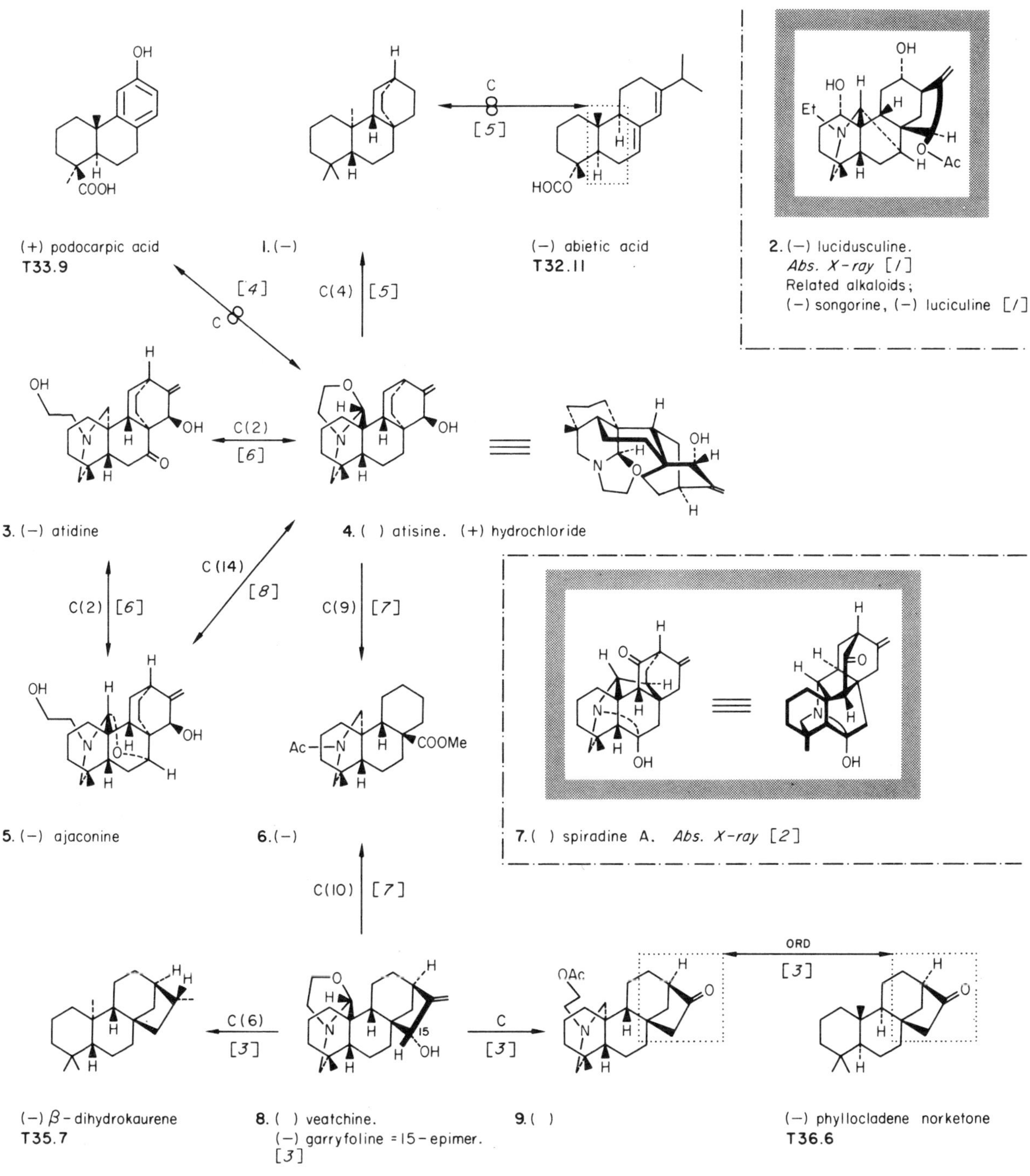

Diterpene alkaloids; Garrya group. K[32]

(+) podocarpic acid
T33.9

1. (−)

(−) abietic acid
T32.11

2. (−) lucidusculine.
Abs. X-ray [1]
Related alkaloids;
(−) songorine, (−) luciculine [1]

3. (−) atidine

4. ( ) atisine. (+) hydrochloride

5. (−) ajaconine

6. (−)

7. ( ) spiradine A. Abs. X-ray [2]

(−) β-dihydrokaurene
T35.7

8. ( ) veatchine.
(−) garryfoline = 15-epimer.
[3]

9. ( )

(−) phyllocladene norketone
T36.6

1. A. Yoshino and Y. Iitaka, *Acta Cryst.*, 1966, **21**, 57.
2. G. Goto, K. Sasaki, and N. Sakabe, *Tetrahedron Letters*, 1968, 1369.
3. H. Vorbrueggen and C. Djerassi, *J. Amer. Chem. Soc.*, 1962, **84**, 2990.
4. N. N. Girotra and L. H. Zalkow, *Tetrahedron*, 1965, **21**, 101.
5. W. A. Ayer, C. E. McDonald, and G. G. Iverach, *Tetrahedron Letters*, 1963, 1095.
6. S. W. Pelletier, *J. Amer. Chem. Soc.*, 1965, **87**, 799.
7. S. W. Pelletier and D. M. Locke, *J. Amer. Chem. Soc.*, 1965, **87**, 761.
8. D. Dvornik and O. E. Edwards, *Chem. and Ind.*, 1957, 952.

K33 Diterpene alkaloids; aconitine group.

1. (+) demethanolaconinone Abs. X-ray [1][3]
2. (+) aconitine [2]
3. (−) des(oxymethylene)-lycoctonine. Abs. X-ray [3]
4. (+) lycoctonine (X=CH$_3$, Y=H) (+) browniine (X=H, Y=CH$_3$)
5. (+) delphinine Rel. X-ray [2]
6. ( )
7. (−) delnudine Abs. X-ray [5]
8. (+) heteratisine. Rel X-ray [9]
9. (−) pyroheteratisine. By ORD (enone chromophore) [8]
10. (−) miyaconitine hydrobromide Abs. X-ray [6]
11. (+) lappaconine. Abs. X-ray [7]

1. M. Przybylska, *Acta Cryst.*, 1961, **14**, 429.
2. K. B. Birnbaum, K. Weisner, E. W. K. Jay, and L. Jay, *Tetrahedron Letters*, 1971, 867.
3. M. Przybylska and L. Marion, *Canad. J. Chem.*, 1959, **37**, 1843.
4. K. Weisner, D. L. Simmons & L. R. Fowler, *Tetrahedron Letters*, 1959, **no. 18**, 1.
5. K. B. Birnbaum, *Acta Cryst.*, 1971, **B27**, 1169.
6. H. Shimanouchi, Y. Sasada, and T. Takeda, *Tetrahedron Letters*, 1970, 2327.
7. G. I. Birnbaum, *Acta Cryst.*, 1970, **B26**, 755.
8. R. Aneja and S. W. Pelletier, *Tetrahedron Letters*, 1965, 215.
9. M. Przybylska, *Canad. J. Chem.*, 1963, **41**, 2911.

K 34

Steroidal alkaloids
See also **T49.4, T49.7, T49.8**.

A number of alkaloids are known which are best described as simple amino- or alkylamino steroids. Examples are funtumine (3a-amino-5a-pregnan-20-one) and paravallarine (3β-dimethylaminopregn-5-en-20-ol-18-oic acid 18→20 lactone). The structures of these have been proved by simple chemical conversion to or from nitrogen-free steroids. Only the more complex type of steroidal alkaloid is included here.

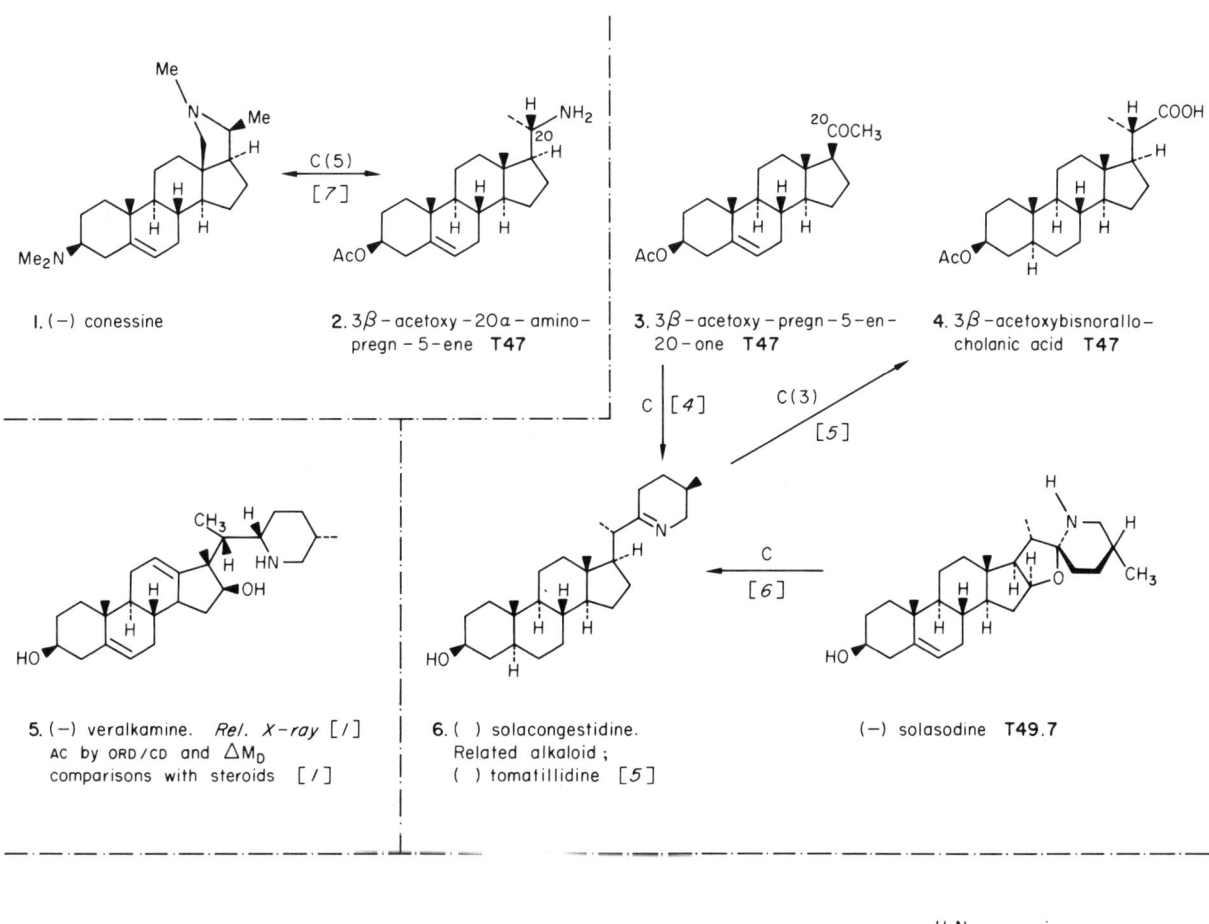

1. (−) conessine
2. 3β-acetoxy-20α-amino-pregn-5-ene **T47**
3. 3β-acetoxy-pregn-5-en-20-one **T47**
4. 3β-acetoxybisnorallo-cholanic acid **T47**
5. (−) veralkamine. *Rel. X-ray* [*1*] AC by ORD/CD and ΔM_D comparisons with steroids [*1*]
6. ( ) solacongestidine. Related alkaloid; ( ) tomatillidine [*5*]
   (−) solasodine **T49.7**

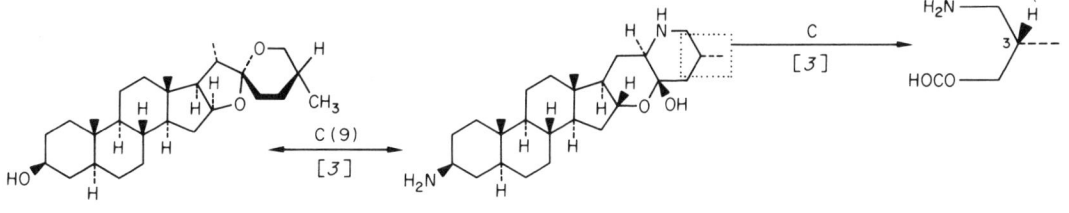

tigogenin **T49.9**
7. (+) solanocapsine. *Rel. X-ray* [*2*]
(R)-(−) 4-amino-3-methylbutyric **A27.16**

1. J. Tomko, A. Vassova, G. Adam, K. Schreiber, and E. Höhne, *Tetrahedron Letters*, 1967, 3907.
2. E. Höhne, H. Ripperger and K. Schreiber, *Tetrahedron*, 1970, **26**, 3569.
3. K. Schreiber and H. Ripperger, *Annalen*, 1962, **655**, 114.
4. K. Schreiber and G. Adam, *Tetrahedron*, 1964, **20**, 1707.
5. E. Bianchi, C. Djerassi, H. Budzikiewitz, and Y. Sato, *J. Org. Chem.*, 1965, **30**, 754.
6. G. Kusano, N. Aimi, and Y. Sato, *J. Org. Chem.*, 1970, **35**, 2624.
7. E. J. Corey and W. R. Hertler, *J. Amer. Chem. Soc.*, 1958, **80**, 2903.

K35

Steroidal alkaloids (contd.); Cevine and jervine groups.

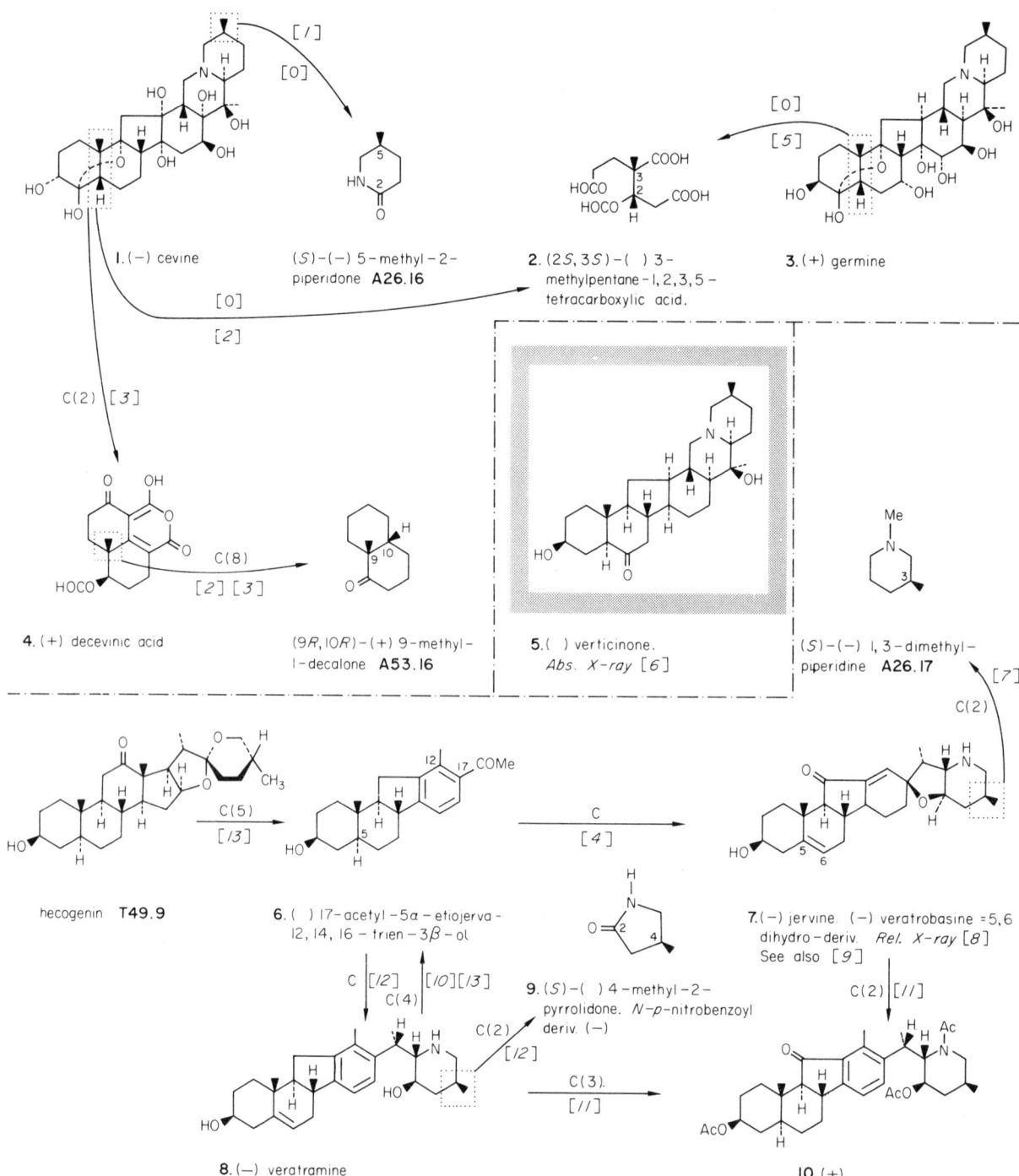

1. (−) cevine
2. (2S,3S)-( ) 3-methylpentane-1,2,3,5-tetracarboxylic acid.
3. (+) germine
4. (+) decevinic acid
   (S)-(−) 5-methyl-2-piperidone A26.16
   (9R,10R)-(+) 9-methyl-1-decalone A53.16
5. ( ) verticinone. Abs. X-ray [6]
   (S)-(−) 1,3-dimethylpiperidine A26.17
6. ( ) 17-acetyl-5α-etiojerva-12,14,16-trien-3β-ol
   hecogenin T49.9
7. (−) jervine. (−) veratrobasine =5,6 dihydro-deriv. Rel. X-ray [8] See also [9]
8. (−) veratramine
9. (S)-( ) 4-methyl-2-pyrrolidone. N-p-nitrobenzoyl deriv. (−)
10. (+)

1. L. C. Craig and W. A. Jacobs, *J. Biol. Chem.*, 1941, **141**, 253.
2. F. Gäutschi, O. Jeger, V. Prelog, and R. B. Woodward, *Helv. Chim. Acta*, 1955, **38**, 296.
3. L. C. Craig and W. A. Jacobs, *J. Biol. Chem.*, 1940, **134**, 123; F. Gäutschi, O. Jeger, V. Prelog, and R. B. Woodward, *Helv. Chim. Acta*, 1954, **37**, 2280.
4. T. Masamune, M. Tagasugi, A. Murai, and K. Kobayashi, *J. Amer. Chem. Soc.*, 1967, **89**, 4521.
5. S. M. Kupchan and C. R. Narayanan, *J. Amer. Chem. Soc.*, 1959, **81**, 1913.
6. S. Itô, Y. Fukazawa, T. Okuda, and Y. Iitaka, *Tetrahedron Letters*, 1968, 5373.
7. S. Okuda, K. Tsuda, and K. Kataoka, *Chem. and Ind.*, 1961, 512.
8. G. N. Reeke, J. L. Vincent, and W. N. Lipscomb, *J. Amer. Chem. Soc.*, 1968, **90**, 1663.
9. S. M. Kupchan and M. I. Suffness, *J. Amer. Chem. Soc.*, 1968, **90**, 2370.
10. R. W. Franck and W. S. Johnson, *Tetrahedron Letters*, 1963, 543.
11. O. Wintersteiner and N. Hosanky, *J. Amer. Chem. Soc.*, 1952, **74**, 4474.
12. W. S. Johnson, H. A. P. de Jongh, C. E. Coverdale, J. W. Scott, and U. Burkhardt, *J. Amer. Chem. Soc.*, 1967, **89**, 4523.
13. H. Mitsuhashi and K. Shibata, *Tetrahedron Letters*, 1964, 2281; S. Masamune, K. Kobayashi, M. Takasugi, Y. Mori, and A. Murai, *Tetrahedron*, 1968, **24**, 3461.

Steroidal alkaloids (contd.); buxus group. K36

1. ( ) buxidienine F
2. ( ) cyclobuxidine F (X=O)
   ( ) dihydrocyclomicrophylline F (X=H₂)
3. ( ) cyclobuxine D
4. (−) cyclobuxosuffrine
5. ( ) buxenine G. *Abs. X-ray* [1]
6. (+) cyclobuxoxazine A.
   (+) cycloeucalenol T50.1
7. ( ) cycloprotobuxine C
   (+) cycloartenol T50.2
8. (−) solaphyllidine. *Abs. X-ray* [2]
9. ( ) batrachotoxinin A *Abs. X-ray* [3]

1. R. T. Puckett, G. A. Sim, and M. G. Waite, *J. Chem. Soc. (B)*, 1971, 935.
2. A. Usubilliga, C. Seelkopf, I. L. Karle, J. W. Daly, and B. Witkop, *J. Amer. Chem. Soc.*, 1970, **92**, 700.
3. R. D. Gilardi, *Acta Cryst.*, 1970, **B26**, 440.
4. F. Khoung-Huu, D. Herlem-Gaulier, Q. Khoung-Huu, E. Stanislas, and R. Goutarel, *Tetrahedron*, 1966, **22**, 3321.
5. T. Nakano and S. Terao, *J. Chem. Soc.*, 1965, 4512.
6. K. S. Brown and S. M. Kupchan, *J. Amer. Chem. Soc.*, 1964, **86**, 4424.
7. T. Nakano and S. Terao, *J. Chem. Soc.*, 1965, 4537.
8. T. Nakano and Z. Votický, *J. Chem. Soc. (C)*, 1970, 590.
9. J. P. Calame and D. Arigoni, *Chimia*, 1964, **18**, 185.

- Y -

Miscellaneous Natural Products

Introductory Notes to Chapter Y

Chapter Y contains natural products not covered by Chapters T or K. The material is divided into the following main categories:

1. *Compounds containing C,H,O and halogens only*
   (a) Open-chain† and O-heterocyclic compounds pp. **Y1-Y15**.
   (b) Carbocyclic compounds pp. **Y15-Y19**.

2. *Compounds containing N, S and P.* pp. **Y20-Y30**

---

† 'Open-chain' here means that the *chiral centres* of the molecule are in an open-chain portion, and not necessarily that the molecule is completely acyclic.

Rotenoids and coumaran derivatives. ORD of rotenoids [4] [5]

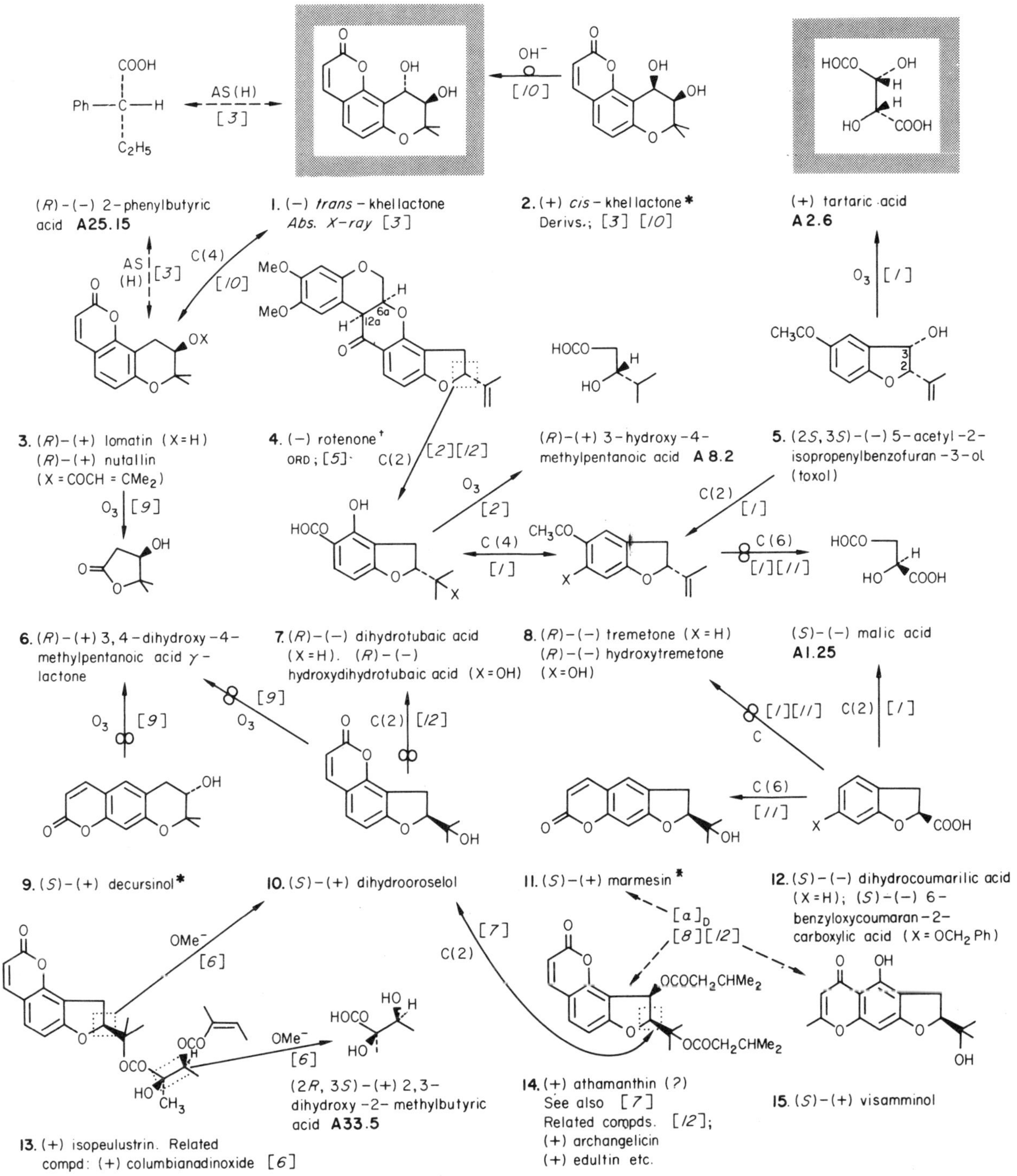

1. W. A. Bonner, N. I. Burke, W. E. Fleck, R. K. Hill, J. A. Joule, B. Sjöberg, and J. H. Zalkow, *Tetrahedron*, 1964, 20, 1419.
2. G. Büchi, J. S. Kaltenbronn, L. Crombie, P. J. Godin, and D. A. Whiting, *Proc. Chem. Soc.*, 1960, 274.
3. H. Bernotat-Wulf, A. Niggli, L. Uhlrich, and H. Schmid, *Helv. Chim. Acta*, 1969, 52, 1165.
4. C. Djerassi, W. D. Ollis, and R. C. Russell, *J. Chem. Soc.*, 1961, 1448.
5. W. D. Ollis, C. A. Rhodes, and I. O. Sutherland, *Tetrahedron*, 1967, 23, 4741.
6. B. E. Neilsen and J. Lemmich, *Acta Chem. Scand.*, 1965, 19, 1810.
7. E. Lemmich, J. Lemmich, and B. E. Neilsen, *Acta Chem. Scand.*, 1970, 24, 2893.
8. W. Bencze, J. Eisenbeiss, and H. Schmid, *Helv. Chim. Acta*, 1956, 39, 923.
9. J. Lemmich and B. E. Neilsen, *Tetrahedron Letters*, 1969, 3.
10. J. Lemmich, P. A. Pedersen, and B. E. Neilsen, *Tetrahedron Letters*, 1969, 3365, and references therein.
11. I. Harada, Y. Hirose, and M. Nakazaki, *Tetrahedron Letters*, 1968, 5463.
12. B. E. Neilsen and J. Lemmich, *Acta Chem. Scand.*, 1964, 18, 2111.

†Other natural rotenoids have been compared with (−) rotenone by ORD [4] [5]. All natural rotenoids examined appear to have the same configuration as rotenone at C6a and C12a. This configuration is characterised by +ve RD at long wavelengths and −ve RD at shorter wavelengths [5].

Y[2]   Isochromanones; phthalans.

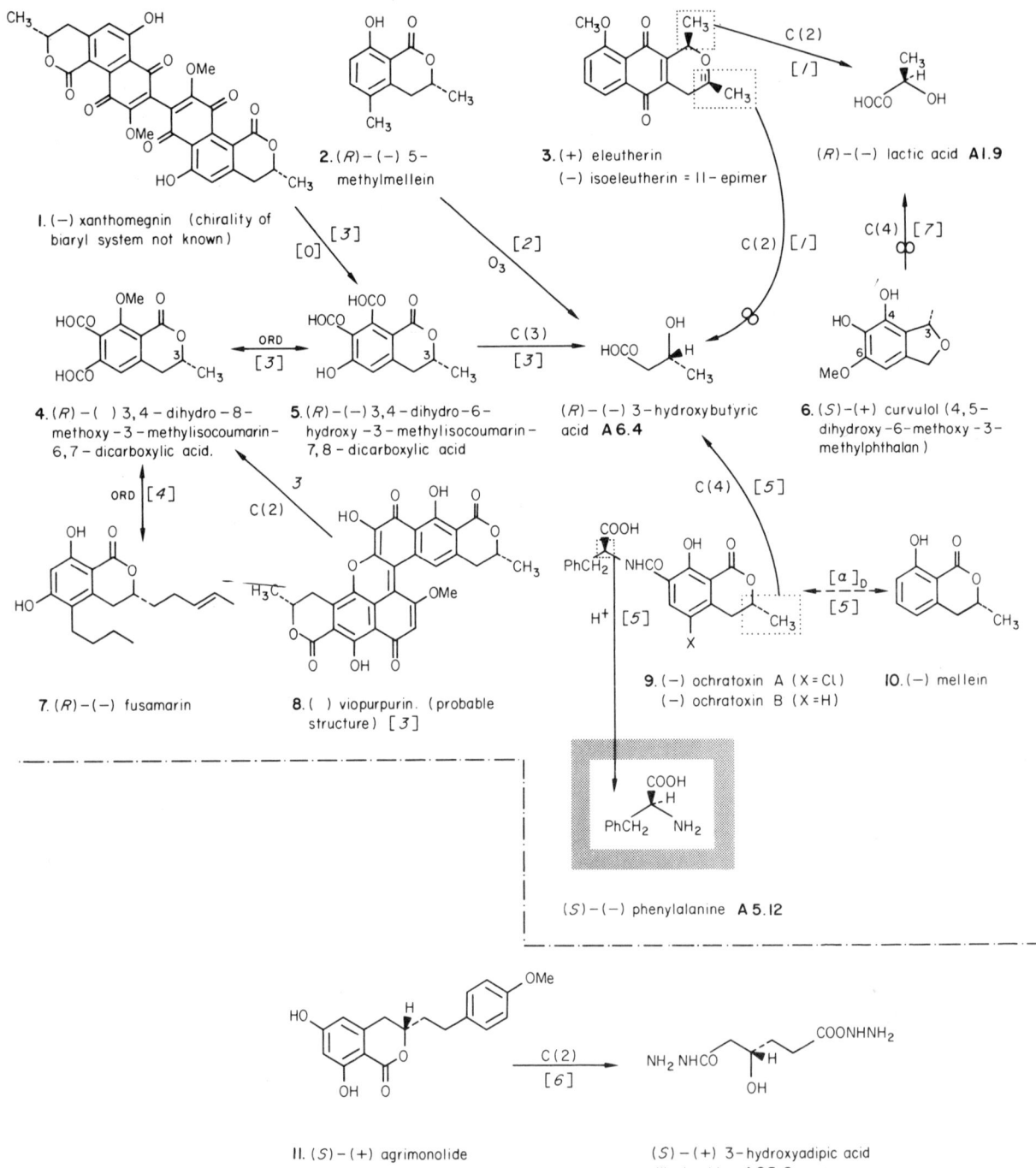

1. H. Schmid and A. Ebnöther, *Helv. Chim. Acta*, 1951, **34**, 1041.
2. A. Ballio, S. Barcellona, and B. Santurbano, *Tetrahedron Letters*, 1966, 3723.
3. A. S. Ng, G. Just, and F. Blank, *Canad. J. Chem.*, 1969, **47**, 1223.
4. Y. Suzuki, *Agr. Biol. Chem. (Japan)*, 1970, **34**, 760.
5. K. J. Van der Merwe, P. S. Steyn, and L. Fourie, *J. Chem. Soc.*, 1965, 7083.
6. H. Arakawa, N. Torimoto, and Y. Masui, *Tetrahedron Letters*, 1968, 4115.
7. A. A. Qureshi, R. W. Rickards, and A. Kamal, *Tetrahedron*, 1967, **23**, 3801.

Flavanoids and isoflavanoids; pterocarpans.

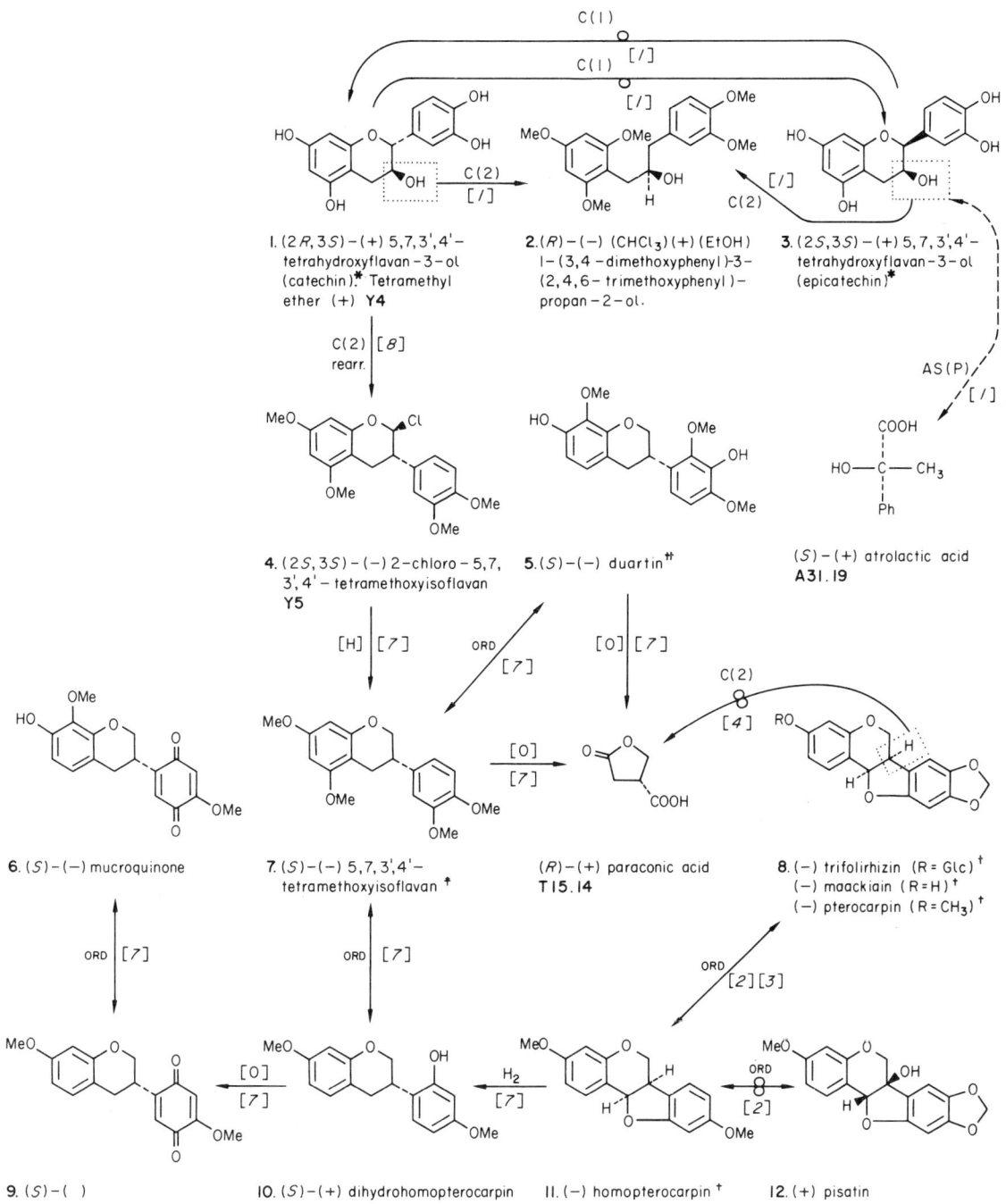

ORD/CD of flavonoids and their glycosides; [5]
A number of dimeric flavonoid types also occur in nature [6]

1. A. J. Birch, J. W. Clark-Lewis, and A. V. Robertson, *J. Chem. Soc.*, 1957, 3586.
2. P. B. Hulbert, Ph.D. Thesis, Westfield College (Univ. of London), 1969.
3. A. Pelter and P. I. Amenechi, *J. Chem. Soc. (C)*, 1969, 887.
4. S. Itô, Y. Fujise, and A. Mori, *Chem. Comm.*, 1965, 595.
5. W. Gaffield and A. C. Waiss, *Chem. Comm.*, 1968, 29; W. Gaffield, *Tetrahedron*, 1970, 26, 4093.
6. S. E. Drewes, D. G. Roux, J. Feeney, and S. H. Eggers, *Chem. Comm.*, 1966, 368; F. D. Monache, I. D. d'Albuquerque, F. Ferrari, and G. B. M. Bettolo, *Tetrahedron Letters*, 1967, 4211.
7. K. Kurosawa, W. D. Ollis, B. T. Redman, I. O. Sutherland, O. R. Gottlieb, and H. M. Alves, *Chem. Comm.*, 1968, 1265, and references therein.
8. K. Weinges, *Proc. Chem. Soc.*, 1964, 138; J. W. Clark-Lewis, I. Dainis, and G. C. Ramsay, *Austral. J. Chem.*, 1965, 18, 1035.

†Other pterocarpinoids have been compared with those shown by chiroptical methods [2] [3]. It has been suggested [4] that all laevorotatory pterocarpinoids have the (3R, 4R) configuration, and vice-versa.

††Other isoflavans have been compared with those shown by chiroptical methods. Care is necessary, however, in the interpretation of the spectra [7].

Y⁴ Flavanoids and isoflavanoids (contd.); neoflavonoids

1. (+) peltogynol. See also [5] Related compds. [3]
2. (+) tri-O-methylpeltogynone
3. (2R,3R)-(+) 3,7,3',4'-tetrahydroxyflavan-4-one (fustin)*
4. (2R,3S,4R)-(+) 3,4,7,3',4'-pentahydroxyflavan. (mollisacacidin)†* Stereoisomers; [1]
5. (+) peltogynol B
6. (2R,3S)-(−) 3',4',7-trihydroxy-flavan-3-ol (fisetinidol)
   (+) catechin Y3.1
7. (S)-(+) 2-(p-methoxyphenyl)-butyric acid
   (R)-(−) 2-phenylbutyric acid A25.15
8. (+) taxifolin (dihydroquercetin).
9. (S)-(−) 4,4'-dimethoxydalbergione. ORD; see also [8]
10. (R) 4-methoxydalbergione (+) (CHCl₃) (−) (dioxan). ORD; see also [8]
11. (2S,3S)-( ) 2,3-dihydroxyglutaric acid
12. (2S,3S)-( ) pentane-1,2,3,5-tetraol (2-desoxy-D-adonitol). Urethane (−)
    D (−) ribose C1.9

1. S. E. Drewes and D. G. Roux, *Biochem. J.*, 1965, **96**, 68.
2. E. Hardegger, H. Gempeler, and A. Züst, *Helv. Chim. Acta*, 1957, **40**, 1819.
3. S. E. Drewes and D. G. Roux, *J. Chem. Soc. (C)*, 1966, 1644.
4. J. W. Clark-Lewis and G. F. Katekar, *Proc. Chem. Soc.*, 1960, 345.
5. C. H. Hassall and J. Weatherston, *J. Chem. Soc.*, 1965, 2844.
6. S. E. Drewes and D. G. Roux, *Biochem. J.*, 1964, **90**, 343, and references therein.
7. W. B. Eyton, W. D. Ollis, I. O. Sutherland, O. R. Gottlieb, M. T. Magalhaes, and L. M. Jackman, *Tetrahedron*, 1966, **21**, 2683.
8. D. M. X. Donnelly, B. J. Nangle, P. B. Hulbert, W. Klyne, and R. J. Swan, *J. Chem. Soc. (C)*, 1967, 2450.
9. J. W. Clark-Lewis and W. Korytnyk, *J. Chem. Soc.*, 1958, 2367.

†A similar series of correlations has been carried out in the melacacidin (3',4',7,8-tetrahydroxyflavan-3,4-diol) series [4]. For a tabulation of other flavandiols, see [5]. For the ORD/CD of flavanones and 3-hydroxyflavones, see [6].

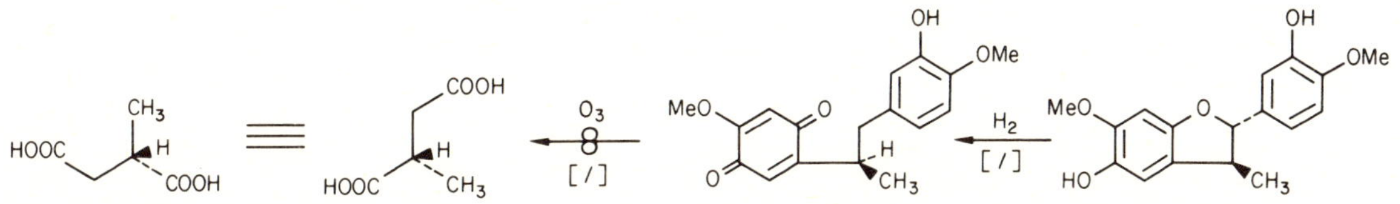

1,2-diphenylpropane types (angolensin etc.).

(S)-(−) methylsuccinic acid **A27.21**

1. (R)-(+) 1-(3-hydroxy-4-methoxyphenyl)-2-(4-methoxy-2,5-benzoquinonyl)-propane

2. (−) melanoxin

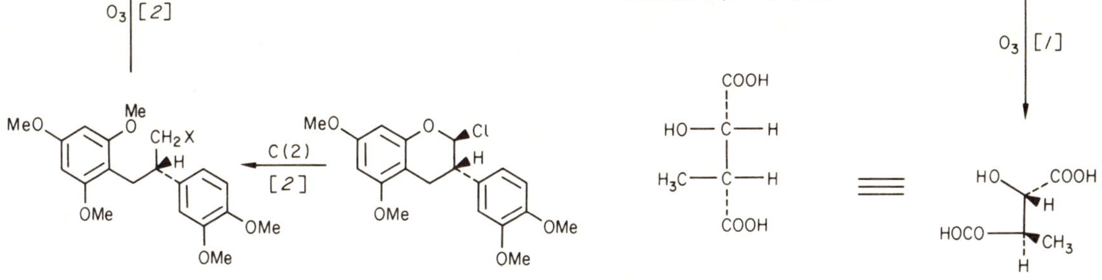

3. (S)-(+) 2-(3,4-dimethoxyphenyl)-1-(2,4,6-trimethoxyphenyl)-propane (X = H);
(S)-(+) 2-(3,4-dimethoxyphenyl)-3-(2,4,6-trimethoxyphenyl)-propan-1-ol (X = OH)

(2S,3S)-(−) 2-chloro-5,7,3',4',-tetramethoxyisoflavan **Y3.3**

4. (2S,3R)-(−) 2-hydroxy-3-methylsuccinic (β-methylmalic) acid

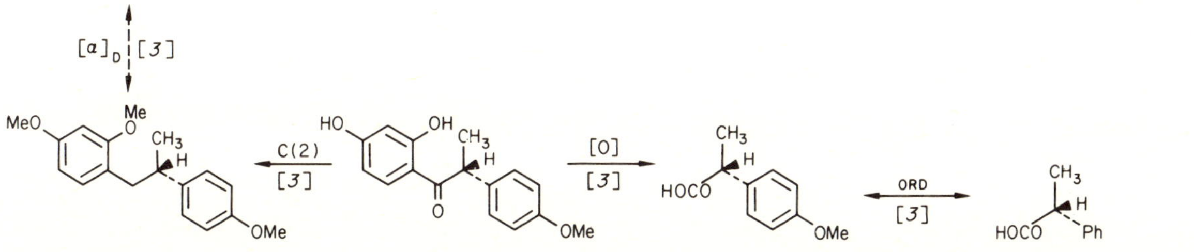

5. (S)-(+) 1-(2,4-dimethoxyphenyl)-2-(p-methoxyphenyl)-propane

6. (R)-(−) angolensin

7. (R)-(−) 2-(p-methoxyphenyl)-propionic acid

(R)-(−) 2-phenylpropionic acid **A41.10**

1. B. J. Donnelly, D. M. X. Donnelly, A. M. O'Sullivan, and J. P. Prendergast, *Tetrahedron*, 1969, 25, 4409.
2. J. W. Clark-Lewis, I. Dainis, and G. C. Ramsay, *Austral. J. Chem.*, 1965, 18, 1035.
3. W. D. Ollis, M. V. J. Ramsay, and I. O. Sutherland, *Austral. J. Chem.*, 1965, 18, 1787; J. W. Clark-Lewis and R. W. Jemison, ibid., 1791.

Y[6]   Flavanones and isoflavanones; Leucodrin, canescins.

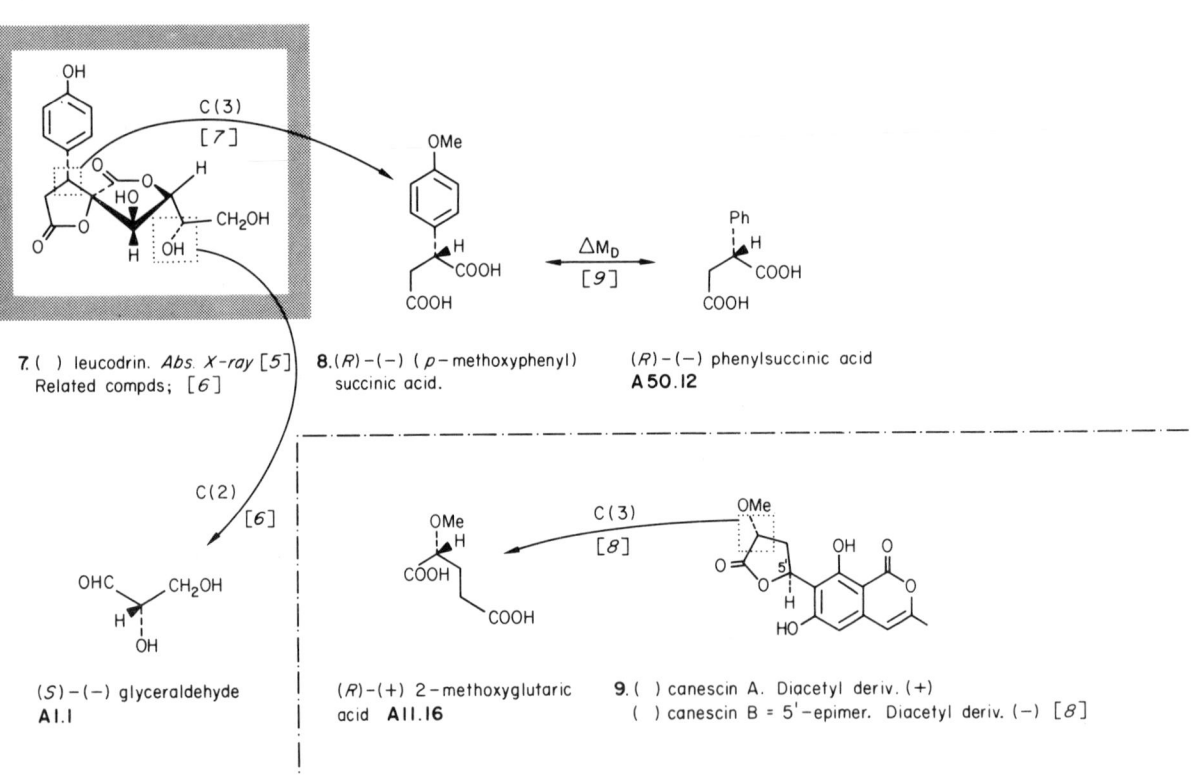

1. (R)-(+) phyllodulcin
2. (S)-(−) liquiritigenin
3. (R)-(+) 7-hydroxyflavan-4-one
4. (R)-(+) flavan-7-ol
5. (S)-(−) hesperetin
   (S)-(−) malic acid A I.25
6. (R)-(+) sakuranetin (R=CH$_3$)
   (R)-(+) naringenin (R=H)*
7. ( ) leucodrin. Abs. X-ray [5]
   Related compds; [6]
8. (R)-(−) (p-methoxyphenyl) succinic acid.
   (R)-(−) phenylsuccinic acid   A 50.12
(S)-(−) glyceraldehyde A I.1
(R)-(+) 2-methoxyglutaric acid A II.16
9. ( ) canescin A. Diacetyl deriv. (+)
   ( ) canescin B = 5'-epimer. Diacetyl deriv. (−) [8]

1. H. Arakawa and M. Nakazaki, *Chem. and Ind.*, 1959, 671.
2. G. Cardillo, L. Merlini, G. Nasini, and P. Salvadori, *J. Chem. Soc. (C)*, 1971, 3967.
3. H. Arakawa and M. Nakazaki, *Chem. and Ind.*, 1960, 73.
4. E. Hardegger and H. Braunschweiger, *Helv. Chim. Acta*, 1961, 44, 1413.
5. R. D. Diamond and D. Rogers, *Proc. Chem. Soc.*, 1964, 63.
6. G. W. Perold and H. K. L. Hundt, *Chem. Comm.*, 1970, 712.
7. G. W. Perold and K. G. R. Pachler, *J. Chem. Soc. (C)*, 1966, 1918.
8. A. J. Birch, J. H. Birkinshaw, P. Chaplen, L. Mo, A. H. Manchanda, A. Pelter, and M. Riano-Martin, *Austral. J. Chem.*, 1969, 22, 1933.
9. M. Naps and I. B. Johns, *J. Amer. Chem. Soc.*, 1940, 62, 2450.

Lignans
Review; [9]

On this page only, Ar = 4-hydroxy-3-methoxyphenyl; Ar' = 3,4-dimethoxyphenyl.

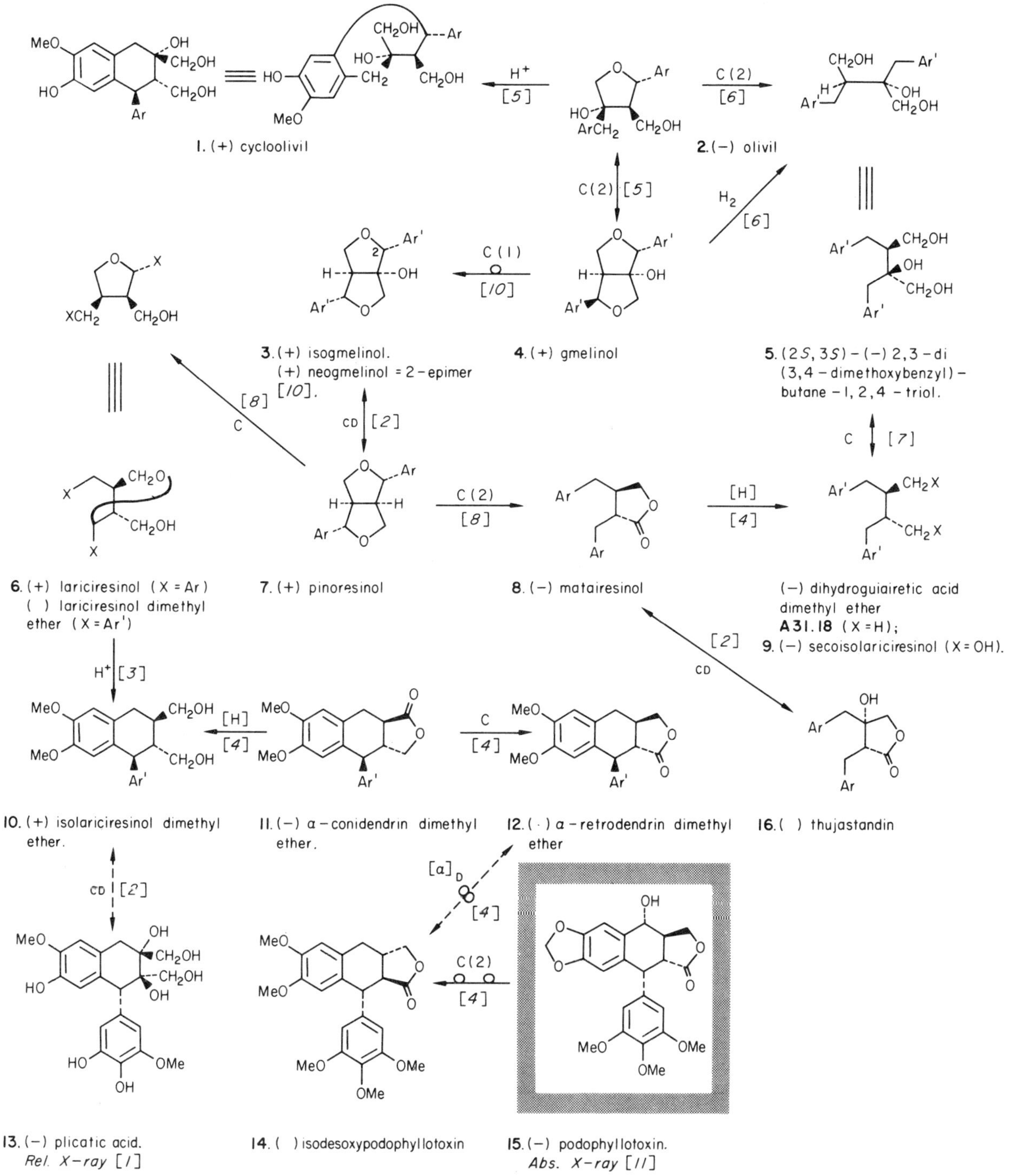

1. (+) cycloolivil
2. (−) olivil
3. (+) isogmelinol. (+) neogmelinol = 2-epimer
4. (+) gmelinol
5. (2S,3S)-(−) 2,3-di (3,4-dimethoxybenzyl)-butane-1,2,4-triol.
6. (+) lariciresinol (X = Ar)
   ( ) lariciresinol dimethyl ether (X = Ar')
7. (+) pinoresinol
8. (−) matairesinol

(−) dihydroguaiaretic acid dimethyl ether
A31.18 (X = H);
9. (−) secoisolariciresinol (X = OH).

10. (+) isolariciresinol dimethyl ether.
11. (−) α-conidendrin dimethyl ether.
12. (·) α-retrodendrin dimethyl ether
16. ( ) thujastandin
13. (−) plicatic acid. Rel. X-ray [1]
14. ( ) isodesoxypodophyllotoxin
15. (−) podophyllotoxin. Abs. X-ray [11]

1. J. A. F. Gardner, E. P. Swan, S. A. Sutherland, and H. MacLean, *Canad. J. Chem.*, 1966, 44, 52.
2. P. B. Hulbert, Ph.D. Thesis, Westfield College (Univ. of London), 1969.
3. R. D. Haworth and W. Kelly, *J. Chem. Soc.*, 1937, 384.
4. A. W. Schrecker and J. L. Hartwell, *J. Amer. Chem. Soc.*, 1955, 77, 432.
5. D. C. Ayres and S. E. Mhasalkar, *Tetrahedron Letters*, 1964, 335.
6. K. Freudenberg and K. Weinges, *Tetrahedron Letters*, 1962, 1077.
7. G. Traverso, *Gazzetta*, 1962, 90, 792.
8. R. D. Haworth, *J. Chem. Soc.*, 1942, 448.
9. K. Weinges and R. Spänig in 'Oxidative Coupling of Phenols' (W. I. Taylor and A. R. Battersby, Eds.), Arnold, 1967.
10. A. J. Birch, P. L. Macdonald, and A. Pelter, *J. Chem. Soc. (C)*, 1967, 1968.
11. T. J. Petcher, H. P. Weber, M. Kuhn, and A. von Wartburg, *J. Chem. Soc., Perkin II*, 1973, 288.

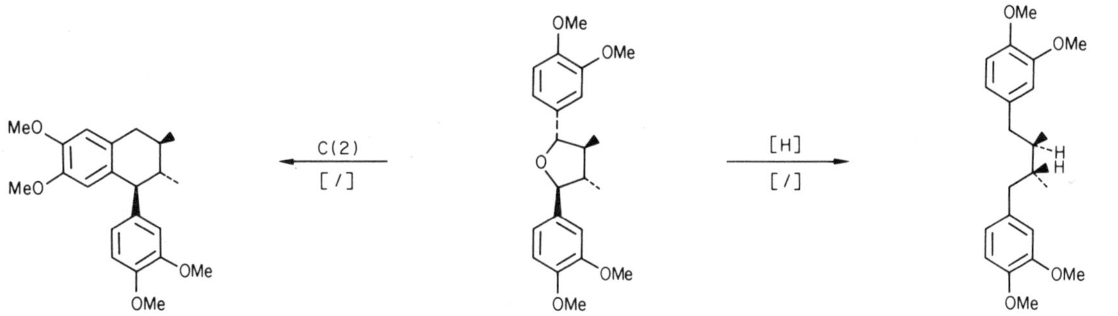

Y8   Lignans (contd.).

1. (−) galbulin

2. (−) galbegin. *Meso* − compounds of this type have also been characterized [/]

(−) dihydroguiaretic acid dimethyl ether A31.18

Notes on lignan stereochemistry†

The evidence on which the absolute configuration of a given lignan is assigned frequently consists of a combination of chemical interconversions within and between classes of compound and optical comparisons. The four main classes of lignan are:

(1) Open-chain types (diarylbutanes), e.g. dihydroguiaretic acid A31.8, secoisolariciresinol Y7.10; these have the general formula $ArCH_2CH(CH_2X)CH(CH_2X)CH_2Ar$. All have the *threo* configuration, leading to optical activity.

(2) Furanoid lignans ('monoepoxylignans'). Cyclization of diarylbutane lignans can occur in two ways, to give types exemplified by lariciresinol Y7.9 and galbegin Y8.2. One or two new chiral centres, respectively, are created, and their configurations may be determined relative to the two pre-existing centres by spectroscopic methods, equilibration experiments, etc.

(3) Cyclolignans (phenyltetralins). The parent of the cyclolignans is 2,3-dimethyl-1-phenyltetralin, for which 8 stereoformulae can be written

(1a) (1S, 2S, 3R)   (1b) (1R, 2R, 3S)   (2a) (1S, 2R, 3S)   (2b) (1R, 2S, 3R)

---

1. A. J. Birch, B. Milligan, E. Smith, and R. N. Speake, *J. Chem. Soc.*, 1958, 4471.
2. D. C. Ayres, personal communication.

†The help of Dr. D. C. Ayres in the preparation of these notes is gratefully acknowledged.

(3a) (1S, 2S, 3S)   (3b) (1R, 2R, 3R)   (4a) (1S, 2R, 3R)   (4b) (1R, 2S, 3S)

Lignans derived from (1a), (1b), (2b), (3a), (3b), and (4b) are known to occur naturally. Assignment of chirality at the 1-position may be carried out by ORD/CD [1], from which the complete stereochemistry may be adduced by methods such as nmr. Alternatively evidence from cyclisation reactions such as lariciresinol → isolariciresinol can be used, but caution is necessary in the interpretation of this type of cyclisation because epimerisation at C_2 may occur [2].

(4) *Bisepoxylignans* (furanofurans), e.g. pinoresinol **Y7.7**. The bisepoxylignan skeleton contains four asymmetric centres, but since steric restrictions demand a *cis*-8,8′ fusion and since almost all natural bisepoxylignans have both aryl groups identical, only six stereochemical types are possible.

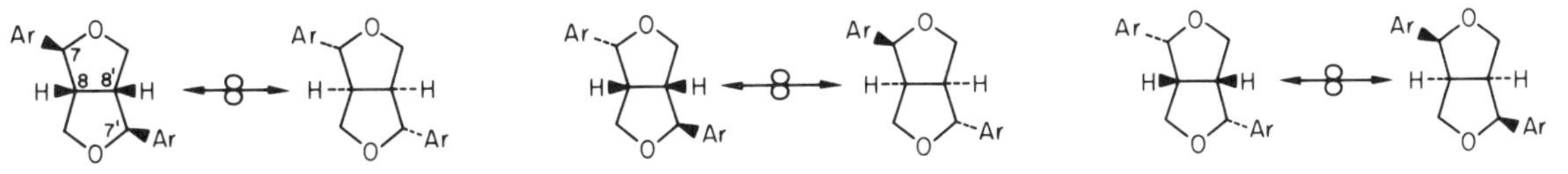

(1a)   (1b)   (2a)   (2b)   (3a)   (3b)

Normal series   *epi* series   *dia* series

Structure determination in this class is generally by ring opening to a furanoid lignan, as shown for pinoresinol → lariciresinol **Y7.7** → **Y7.6**.

Note added in proof: The first unambiguous X-ray determination of the absolute configuration of a lignan, podophyllotoxin, has now been carried out (T. J. Petcher, H. P. Weber, M. Kuhn, and A. von Wartburg, *J. Chem. Soc., Perkin II*, 1973, 288). This lignan has been related to isolariciresinol **Y7.5** by rotational arguments (A. W. Schrecker and J. L. Hartwell, *J. Amer. Chem. Soc.*, 1955, 77, 432) and the determination therefore lends support to the ACs shown on **Y7/Y8**. See **Y7.15**.

Y⁹ 1,3-diarylpentane types; gambogic acid group.

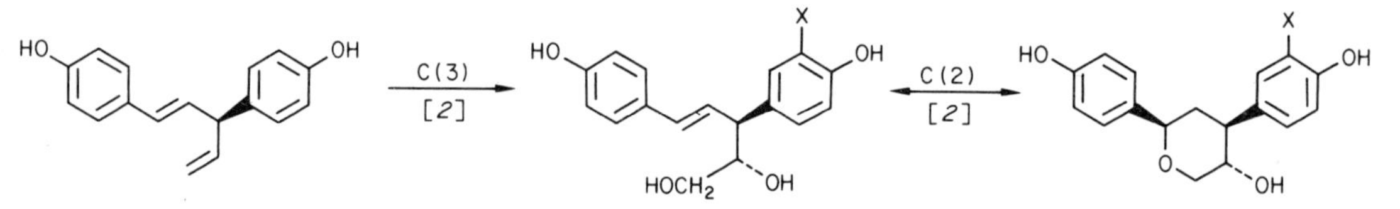

1. (S)-( ) hinokiresinol. Dimethyl ether (+)

2. (−) agatharesinol (X = H)
( ) sequirin C (X = OH) [/]

3. ( ) sugiresinol (X = H) (dimethyl ether (−))
( ) sequirin B (X = OH) [/]

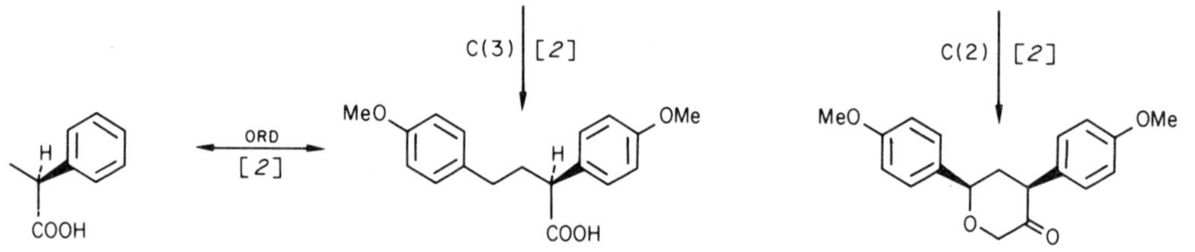

(S)-(+) 2-phenylpropionic acid
A41.10

4. (S)-( ) 2,4-bis(p-methoxyphenyl)-butyric acid

5. ( ) sugiresinone dimethyl ether. AC by ORD [2]

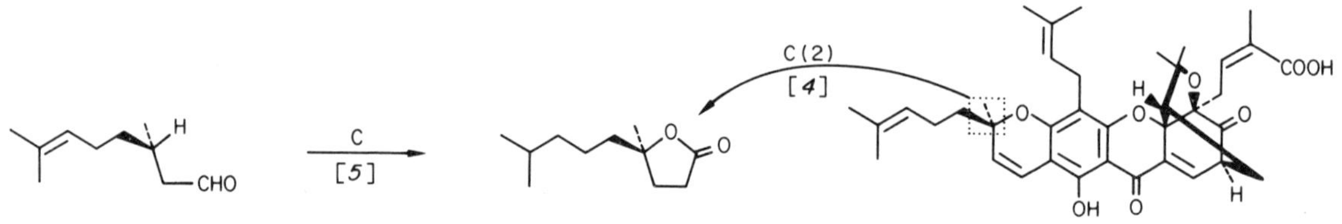

(R)-(+) citronellal
T1.4

6. (R)-(+) 4,8-dimethyl-4-hydroxynonanoic acid lactone

7. (−) gambogic acid. Rel. X-ray of (−) morellin (a similar natural product which does not appear to have been correlated with gambogic acid) [3]

---

1. N. A. R. Hatam and D. A. Whiting, *J. Chem. Soc. (C)*, 1969, 1921.
2. C. R. Enzell, Y. Hirose, and B. R. Thomas, *Tetrahedron Letters*, 1967, 793.
3. G. Kartha, G. N. Ramachandran, H. B. Bhat, P. M. Nair, V. K. V. Raghavan, and K. Venkataraman, *Tetrahedron Letters*, 1963, 459.
4. G. Cardillo and L. Merlini, *Tetrahedron Letters*, 1967, 2529.
5. I. Orban, K. Schaffner, and O. Jeger, *J. Amer. Chem. Soc.*, 1963, 85, 3033.

Aphid pigments.

1. ( ) neriaphin
2. ( ) xanthodactynaphin –*jc* – I.
3. ( ) rhododactynaphin –*jc* – I.
4. ( ) protoaphin –*fb*.
5. ( ) "quinone A"
(*R*)-(+) dilactic acid A I.5
6. ( ) xanthoaphin –*fb*.
7. ( ) chrysoaphin –*fb*.
8. ( ) erythroaphin –*fb*

Pigments of the *-jc-2, -sl* and *-tt* series are epimeric with those shown [*1*] [*3*].

N.B. the protoaphin molecule contains a biaryl system of axial chirality, but its configuration is unknown [*1*].

The suffixes *-fb, -jc, -sl* and *-tt* refer to the aphid species from which the relevant pigments are obtained, e.g. *-fb* pigments from *Aphis fabae* [*1*].

1. A. J. Banks, D. W. Cameron, and J. C. A. Craik, *J. Chem. Soc. (C)*, 1969, 627 and references therein.
2. K. S. Brown, D. W. Cameron, and U. Weiss, *Tetrahedron Letters*, 1969, 471.
3. J. C. Bowie and D. W. Cameron, *J. Chem. Soc. (C)*, 1967, 712.

Y11 Polyether antibiotics; azaphilone mould metabolites.

1. §( ) nigericin (antibiotic X-464) [1] Abs. X-ray [2]

2. ( ) antibiotic X-537A. Abs. X-ray [4]. Rel. X-ray [7]

3. ( ) antibiotic X-206. Abs. X-ray [3]

4. (R)-(−) 6-(3-formylbutyl)-2-methoxy-3-methylbenzoic acid, methyl ester

5. §( ) grisorixin [5]. Abs. X-ray [6]

(R)-(−) 2-methyl-butanal A27.5

6. (S)-( ) aposclerotioramine
Related compds. [10]

7. (+) sclerotiorin.
(−) sclerotiorin = 7-epimer
(*not* enantiomer) [9]
Other related compds. [10]

1. J. F. Blount and J. W. Westley, personal communication.
2. M. Shiro and H. Koyama, *J. Chem. Soc. (B)*, 1970, 243.
3. J. F. Blount and J. W. Westley, *Chem. Comm.*, 1971, 927.
4. S. M. Johnson, J. Herrin, S. J. Liu, and I. C. Paul, *J. Amer. Chem. Soc.*, 1970, 92, 4428.
5. M. Alléaume, personal communication.
6. M. Alléaume and D. Hickel, *Chem. Comm.*, 1971, 1423.
7. S. M. Johnson, J. Herrin, S. J. Liu, and I. C. Paul, *Chem. Comm.*, 1970, 72.
8. J. W. Westley, R. H. Evans, T. Williams, and A. Stempel, *Chem. Comm.*, 1970, 71.
9. G. A. Ellestad and W. B. Whalley, *J. Chem. Soc.*, 1965, 7260.
10. F. C. Chen, P. S. Manchand, and W. B. Whalley, *J. Chem. Soc. (C)*, 1971, 3577.

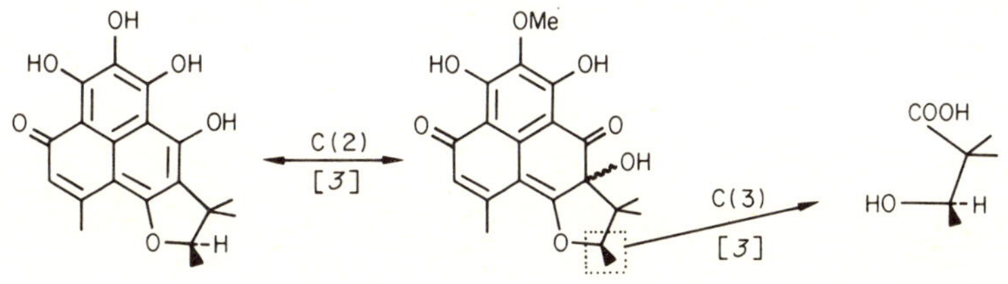

Atrovenetin and blancoic acid groups.

1. (R)-(−) atrovenetin
2. (+) herqueinone
(R)-(−) 2,2-dimethyl-3-hydroxybutyric acid **A12.14**

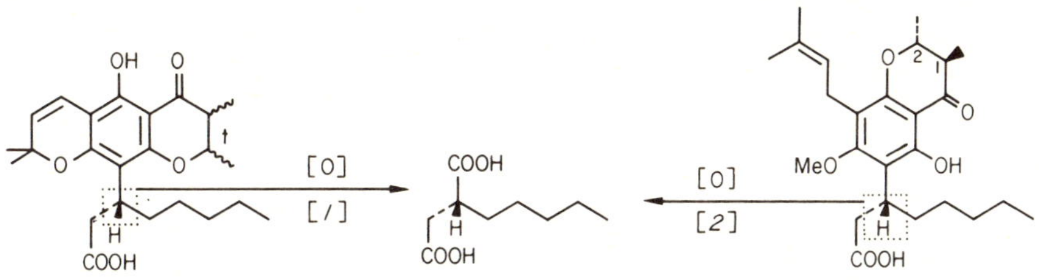

3. ( ) blancoic acid
(R)-(+) pentylsuccinic acid **A28.9**
4. (+) papuanic acid
(−) isopapuanic acid = 2-epimer [2]
Rel. X-ray [2]

1. G. H. Stout and K. D. Sears, *J. Org. Chem.*, 1968, **33**, 4185.
2. G. H. Stout, G. K. Hickernell, and K. D. Sears, *J. Org. Chem.*, 1968, **33**, 4191.
3. J. S. Brooks and G. A. Morrison, *Chem. Comm.*, 1971, 1359.

Y13 Aflatoxins; citrinin group of mould metabolites.

1. (−) aflatoxin G1. Rel. X-ray [1]
2. (−) aflatoxin B1. Abs. X-ray [2]
3. (−) sterigmatocystin
4. (+) pulvilloric acid
5. (S)-(−) 1-(3,5-dimethoxyphenyl)-heptan-2-ol
6. ( ) tetraacetoxydothistromin Abs. X-ray [5]
7. (−) citrinin. Rel. X-ray [7]
8. (2R,3S)-(−) 3-(3,5-dimethoxy-2-methylphenyl)-butan-2-ol
9. (2R,3S)-(−) 3-(3,5-dihydroxy-2-methylpheny)-butan-2-ol
10. (R)-(−) 2-(3,5-dimethoxy-2-methylphenyl)-butane

(S)-(+) 2-methylbutyric acid **A29.11**

(S)-(+) atrolactic acid **A31.19**

(R)-(−) 2-phenylbutane **A49.15**

1. K. K. Cheung and G. A. Sim, *Nature*, 1967, **201**, 1185.
2. T. E. van Soest and A. F. Peerdeman, *Acta Cryst.*, 1970, **B26**, 1940.
3. S. Brechbühler, G. Büchi, and G. Milne, *J. Org. Chem.*, 1967, **32**, 2641.
4. J. S. E. Holker and J. Mulheirn, *Chem. Comm.*, 1968, 1576.
5. C. A. Bear, J. M. Waters, T. N. Waters, R. T. Gallagher, and R. Hodges, *Chem. Comm.*, 1970, 1705.
6. G. C. Barrett, J. F. W. McOmie, S. Nakajima, and S. W. Tanenbaum, *J. Chem. Soc. (C)*, 1969, 1068.
7. O. R. Rodig, M. Shiro, and Q. Fernando, *Chem. Comm.*, 1971, 1553.
8. P. P. Mehta and W. B. Whalley, *J. Chem. Soc.*, 1963, 3777.
9. R. K. Hill and L. A. Gardella, *J. Org. Chem.*, 1964, **29**, 766.

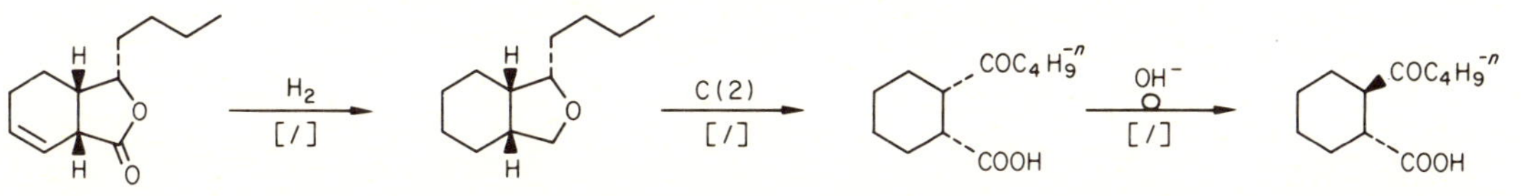

Phthalides; macrocyclic lactones (brefeldin, zearalenone). Y[14]

1. (−) cnidilide
2. (−) β-dihydrosedanolide
3. ( )
4. ( ), 2,4-DNP deriv. (−)

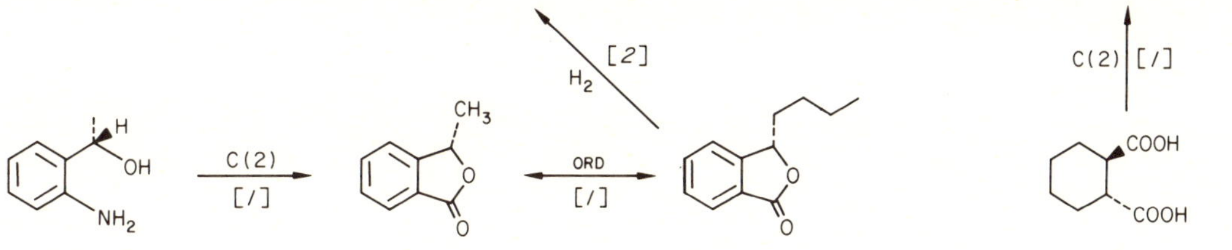

(S)-(−) 1-(2-aminophenyl) ethanol **A22.6**

5. (S)-(−) 3-methylphthalide
6. (S)-(−) 3-butylphthalide

(1R, 2R)-(−) cyclohexane-1,2-dicarboxylic acid **A37.6**

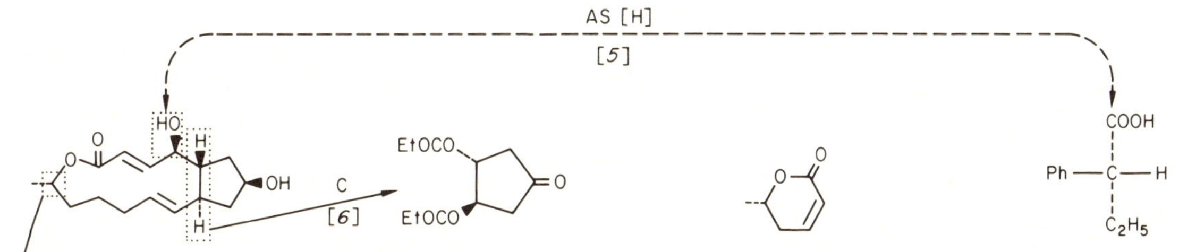

7. (+) brefeldin A. Rel. X-ray [5]
8. (1R, 2R)-(−) 4-oxocyclopentane-1,2-dicarboxylic acid, diethyl ester
9. (S)-(+) sorbinol (5-hydroxy-hex-2-enoic acid lactone)

(R)-(−) 2-phenylbutyric acid **A25.15**

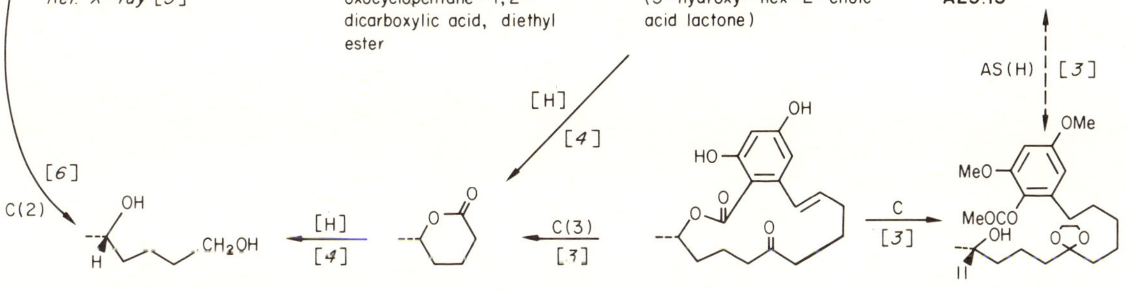

(S)-(+) hexane-1,5-diol **A18.17**
10. (S)-(−) 5-hydroxyhexanoic acid lactone
11. (S)-(−) zearalenone
12. (S)-(+)

1. U. Nagai, T. Shishido, R. Chiba, and H. Mitsuhashi, *Tetrahedron*, 1965, **21**, 1701, and references therein.
2. D. H. R. Barton and J. X. de Vries, *J. Chem. Soc.*, 1963, 1916.
3. C. H. Kuo, D. Taub, R. D. Hoffsommer, N. L. Wendler, W. H. Urry, and G. Mullenbach, *Chem. Comm.*, 1967, 761.
4. R. Kuhn and K. Kum, *Chem. Ber.*, 1962, **95**, 2009.
5. H. P. Weber, D. Hauser, and H. P. Sigg, *Helv. Chim. Acta*, 1971, **54**, 2763.
6. H. P. Sigg, *Helv. Chim. Acta*, 1964, **47**, 1401.

Y15 Miscellaneous natural products containing C, H, O, halogens only.

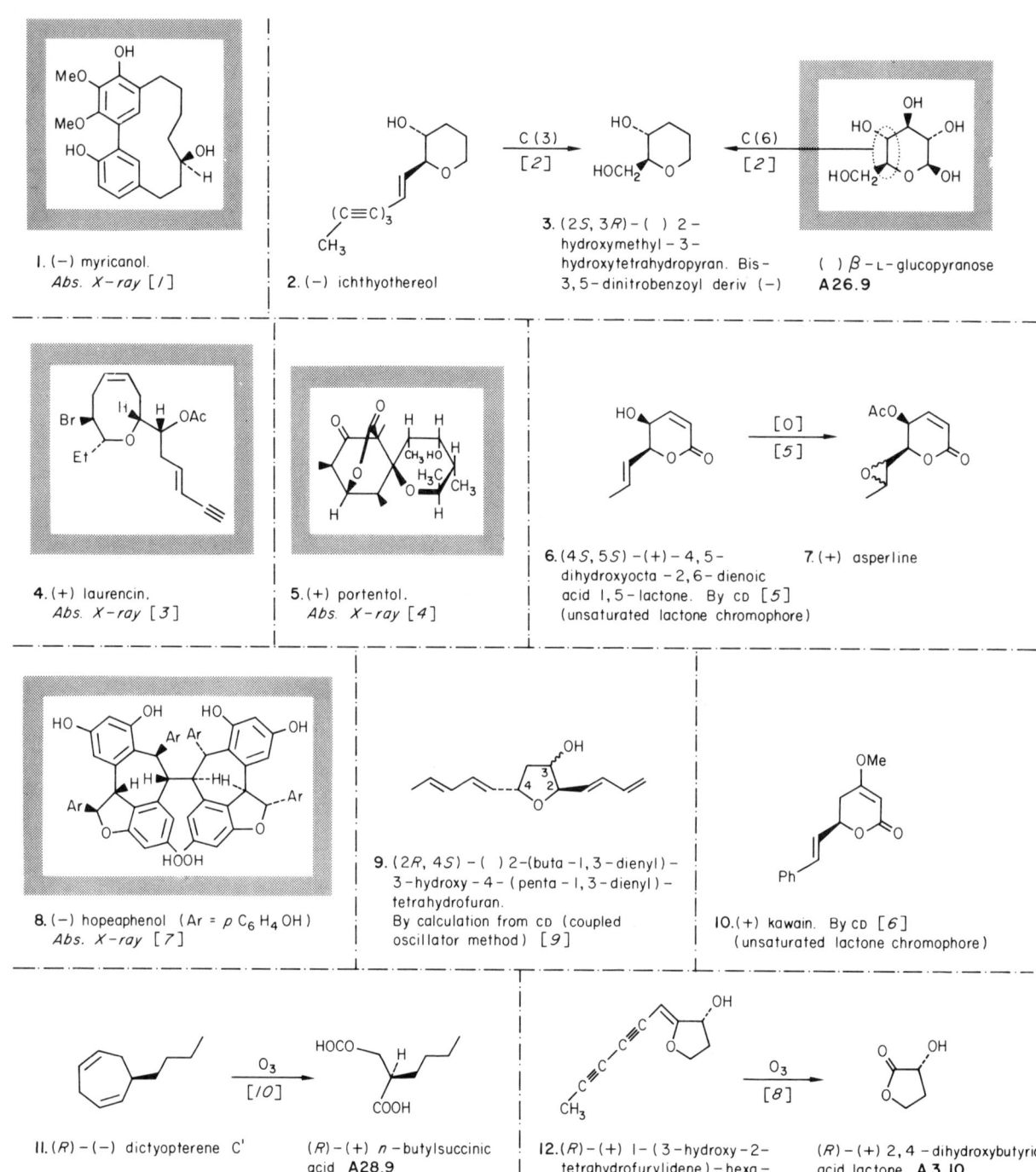

1. (−) myricanol. Abs. X-ray [1]
2. (−) ichthyothereol
3. (2S, 3R)-( ) 2-hydroxymethyl-3-hydroxytetrahydropyran. Bis-3,5-dinitrobenzoyl deriv (−)
( ) β-L-glucopyranose A26.9
4. (+) laurencin. Abs. X-ray [3]
5. (+) portentol. Abs. X-ray [4]
6. (4S, 5S)-(+)-4,5-dihydroxyocta-2,6-dienoic acid 1,5-lactone. By CD [5] (unsaturated lactone chromophore)
7. (+) asperline
8. (−) hopeaphenol (Ar = p C$_6$H$_4$OH) Abs. X-ray [7]
9. (2R, 4S)-( ) 2-(buta-1,3-dienyl)-3-hydroxy-4-(penta-1,3-dienyl)-tetrahydrofuran. By calculation from CD (coupled oscillator method) [9]
10. (+) kawain. By CD [6] (unsaturated lactone chromophore)
11. (R)-(−) dictyopterene C' → (R)-(+) n-butylsuccinic acid A28.9
12. (R)-(+) 1-(3-hydroxy-2-tetrahydrofurylidene)-hexa-2,4-diyne → (R)-(+) 2,4-dihydroxybutyric acid lactone A 3.10

1. M. J. Begley and D. A. Whiting, *Chem. Comm.*, 1970, 1207.
2. C. Chin, M. C. Cutler, E. R. H. Jones, J. Lee, S. Safe, and V. Thaller, *J. Chem. Soc. (C)*, 1970, 314.
3. A. F. Cameron, K. K. Cheung, G. Ferguson, and J. M. Robertson, *J. Chem. Soc. (B)*, 1969, 559.
4. G. Ferguson and I. R. Mackay, *Chem. Comm.*, 1970, 665.
5. R. H. Evans, G. A. Ellestad, and M. P. Kunstmann, *Tetrahedron Letters*, 1969, 1791.
6. G. Snatzke and R. Hänsel, *Tetrahedron Letters*, 1968, 1797.
7. P. Coggon, A. T. McPhail, and S. C. Wallwork, *J. Chem. Soc. (B)*, 1970, 884.
8. R. K. Bentley, E. R. H. Jones, and V. Thaller, *J. Chem. Soc. (C)*, 1969, 1096.
9. S. F. Mason and G. W. Vane, *Chem. Comm.*, 1967, 598.
10. J. A. Pettus and R. E. Moore, *J. Amer. Chem. Soc.*, 1971, 93, 3087.

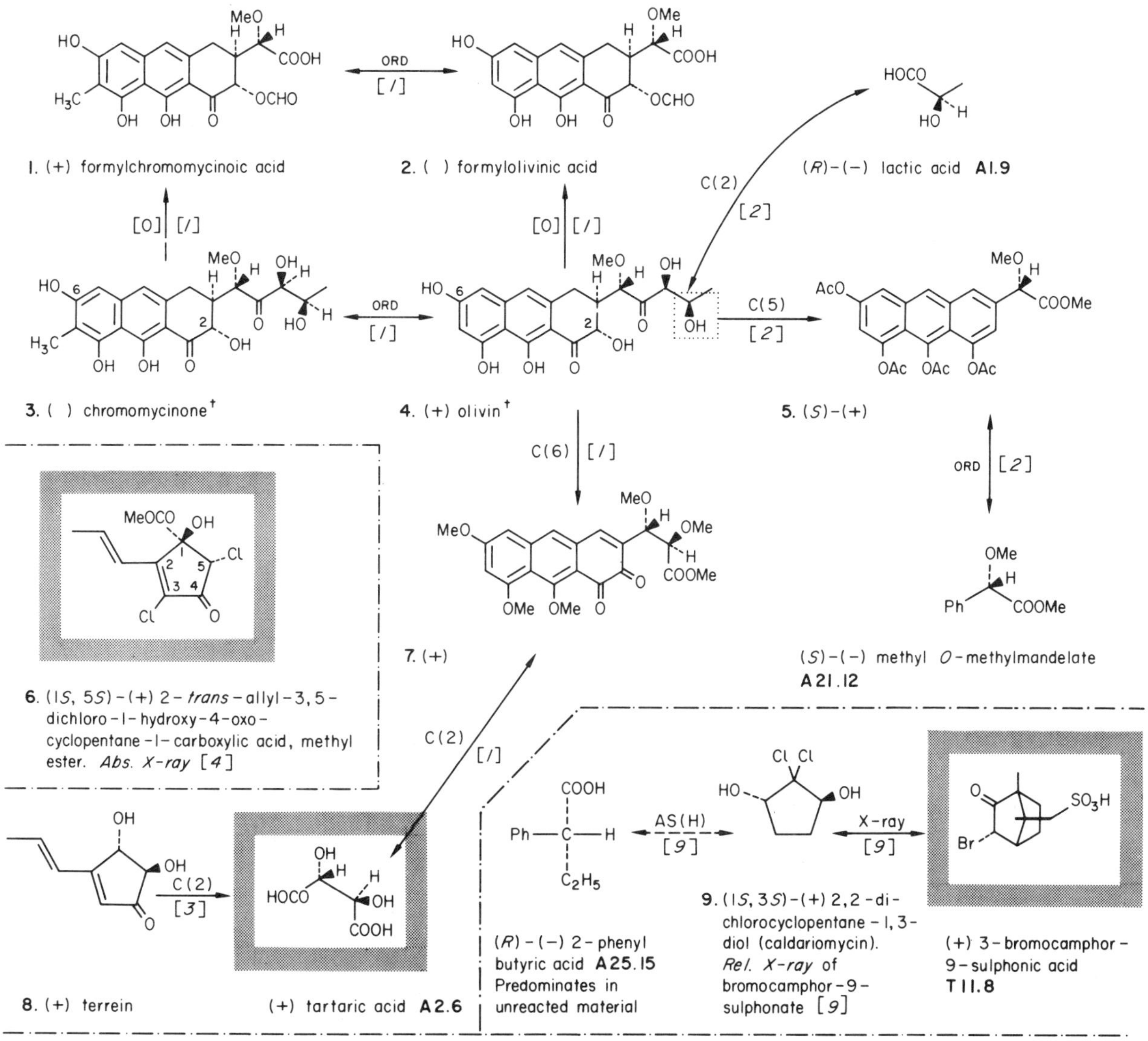

Aureolic acid group antibiotics; cyclopentanoid antibiotics; prostaglandins.

Y16

1. (+) formylchromomycinoic acid
2. (−) formylolivinic acid
3. (−) chromomycinone†
4. (+) olivin†
5. (S)-(+)
6. (1S, 5S)-(+) 2-*trans*-allyl-3,5-dichloro-1-hydroxy-4-oxo-cyclopentane-1-carboxylic acid, methyl ester. Abs. X-ray [4]
7. (+)
8. (+) terrein

(R)-(−) lactic acid A1.9

(S)-(−) methyl O-methylmandelate A21.12

(+) tartaric acid A2.6

(R)-(−) 2-phenyl butyric acid A25.15 Predominates in unreacted material

9. (1S, 3S)-(+) 2,2-dichlorocyclopentane-1,3-diol (caldariomycin). Rel. X-ray of bromocamphor-9-sulphonate [9]

(+) 3-bromocamphor-9-sulphonic acid T11.8

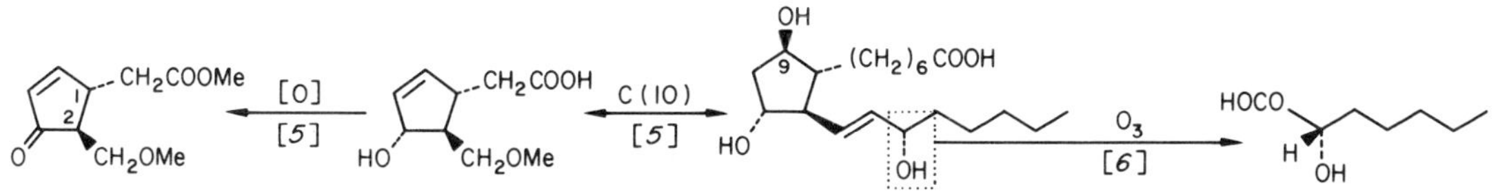

10. (1R, 2R)-(−) 2-methoxymethyl-3-oxocyclopent-4-ene-1-acetic acid. By CD [5]
11. (+)
12. (−) prostaglandin-F1β. Rel. X-ray [8]
    (−) prostaglandin-F1α = 9-epimer
    (−) prostaglandin E₁ = 9-(oxodesoxy) deriv.
    Other prostaglandins related. (See also [7])

(S) 2-hydroxyheptanoic acid, (−) in NaOH A24.15

1. Yu. A. Berlin, M. N. Kolosov, and L. A. Piotrovich, *Tetrahedron Letters*, 1970, 1329.
2. G. P. Bakhaeva, Yu. A. Berlin, O. A. Chuprunova, M. N. Kolosov, G. Yu. Peck, L. A. Piotrovich, M. M. Shemyakin, and I. V. Vasina, *Chem. Comm.*, 1967, 10.
3. D. H. R. Barton and E. Miller, *J. Chem. Soc.*, 1955, 1028.
4. W. J. McGahren, J. H. van den Hende, and L. A. Mitscher, *J. Amer. Chem. Soc.*, 1969, 91, 157.
5. E. J. Corey, T. K. Schaaf, W. Huber, U. Koelliker, and M. N. Weinshenker, *J. Amer. Chem. Soc.*, 1970, 92, 397; E. J. Corey, R. Noyori, and T. K. Schaaf, ibid., 2586.
6. D. H. Nugteren, D. A. Van Dorp, S. Bergström, K. Hamberg, and B. Samuelson, *Nature*, 1966, 212, 38.
7. E. J. Corey, H. Shirahama, H. Yamamoto, S. Terashima, A. Venkateswarlu, and T. K. Schaaf, *J. Amer. Chem. Soc.*, 1971, 93, 1491.
8. S. Abrahamsson, *Acta Cryst.*, 1963, 16, 409.
9. S. M. Johnson, I. C. Paul, K. L. Rinehart, and R. Srinavasan, *J. Amer. Chem. Soc.*, 1968, 90, 136.

†The antibiotics of the aureolic acid group (chromomycins, olivomycins etc.) are O-glycosides of olivin and chromomycinone at C_2 and C_6.

Y17 Anthracenes; ergot pigments.

1. (+) catalponol. By CD (aromatic chirality rule) [1]
2. ( ) bostrychin. Abs. X-ray [2]
3. ( ) purpurogenone. Abs. X-ray [9]
4. (R)-(+) cacalol [7]. ORD [7]
5. (+) griseofulvin. Rel. X-ray [3]
6. (+) ergoflavin (ergochrome CC) Abs. X-ray [6]

(R)-(+) 3-methyladipic acid A 26.15

(S)-(-) methylsuccinic acid A 27.21

The ergot pigments are derived from the following four structural units **Y17A-D** by dimerisation and cross-dimerisation at the 2-positions [10]. Several of the pigments other than ergoflavin have been degraded to (R) or (S) methylsuccinic acid and the full stereochemistry worked out by a combination of spectroscopis methods and rotational arguments [5] [10].

A   B   C   D

1. H. Inouye, T. Okuda, and T. Hayashi, *Tetrahedron Letters*, 1971, 3615.
2. A. Takenaka, A. Furusaki, T. Watanabé, T. Noda, T. Take, T. Watanabe, and J. Abe, *Tetrahedron Letters*, 1968, 6091
3. W. A. C. Brown and G. A. Sim, *J. Chem. Soc.*, 1963, 1050.
4. J. F. Grove, J. MacMillan, T. P. C. Mulholland, and J. Zealley, *J. Chem. Soc.*, 1952, 3967.
5. B. Frank, G. Baumann, and U. Ohnsorge, *Tetrahedron Letters*, 1965, 2031.
6. A. T. McPhail, G. A. Sim, J. D. M. Asher, J. M. Robertson, and J. V. Silverton, *J. Chem. Soc. (B)*, 1966, 18.
7. P. Joseph-Nathan and M. P. Gonzalez, *Canad. J. Chem.*, 1969, 47, 2465.
8. P. Joseph-Nathan, J. J. Morales, and J. Romo, *Tetrahedron*, 1966, 22, 301.
9. T. J. King, J. C. Roberts, and D. J. Thompson, *Chem. Comm.*, 1970, 1499.
10. J. W. Hooper, W. Marlow, W. B. Whalley, A. D. Borthwick, and R. Bowden, *Chem. Comm.*, 1971, 111.

Further anthracene types and other polycyclic natural products.

Y18

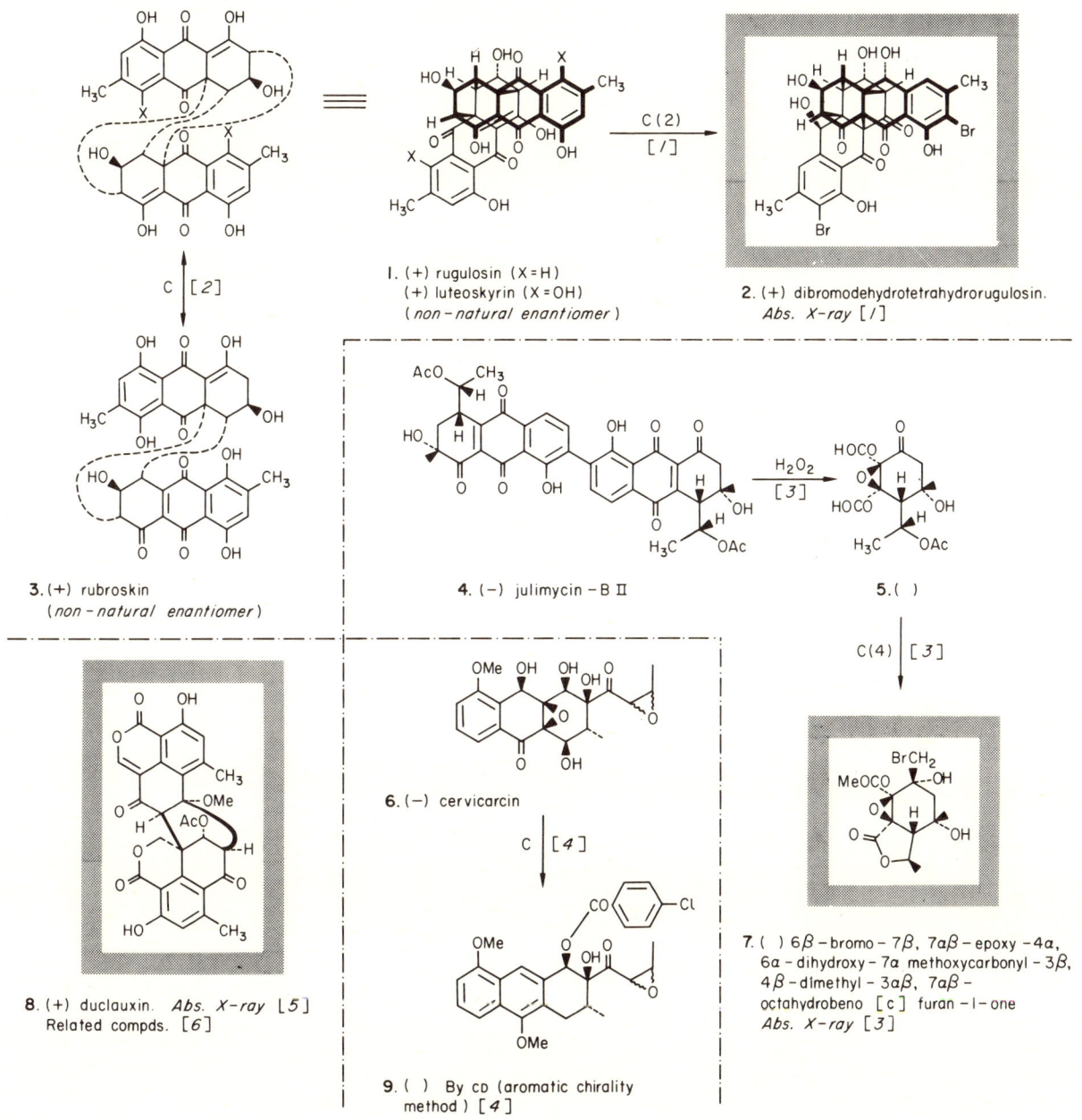

1. N. Kobayashi, Y. Iitaka, U. Sankawa, Y. Ogihara, and S. Shibata, *Tetrahedron Letters*, 1968, 6135.
2. U. Sankawa, S. Seo, N. Kabayashi, Y. Ogihara, and S. Shibata, *Tetrahedron Letters*, 1968, 5557.
3. H. Nakai, M. Shiro, and H. Koyama, *J. Chem. Soc. (B)*, 1969, 498.
4. S. Marumo, N. Harada, K. Nakanishi, and T. Nishida, *Chem. Comm.*, 1970, 1693.
5. Y. Ogihara, Y. Iitaka, and S. Shibata, *Tetrahedron Letters*, 1965, 1289.
6. Y. Ogihara, O. Tanaka, and S. Shibata, *Tetrahedron Letters*, 1966, 2867.

Y19 Cyclohexane derivatives; crotepoxide-senepoxide group, trisporic acids etc.

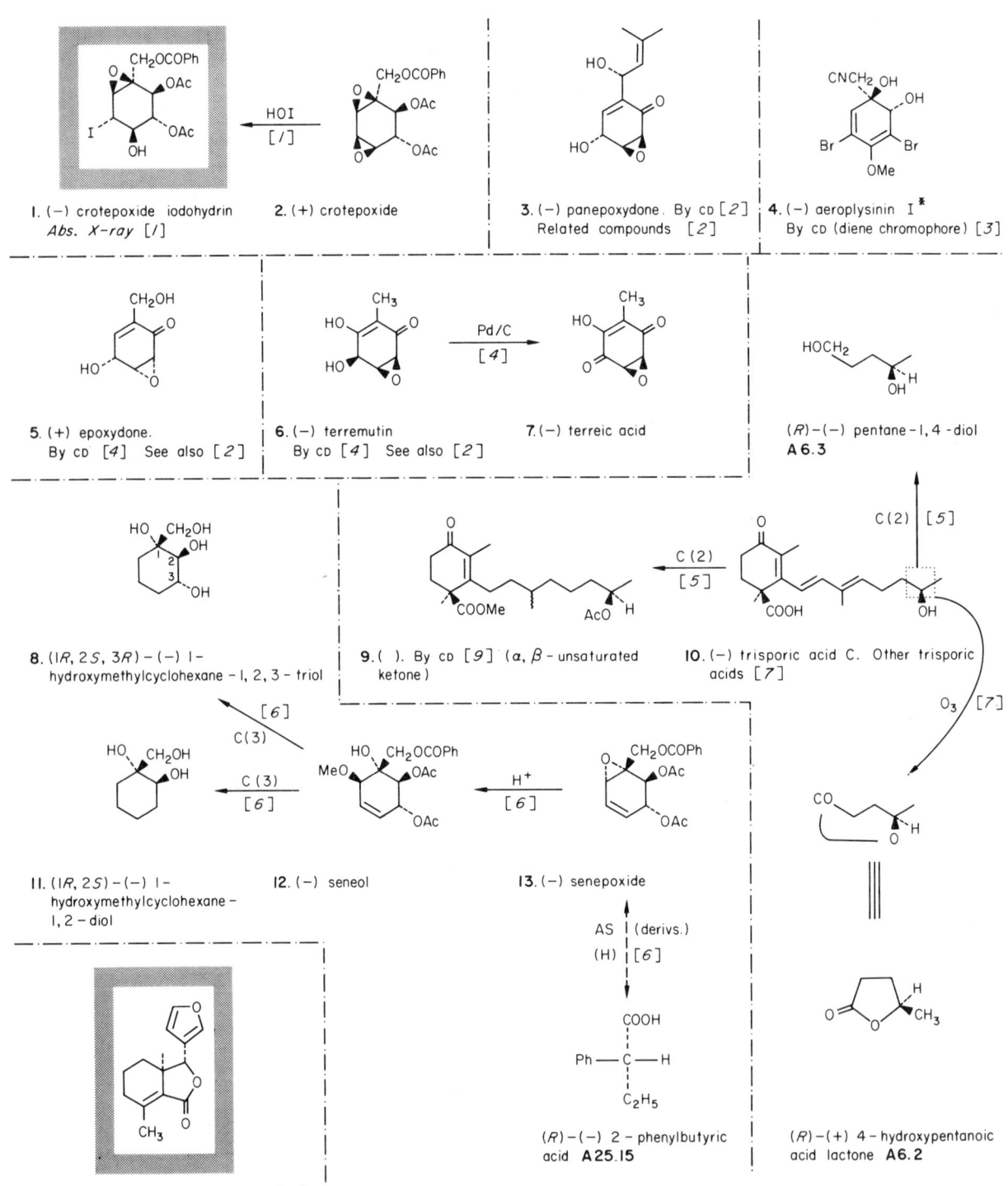

1. S. M. Kupchan, R. J. Hemingway, P. Coggon, A. T. McPhail, and G. A. Sim, *J. Amer. Chem. Soc.*, 1968, **90**, 2982.
2. Z. Kis, A. Closse, H. P. Sigg, L. Hruban, and G. Snatzke, *Helv. Chim. Acta*, 1970, **53**, 1577.
3. W. Fulmor, G. E. van Lear, G. O. Morton, and R. D. Mills, *Tetrahedron Letters*, 1970, 4551.
4. M. W. Miller, *Tetrahedron*, 1968, **24**, 4839.
5. J. D. Bu'Lock, D. J. Austin, G. Snatzke, and L. Hruban, *Chem. Comm.*, 1970, 255.
6. R. Hollands, D. Becher, A. Gaudemer, and J. Polonsky, *Tetrahedron*, 1968, **24**, 1633.
7. T. Resche, *Tetrahedron Letters*, 1969, 3435.
8. P. Coggon, A. T. McPhail, R. Storer, and D. W. Young, *Chem. Comm.*, 1969, 828.

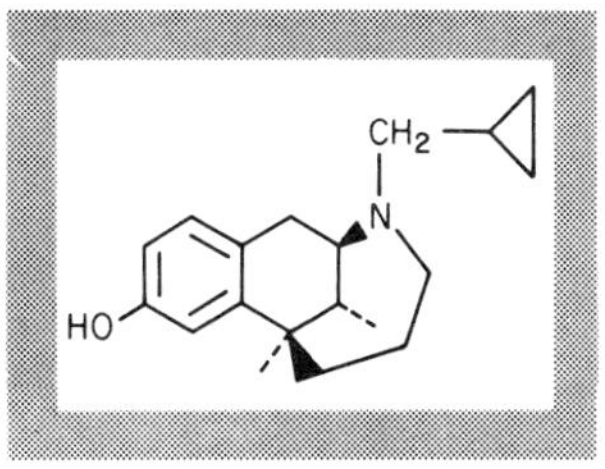

Miscellaneous nitrogen and phosphorus compounds.

1. (−) cyclazocine (synthetic compound) Abs. X-ray [1]

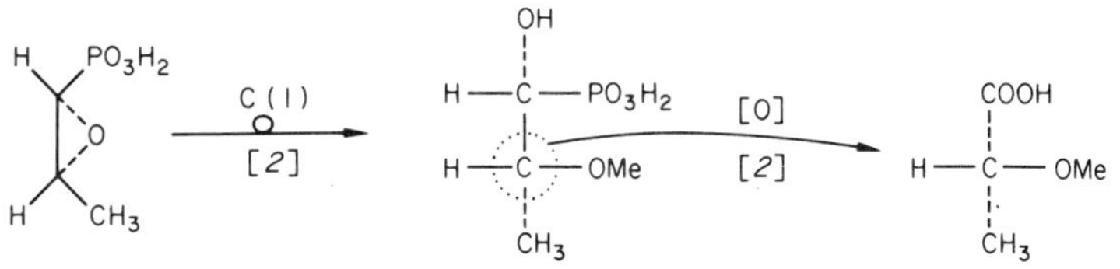

2. (−) phosphonomycin ((1R, 2S) − 1,2 − epoxypropylphosphonic acid)

3. (1R, 2R) − (−) 1−hydroxy−2−methoxypropylphosphonic acid

(R) − (+) O − methyllactic acid A 1.8

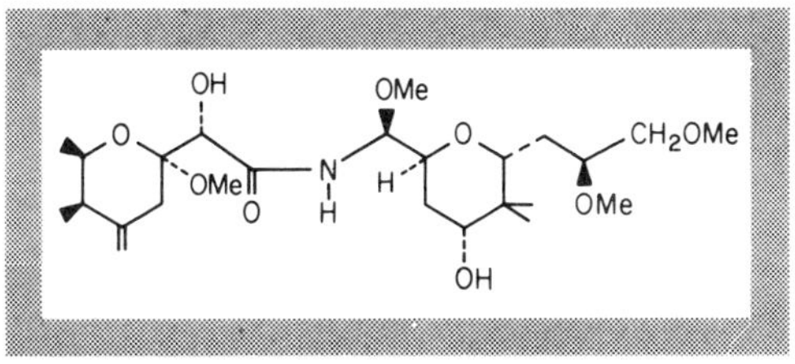

4. ( ) pederin. Abs. X-rays [3] [4]

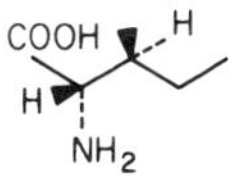

(+) isoleucine A 27.6

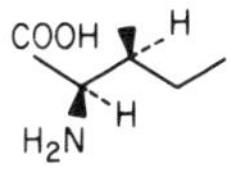

(−) alloisoleucine A 27.2

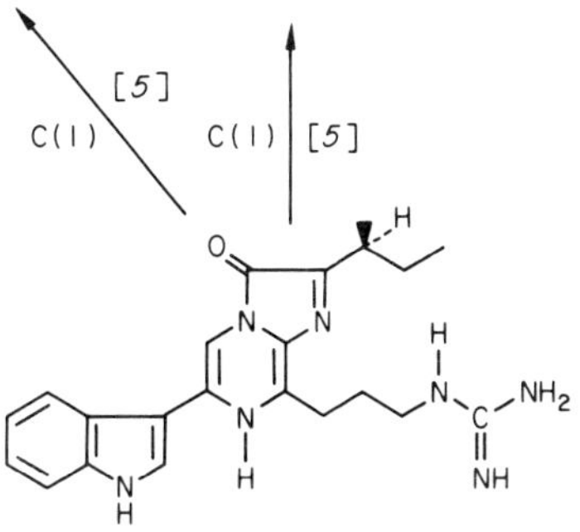

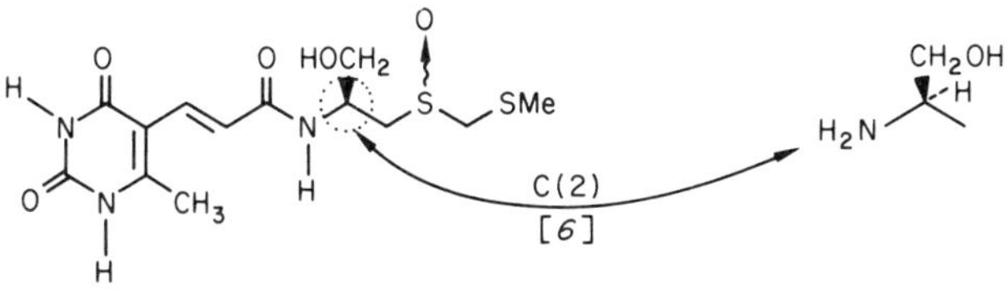

5. ( ) sparsomycin

(R) − (−) 2−aminopropan−1−ol A 19.13

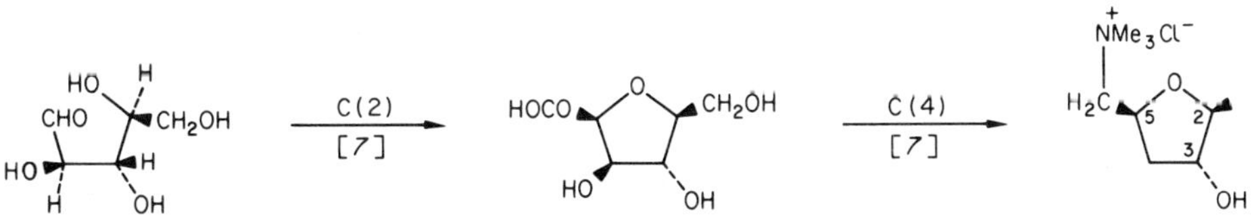

6. (S) − ( ) cypridina luciferin

L (+) arabinose C 1.8

7. L ( ) chitaric acid

8. (+) muscarine. Rel. X-ray [8] stereoisomers; [9]

---

1. I. L. Karle, R. D. Gilardi, A. V. Fratini, and J. Karle, *Acta Cryst.*, 1969, **B25**, 1469.
2. B. G. Christensen, W. J. Leanza, T. R. Beattie, A. A. Patchett, B. H. Arison, R. E. Ormond, F. A. Kuehl, G. Albers-Schonberg, and O. Jardetzky, *Science*, 1969, **166**, 123.
3. A. Furusaki, T. Watanabé, T. Matsumoto, and M. Yanagiya, *Tetrahedron Letters*, 1968, 6301.
4. A. B. Corradi, A. Mangia, M. Nardelli, and G. Pelizzi, *Gazzetta*, 1971, **101**, 591.
5. Y. Kishi, T. Goto, Y. Hirata, O. Shimomura, and F. H. Johnson, *Tetrahedron Letters*, 1966, 3427.
6. P. F. Wiley and F. A. MacKellar, *J. Amer. Chem. Soc.*, 1970, **92**, 417.
7. C. H. Eugster, *Adv. Organic. Chem.*, II, 427.
8. F. Jellinek, *Acta Cryst.*, 1957, **10**, 277.
9. C. H. Eugster and E. Schleusener, *Helv. Chim. Acta*, 1969, **52**, 708; H. Bollinger and C. H. Eugster, *Helv. Chim. Acta*, 1971, **54**, 2704.

Y21 Lincomycin group antibiotics; actidione.

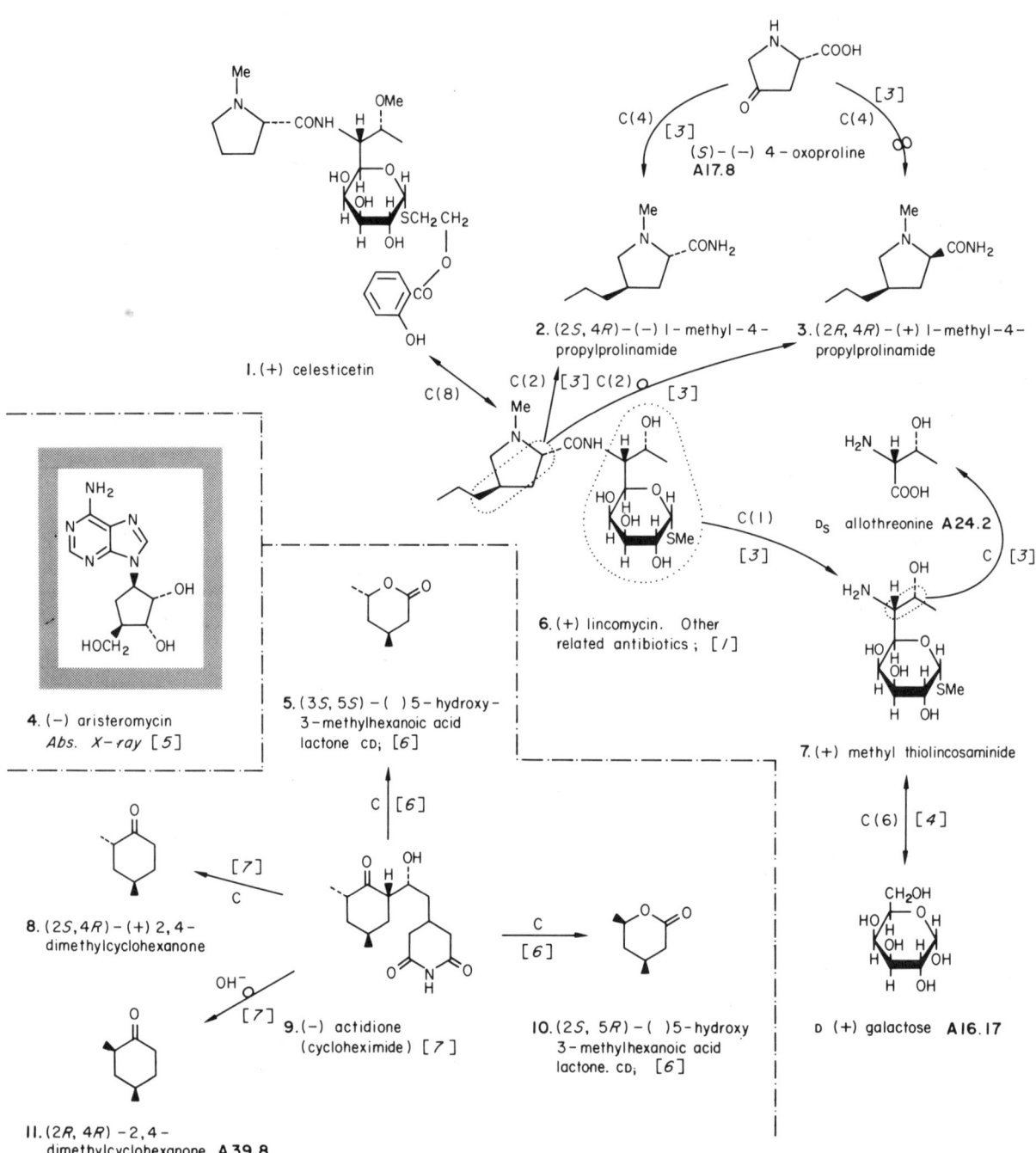

1. A. D. Argondelis, J. A. Fox, D. J. Mason, and T. E. Eble, *J. Amer. Chem. Soc.*, 1964, **86**, 5044.
2. H. Hoeksma, *J. Amer. Chem. Soc.*, 1964, **86**, 4224.
3. B. J. Magerlein, R. D. Birkenmeyer, R. R. Herr, and F. Kagan, *J. Amer. Chem. Soc.*, 1967, **89**, 2459, and references therein.
4. W. Schroeder, R. Bannister, and H. Hoeksma, *J. Amer. Chem. Soc.*, 1967, **89**, 4448.
5. T. Kishi, M. Muroi, T. Kusaka, M. Nishikawa, K. Kamiya, and K. Mizuno, *Chem. Comm.*, 1967, 852.
6. F. I. Carroll, A. Sobti, and R. Meck, *Tetrahedron Letters*, 1971, 405.
7. E. C. Kornfeld, R. G. Jones, and T. V. Parke, *J. Amer. Chem. Soc.*, 1949, **71**, 150; F. Johnson, N. A. Starkovsky, A. C. Paton, and A. A. Carlson, ibid., 1966, **88**, 149, and references therein.

Viomycin, mycobactin.

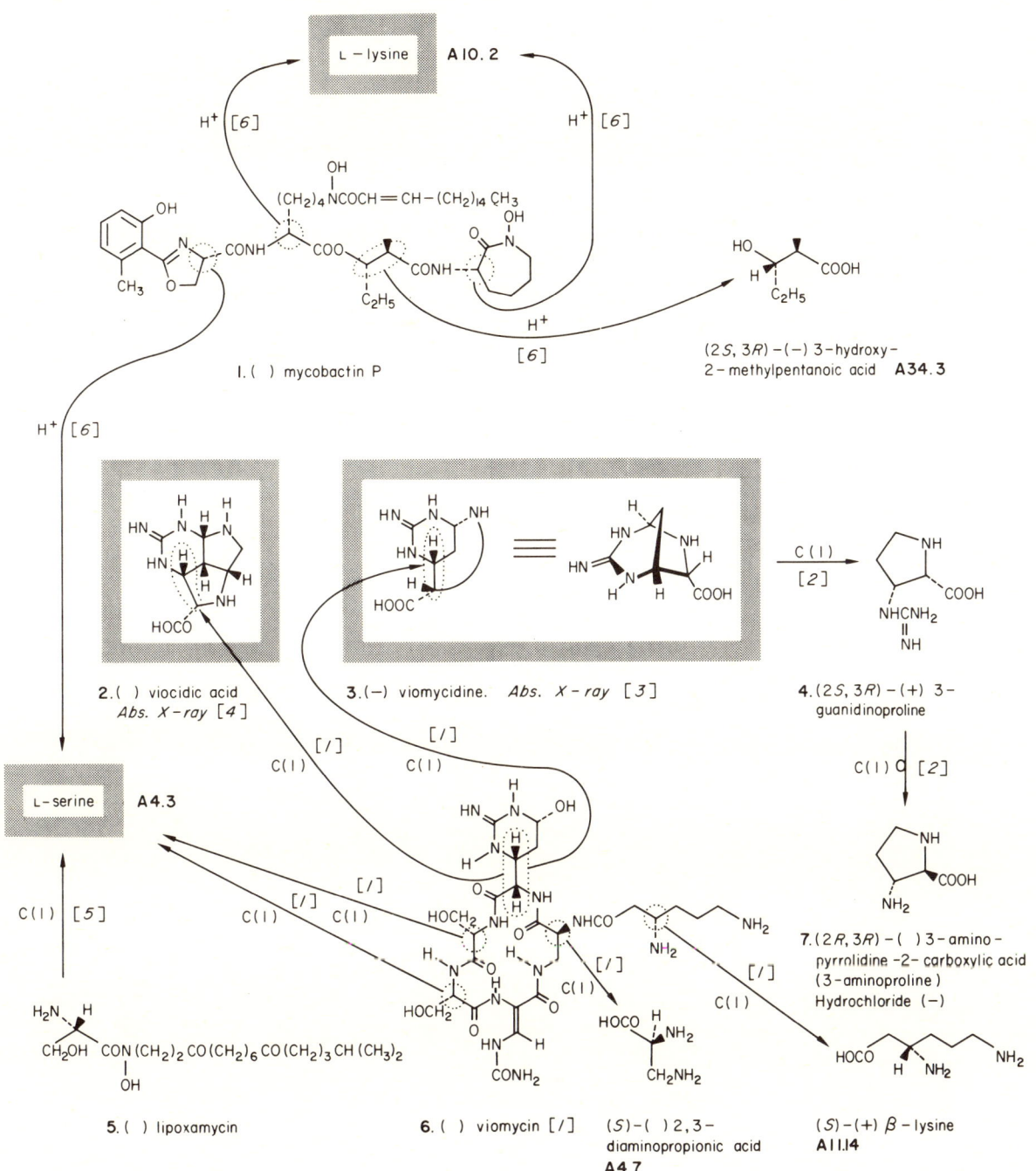

1. B. W. Bycroft, D. Cameron, L. R. Croft, A. Hassanali-Walji, A. W. Johnson, and T. Webb, *Experientia*, 1971, **27**, 501, and references therein.
2. C. Gallina, C. Marta, C. Colombo, and A. Romeo, *Tetrahedron*, 1971, **27**, 4681.
3. J. C. Floyd, J. A. Bertrand, and J. R. Dyer, *Chem. Comm.*, 1968, 998.
4. P. Coggon, *J. Chem. Soc. (B)*, 1970, 838.
5. H. A. Whaley, *J. Amer. Chem. Soc.*, 1971, **93**, 3767.
6. G. A. Snow, *Biochem. J.*, 1965, **94**, 160; *J. Chem. Soc.*, 1954, 2588, 4080.

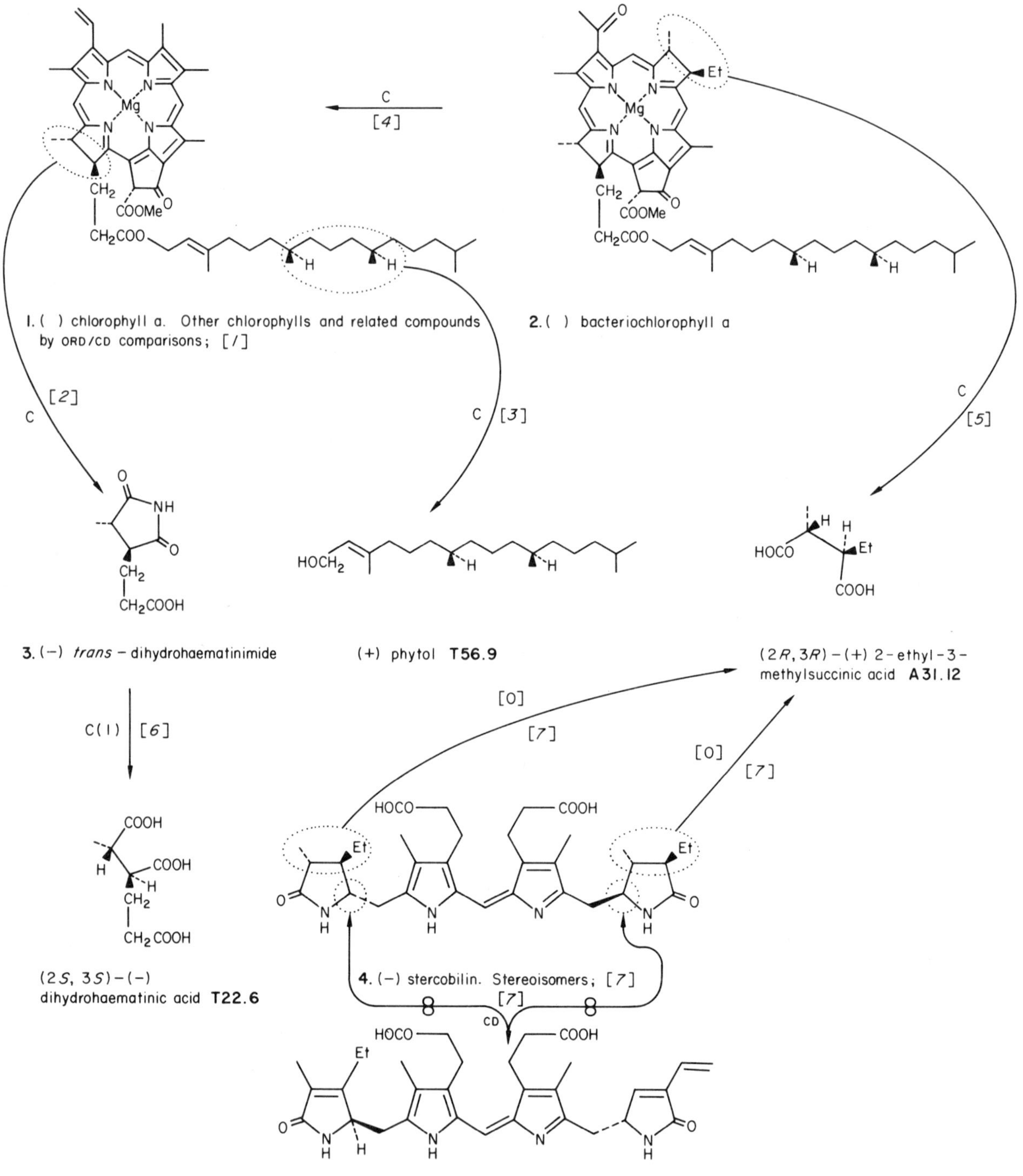

Y 23 Porphyrins; chlorophylls, bile pigments.

1. ( ) chlorophyll a. Other chlorophylls and related compounds by ORD/CD comparisons; [/]

2. ( ) bacteriochlorophyll a

3. (−) *trans* − dihydrohaematinimide

(+) phytol T56.9

(2R, 3R) − (+) 2-ethyl-3-methylsuccinic acid A31.12

(2S, 3S) − (−) dihydrohaematinic acid T22.6

4. (−) stercobilin. Stereoisomers; [7]

5. (+) urobilin

1. H. Wolf and H. Scheer, *Annalen*, 1971, **745**, 87.
2. G. E. Ficken, R. B. Johns, and R. P. Linstead, *J. Chem. Soc.*, 1956, 2272.
3. R. Willstätter and F. Hocheder, *Annalen*, 1907, **354**, 205.
4. H. Fisher, H. Mittenzwei, and D. B. Héver, *Annalen*, 1940, **545**, 154.
5. H. Brockmann, *Angew. Chem. Internat. Edn.*, 1968, 7, 222.
6. I. Fleming, *Nature*, 1967, **216**, 151.
7. H. Brockmann, G. Knobloch, H. Plieninger, K. Ehl, J. Ruppert, A. Moscowitz, and C. J. Watson, *Proc. Nat. Acad. Sci. USA*, 1971, **68**, 2141.

Y24
Vitamin B₁₂ (review; [5]); gliotoxin group of fungal metabolites.

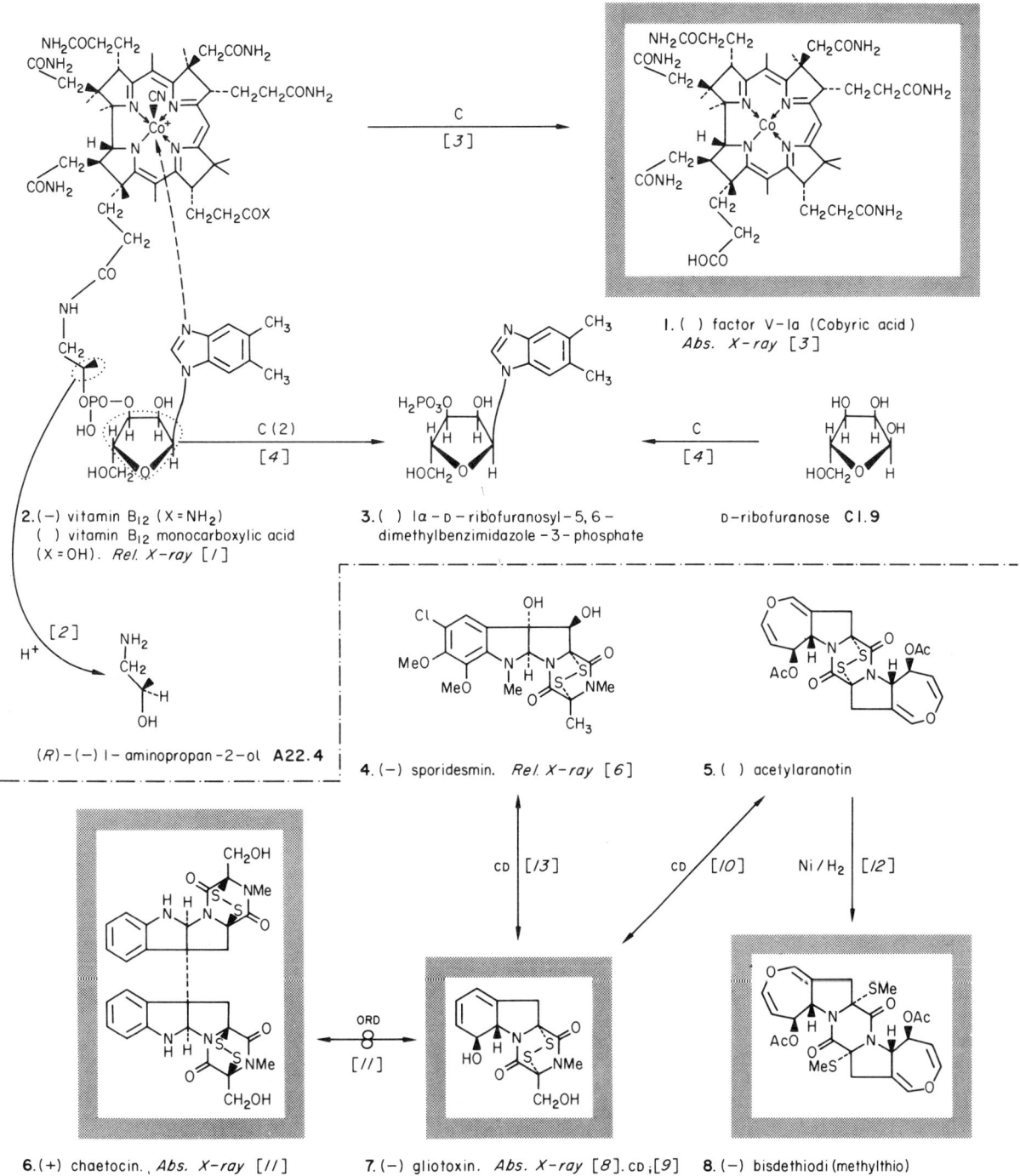

1. ( ) factor V-1a (Cobyric acid) Abs. X-ray [3]
2. (−) vitamin B₁₂ (X=NH₂) ( ) vitamin B₁₂ monocarboxylic acid (X=OH). Rel. X-ray [1]
3. ( ) 1α-D-ribofuranosyl-5,6-dimethylbenzimidazole-3-phosphate
   D-ribofuranose Cl.9
   (R)-(−)1-aminopropan-2-ol A22.4
4. (−) sporidesmin. Rel. X-ray [6]
5. ( ) acetylaranotin
6. (+) chaetocin. Abs. X-ray [11]
7. (−) gliotoxin. Abs. X-ray [8]. CD [9]
8. (−) bisdethiodi(methylthio) acetylaranotin. Abs. X-ray [7]

1. C. K. Nockolds, T. N. M. Waters, S. Ramaseshan, J. M. Waters, and D. C. Hodgkin, *Nature*, 1967, **214**, 129.
2. D. E. Wolf, W. H. Jones, J. Valiant, and K. Folkers, *J. Amer. Chem. Soc.*, 1950, **72** 2820.
3. D. C. Hodgkin in *Crystallography and Crystal Perfection* (G. N. Ramachandran, Ed.), Academic Press, 1963, p. 237.
4. E. A. Kaczka, D. Heyl, W. H. Jones, and K. Folkers, *J. Amer. Chem. Soc.*, 1952, **74**, 5549.
5. H. W. Moore and K. Folkers in *The Vitamins*, vol. II (Sebrell & Harris, Eds.), Academic Press, 1968, p. 121.
6. J. Fridrichsons and A. McL. Mathieson, *Acta Cryst.*, 1965, **18**, 1043.
7. J. W. Moncrief, *J. Amer. Chem. Soc.*, 1968, **90**, 6517.
8. A. F. Beecham, J. Fridrichsons, and A. McL. Mathieson, *Acta Cryst.*, 1967, **23**, 439.
9. A. F. Beecham and A. McL. Mathieson, *Tetrahedron Letters*, 1966, 3139.
10. R. Nagarajan, N. Neuss, and M. M. Marsh, *J. Amer. Chem. Soc.*, 1968, **90**, 6518.
11. D. Hauser, H. P. Weber, and H. P. Sigg, *Helv. Chim. Acta*, 1970, **53**, 1061.
12. R. Nagarajan, L. L. Huckstep, D. H. Lively, D. C. DeLong, M. M. Marsh, and N. Neuss, *J. Amer. Chem. Soc.*, 1968, **90**, 2980.
13. H. Hermann, R. Hodges, and A. Taylor, *J. Chem. Soc.*, 1964, 4315.

Y25 Macrolide antibiotics.

1. L-(−) oleandrose (2,6-dideoxy-L-arabinohexose 3-O-methyl ether) C1
2. (−) oleandomycin [1]
3. D-( ) desosamine (3,4,6-trideoxy-3-dimethylamino-D-xylohexose) C1
4. ( ) erythromycin A. Abs. X-ray [3]
5. (−) leucomycin A₃ [7]
6. ( ) demycarosyl-leucomycin A. Abs. X-ray [8]

(2S,3R)-( ) 3-acetoxy-2-methylbutyric acid A32.17

(S)-(−) 2-methylsuccinic acid A27.21

(R)-(+) 2-methyl-4-oxopentanoic acid A28.3

---

1. W. D. Celmer, *J. Amer. Chem. Soc.*, 1965, **87**, 1797, 1799.
2. H. Els, W. D. Celmer, and K. Murai, *J. Amer. Chem. Soc.*, 1958, **80**, 3777.
3. D. R. Harris, S. G. McGeachin, and H. H. Mills, *Tetrahedron Letters*, 1965, 679, and references therein.
4. F. A. Hochstein, H. Els, W. D. Celmer, B. L. Shapiro, and R. B. Woodward, *J. Amer. Chem. Soc.*, 1960, **82**, 3225.
5. C. Djerassi, O. Halpern, D. I. Wilkinson, and E. J. Eisenbraun, *Tetrahedron*, 1958, **4**, 369.
6. W. D. Celmer, *J. Amer. Chem. Soc.*, 1966, **88**, 5028, and references therein.
7. S. Omura, A. Nakagawa, M. Katagiri, T. Hata, M. Hiramatsu, T. Kimura, and H. Naya, *Chem. Pharm. Bull. Japan*, 1970, **18**, 1501.
8. M. Hiramatsu, A. Furusaki, T. Noda, K. Naya, Y. Tomiie, I. Nitta, T. Watanabe, T. Take, and J. Abe, *Bull. Chem. Soc. Japan*, 1967, **40**, 2982.

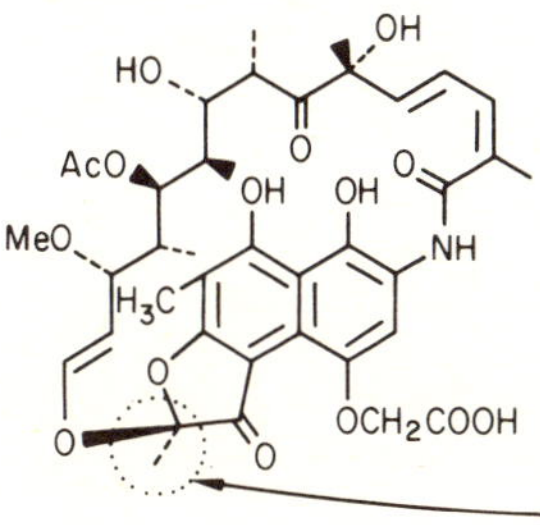

Macrolide antibiotics (contd.).

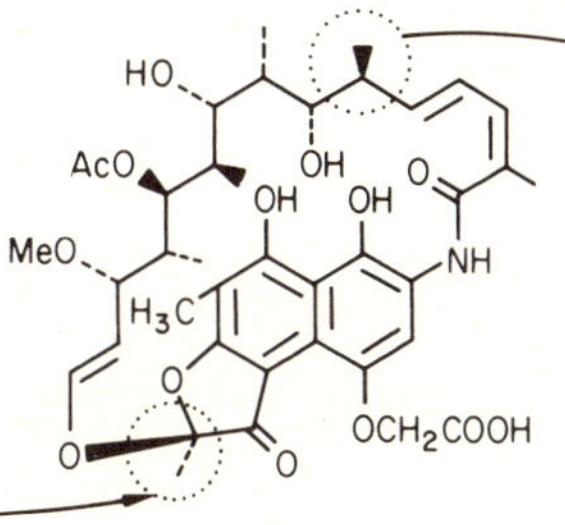

1. (+) rifamycin Y. *Rel. X-ray* [1]
2. (−) rifamycin B. *Rel. X-ray* [2]

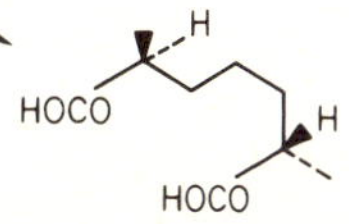

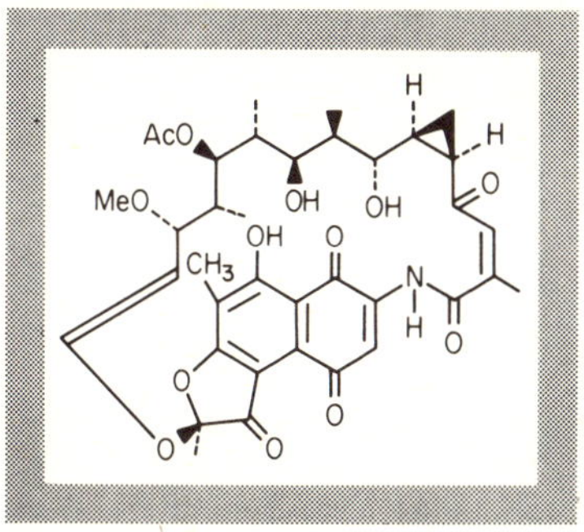

(2S, 6S)−(+) 2,6-dimethylheptanedioic acid A32.8

4. (+) tolpomycinone. *Abs. X-ray* [5]

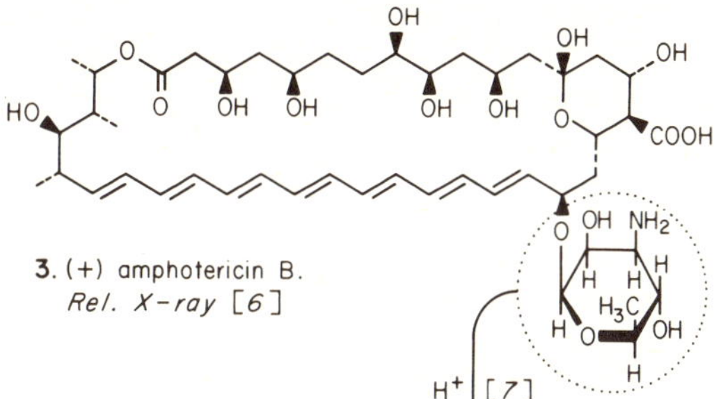

3. (+) amphotericin B. *Rel. X-ray* [6]

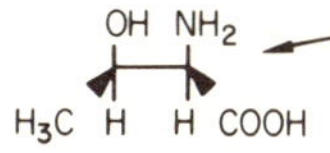

(−) allothreonine A24.2

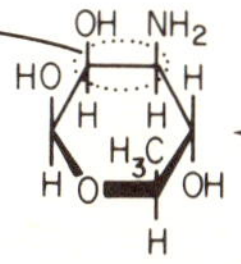

5. ( ) mycosamine (3-amino-3,6-dideoxy-β-D-mannopyranoside). A26.9

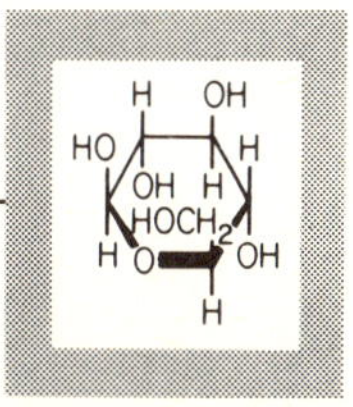

β-D-glucopyranose

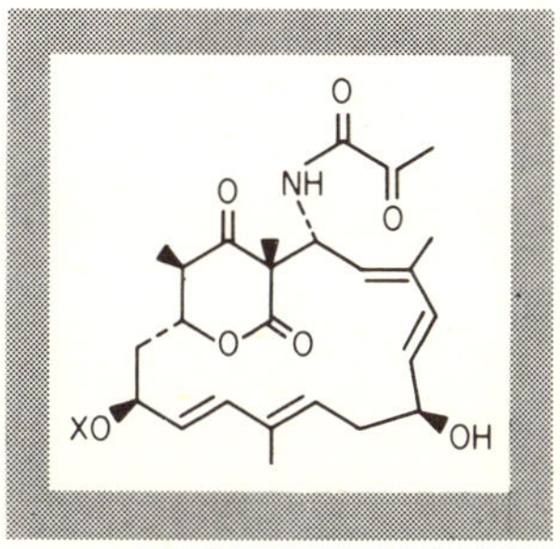

6. (−) bundlin A (X = H). *Abs. X-ray* [9]
   (−) bundlin B (X = Ac) *Abs. X-ray* [10]

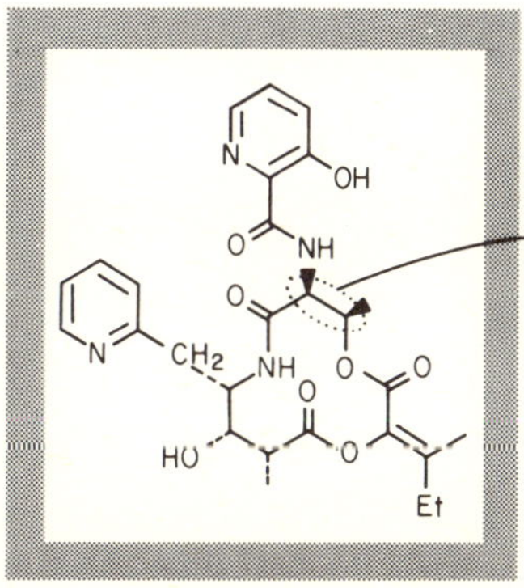

7. ( ) pyridomycin. *Abs. X-ray* [11]

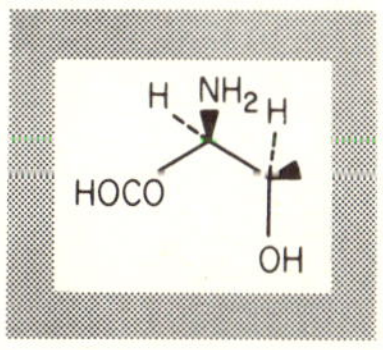

(−) threonine A8.16

1. M. Brufani, W. Fedeli, G. Giacomello, and A. Vaciago, *Experientia*, 1967, **23**, 508.
2. M. Brufani, W. Fedeli, G. Giacomello, and A. Vaciago, *Experientia*, 1964, **20**, 339.
3. J. Leitich, V. Prelog, and P. Sensi, *Experientia*, 1967, **23**, 505.
4. J. Leitich, W. Oppolzer, and V. Prelog, *Experientia*, 1964, **20**, 336.
5. K. Kamiya, T. Sugino, Y. Wada, M. Nishikawa, and T. Kishi, *Experientia*, 1969, **25**, 901.
6. W. Mechlinksi, C. P. Schaffner, P. Ganis, and G. Avitabile, *Tetrahedron Letters*, 1970, 3873.
7. M. H. von Saltza, J. Reid, J. D. Dutcher, and O. Wintersteiner, *J. Amer. Chem. Soc.*, 1961, **83**, 2785; *J. Org. Chem.*, 1963, **28**, 999.
8. K. Maeda, *J. Antibiotics (Japan)*, 1957, **10A**, 94.
9. M. Uramoto, N. Otake, Y. Ogawa, H. Yonehara, F. Marumo, and Y. Saito, *Tetrahedron Letters*, 1969, 2249.
10. K. Kamiya, S. Harada, Y. Wada, M. Nishikawa, and T. Kishi, *Tetrahedron Letters*, 1969, 2245.
11. G. Koyama, Y. Iitaka, K. Maeda, and H. Umezawa, *Tetrahedron Letters*, 1967, 3587.

Y27 Further antibiotics; boromycin, cytochalasin (phomin) group etc.

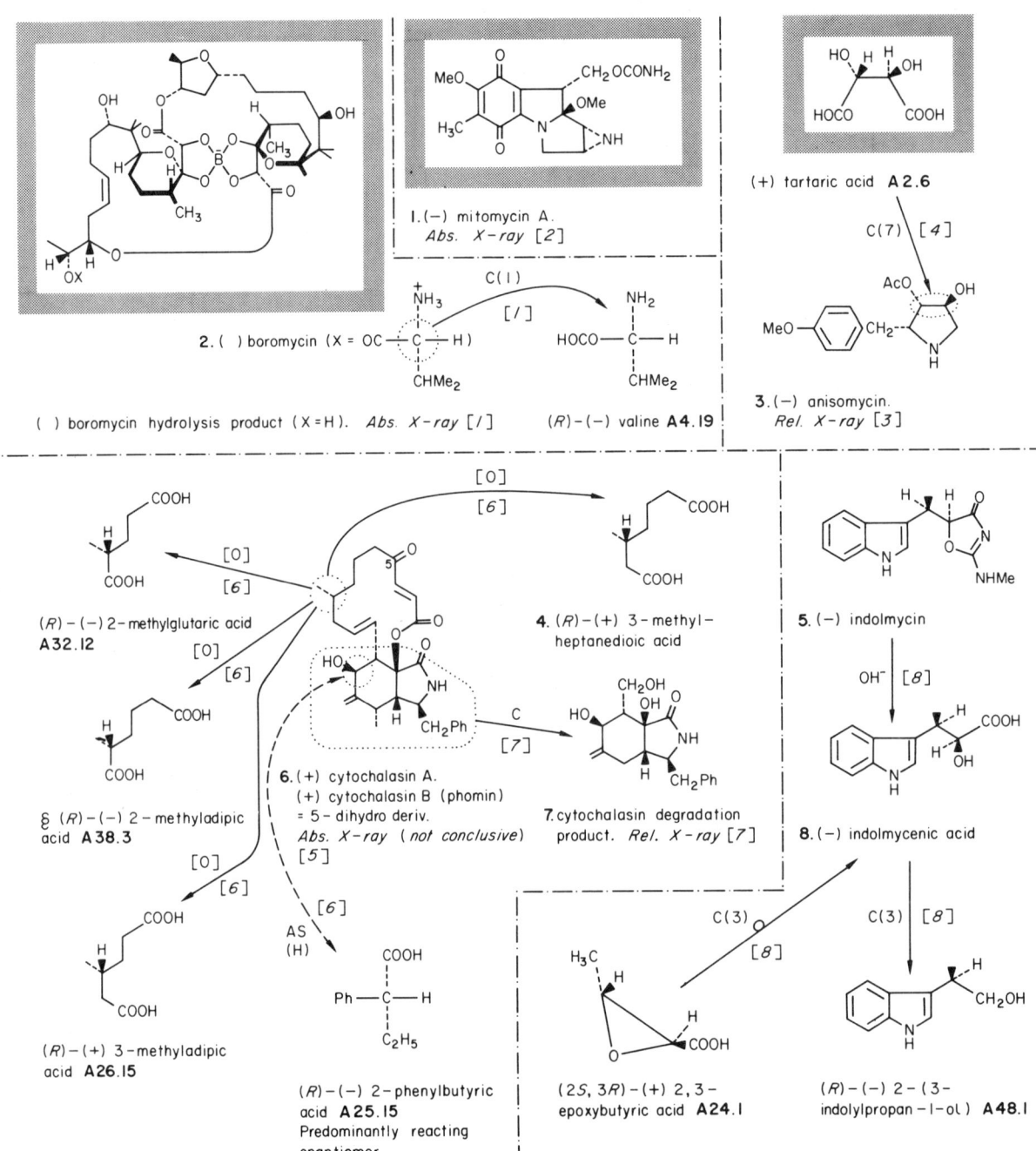

1. J. D. Dunitz, D. M. Hawley, D. Miklos, D. N. J. White, Yu. Berlin, R. Marusić, and V. Prelog, *Helv. Chim. Acta*, 1971, **54**, 1709.
2. A. Tulinsky and J. H. van den Hende, *J. Amer. Chem. Soc.*, 1967, **89**, 2905.
3. J. P. Schaefer and P. J. Wheatley, *Chem. Comm.*, 1967, 578.
4. C. M. Wong, J. Buccini, I. Chang, J. T. Raa, and R. Schwenk, *Canad. J. Chem.*, 1969, **47**, 2421.
5. G. M. McLaughlin, G. A. Sim, J. R. Kiechel, and Ch. Tamm, *Chem. Comm.*, 1970, 1398.
6. D. C. Aldridge, J. J. Armstrong, R. N. Speake, and W. B. Turner, *J. Chem. Soc. (C)*, 1967, 1667; W. Rothweiler and Ch. Tamm, *Helv. Chim. Acta*, 1970, **53**, 696.
7. G. A. Sim and Ch. Tamm, unpublished work quoted by W. Rothweiler and Ch. Tamm, *Helv. Chim. Acta*, 1970, **53**, 696.
8. T. H. Chan and R. K. Hill, *J. Org. Chem.*, 1970, **35**, 3519.

Tetracyclines; anthracyclinones.

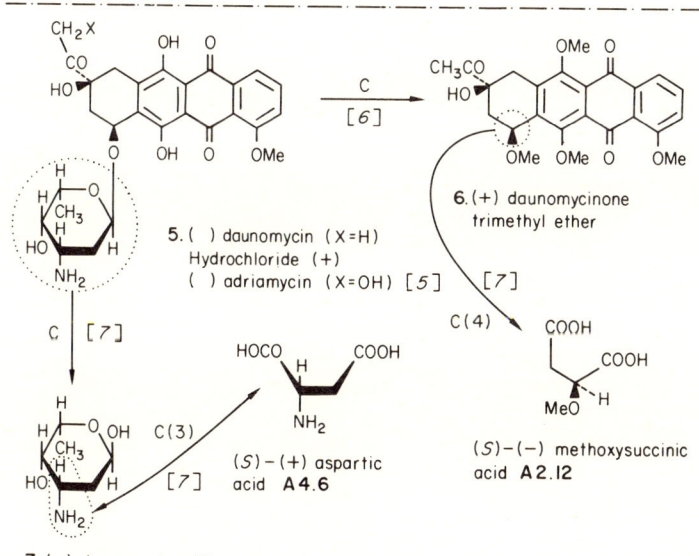

1. ≨ (−) oxytetracycline (terramycin) [1]
2. ( ) aureomycin (chlortetracycline) (X = Cl). Rel. X-ray [2]
   ( ) achromycin (X = H). Abs. X-ray [3]
3. ( ) pillaromycin A
4. ( ) pillaronone. Abs. X-ray [3]

(R)-(+) 4-chloro-7-methoxy-3-methylphthalide-3-carboxylic acid   A 51.4

N.B. pillaromycin is the only known example of a tetracycline antibiotic having the antipodal-type configuration at the C/D ring junction relative to the other members of the group [3].

5. ( ) daunomycin (X = H) Hydrochloride (+)
   ( ) adriamycin (X = OH) [5]
6. (+) daunomycinone trimethyl ether
7. (−) daunosamine (3-amino-2,3,6-trideoxy-L-lyxohexose)

(S)-(+) aspartic acid   A 4.6

(S)-(−) methoxysuccinic acid   A 2.12

Other antibiotics of the anthracyclinone group are derived from aglycones of the daunomycinone type by glycoside formation with various sugars [6]. The available evidence for their absolute configurations has been summarized [6].

---

1. M. S. von Wittenau, R. K. Blackwood, L. H. Conover, R. H. Glauert, and R. B. Woodward, *J. Amer. Chem. Soc.*, 1965, 87, 134.
2. J. Donohue, J. D. Dunitz, K. N. Trueblood, and M. S. Webster, *J. Amer. Chem. Soc.*, 1963, 85, 851.
3. K. Kamiya, M. Asai, Y. Wada, and M. Nishikawa, *Experientia*, 1971, 27, 363.
4. V. N. Dobrynin, A. I. Gurevich, M. G. Karapetyan, M. N. Kolosov, and M. M. Shemyakin, *Tetrahedron Letters*, 1962, 901.
5. F. Arcamone, G. Franceschi, S. Penco, and A. Selva, *Tetrahedron Letters*, 1969, 1007.
6. H. Brockmann, H. Brockmann and J. Niemeyer, *Tetrahedron Letters*, 1968, 4719.
7. F. Arcamone, G. Cassinelli, G. Francesci, P. Orezzi, and R. Mondelli, *Tetrahedron Letters*, 1968, 3353; *J. Amer. Chem. Soc.*, 1964, 86, 5335.

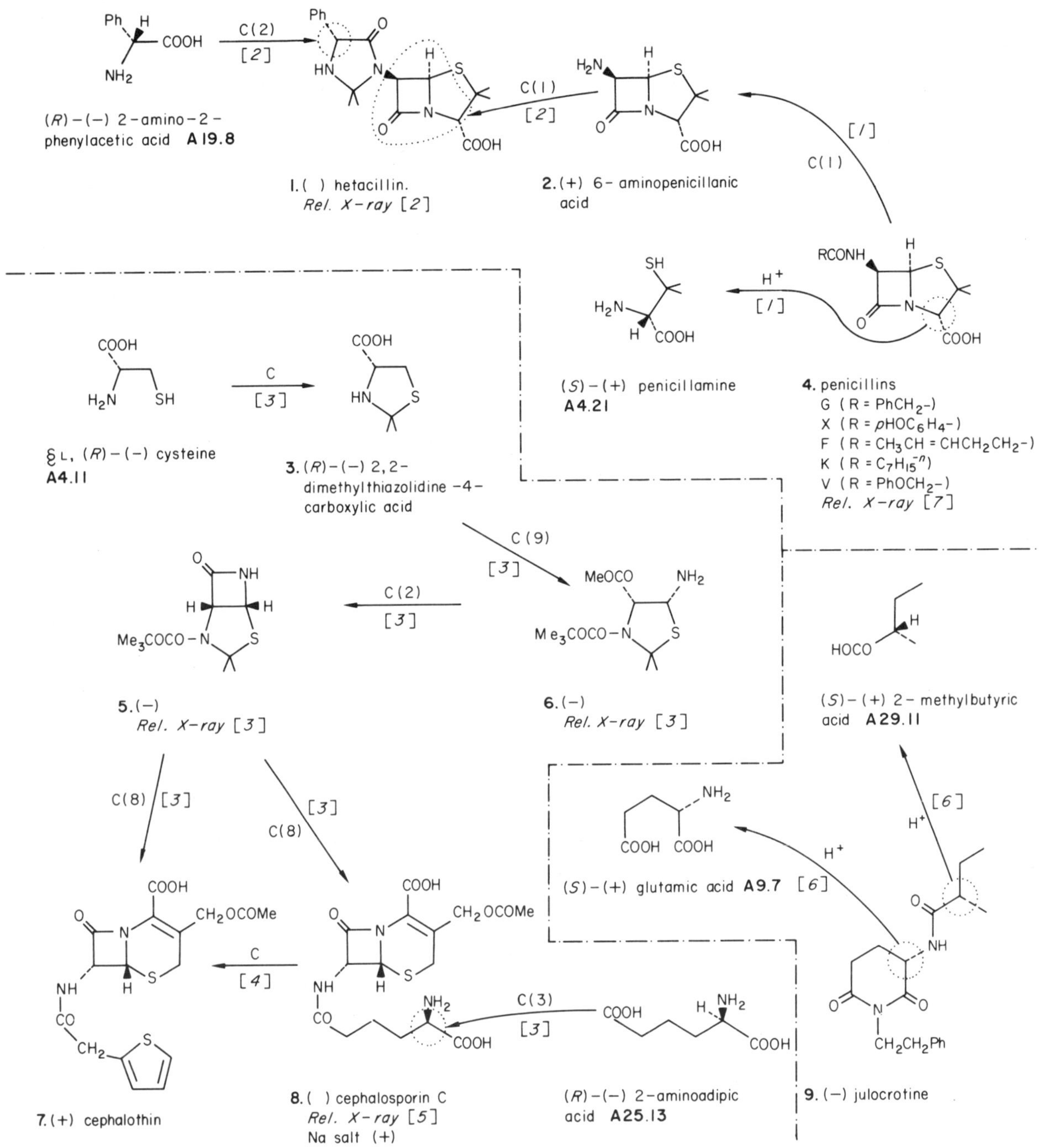

1. A. Heusner, *Z. Naturforsch*, 1946, **1**, 171.
2. G. A. Hardcastle, D. A. Johnson, C. A. Panetta, A. I. Scott, and S. A. Sutherland, *J. Org. Chem.*, 1966, **31**, 897.
3. R. B. Woodward, K. Heusler, J. Gosteli, P. Naegeli, W. Oppolzer, R. Ramage, S. Ranganathan, and H. Vorbrüggen, *J. Amer. Chem. Soc.*, 1966, **88**, 852.
4. R. R. Chauvette, E. H. Flynn, B. G. Jackson, E. R. Lavagnino, R. B. Morin, R. A. Mueller, R. P. Pioch, R. W. Roeske, C. W. Ryan, J. L. Spencer, and E. Van Heyningen, *J. Amer. Chem. Soc.*, 1962, **84**, 3402.
5. D. C. Hodgkin and E. N. Maslen, *Biochem. J.*, 1961, **79**, 393.
6. T. Nakano, C. Djerassi, R. A. Corral, and O. O. Orazi, *Tetrahedron Letters*, 1959, **no. 14**, 8.
7. D. C. Crowfoot, C. W. Bunn, B. W. Rogers-Low, and A. Turner-Jones, in *The Chemistry of Penicillin* (H. T. Clarke, J. R. Johnson, and R. Robinson, Eds.), Princeton Univ. Press, 1949, p. 310.

Y30
Biotin; miscellaneous antibiotics.

1. (+) biotin. Abs. *X-ray* [*1*] ORD/CD;[*4*]
2. (+) dethiobiotin
3. ( ) 4,6-*O*-benzylidene-3-deoxy-α-D-*ribo*-hexopyranoside
α-D-glucopyranose **A26.9**

4. ( ) actinobolin. Sulphate (+)
(S)-(+) alanine **A1.18**
5. (R)-(+) 3-(1-hexenylazoxy)-1-hydroxybutan-2-one (antibiotic LL-BH-872α)
(−) threonine **A8.16**

(R)-(+) 2-methylpentan-3-ol **A8.6**
(R)-(−) phenylethylenediamine **A19.7**
6. (R)-(−) 2-mercapto-4-phenylimidazoline
7. (R)-(+) tetramisole

8. (−) piericidin A (small rotation)
9. (+) abikoviromycin
10. (−). AC by ΔM_D of 3,5-dinitrobenzoate formation [*9*]

---

1. J. Trotter and J. A. Hamilton, *Biochemistry*, 1966, **5**, 713.
2. V. du Vigneaud, D. B. Melville, K. Folkers, D. E. Wolf, R. Mozingo, J. C. Keresztesy, and S. A. Harris, *J. Biol. Chem.*, 1942, **146**, 475.
3. H. Kuzuhara, H. Ohrui, and S. Emoto, *Tetrahedron Letters*, 1970, 1185.
4. N. M. Green, W. P. Mose, and P. M. Scopes, *J. Chem. Soc. (C)*, 1970, 1330.
5. F. J. Antosz, D. M. Nelson, D. L. Herald, and M. E. Munk, *J. Amer. Chem. Soc.*, 1970, **92**, 4933.
6. W. J. McGahren and M. P. Kunstmann, *J. Amer. Chem. Soc.*, 1969, **91**, 2808.
7. N. Takahashi, A. Suzuki, Y. Kimura, S. Miyamota, and S. Tamura, *Tetrahedron Letters*, 1967, 1961.
8. A. H. M. Raeymaekers, L. F. C. Roevens, and P. A. J. Janssen, *Tetrahedron Letters*, 1967, 1467.
9. A. I. Gurevich, M. N. Kolosov, V. G. Korobko, and V. V. Onoprienko, *Tetrahedron Letters*, 1968, 2209.

-D-
Compounds with Chirality due to Isotopic substitution

Introductory Notes to Chapter D

Scope

Chapter D consists of compounds where the chirality is due solely to substitution of a hydrogen atom by deuterium or tritium (or, in one case, by substitution of ^{16}O by ^{18}O). Compounds containing stereospecifically incorporated deuterium or tritium in addition to other centres of chirality, of which many are known, are not specifically treated.

Application of (R,S) nomenclature

A sub-rule of the Cahn-Ingold-Prelog rules states that heavier isotopes take precedence over lighter. The order of priority of hydrogen isotopes is thus $^3H > {}^2H > {}^1H$.

Nomenclature

The IUPAC system of denoting isotopic substitution has been used in preference to the Chemical Abstracts system. Thus CH_3CHDOH is [1-^2H] ethanol rather than ethanol-1-*d*.

Brewster's rules

Brewster's rules (J. H. Brewster, *J.A.C.S.*, 1959, 81, 5475) are a set of semi-empirical rules relating molecular chirality to sign of molecular rotation. For a molecule containing a chiral centre Cabde, the sign of rotation is predicted from the polarizabilities of the four groups abde. This rule has been used to predict the absolute configurations of a number of deutero-compounds where d = ^2H and e = ^1H.

D[1]

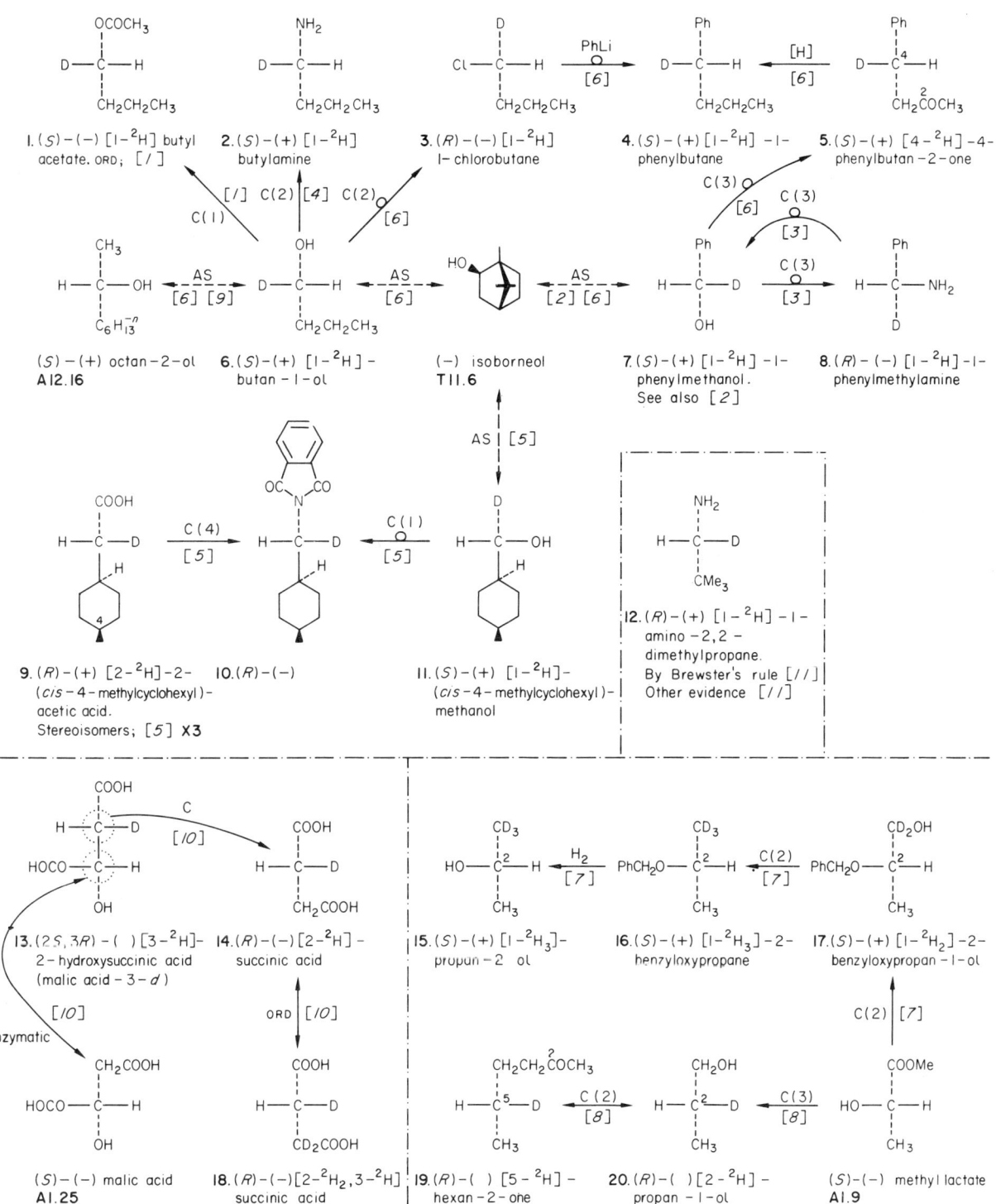

1. L. Verbit, *J. Amer. Chem. Soc.*, 1967, **89**, 167.
2. D. Nasipuri, C. K. Ghosh, and R. J. L. Martin, *J. Org. Chem.*, 1970, **35**, 657.
3. A. Streitweiser and J. R. Wolfe, *J. Org. Chem.*, 1969, **28**, 3263.
4. A. Streitweiser and W. D. Schaeffer, *J. Amer. Chem. Soc.*, 1956, **78**, 5597.
5. H. Gerlach, *Helv. Chim. Acta*, 1966, **49**, 1291.
6. A. Streitweiser, J. R. Wolfe, and W. D. Schaeffer, *Tetrahedron*, 1959, **6**, 338.
7. K. Mislow, R. E. O'Brien, and H. Schaefer, *J. Amer. Chem. Soc.*, 1962, **84**, 1940.
8. M. M. Green, J. G. McGrew, and J. M. Moldowan, *J. Amer. Chem. Soc.*, 1972, **93**, 6700.
9. H. R. Levy, F. A. Loewus, and B. Vennesland, *J. Amer. Chem. Soc.*, 1957, **79**, 2949.
10. G. Popják and J. W. Cornforth, *Biochem. J.*, 1966, **101**, 553.
11. W. Meister and D. J. Cram, *J. Amer. Chem. Soc.*, 1967, **89**, 5288.

D²

1. (S)-( ) [2-²H] - glycollic acid. Abs neutron diffraction [2]
2. (S)-( ) [2-³H₁]- glycollic acid
3. (S)-( ) [1-³H₁] - ethane -1,2-diol
4. (R)-( ) [²H₁, ³H₁]- acetic acid
5. (R)-( ) [²H₁, ³H₁]- acetophenone
6. (S)-( ) [2-³H] -2- aminoacetic acid (2-tritioglycine) — (S)-(+) alanine A1.18
7. (R)-(+) [1-²H] -1- aminoethane. By Brewster's rules [4] Other evidence [4]
8. (1R, 2R)-( ) [²H₁, ³H₁] - 1- phenylethanol
9. (1S, 3S)-(+) [3-²H]-1- isopropenylcyclopentane
10. (S) [3-²H] - cyclopentanone. No detectable rotation [6]
11. (R)-(+) benzyl-p-tolylsulphoxide Z7
12. (S)-(-)(?) benzyl-p-tolyl - ¹⁶O ¹⁸O sulphone

(1R, 3S)-(+) 3- isopropenylcyclopentanol T9.2

(2R, 3R)-(+) 2,3-epoxybutane A13.8

13. (2R, 3S)-(-) [3-²H] butan-2-ol
14. (2S, 3S)-(+) [3-²H] 2-bromobutane
15. (S)-(-) [2-²H] butane
16. (S)-( ) [2-³H] propionic acid
17. (2R, 3S)-( ) [3-³H] butan-2-ol
18. (S)-(-) [1-²H] ethanol
19. (S)-( ) [3-²H] butan-2-one

1. J. Lüthy, J. Rétey, and D. Arigoni, *Nature*, 1969, **221**, 1213.
2. C. K. Johnson, E. J. Gabe, M. R. Taylor, and I. A. Rose, *J. Amer. Chem. Soc.*, 1965, **87**, 1802.
3. M. Akhtar and P. M. Jordan, *Tetrahedron Letters*, 1969, 875.
4. W. Meister, R. D. Guthrie, J. L. Maxwell, D. A. Jaeger, and D. J. Cram, *J. Amer. Chem. Soc.*, 1969, **91**, 4452.
5. J. W. Cornforth, J. W. Redmond, H. Eggerer, W. Buckel, and C. Gutschow, *Nature*, 1969, **221**, 1212.
6. C. Djerassi and B. Tursch, *J. Amer. Chem. Soc.*, 1961, **83**, 4609.
7. C. J. M. Stirling, *J. Chem. Soc.*, 1963, 5741.
8. H. Weber, J. Seibl, and D. Arigoni, *Helv. Chim. Acta*, 1966, **49**, 741.
9. J. Rétey and F. Lynen, *Biochem. Z.*, 1965, **342**, 256.
10. G. K. Helmcamp, C. D. Joel, and H. Sharman, *J. Org. Chem.*, 1956, **21**, 844.

-X-
Compounds containing Chiral Axes, Planes, etc.

Introductory Notes to Chapter X

Scope

Chapter X contains compounds having axes and planes of chirality.
Establishment of the absolute configuration of compounds having axial or planar chirality by comparison with a compound containing a chiral centre is a problem of some difficulty. A number of different approaches have been employed; many of these entail a form of asymmetric synthesis. A recent review is by G. Krow in *Topics stereochem.*, 1970, 5, 31. In the 'Atlas' it is not possible to portray these various types of asymmetric synthesis in detail, but the user should have no difficulty in consulting the original publications.

Arrangement

Axes of chirality
    Allenes, alkylidenecycloalkanes and related compounds **X1-X4**
    Spiro compounds **X4-X5**

Axes of chirality; compounds showing atropisomerism
    Biaryls **X5-X7**
    Helicenes and related compounds **X7**

Planes of chirality
    Cyclophanes **X8**
    Metallocenes **X8-X9**

Miscellaneous compounds of C_2, C_3, D_2, and D_3 symmetry **X10-X11**

Allenes, alkylidenecycloalkanes and related compounds X[1]

Lowe's rule for allenes. 'An allene is viewed along its orthogonal axis with the more polarizable substituent in the vertical axis uppermost. If the more polarizable substituent in the horizontal axis is to the right, then this enantiomer will be dextrorotatory and vice-versa' [4] [5] [6]. (This rule is not obeyed by cycloocta-1,2-diene X2.2.)

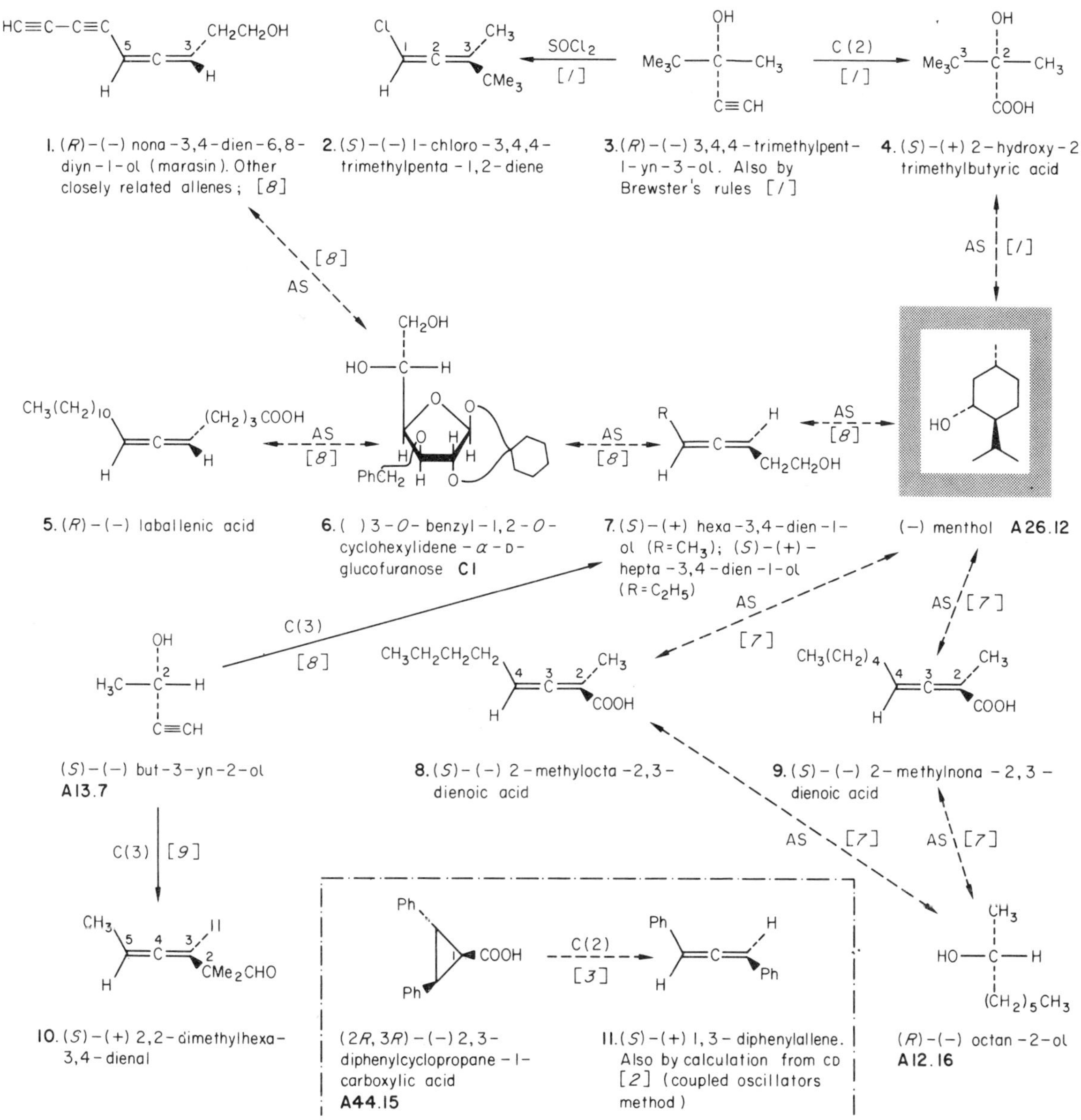

1. (R)-(−) nona-3,4-dien-6,8-diyn-1-ol (marasin). Other closely related allenes; [8]
2. (S)-(−) 1-chloro-3,4,4-trimethylpenta-1,2-diene
3. (R)-(−) 3,4,4-trimethylpent-1-yn-3-ol. Also by Brewster's rules [/]
4. (S)-(+) 2-hydroxy-2,3,3-trimethylbutyric acid
5. (R)-(−) laballenic acid
6. ( ) 3-O-benzyl-1,2-O-cyclohexylidene-α-D-glucofuranose C1
7. (S)-(+) hexa-3,4-dien-1-ol (R=CH₃); (S)-(+)-hepta-3,4-dien-1-ol (R=C₂H₅)
8. (S)-(−) 2-methylocta-2,3-dienoic acid
9. (S)-(−) 2-methylnona-2,3-dienoic acid
10. (S)-(+) 2,2-dimethylhexa-3,4-dienal
11. (S)-(+) 1,3-diphenylallene. Also by calculation from cD [2] (coupled oscillators method)

(S)-(−) but-3-yn-2-ol A13.7

(2R,3R)-(−) 2,3-diphenylcyclopropane-1-carboxylic acid A44.15

(−) menthol A26.12

(R)-(−) octan-2-ol A12.16

1. R. J. D. Evans and S. R. Landor, *J. Chem. Soc.*, 1965, 2553.
2. S. F. Mason and G. W. Vane, *Tetrahedron Letters*, 1965, 1593.
3. W. M. Jones and J. W. Wilson, *Tetrahedron Letters*, 1965, 1587.
4. G. Lowe, *Chem. Comm.*, 1965, 411.
5. G. Krow, *Topics Stereochem.*, 1970, 5, 31.
6. P. Crabbé, E. Velarde, H. W. Anderson, S. D. Clark, W. R. Moore, A. F. Drake, and S. F. Mason, *Chem. Comm.*, 1971, 1261.
7. I. Tömösközi and H. J. Bestmann, *Tetrahedron Letters*, 1964, 1293.
8. R. J. D. Evans, S. R. Landor, and J. P. Regan, *Chem. Comm.*, 1965, 397; S. R. Landor, B. J. Miller, J. P. Regan, and A. R. Tatchell, ibid., 1966, 585.
9. E. R. H. Jones, J. D. Loder, and M. C. Whiting, *Proc. Chem. Soc.*, 1960, 180.

X² Allenes and related compounds (contd.)

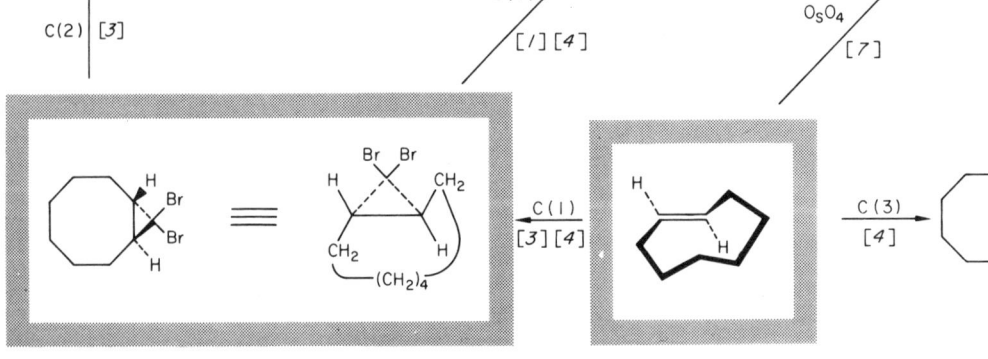

1. (S)-(+) penta-2,3-diene (1,3-dimethylallene)
2. (R)-(+) cycloöcta-1,2-diene
3. (1R,9R)-(+) bicyclo-[7.1.0.] decane
4. (R)-(+) cyclonona-1,2-diene. (does not obey Lowe's rule [/])
5. (S)-(+) cycloöct-1-en-3-ol O-methyl (+) O-acetyl (+)
6. (S)-(−) 3-acetoxycycloöctyne
7. (1R,8R)-(+) bicyclo-[6.1.0.] nonane
8. (1S,8S)-(+) 9,9-dibromobicyclo [6.1.0.] nonane. Abs. X-ray [3]
9. (R)-(−) cycloöctene. Abs. X-ray [2]
10. (1S,8S)-(+) 9-phenyl-9-azabicyclo [6.1.0.] nonane

(+) α-pinene T8.4
(1R,2R)-(−) 1,2-dimethylcyclopropane A44.1
(1S,2S)-(+) cycloöctane-1,2-diol A3.16

1. W. R. Moore, *J. Amer. Chem. Soc.*, 1971, **93**, 4932.
2. P. C. Manor, D. P. Shoemaker, and A. S. Parkes, *J. Amer. Chem. Soc.*, 1970, **92**, 5261.
3. R. D. Bach, U. Mazur, R. N. Brummel, and L.-H. Lin, *J. Amer. Chem. Soc.*, 1971, **93**, 7120.
4. T. Aratani, Y. Nakanishi, and H. Nozaki, *Tetrahedron*, 1970, **26**, 4339.
5. W. M. Jones and J. M. Walbrick, *Tetrahedron Letters*, 1968, 5229.
6. W. L. Waters and M. C. Caserio, *Tetrahedron Letters*, 1968, 5233.
7. A. C. Cope and A. S. Mehta, *J. Amer. Chem. Soc.*, 1964, **86**, 5626.

Allenes and related compounds (contd.) X³

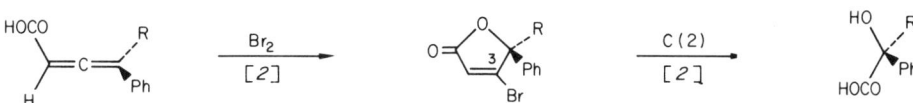

1. (R)-(−) pentadiendioic (glutinic) acid
2. (+) glutinic acid cyclopentadiene adduct
(+) norcamphor **A36.11**

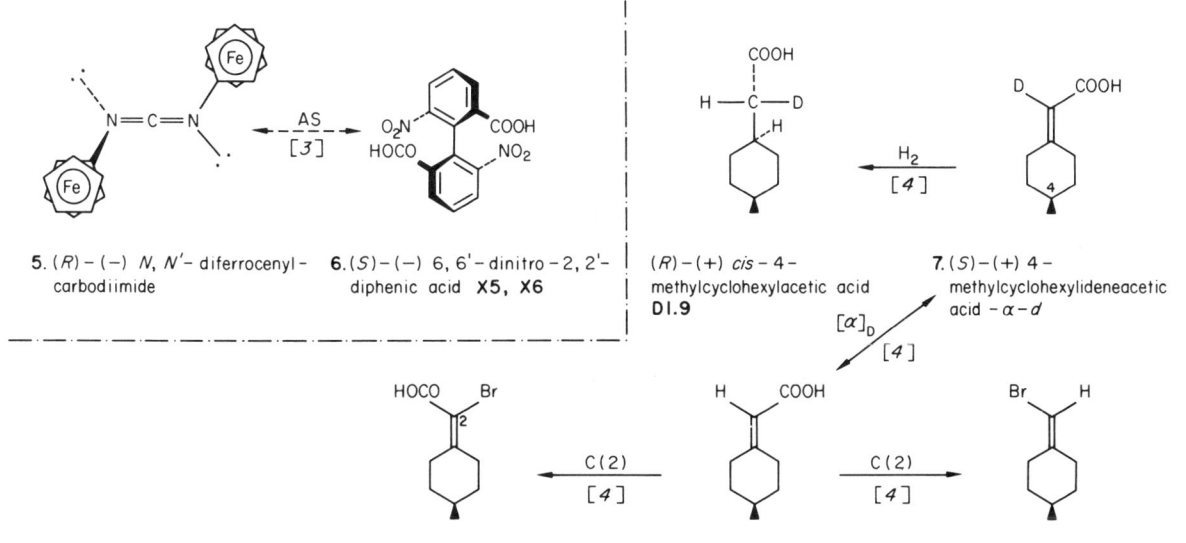

3. (S)-(+) phenyl alkyl allenic acids (R=H, CH₃, C₂H₅, CHMe₂, CMe₃)

4. (S)-(+) 3-bromo-4-alkyl-4-phenylcrotonolactones (R=H, CH₃, C₂H₅, CHMe₂, CMe₃)

(S)-(+) mandelic acid (R=H) **A21.12**
(S)-alkylmandelic acids **A31.19, A51.14, A51.1, A51.19** (R = CH₃, C₂H₅, CHMe₂, CMe₃)

5. (R)-(−) N,N'-diferrocenyl-carbodiimide

6. (S)-(−) 6,6'-dinitro-2,2'-diphenic acid **X5, X6**

(R)-(+) cis-4-methylcyclohexylacetic acid **D1.9**

7. (S)-(+) 4-methylcyclohexylideneacetic acid -α-d

8. (S)-(+) 2-bromo-2-(4-methylcyclohexylidene) acetic acid

9. (S)-(+) 4-methylcyclohexylideneacetic acid

10. (R)-(−) (4-methylcyclohexylidene) bromomethane

---

1. W. C. Agosta, *J. Amer. Chem. Soc.*, 1964, **86**, 2638.
2. K. Shingu, S. Hagishita, and M. Nakagawa, *Tetrahedron Letters*, 1967, 4371.
3. K. Schlögl and H. Mechtler, *Angew. Chem. Internat. Edn.*, 1966, **5**, 596.
4. H. Gerlach, *Helv. Chim. Acta*, 1966, **49**, 1291.

X⁴ (i) Alkylidenecycloalkanes; (ii) Spiro compounds

(R)-(+) 3-methylcyclohexanone A38.4 → [PhCHO, */*] **1. (R)-(−) 2-benzylidene-5-methylcyclohexanone (E-isomer). (Z)-isomer (−) [/]** → [C(1), */*] **2. (S)-(+) 1-benzylidene-4-methylcyclohexane**

3. (2S,6R)(Z) or (2R,6S)(E)-(+) 2,6-diphenyl-1-methyl-4-piperidone oxime → [H⁺, [2]] **4. (−)** → [C(2), [2]] **(R)-(−) 2-methylamino-2-phenylacetic acid A19.8**

Spiro-compounds
Lowe's rule has been applied to spirans, but fails for 2,7-diazaspiro-[4.4] nonane, ditosyl deriv X 4.8. This failure has been discussed [3] [4].

5. (R)-(−) 4-carbethoxy-2-pyrrolidone-4-acetic acid → [C(4), [5]] **6. (R)-(+) 2-aza-7-thiaspiro [4.4] nonane, methanesulphonyl deriv.**

5 → [C(4), [5]] **8. § (R)-(−) 2,7-diazaspiro [4.4] nonane, ditosyl deriv.†**

6 ↕ [C(4), [5]] **(R)-(+) 2-ethyl-2-methylsuccinic acid A55.5**

7. § (S)-(−) 4,4'-dimethoxy-1,1',3,3'-tetrahydrospiro-[isoindole-2,2'-isoindolium] bromide → [NaH, [3]] **9. (S)-(−) 4,8-dimethoxy-5,7,12,12a-tetrahydroisoindolo[2,1-b] isoquinoline**

(S)-(+) aspartic acid A4.6 ⇌ [C(2), [3]] **9**

---

1. J. H. Brewster and J. E. Privett, *J. Amer. Chem. Soc.*, 1966, **88**, 1419.
2. G. G. Lyle and E. T. Pelosi, *J. Amer. Chem. Soc.*, 1966, **88**, 5276.
3. J. H. Brewster and R. S. Jones, *J. Org. Chem.*, 1969, **34**, 354.
4. H. Wynberg and J. P. M. Houbiers, *J. Org. Chem.*, 1971, **36**, 834.
5. G. Krow and R. K. Hill, *Chem. Comm.*, 1968, 430.
6. G. Krow, *Topics Stereochem.*, 1970, **5**, 31; R. S. Cahn, C. K. Ingold, and V. Prelog, *Angew. Chem. Internat. Edn.*, 1966, **5**, 385.

†Redesignated here as centrally chiral molecules with consequent reversal of the original (R,S) assignment [6].

Spiro compounds (contd.); biaryls

(R)-(−) 2-phenylbutyric acid **A25.15**

1. (1R, 5S, 6R)-(−) spiro-[4.4] nonane-1,6-diol
2. (S)-(−) spiro-[4.4] nonane-1,6-dione
3. (R)-(−) 3,3,3',3'-tetramethyl-bis-1,1'-spiroindane. Related compds; [2]
4. (R)-(−) 7,7' dibromo 6,6' dihydroxy-3,3,3',3',5,5'-hexamethylbis-1,1'-spiroindane. Abs. X-ray [2]

(S)-(+) 9,10-dihydro-3-hydroxy-4-methoxy-2,5,8-trimethylphenanthrene-1-acetic acid **T53.14**

5. (R)-(−) 9,10-dihydro-4,5-dimethylphenanthrene
6. (R)-(−) 2,2'-bis(bromomethyl)-6,6'-dimethylbiphenyl
7. (R)-(+) dimethyldibenz-1,3-cycloheptadiene-6-one

(R)-(+) 6,6'-dinitro-2,2'-diphenic acid **X3.6**

8. (R)-(+) 2,2'-diamino-6,6'-dimethylbiphenyl. Abs. X-ray [6]
9. (R)-(−) 2,2'-dimethylbiphenyl

10. (R)-(+) 4,4'-dinitro-3,3'-bithienyl-2,2'-dicarboxylic acid
11. (R)-(+) 2,2'-dinitro-3,3'-bithienyl-4,4'-dicarboxylic acid
12. (R)-(−) 6,6'-dimethyl-2,2'-diphenic acid **X6, X10**
13. (R)-(−) 1,2-dimethyl-6,7-diphenyldibenzo [e,g] [1,4] diazocine. Other related heterocylic 2,2' bridged biaryls; [3] [7]

1. H. Gerlach, *Helv. Chim. Acta*, 1968, **51**, 1587.
2. S. Hagishita, *Bull. Chem. Soc. Japan*, 1971, **44**, 496.
3. K. Mislow, *Angew. Chem.*, 1958, **70**, 683.
4. K. Mislow, M. A. W. Glass, R. E. O'Brien, P. Rutkin, D. H. Steinberg, J. Weiss, and C. Djerassi, *J. Amer. Chem. Soc.*, 1962, **84**, 1455; K. Mislow, E. Bunnenberg, R. Records, K. Wellman, and C. Djerassi, *J. Amer. Chem. Soc.*, 1963, **85**, 1342 and references therein.
5. T. R. Hollands, P. de Mayo, M. Nisbet, and P. Crabbé, *Canad. J. Chem.*, 1965, **43**, 3008.
6. L. H. Pignolet, R. P. Taylor, and W. De W. Horrocks, *Chem. Comm.*, 1968, 1443.
7. J. M. Insole, *J. Chem. Soc. (C)*, 1971, 1712 and references therein.
8. S. Gronowitz and P. Gustafson, *Arkiv Kemi*, 1963, **20**, 289; S. Gronowitz, ibid., 1965, **23**, 307.
9. W. Thielacker and H. Böhm, *Angew. Chem. Internat. Edn.*, 1967, **6**, 251.
10. K. Mislow and H. B. Hopps, *J. Amer. Chem. Soc.*, 1962, **84**, 3018.

6 Biaryls (contd.)

1. (R)-(−) 4',1"-dinitro-dibenz-1,3-cycloheptadien-6-one
2. (R)-(+) dinaphthocycloheptadienone
3. (S)-(−) 2-hydroxy-1,1'-binaphthyl

(R)-(+) 6,6'-dinitro-2,2'-diphenic acid **X3.6**
(S)-(+) pinacolyl alcohol **A12.4**
(S)-(+) octan-2-ol **A12.16**
(S)-(+) atrolactic acid **A31.19**

4. (R)-(−) 9,10-dihydrodinaphtho-2,3'-3,4: 2",3"-5,6-phenanthrene
5. (R)-(+) 1,1'-binaphthyl-2,2'-dicarboxylic acid
6. (R)-(+) 2,2'-dihydroxy-1,1'-binaphthyl-3,3'-dicarboxylic acid dimethyl ester. Abs. X-ray [2]
7. (R)-(−) 1,1'-binaphthyl. Also by CD [8] Other derivs; [2]

8. (R)-(−) 2,2'-dimethyl-1,1'-bianthryl. Also by calculation from CD [3] Related compds; [4]
9. (R)-(−) 2,2'-dimethyl-1,1'-bisanthraquinonyl
(R)-(−) 6,6'-dimethyl-2,2'-diphenic acid **X5.12**
10. (R)-(−) 2,2'-dimethyl-1,1'-binaphthyl

1. P. Newman, P. Rutkin, and K. Mislow, *J. Amer. Chem. Soc.*, 1958, 80, 465.
2. H. Akimoto, T. Shioiri, Y. Iitaka, and S. Yamada, *Tetrahedron Letters*, 1968, 97.
3. R. Grinter and S. F. Mason, *Trans. Faraday Soc.*, 1964, 60, 274.
4. G. M. Badger, R. J. Drewer, and G. E. Lewis, *J. Chem. Soc.*, 1962, 4268.
5. S. Yamada and H. Akimoto, *Tetrahedron Letters*, 1968, 3967.
6. J. A. Berson and M. A. Greenbaum, *J. Amer. Chem. Soc.*, 1958, 80, 653.
7. K. Mislow, V. Prelog, and H. Scherrer, *Helv. Chim. Acta*, 1958, 41, 1410; K. Mislow and F. A. McGinn, *J. Amer. Chem. Soc.*, 1958, 80, 6036.
8. P. A. Browne, M. M. Harris, R. Z. Mazengo, and S. Singh, *J. Chem. Soc. (C)*, 1971, 3990.

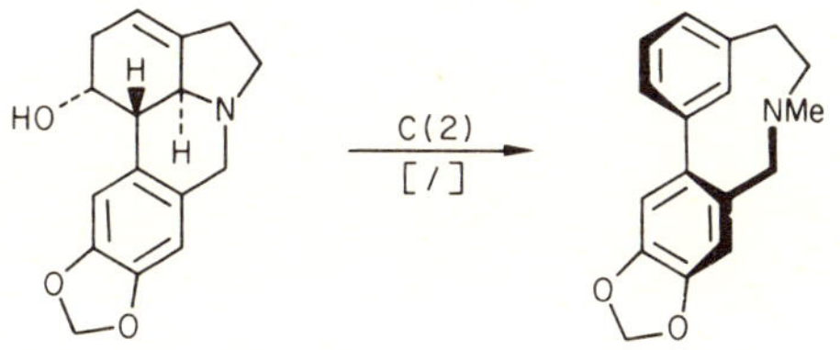

(i) miscellaneous bridged biaryls (see also **K27**) (ii) helicenes

1. ( ) caranine **K6.3** → 1. (S)-(−) caranine anhydromethine ORD/CD; [1] C(2) [1]

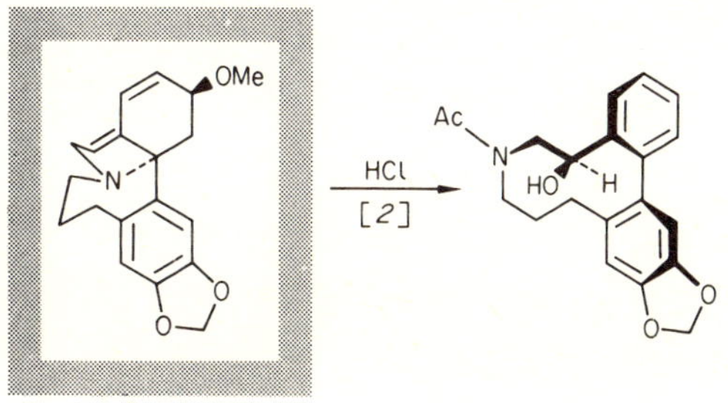

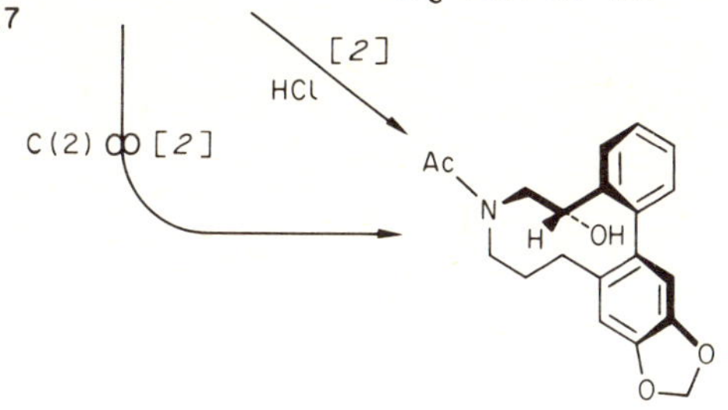

2. § (aS,7R)-(+)† via HCl [2] from (+) schellhammeridine **K7.7**

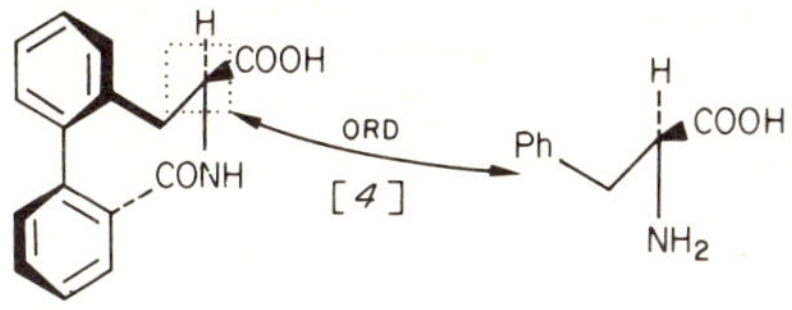

3. ( ) By ORD [4] → (S)-( ) phenylalanine **A5.12**

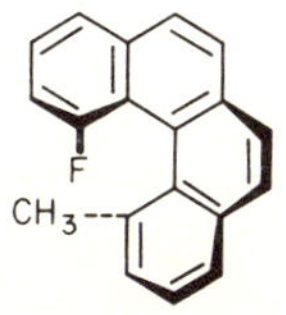

4. § (aS,7S)-(+)† via C(2) CD [2], HCl [2]

5. (S)-(or P)-(+) 1-fluoro-12-methylbenzo[c]phenanthrene. By calculation from CD [5]‡

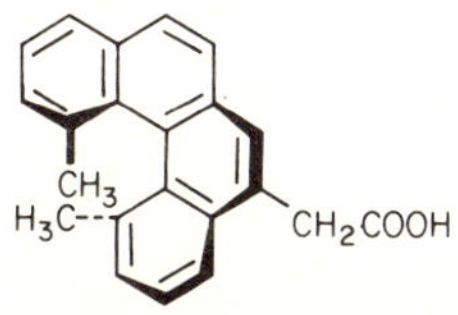

6. (S)-(or P)-(+) 1,12-dimethylbenzo[c]phenanthrene-5-acetic acid. By calculation from CD [5]‡

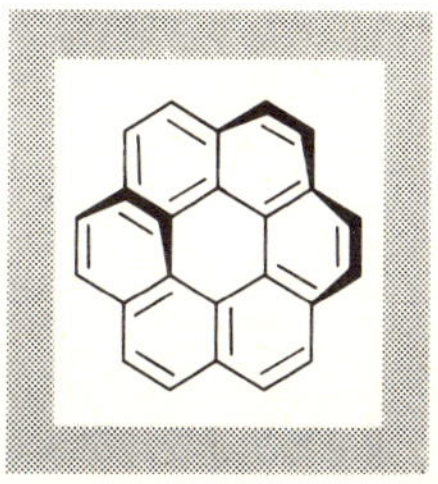

7. § (R)-(or M)-(−) [6]-helicene. Abs. X-ray [7]. Also by calculation from CD [6]‡

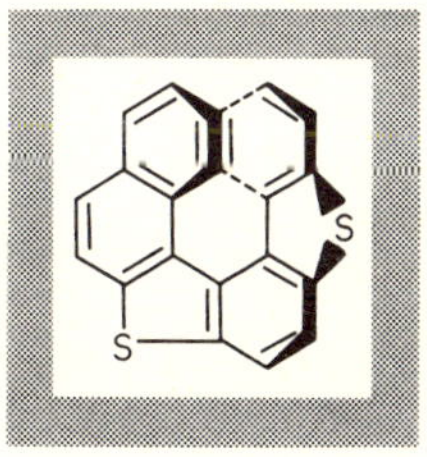

8. (R)-(or M)-(−) benzo[d]naphtho-[1,2-d']benzo[1,2-b; 4,3-b']dithiophene. Abs. X-ray [3]. Other heterohelicenes by calculation from CD [3] [8]

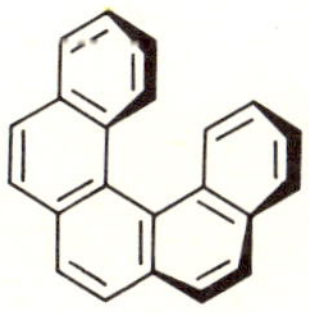

9. (R)-(or M)-(−) [5] helicene. By calculation from CD [6]‡

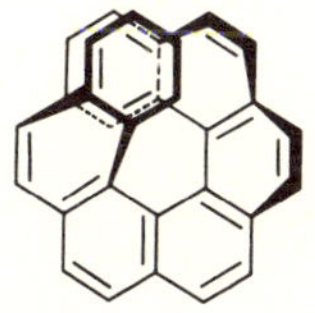

10. (R)-(or M)-(−) [7] helicene. By calculation from CD [6]‡

1. E. W. Warnhoff and S. V. Lopez, *Tetrahedron Letters*, 1967, 2723.
2. S. R. Johns, J. A. Lamberton, A. A. Sioumis, and H. Suares, *Chem. Comm.*, 1969, 646.
3. M. B. Groen, G. Stulen, G. J. Visser, and H. Wynberg, *J. Amer. Chem. Soc.*, 1970, 92, 7218.
4. B. Belleau and R. Chevalier, *J. Amer. Chem. Soc.*, 1968, 90, 6864.
5. C. M. Kemp and S. F. Mason, *Chem. Comm.*, 1965, 559.
6. A. Brown, C. M. Kemp, and S. F. Mason, *J. Chem. Soc. (A)*, 1971, 751, 756.
7. D. A. Lightner, D. T. Hefelfinger, G. W. Frank, T. W. Powers, and K. N. Trueblood, *Nature (Physical Science)*, 1971, 232, (no. 32), 124.
8. M. B. Groen and H. Wynberg, *J. Amer. Chem. Soc.*, 1971, 93, 2968.

†Chirality specified according to the revised (1966) Cahn-Ingold-Prelog rules (*Angew. Chem. Internat. Edn.*, 1966, 5, 385).

‡The recent X-ray determination of the absolute configuration of [6]-helicene **X8.3** [7] has validated the method used [1] for calculation of the absolute configurations of other helicene systems.

Cyclophanes (For application of (R, S) nomenclature, see [3])

(R)-(−) 2-phenylbutyric acid
A25.15

1. (1R)-(−)1-hydroxy-1,2,3,4-tetrahydrobenzo [d][2.2] paracyclophane

2. (S)-(+) [2.2] paracyclophane carboxylic acid. Other [2.2] paracyclophanes; [5] CD: [5][6]

(R)-(+) 1-phenylethylamine
A19.14

3. (S)-(+) 2-methyl-[2.2] paracyclophane

4. (S)-(+) 12-methyl-[2.2] metaparacyclophane

Metallocenes (Review; [1])

Application of the (R,S) system to metallocenes

The most widely employed system for specification of metallocene chirality was put forward by Schlögl in consultation with Cahn and Prelog [1]. The bond from the central metal atom to the ring carbon atom under consideration is treated as a formal single bond. The carbon atom is then considered as a chiral centre and (R,S) nomenclature is applied in the usual way.

(1S, 2R) 2-methylferrocene 1-carboxylic acid

order of precedence
Fe > C_1 > C_3 > CH_3
chirality (2R)

A recent publication [2] has suggested the readoption of an earlier proposal [1] according to which the molecule is treated overall as a case of planar chirality (cf. cyclophanes). In view of its widespread use, however, the Schlögl-Cahn-Prelog system is employed here [4].

1. K. Schlögl, *Topics Stereochem.*, 1967, 39.
2. D. Marquarding, H. Klusacek, G. Gokel, P. Hoffmann, and I. Ugi, *J. Amer. Chem. Soc.*, 1970, **92**, 5389.
3. R. S. Cahn, C. K. Ingold, and V. Prelog, *Angew. Chem. Internat. Edn.*, 1966, **5**, 385.
4. K. Schlögl, Personal communication.
5. H. Falk, P. Reich-Rohrwig, and K. Schlögl, *Tetrahedron*, 1970, **26**, 511.
6. M. J. Nugent and O. E. Weigang, *J. Amer. Chem. Soc.*, 1969, **91**, 4556.
7. H. Falk and K. Schlögl, *Angew. Chem. Internat. Edn.*, 1968, 7, 383.
8. M. H. Dalton, R. E. Gilman, and D. J. Cram, *J. Amer. Chem. Soc.*, 1971, **93**, 2329.

8. H. Gowal and K. Schlögl, *Monatsh*, 1968, **99**, 267. ▶
9. K. Bauer, H. Falk, and K. Schlögl, *Angew. Chem. Internat. Edn.*, 1969, **8**, 135.
10. H. Falk and K. Schlögl, *Tetrahedron*, 1966, **22**, 3047 and references therein.
11. G. Haller and K. Schlögl, *Monatsh*, 1967, **98**, 2044.
12. H. Falk and K. Schlögl, *Monatsh*, 1971, **102**, 33 and references therein.
13. H. Falk and K. Schlögl, *Monatsh*, 1965, **96**, 266.
14. S. G. Cottis, H. Falk, and K. Schlögl, *Tetrahedron Letters*, 1965, 2857.

†Compounds X9.1 and X9.2 are examples of compounds having pure planar chirality independently of any conformation contribution from substituents [10]. A correlation, analogous to Brewster's Rules, between absolute configuration and rotation sign, has been put forward for this type of compound [10].

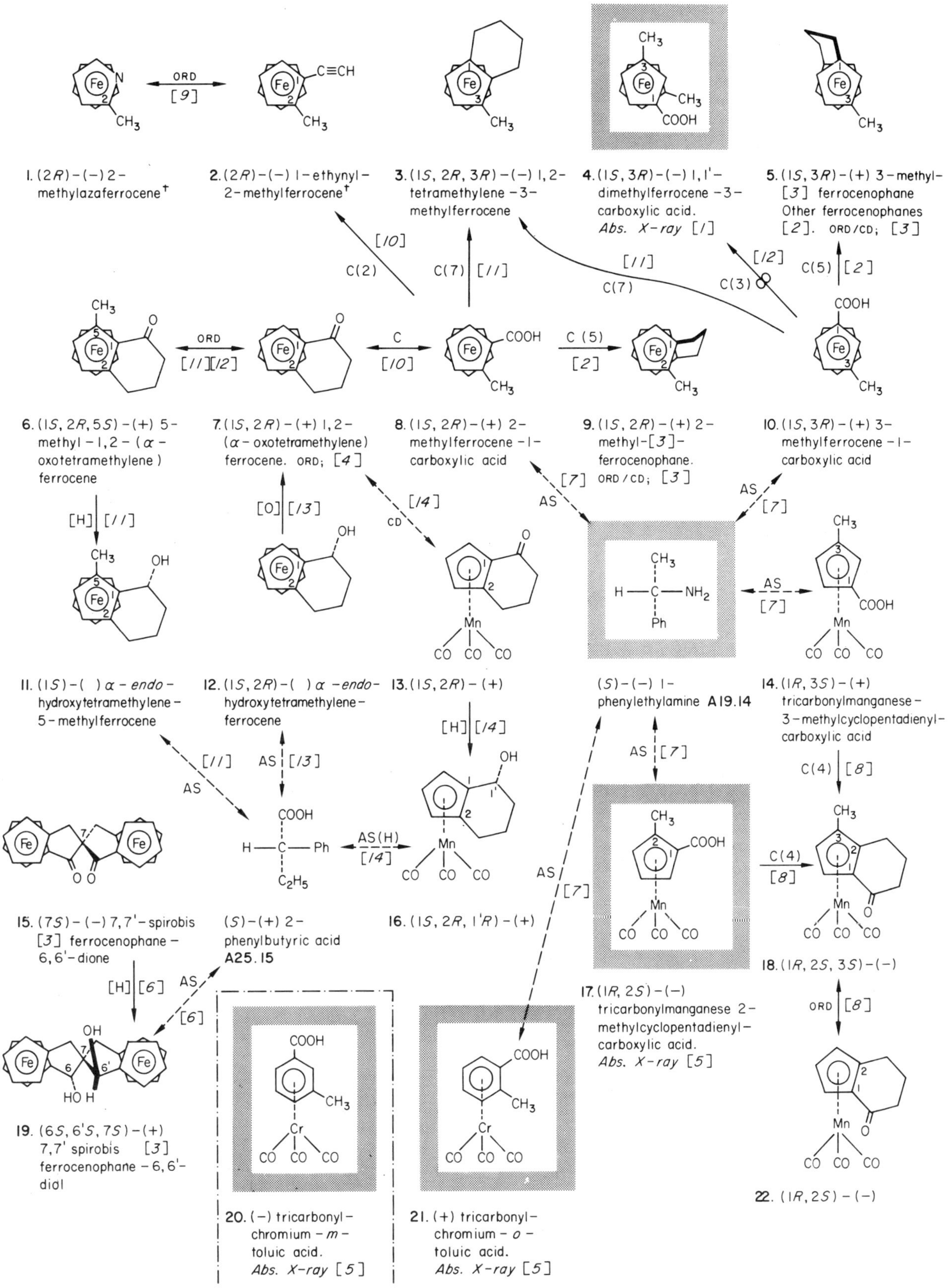
ORD [9]
1. (2R)-(−) 2-methylazaferrocene†
2. (2R)-(−) 1-ethynyl-2-methylferrocene†
3. (1S, 2R, 3R)-(−) 1,2-tetramethylene-3-methylferrocene
4. (1S, 3R)-(−) 1,1'-dimethylferrocene-3-carboxylic acid. Abs. X-ray [1]
5. (1S, 3R)-(+) 3-methyl-[3] ferrocenophane Other ferrocenophanes [2]. ORD/CD; [3]
C(2) [10]
C(7) [11]
C(3) [12]
C(5) [2]
ORD [11][12]
C [10]
C (5) [2]
6. (1S, 2R, 5S)-(+) 5-methyl-1,2-(α-oxotetramethylene) ferrocene
7. (1S, 2R)-(+) 1,2-(α-oxotetramethylene) ferrocene. ORD; [4]
8. (1S, 2R)-(+) 2-methylferrocene-1-carboxylic acid
9. (1S, 2R)-(+) 2-methyl-[3]-ferrocenophane. ORD/CD; [3]
10. (1S, 3R)-(+) 3-methylferrocene-1-carboxylic acid
[H] [11]
[O] [13]
CD [14]
AS [7]
AS [7]
AS [7]
11. (1S)-( ) α-endo-hydroxytetramethylene-5-methylferrocene
12. (1S, 2R)-( ) α-endo-hydroxytetramethylene-ferrocene
13. (1S, 2R)-(+)
(S)-(−) 1-phenylethylamine A19.14
14. (1R, 3S)-(+) tricarbonylmanganese-3-methylcyclopentadienyl-carboxylic acid
AS [11]
AS [13]
[H] [14]
AS [7]
C(4) [8]
AS(H) [14]
AS [7]
C(4) [8]
15. (7S)-(−) 7,7'-spirobis [3] ferrocenophane-6,6'-dione
(S)-(+) 2-phenylbutyric acid A25.15
16. (1S, 2R, 1'R)-(+)
17. (1R, 2S)-(−) tricarbonylmanganese 2-methylcyclopentadienyl-carboxylic acid. Abs. X-ray [5]
18. (1R, 2S, 3S)-(−)
[H] [6]
AS [6]
ORD [8]
19. (6S, 6'S, 7S)-(+) 7,7' spirobis [3] ferrocenophane-6,6'-diol
20. (−) tricarbonylchromium-m-toluic acid. Abs. X-ray [5]
21. (+) tricarbonylchromium-o-toluic acid. Abs. X-ray [5]
22. (1R, 2S)-(−)

1. O. L. Carter, A. T. McPhail, and G. A. Sim, *J. Chem. Soc. (A)*, 1967, 365.
2. H. Falk, O. Hofer, and K. Schlögl, *Monatsh*, 1969, **100**, 624.
3. H. Falk and O. Hofer, *Monatsh*, 1969, **100**, 1540.
4. H. Falk and K. Schlögl, *Monatsh*, 1968, **99**, 279.
5. M. A. Bush, T. A. Dullforce, and G. A. Sim, *Chem. Comm.*, 1969, 1491.
6. H. Falk, W. Fröstl, and K. Schlögl, *Monatsh*, 1971, **102**, 1270.
7. H. Falk and K. Schlögl, *Monatsh*, 1968, **99**, 578.

References continued on page 224

X 10 Compounds of high (C_2, C_3, D_2 and D_3) symmetry

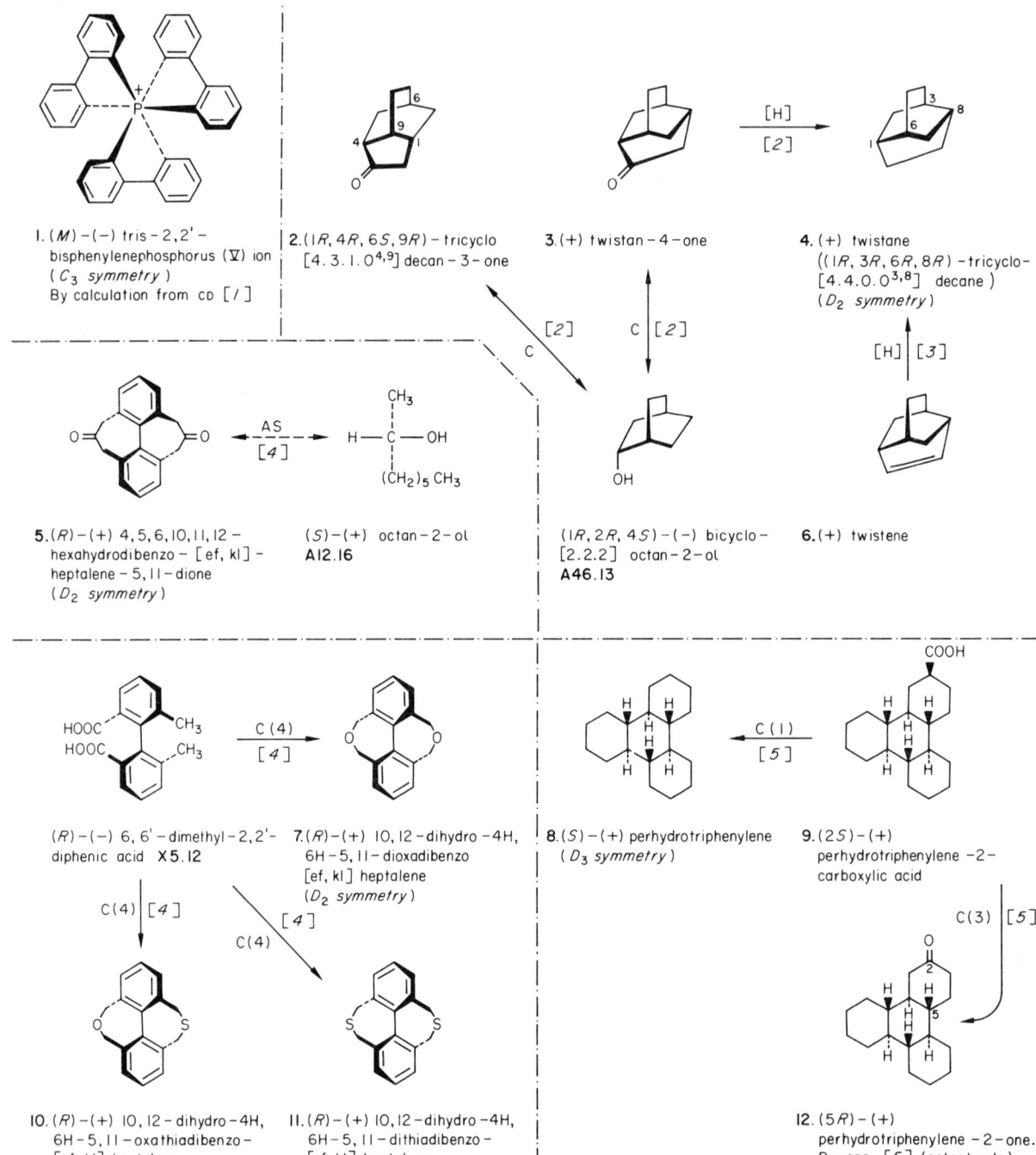

1. (M)-(−) tris-2,2'-bisphenylenephosphorus (V) ion (C_3 symmetry) By calculation from CD [1]
2. (1R,4R,6S,9R)-tricyclo [4.3.1.0^{4,9}] decan-3-one
3. (+) twistan-4-one
4. (+) twistane ((1R,3R,6R,8R)-tricyclo-[4.4.0.0^{3,8}] decane) (D_2 symmetry)
5. (R)-(+) 4,5,6,10,11,12-hexahydrodibenzo-[ef,kl]-heptalene-5,11-dione (D_2 symmetry)
   (S)-(+) octan-2-ol A12.16
   (1R,2R,4S)-(−) bicyclo-[2.2.2] octan-2-ol A46.13
6. (+) twistene
7. (R)-(+) 10,12-dihydro-4H,6H-5,11-dioxadibenzo [ef,kl] heptalene (D_2 symmetry)
   (R)-(−) 6,6'-dimethyl-2,2'-diphenic acid X5.12
8. (S)-(+) perhydrotriphenylene (D_3 symmetry)
9. (2S)-(+) perhydrotriphenylene-2-carboxylic acid
10. (R)-(+) 10,12-dihydro-4H, 6H-5,11-oxathiadibenzo-[ef,kl] heptalene (C_2 symmetry)
11. (R)-(+) 10,12-dihydro-4H,6H-5,11-dithiadibenzo-[ef,kl] heptalene (D_2 symmetry)
12. (5R)-(+) perhydrotriphenylene-2-one. By ORD [5] (octant rule)

1. D. Hellwinkel and S. F. Mason, *J. Chem. Soc. (B)*, 1970, 640.
2. M. Tichý, *Tetrahedron Letters*, 1972, 2001.
3. M. Tichý and J. Sicher, *Tetrahedron Letters*, 1969, 4609.
4. K. Mislow, M. A. W. Glass, H. B. Hopps, E. Simon, and G. H. Wahl, *J. Amer. Chem. Soc.*, 1964, **86**, 1710.
5. M. Farina and G. Audisio, *Tetrahedron*, 1970, **26**, 1839.

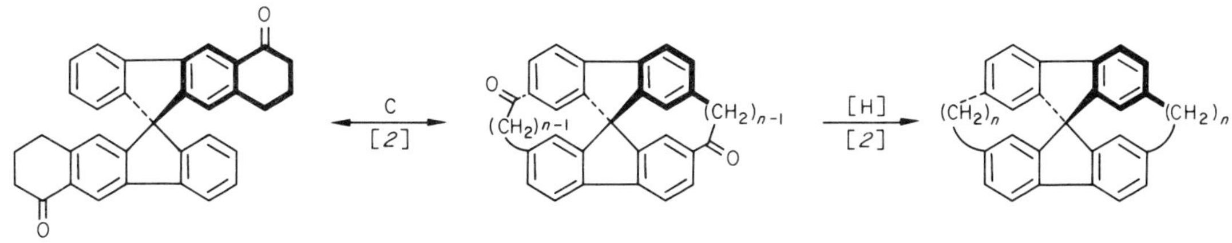

Compounds of high symmetry (contd.); vespiranes

1. (R)-(+) 2,3,2',3'-bis(α-oxo-tetramethylene)-9,9'-spirobifluorene. By CD [1] Related compds. [1] [2]

2. (R)-(−) [6,6]-vespirone (n = 6)
   (R)-(−) [7,7]-vespirone (n = 7)
   (R)-(−) [8,8]-vespirone (n = 8)
   By CD [1]
   Related compds. [1] [2]

3. (R)-(−) [6,6]-vespirene (n = 6)
   (R)-(−) [7,7]-vespirene (n = 7)
   (R)-(−) [8,8]-vespirene (n = 8)
   (D_2 symmetry) By CD [1]
   Related compds. [1] [2]

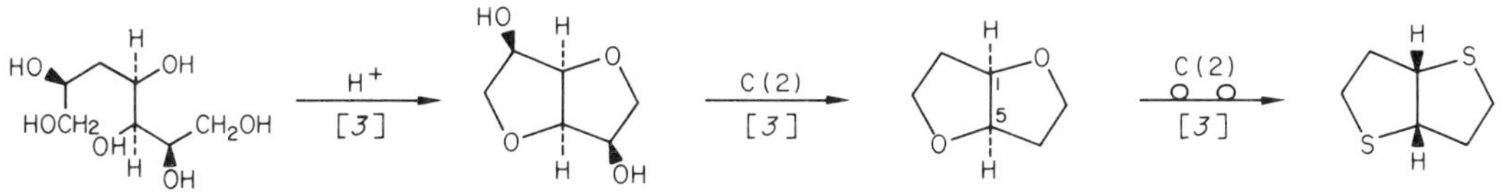

D (−) mannitol A II.1

4. D (+) isomannide

5. (1R,5R)-(+) 2,6-dioxabicyclo-[3.3.0] octane

6. (1S,5S)-(−) 2,6-dithiabicyclo-[3.3.0] octane

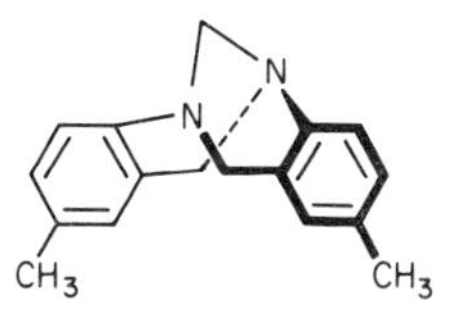

7. (+) Tröger's base. By calculation from CD [4] (coupled oscillators)

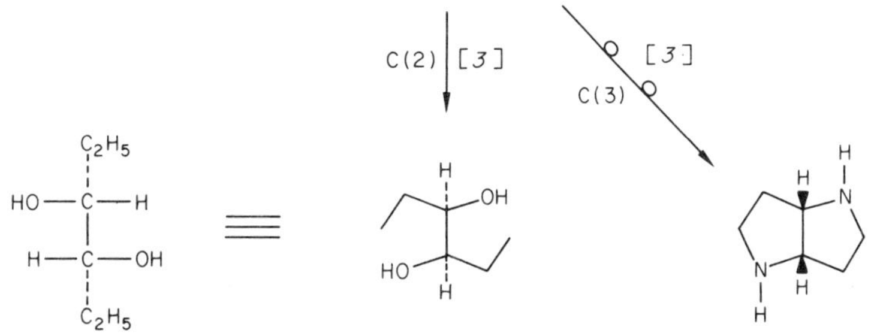

8. (3R,4R)-(+) hexane-3,4-diol

9. (1S,5S)-(−) 2,6-diazabicyclo-[3.3.0] octane

1. G. Haas, P. B. Hulbert, W. Klyne, V. Prelog, and G. Snatzke, *Helv. Chim. Acta*, 1971, **54**, 491.
2. G. Haas and V. Prelog, *Helv. Chim. Acta*, 1969, **52**, 1202.
3. A. C. Cope and T. Y. Shen, *J. Amer. Chem. Soc.*, 1956, **78**, 5916.
4. S. F. Mason, G. W. Vane, K. Schofield, R. J. Wells, and J. S. Whitehurst, *J. Chem. Soc. (B)*, 1967, 553.

-Z-

Compounds containing Chiral Atoms other than Carbon

Introductory Notes to Chapter Z

Chapter Z deals with chiral compounds containing silicon, germanium, tin, lead, nitrogen, phosphorus, arsenic, and sulphur.

Arrangement

(1) Compounds containing a silicon, germanium, tin, or lead atom directly attached to a chiral carbon atom—page **Z1**.
(2) Compounds containing a chiral silicon or germanium atom—pages **Z1-Z2**.
(3) Compounds containing a chiral nitrogen, phosphorus or arsenic atom—pages **Z3-Z6**.
(4) Compounds containing a chiral sulphur atom—pages **Z7-Z8**.
For compounds containing a nitrogen or sulphur atom directly attached to a chiral carbon atom, see Chapter *A*.

Abbreviations

For clarity, the following abbreviations are used throughout section Z.

Mcn = (-) menthyl,

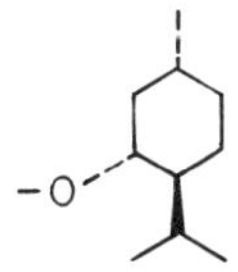

Note that since the menthyl group contains three chiral centres, forms such as

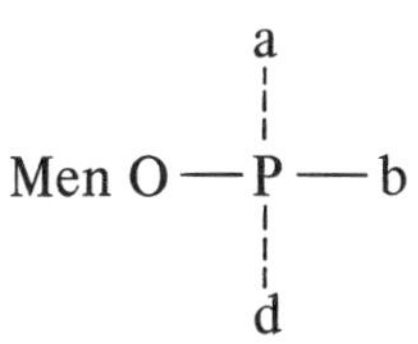

Men O—P—b   and   b—P—O Men

(with substituents a and d)

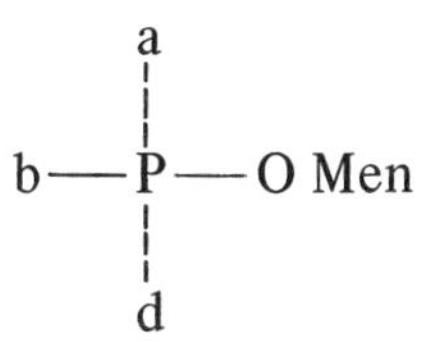

are diastereoisomers and not enantiomers.

Np = naphthyl, $C_{10}H_7$.

Specification of chirality at phosphorus and sulphur

The Cahn-Ingold-Prelog rules contain a sub-rule which states that d orbital expansion is disregarded in assigning chirality. Hence, the SO bond in sulphoxides and the PO, PS etc. bond in P(V) compounds is regarded as a formal single bond. For example in the compound

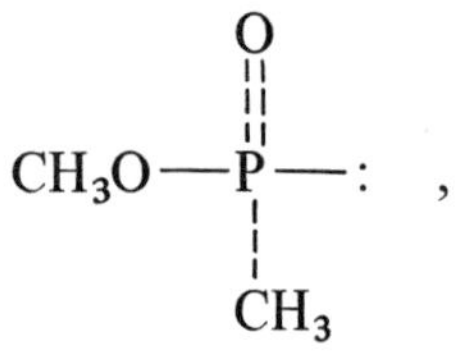

$$\text{CH}_3\text{O}-\overset{\overset{\displaystyle O}{\|}}{\underset{\underset{\displaystyle \text{CH}_3}{|}}{P}}-:\quad,$$

the order of precedence of the four substituents is $\text{OCH}_3 > =\text{O} > \text{CH}_3 >$ lone pair and the chirality is (S).

In dealing with compounds containing chiral P,S or other hetero-atoms, a recent development has been the introduction of P S, etc. as a subscript to denote the chirality at the hetero-atom as distinct from that at other chiral centres in the molecule, for example

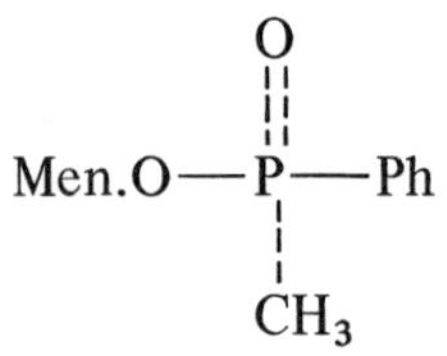

$$\text{Men.O}-\overset{\overset{\displaystyle O}{\|}}{\underset{\underset{\displaystyle \text{CH}_3}{|}}{P}}-\text{Ph} \quad = \quad (R)\text{p - menthyl methylphenyl phosphinate}$$

Recent developments in phosphorus chirality

The study of phosphorus reaction mechanisms is still at a relatively early stage, and many aspects of the course of displacement reactions at chiral phosphorus are not yet completely understood. A recent publication (J. Donohue, N. Mandel, W. B. Farnham, R. K. Murray, K. Mislow, and H. P. Benshop, *J. Amer. Chem. Soc.*, 1971, **93**, 3792) has demonstrated that a number of earlier correlations are of doubtful validity, and pending clarification anticipated as a result of future work, all of these have been omitted.

Z¹
Silicon, germanium, tin and lead

(a) Compounds containing a silicon, germanium, tin or lead atom directly attached to a chiral carbon atom.

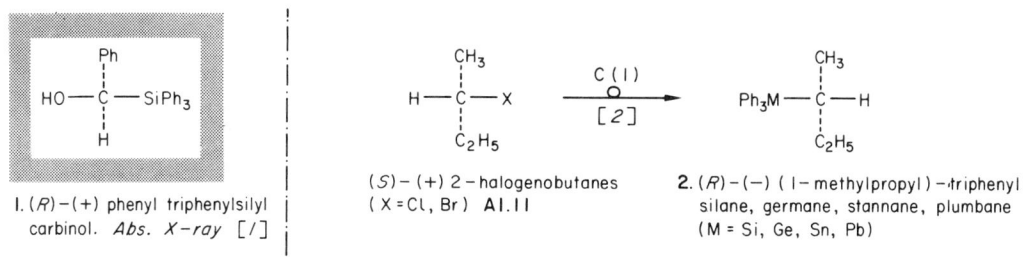

1. (R)-(+) phenyl triphenylsilyl carbinol. Abs. X-ray [1]

(S)-(+) 2-halogenobutanes (X = Cl, Br) A1.11

2. (R)-(−) (1-methylpropyl)-triphenyl silane, germane, stannane, plumbane (M = Si, Ge, Sn, Pb)

(b) Compounds containing a chiral silicon or germanium atom. (Reviews; [3] [4].) See [3] for further compounds not given here.

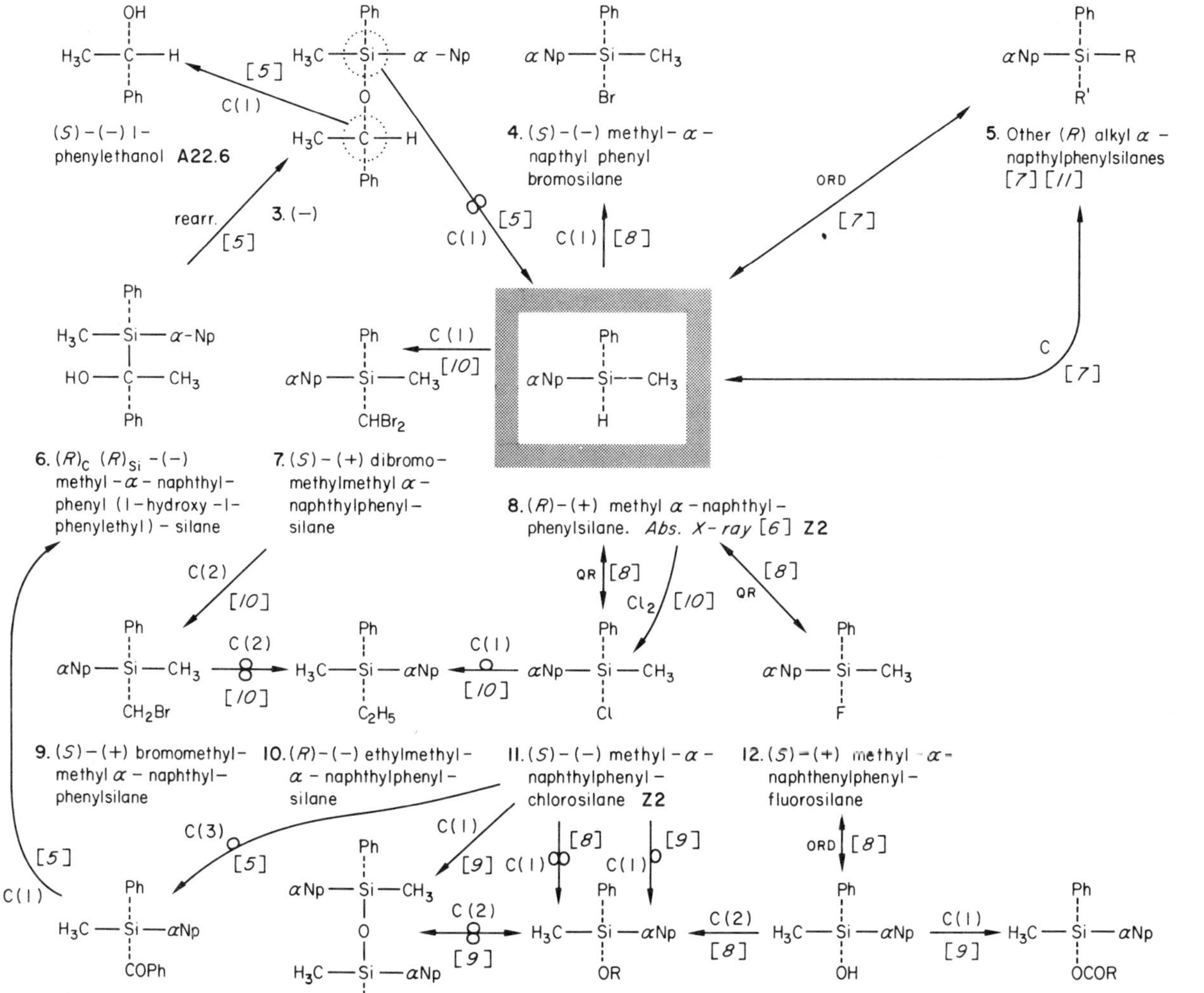

(S)-(−) 1-phenylethanol A22.6

4. (S)-(−) methyl-α-napthyl phenyl bromosilane

5. Other (R) alkyl α-napthylphenylsilanes [7] [11]

3. (−)

6. (R)$_C$ (R)$_{Si}$ -(−) methyl-α-naphthyl-phenyl (1-hydroxy-1-phenylethyl)-silane

7. (S)-(+) dibromo-methylmethyl α-naphthylphenyl-silane

8. (R)-(+) methyl α-naphthyl-phenylsilane. Abs. X-ray [6] Z2

9. (S)-(+) bromomethyl-methyl α-naphthyl-phenylsilane

10. (R)-(−) ethylmethyl-α-naphthylphenyl-silane

11. (S)-(−) methyl-α-naphthylphenyl-chlorosilane Z2

12. (S)-(+) methyl-α-naphthenylphenyl-fluorosilane

13. (R)-(+) benzoylmethyl-α-naphthylphenyl-silane

14. (S)-(+)

15. (R)-alkoxymethyl-α-naphthylphenylsilanes R = CH₃ (−) R = CMe₃ (+) R = cyclohexyl (+) Z2

16. (R)-(−) methyl-α-naphthylphenylsilanol. K salt (+)

17. (R)-(−) methyl-α-naphthylphenylacyloxy-silanes (R = CH₃, Ph, p-C₆H₄NO₂)

1. K. T. Black and H. Hope, *J. Amer. Chem. Soc.*, 1971, **93**, 3053.
2. F. R. Jensen and D. D. Davis, *J. Amer. Chem. Soc.*, 1971, **93**, 4047.
3. L. H. Sommer, 'Stereochemistry, Mechanism and Silicon', McGraw-Hill, 1965.
4. B. J. Aylett, *Progr. Stereochem.*, 1969, **4**, 213.
5. A. G. Brook and W. W. Limburg, *J. Amer. Chem. Soc.*, 1963, **85**, 833.
6. T. Ashida, R. Pepinsky and Y. Okaya, *Acta Cryst.*, 1963, **16** (suppl.), A48, abstr. 5.8.
7. R. J. P. Corriu, G. F. Lanneau, and M. Leard, *Chem. Comm.*, 1971, 1365.
8. L. H. Sommer, C. L. Frye, G. A. Parker, and K. W. Michael, *J. Amer. Chem. Soc.*, 1964, **86**, 3271.
9. L. H. Sommer, C. L. Frye, and G. A. Parker, *J. Amer. Chem. Soc.*, 1964, **86**, 3276, 3280.
10. A. G. Brook, J. M. Duff, and D. G. Anderson, *J. Amer. Chem. Soc.*, 1970, **92**, 7567.
11. L. H. Sommer, W. D. Korte, and P. G. Rodewald, *J. Amer. Chem. Soc.*, 1967, **89**, 862; L. H. Sommer, K. W. Michael, and W. D. Korte, ibid., 868.

Z² Chirality at Si and Ge (contd.)

(S)-(−) methyl α-naphthenylphenyl-silane Z1.8

1. (R)-(−) dibromomethyl methyl-α-naphthyl-phenyl germane
2. (R)-(−) bromomethyl methyl-α-naphthyl-phenyl germane
3. (R)-(−) ethyl-α-naphthylphenylgermane

(R)-(+) methyl-α-naphthylphenyl-chlorosilane Z1.11

4. (R)-(+) methyl-α-naphthylphenyl-chlorogermane
5. (S)-(−) alkyl-α-naphthylphenyl-germanes (R=Me, Et)
6. (S)-(−) isopropyl-α-naphthylphenylgermane Related compds. [4]
7. (R)-(−) ethylmethyl-α-naphthylphenyl-germane

(R)-(−) methoxy-methyl-α-naphthyl-phenylsilane Z1.15

8. (R)-(−) methoxy-methyl-α-naphthyl-phenylgermane
9. (R)-(+) alkyl-α-naphthylphenyl carboxy germanes (R=Me, Et)
10. (S)-( ) alkyl-α-naphthylphenyl-germyllithiums (R=Me, Et)
11. (R)-(+)

---

1. A. G. Brook and G. J. D. Peddle, *J. Amer. Chem. Soc.*, 1963, **85**, 1869, 2338.
2. A. G. Brook, J. M. Duff, and D. G. Anderson, *J. Amer. Chem. Soc.*, 1970, **92**, 7567.
3. C. Eaborn, R. E. E. Hill, and P. Simpson, *J. Organometall. Chem.*, 1968, **15**, P1.
4. F. Carré and R. Corriu, *J. Organometall. Chem.*, 1970, **25**, 395.

Nitrogen, phosphorus and arsenic
Reviews of phosphorus chirality; [6] [7]

(See also N,N'-diferrocenylcarbodiimide, **X3.5**, Tröger's base **X11.7** and tris-2,2'-bisphenylene-phosphorus (V) ion **X10.1**)

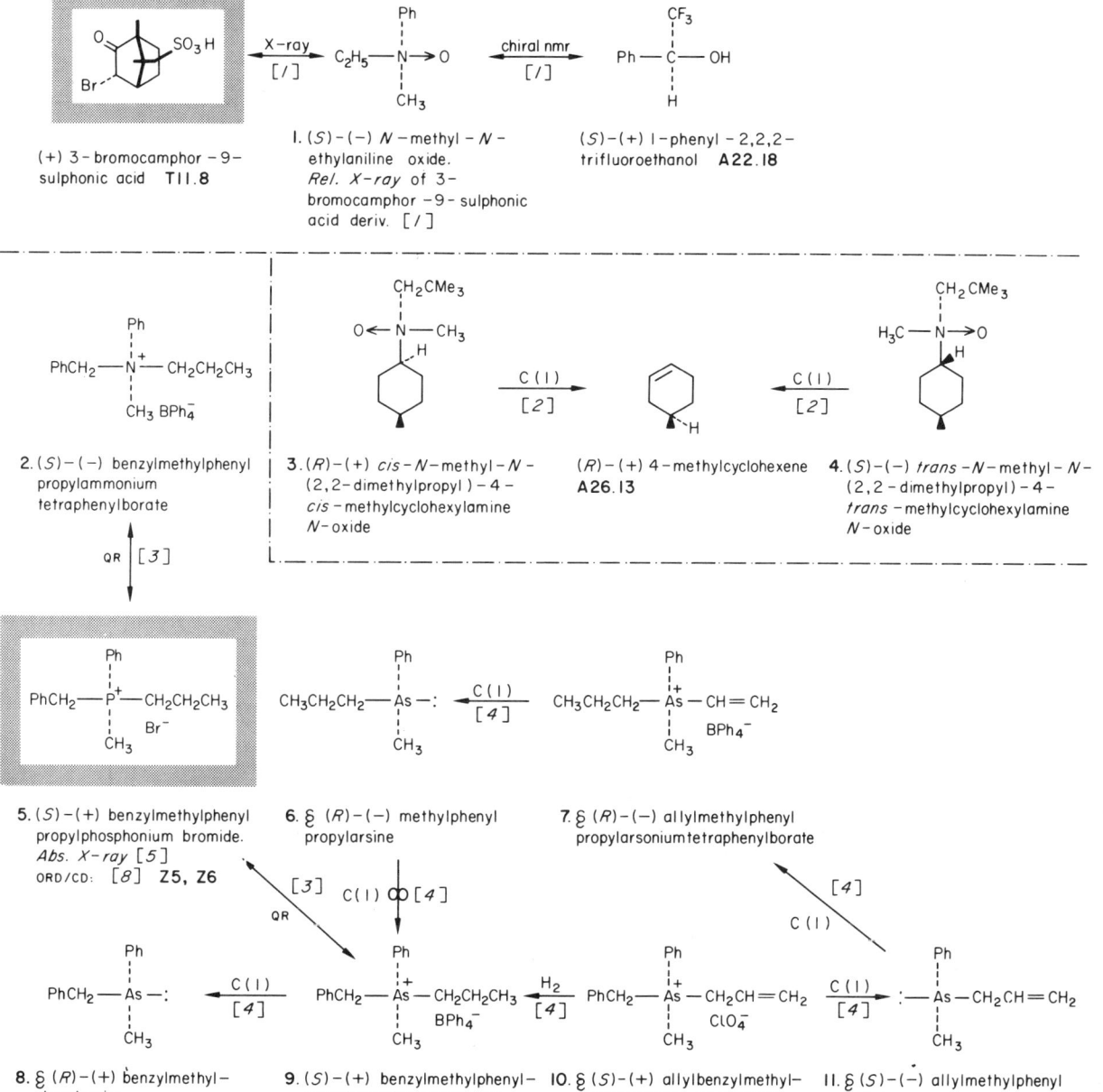

1. (S)-(−) N-methyl-N-ethylaniline oxide. Rel. X-ray of 3-bromocamphor-9-sulphonic acid deriv. [1]

(+) 3-bromocamphor-9-sulphonic acid **T11.8**

(S)-(+) 1-phenyl-2,2,2-trifluoroethanol **A22.18**

2. (S)-(−) benzylmethylphenyl propylammonium tetraphenylborate

3. (R)-(+) cis-N-methyl-N-(2,2-dimethylpropyl)-4-cis-methylcyclohexylamine N-oxide

(R)-(+) 4-methylcyclohexene **A26.13**

4. (S)-(−) trans-N-methyl-N-(2,2-dimethylpropyl)-4-trans-methylcyclohexylamine N-oxide

5. (S)-(+) benzylmethylphenyl propylphosphonium bromide. Abs. X-ray [5] ORD/CD: [8] **Z5, Z6**

6. § (R)-(−) methylphenyl propylarsine

7. § (R)-(−) allylmethylphenyl propylarsonium tetraphenylborate

8. § (R)-(+) benzylmethylphenylarsine

9. (S)-(+) benzylmethylphenyl-propylarsonium tetraphenyl borate. ORD/CD; [8]

10. § (S)-(+) allylbenzylmethylphenylarsonium perchlorate

11. § (S)-(−) allylmethylphenyl arsine

1. W. H. Pirkle, R. L. Muntz, and I. C. Paul, *J. Amer. Chem. Soc.*, 1971, **93**, 2817.
2. S. I. Goldberg and F.-K. Lam, *J. Amer. Chem. Soc.*, 1969, **91**, 5113.
3. L. Horner, H. Winkler, and E. Meyer, *Tetrahedron Letters*, 1965, 789.
4. L. Horner and H. Fuchs, *Tetrahedron Letters*, 1963, 1573.
5. A. F. Peerdeman, J. P. C. Holst, L. Horner, and H. Winkler, *Tetrahedron Letters*, 1965, 811.
6. L. Horner, *Pure Appl. Chem.*, 1964, **9**, 225; *Helv. Chim. Acta, Fasc. Extraord. A. Werner*, 1967, 93.
7. M. J. Gallagher and I. D. Jenkins, *Topics Stereochem.*, 1968, **3**, 1.
8. L. Horner and W.-D. Balzer, *Chem. Ber.*, 1969, **102**, 3542.

Z⁴ Chirality at P (contd.)

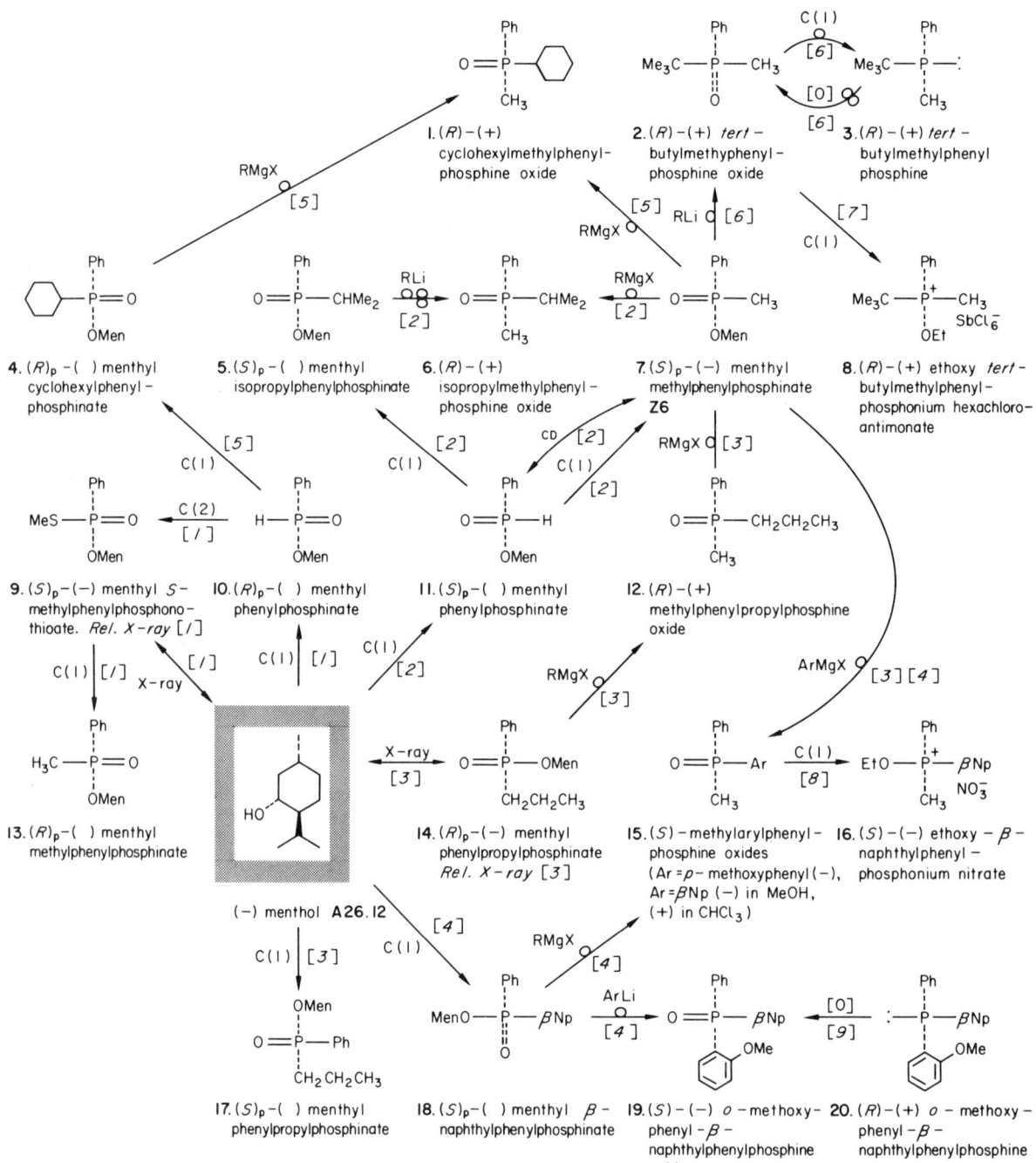

1. J. Donohue, N. Mandel, W. B. Farnham, R. K. Murray, K. Mislow, and H. P. Benschop, *J. Amer. Chem. Soc.*, 1971, **93**, 3792.
2. W. B. Farnham, R. K. Murray, and K. Mislow, *J. Amer. Chem. Soc.*, 1970, **92**, 5809.
3. O. Korpiun and K. Mislow, *J. Amer. Chem. Soc.*, 1967, **89**, 4784; O. Korpiun, R. A. Lewis, J. Chickos, and K. Mislow, ibid., 1968, **90**, 4842 and references therein.
4. R. A. Lewis and K. Mislow, *J. Amer. Chem. Soc.*, 1969, **91**, 7009.
5. W. B. Farnham, R. K. Murray, and K. Mislow, *Chem. Comm.*, 1971, 146.
6. R. A. Lewis, K. Naumann, K. E. DeBruin, and K. Mislow, *Chem. Comm.*, 1969, 1010.
7. L. Horner and W.-D. Balzer, *Chem. Ber.*, 1969, **102**, 3542.
8. G. Zon, K. E. DeBruin, K. Naumann, and K. Mislow, *J. Amer. Chem. Soc.*, 1969, **91**, 7023.
9. K. Naumann, G. Zon, and K. Mislow, *J. Amer. Chem. Soc.*, 1969, **91**, 7012.

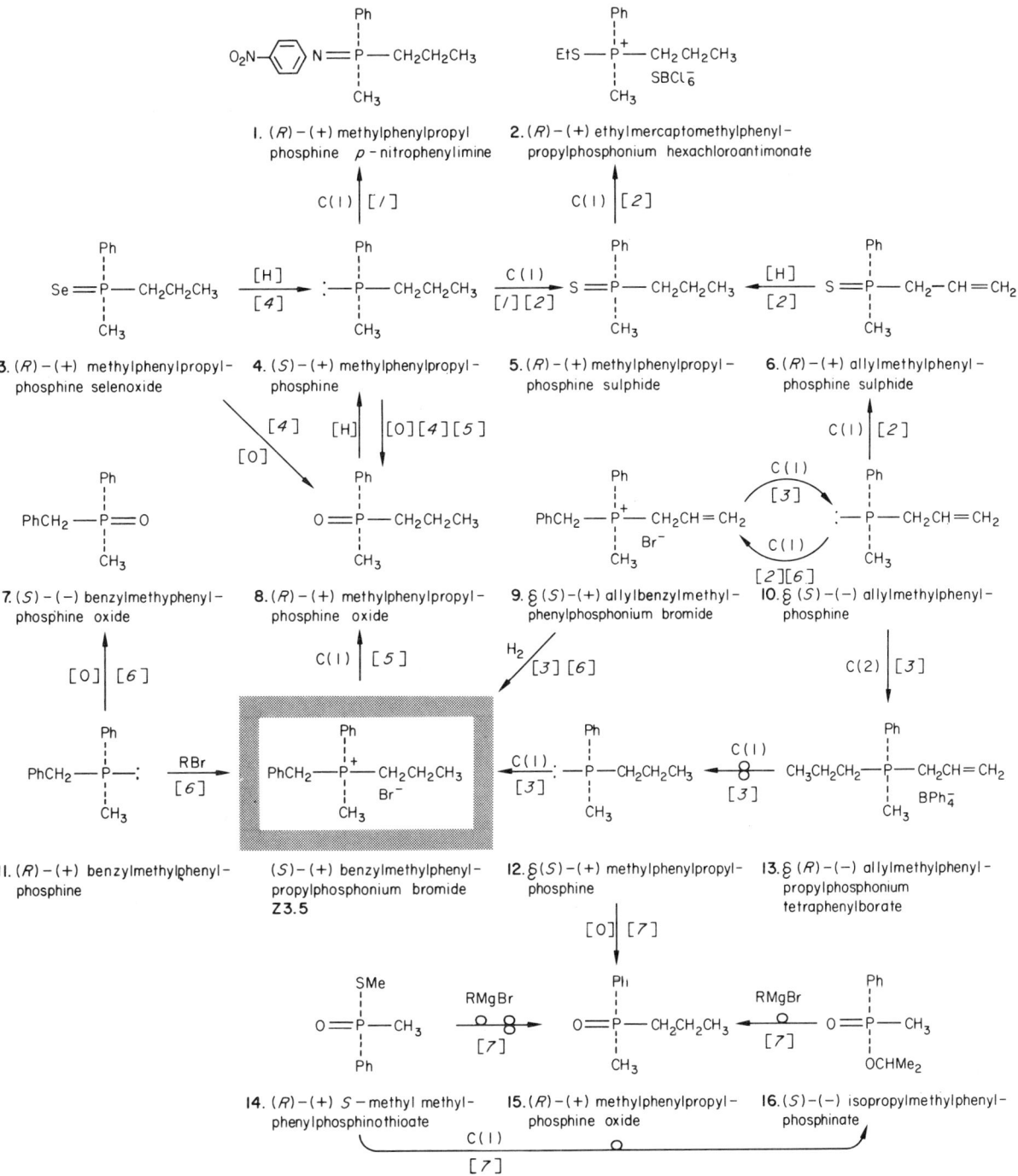

1. L. Horner and H. Winkler, *Tetrahedron Letters*, 1964, 175.
2. G. Zon, K. E. DeBruin, K. Naumann, and K. Mislow, *J. Amer. Chem. Soc.*, 1969, **91**, 7023 and references therein.
3. L. Horner, H. Fuchs, H. Winkler, and A. Rapp, *Tetrahedron Letters*, 1963, 965.
4. W. Stec, A. Okruszek, and J. Michalski, *Angew. Chem. Internat. Edn.*, 1971, **10**, 494.
5. O. Korpiun & K. Mislow, *J. Amer. Chem. Soc.*, 1967, **89**, 4784; O. Korpiun, R. A. Lewis, J. Chickos, and K. Mislow, ibid., 1968, **90**, 4842.
6. K. Naumann, G. Zon, and K. Mislow, *J. Amer. Chem. Soc.*, 1969, **91**, 7012.
7. H. P. Benschop, G. R. van den Berg, and H. L. Boter, *Rec. Trav. Chim.*, 1968, **87**, 387.

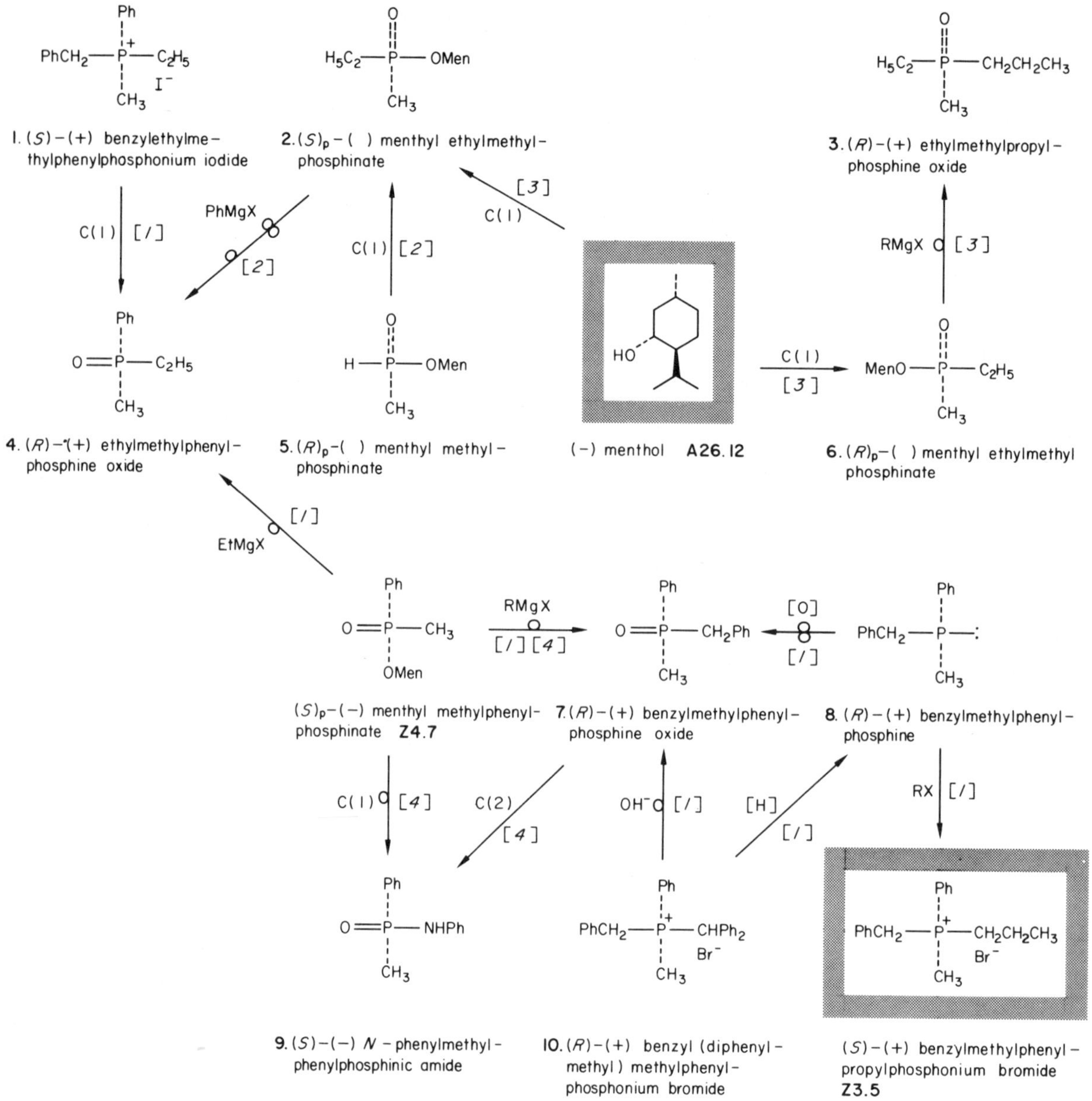

Z⁶ Chirality at P (contd.)

1. $(S)-(+)$ benzylethylmethylphenylphosphonium iodide
2. $(S)_P-(\ )$ menthyl ethylmethylphosphinate
3. $(R)-(+)$ ethylmethylpropylphosphine oxide
4. $(R)-(+)$ ethylmethylphenylphosphine oxide
5. $(R)_P-(\ )$ menthyl methylphosphinate
6. $(R)_P-(\ )$ menthyl ethylmethyl phosphinate

$(-)$ menthol A26.12

$(S)_P-(-)$ menthyl methylphenylphosphinate Z4.7

7. $(R)-(+)$ benzylmethylphenylphosphine oxide
8. $(R)-(+)$ benzylmethylphenylphosphine
9. $(S)-(-)$ N-phenylmethylphenylphosphinic amide
10. $(R)-(+)$ benzyl(diphenylmethyl)methylphenylphosphonium bromide

$(S)-(+)$ benzylmethylphenylpropylphosphonium bromide Z3.5

---

1. O. Korpiun and K. Mislow, *J. Amer. Chem. Soc.*, 1967, **89**, 4784; O. Korpiun, R. A. Lewis, J. Chickos, and K. Mislow, ibid., 1968, **90**, 4842 and references therein.
2. W. B. Farnham, R. K. Murray, and K. Mislow, *Chem. Comm.*, 1971, 605.
3. R. A. Lewis and K. Mislow, *J. Amer. Chem. Soc.*, 1969, **91**, 7009.
4. A. Nudelman and D. J. Cram, *J. Org. Chem.*, 1971, **36**, 335.

Chirality at sulphur†

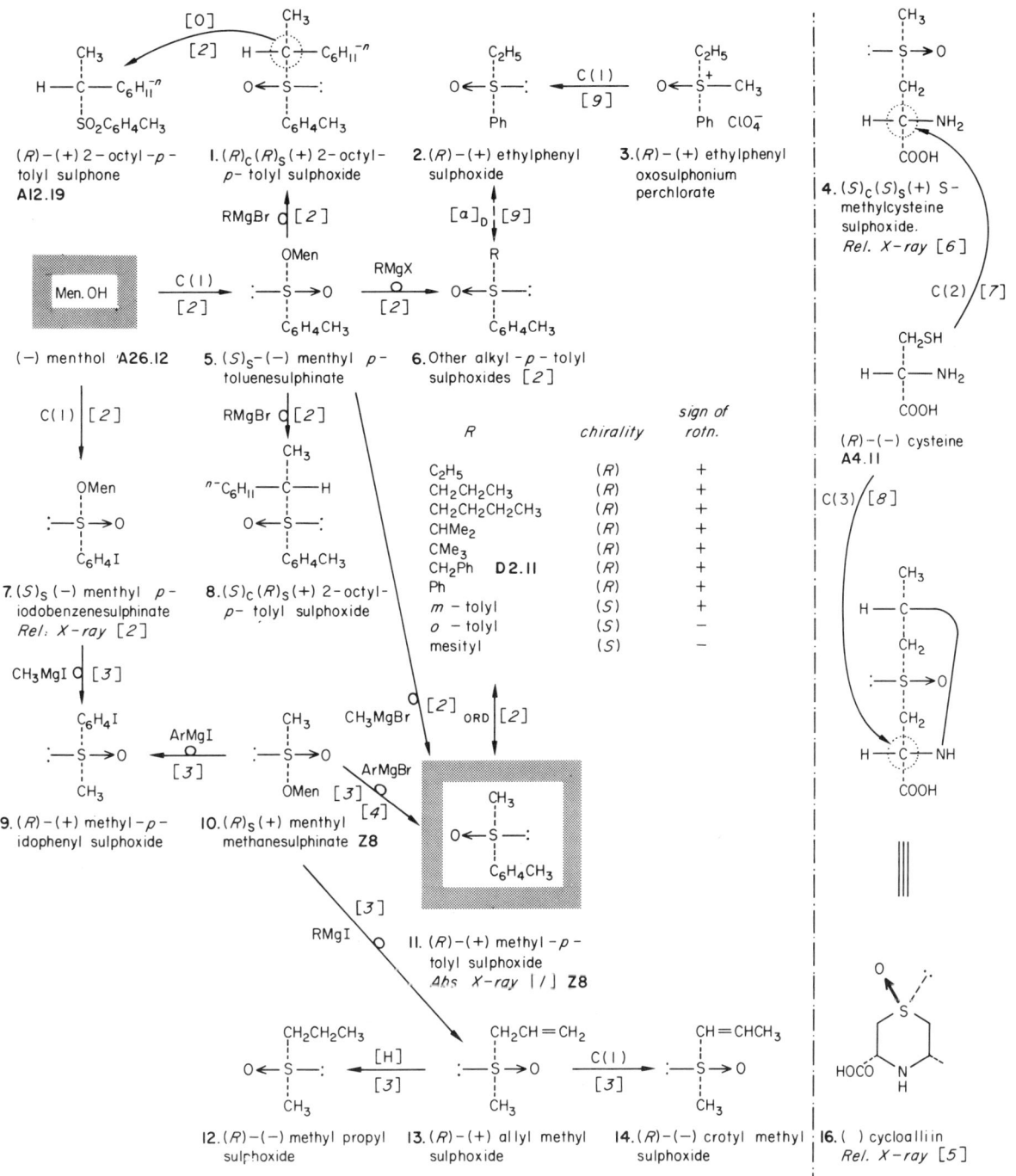

1. H. Hope, U. de la Camp, G. D. Homer, A. W. Messing, and L. H. Sommer, *Angew. Chem. Internat. Edn.*, 1969, **8**, 612.
2. K. Mislow, M. M. Green, P. Laur, J. T. Melillo, T. Simmons, and A. L. Ternay, *J. Amer. Chem. Soc.*, 1965, **87**, 1958.
3. M. Axelrod, P. Bickart, J. Jacobus, M. M. Green, and K. Mislow, *J. Amer. Chem. Soc.*, 1968, **90**, 4835.
4. J. Jacobus and K. Mislow, *J. Amer. Chem. Soc.*, 1967, **89**, 5228.
5. K. J. Palmer and K. S. Lee, *Acta Cryst.*, 1966, **20**, 790.
6. R. Hine, *Acta Cryst.*, 1962, **15**, 635.
7. C. J. Morris and J. F. Thompson, *Chem. and Ind.*, 1955, 951.
8. A. I. Virtanen and E. J. Matikkala, *Acta Chem. Scand.*, 1959, **13**, 623.
9. M. Kobayashi, K. Kamiyama, H. Minato, Y. Oishi, Y. Takeda, and Y. Hattori, *Chem. Comm.*, 1971, 1577.

†*Note added in proof*; for a review, see P. H. Laur in 'Sulphur in Organic and Inorganic Chemistry' (A. Senning, Ed.), Chap. 24 (Marcel Dekker, Inc., 1972).

Z⁸ Chirality at S (contd.)

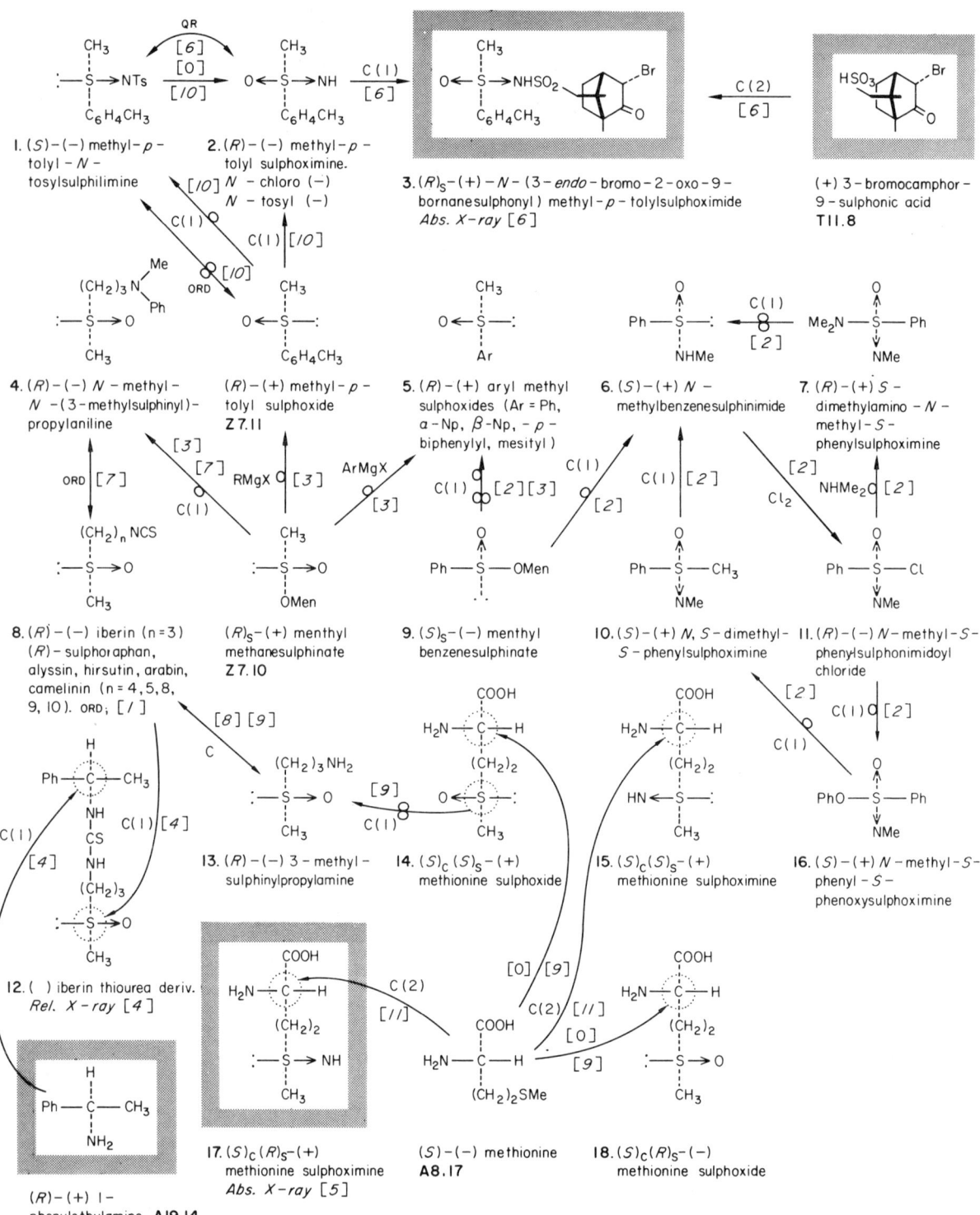

1. W. Klyne, J. Day, and A. Kjaer, *Acta Chem. Scand.*, 1960, **14**, 215.
2. E. U. Jonsson and C. R. Johnson, *J. Amer. Chem. Soc.*, 1971, **93**, 5308.
3. J. Jacobus and K. Mislow, *J. Amer. Chem. Soc.*, 1967, **89**, 5228.
4. K. K. Cheung, A. Kjaer, and G. A. Sim, *Chem. Comm.*, 1965, 100.
5. S. Neidle and D. Rogers, *J. Chem. Soc. (B)*, 1970, 694.
6. D. J. Cram, J. Day, D. R. Rayner, D. M. von Schriltz, D. J. Duchamp, and D. C. Garwood, *J. Amer. Chem. Soc.*, 1970, **92**, 7369.
7. K. H. Andersen, *J. Org. Chem.*, 1964, **29**, 1953.
8. P. Karrer, E. Scheitlin, and H. Siegrist, *Helv. Chim. Acta*, 1950, **33**, 1237; M. Axelrod, P. Bickart, J. Jacobus, M. M. Green, and K. Mislow, *J. Amer. Chem. Soc.*, 1968, **90**, 4835.
9. B. W. Christensen and A. Kjaer, *Chem. Comm.*, 1965, 225.
10. D. R. Rayner, D. M. von Schriltz, J. Day, and D. J. Cram, *J. Amer. Chem. Soc.*, 1968, **90**, 2721.
11. B. W. Christensen and A. Kjaer, *Chem. Comm.*, 1969, 934.

FORMULAE,
AUTHOR AND SUBJECT
INDEXES

# Formulae Index

Isotopically labelled compounds. In this index, deuterium and tritium are given as D and T respectively and are created as heteroatoms for the purpose of indexing.

C_2

$C_2H_2DO_2T$
   [$^2H,^3H$]-acetic acid   D2.4

$C_2H_3DO_3$
   [2-^2H]-glycollic acid   D2.1

$C_2H_3O_3T$
   [2-^3H]-glycollic acid   D2.2

$C_2H_4NO_2T$
   [2-^3H]-2-aminoacetic acid   D2.6

C_2H_5DO
   [1-^2H]-ethanol   D2.18

$C_2H_5O_2T$
   [1-$^3H_1$]-ethane-1,2-diol   D2.3

C_2H_6DN
   [1-^2H]-1-aminoethane   D2.7

C_3

$C_3H_5BrO_2$
   2-Bromopropionic acid   A1.14

$C_3H_5BrO_3$
   3-Bromo-2-hydroxypropionic acid   A1.4

$C_3H_5ClO_2$
   2-Chloropropionic acid   A1.14

$C_3H_5D_3O$
   [1-$^2H_3$]-propan-2-ol   D1.15

$C_3H_5F_3O$
   1,1,1-trifluoropropan-2-ol   A12.2

$C_3H_5IO_2$
   2-Iodopropionic acid   A1.14

$C_3H_5N_3O_2$
   2-Azidopropionic acid   A1.19

$C_3H_5O_2T$
   [2-^3H]-propionic acid   D2.16

$C_3H_6Cl_2$
   1,2-Dichloropropane   A13.10

$C_3H_6ClNO_2$
   2-Amino-3-chloropropionic acid   A4.8

C_3H_6O
   1,2-Epoxypropane   A13.15

$C_3H_6O_2$
   Lactic aldehyde   A1.9

$C_3H_6O_2S$
   Thiolactic acid   A1.15
   2-Mercaptopropionic acid   A12.10

$C_3H_6O_3$
   Lactic acid   A1.9
   Glyceraldehyde   A1.1

$C_3H_6O_3S$
   2-Hydroxy-3-mercaptopropionic acid   A12.6

$C_3H_6O_4$
   Glyceric acid   A1.2

C_3H_6S
   2-Methylthiiran   A14.2

C_3H_7BrO
   1-Bromopropan-2-ol   A14.13

C_3H_7ClO
   2-Chloropropan-1-ol   A13.6

C_3H_7DO
   [2-^2H]-propan-1-ol   D1.20

$C_3H_7NO_2$
   Alanine   A1.18

$C_3H_7NO_2S$
   Cysteine   A4.11

$C_3H_7NO_3$
   Isoserine   A1.3
   Serine   A4.3

$C_3H_7O_4P$
   Phosphonomycin   Y20.2

$C_3H_8N_2O_2$
   2,3-Diaminopropionic acid   A4.7

C_3H_8OS
   2-Mercaptopropan-1-ol   A14.15

$C_3H_8OS_2$
   2,3-Dimercaptopropan-1-ol   A14.9

$C_3H_8O_2$
   Propane-1,2-diol   A14.8

$C_3H_8O_2S$
   3-Mercaptopropane-1,2-diol   A14.6

C_3H_9NO
   1-Aminopropan-2-ol   A22.4
   Alaninol   A19.13

$C_3H_9O_6P$
   Glycerophosphoric acid   A14.11

$C_3H_{10}N_2$
   1,2-Diaminopropane   A19.6

C_4

$C_4H_3D_3O_4$
   [2-$^2H_2$, 3-^2H]-succinic acid   D1.18

$C_4H_4Cl_2O_4$
   2,3-Dichlorosuccinic acid   A2.15

$C_4H_4O_5$
   2,3-Epoxysuccinic acid   A2.10

$C_4H_5BrO_4$
   Bromosuccinic acid   A4.1

$C_4H_5ClO_4$
   Chlorosuccinic acid   A4.1

$C_4H_5ClO_5$
   Chloromalic acids   A2.5, A2.9

$C_4H_5DO_4$
   [2-^2H]-succinic acid   D1.14

$C_4H_5DO_5$
   [3-^2H]-2-hydroxysuccinic acid   D1.13

$C_4H_5NO_3$
   3-Hydroxypyrrolidin-2,5-dione   A3.6

$C_4H_5NO_4$
   3,4-Dihydroxypyrrolidin-2,5-dione   A2.7

C_4H_6O
   But-3-yn-2-ol   A13.7

$C_4H_6O_3$
   2,3-Epoxybutyric acid   A24.1
   Propane-1,2-diol carbonate   A14.5

$C_4H_6O_4$
   Glycerol-1,2-carbonate   A15.7

$C_4H_6O_5$
   Malic acid   A1.25

Formulae Index

$C_4H_6O_2S$
    4-Methylthiete-1,1-dioxide — A22.17

$C_4H_6O_2S_2$
    1,2-Dithiolane-3-carboxylic acid — A18.1

$C_4H_6O_4S$
    Mercaptosuccinic acid — A2.8

$C_4H_7BrO_2$
    2-Bromobutyric acid — A8.12
    3-Bromo-2-methylpropionic acid — A34.9

C_4H_7Cl
    3-Chlorobut-1-ene — A1.20

$C_4H_7ClO_2$
    2-Chlorobutyric acid — A8.12
    3-Chlorobutyric acid — A22.7

C_4H_7DO
    [3-^{2}H]-butan-2-one — D2.19

C_4H_7NO
    3-Hydroxybutyronitrile — A6.4
    4-Methylazetidin-2-one — A19.12

$C_4H_7NO_2$
    3-Hydroxy-2-pyrrolidone — A9.5
    4-Hydroxy-2-pyrrolidone — A2.17
    4-Methyl-2-oxazolidone — A19.16

$C_4H_7NO_4$
    Aspartic acid — A4.6

$C_4H_8Br_2$
    1,3-Dibromobutane — A22.15

C_4H_8BrD
    [3-^{2}H]-2-bromobutane — D2.14

C_4H_8ClD
    [1-^{2}H]-1-chlorobutane — D1.3

$C_4H_8N_2O_3$
    Asparagine — A4.5
    Isoasparagine — A20.17
    Malamide — A1.25

C_4H_8O
    1,2-Epoxybutane — A13.13
    2,3-Epoxybutane — A13.8
    But-1-en-3-ol — A1.6

C_4H_8OS
    Allyl methyl sulphoxide — Z7.13
    Crotyl methyl sulphoxide — Z7.14

$C_4H_8O_2$
    3-Hydroxy-2-methylpropanol — A28.7

$C_4H_8O_2S$
    2-methylthietane-1,1-dioxide — A22.16

$C_4H_8O_2S_2$
    4,5-dihydroxy-1,2-dithiane — A2.3

$C_4H_8O_3$
    2-Hydroxybutyric acid — A1.22
    3-Hydroxybutyric acid — A6.4
    3-Hydroxy-2-methylpropionic acid — A28.12
    2-Methoxypropionic acid — A1.8
    Lactic acid, methyl ester — A1.9

$C_4H_8O_4$
    Threose — C1.11
    Erythrose — C1.12
    2,3-Dihydroxybutyric acids — A24.5, A24.7
    2,4-Dihydroxybutyric acid — A3.10
    Glycerotetrulose — C2.7

$C_4H_8O_5$
    2,3,4-Trihydroxybutyric acid — A3.12

C_4H_9Br
    2-Bromobutane — A1.11

C_4H_9BrO
    1-Bromobutan-2-ol — A1.17
    4-Bromobutan-2-ol — A1.13

C_4H_9Cl
    2-Chlorobutane — A1.11

C_4H_9ClO
    3-Chlorobutan-2-ol — A13.9

C_4H_9D
    [2-^{2}H]-butane — D2.15

C_4H_9DO
    [1-^{2}H]-butan-1-ol — D1.6
    [3-^{2}H]-butan-2-ol — D2.13

C_4H_9IO
    4-Iodobutan-2-ol — A1.13

C_4H_9N
    2,3-Dimethyl-ethyleneimine — A13.1

C_4H_9NO
    1-Aminobut-3-en-2-ol — A8.9

C_4H_9NOS
    Methylcysteine sulphoxide — Z7.4

$C_4H_9NO_2$
    2-Aminobutyric acid — A8.13
    3-Aminobutyric acid — A19.1
    3-Amino-2-methylpropionic acid — A31.7

$C_4H_9NO_3$
    Threonine — A8.16
    2-Amino-3-hydroxy-2-methylpropionic acid — A40.8
    4-Amino-2-hydroxybutyric acid — A9.4
    4-Amino-3-hydroxybutyric acid — A2.16
    Allothreonine — A24.2

C_4H_9OT
    [3-^{3}H]-butan-2-ol — D2.17

$C_4H_{10}DN$
    [1-^{2}H]-butylamine — D1.2

$C_4H_{10}N_2O_2$
    3-Hydroxy-2-methylpropionic acid hydrazide — A28.12
    2,4-Diaminobutyric acid — A9.3

$C_4H_{10}O$
    Butan-2-ol — A1.16

$C_4H_{10}OS$
    Methyl propyl sulphoxide — Z7.12

$C_4H_{10}O_2S_2$
    Dithiothreitol — A2.2

$C_4H_{10}O_2$
    Butane-1,2-diol — A1.21
    Butane-1,3-diol — A1.12
    Butane-2,3-diol — A3.7

$C_4H_{10}O_4$
    Threitol — C1.11
    Erythritol — C1.12

$C_4H_{10}S_2$
    Butane-2,3-dithiol — A13.14

Formulae Index

$C_4H_{11}NO$
  1-Aminobutan-2-ol — A8.10
  1-Aminobutan-3-ol — A6.1
  2-Aminobutan-1-ol — A8.19
  3-Aminobutan-2-ol — A13.4

$C_4H_{11}NOS$
  3-Methylsulphinylpropylamine — Z8.13

$C_4H_{11}NO_2$
  2-Aminobutane-1,4-diol — A4.2

$C_4H_{11}O_5P$
  1-Hydroxy-2-methoxypropylphosphonic acid — Y20.3

$C_4H_{12}N_2$
  2,3-Diaminobutane — A19.2

C_5

$C_5H_4O_4$
  Pentadiendioic acid — X3.1

$C_5H_6O_3$
  2-Hydroxy-2-methylbut-3-ynoic acid — A33.3

$C_5H_6O_4$
  Paraconic acid — T15.14
  2-Hydroxyglutaric acid lactone — A11.13
  Cyclopropane-1,2-dicarboxylic acid — A37.1

$C_5H_7BrO_4$
  2-Bromo-3-methylsuccinic acid — A28.10

$C_5H_7ClO_4$
  2-Chloroglutaric acid — A11.18

C_5H_7DO
  [3-^{2}H]-cyclopentanone — D2.10

$C_5H_7NO_2$
  3,4-Dehydroproline — A17.2

$C_5H_7NO_3$
  Pyroglutamic acid — A9.11
  4-Oxoproline — A17.8

$C_5H_7N_3O_2$ — A11.3
  5-Azido-4-hydroxypentanoic acid lactone

$C_5H_7N_3O_4$
  Azaserine — A4.4

C_5H_7NOS
  Goitrin — A8.11

C_5H_8
  3-Methylcyclobutene — A28.8
  Penta-1,3-diene — X2.1

$C_5H_8Cl_2O_2$
  Caldariomycin — Y16.9

C_5H_8INO
  5-iodomethyl-2-pyrrolidone — A9.13

$C_5H_8O_2$
  4-hydroxypentanoic acid lactone — A12.18
    A44.12, A44.13
  2-Methylcyclopropanecarboxylic acids

$C_5H_8O_2S$
  Thiophane-2-carboxylic acid — A18.2

$C_5H_8O_3$
  Tetrahydrofuran-3-carboxylic acid — A30.8
  4-Hydroxy-2-hydroxymethylbutyric acid 1,4-lactone — A30.7
  2-Hydroxyglutaric acid lactone, methyl ester — A11.13
  4,5-Dihydroxypentanoic acid 1,4-lactone — A11.12

$C_5H_8O_3S$
  2-Mercaptopropionic acid, S-acetyl — A12.10

$C_5H_8O_4$
  2-Methylsuccinic acid — A27.21

$C_5H_8O_4S$
  2-Thiomethylsuccinic acid — A2.8

$C_5H_8O_5$
  2-Hydroxy-2-methylsuccinic acid — A33.12
  2-Hydroxy-3-methylsuccinic acid — Y5.4
  Methoxysuccinic acid — A2.12
  Malic acid, methyl ester — A1.25
  2-Hydroxyglutaric acid — A11.13

$C_5H_8O_6$
  2,3-dihydroxyglutaric acid — Y4.11
  2,4-dihydroxyglutaric acid — A11.20

$C_5H_9BrO_2$
  2-Bromo-3-methylbutyric acid — A4.20

$C_5H_9ClO_2$
  2-Chloro-3-methylbutyric acid — A4.20

$C_5H_9F_3O$
  1,1,1-Trifluoro-2-ethoxypropane — A12.2

C_5H_9N
  2-Methylbutyronitrile — A29.11
  1-Azabicyclo-[3.1.0]-hexane — A17.1

C_5H_9NO
  4-Methyl-2-pyrrolidone — K35.8

C_5H_9NOS
  Sisaustricin — A8.18
  Thiophane-2-carboxylic acid, amide — A18.2

$C_5H_9NO_2$
  Proline — A11.14
  3-Hydroxy-2-piperidone — A11.9
  5-Hydroxy-2-piperidone — A11.6
  2-Amino-2-methylbut-3-enoic acid — A40.14

$C_5H_9NO_2S$
  Iberin — Z8.8

$C_5H_9NO_3$
  β-Carboxybutyramide — A34.11
  4-Hydroxypyrrolidine-2-carboxylic acids — A17.9, A17.11

$C_5H_9NO_3S$
  2-Oxo-4-pyrrolidyl methyl sulphone — A17.12

$C_5H_9NO_4$
  2-Amino-2-methylsuccinic acid — A40.4
  2-Amino-3-methylsuccinic acid — A28.11
  Glutamic acid — A9.7

$C_5H_9NO_5$
  β-Hydroxyglutamic acid — A2.13, A2.14
  γ-Hydroxyglutamic acid — A11.9

$C_5H_9N_3O_2$
  Pyroglutamic acid hydrazide — A9.11

C_5H_{10}
  1,2-Dimethylcyclopropane — A44.1

243

Formulae Index

$C_5H_{10}Br_2$
    1,4-Dibromo-2-methylbutane   A27.15
    2,4-Dibromopentane   A6.17

$C_5H_{10}Cl_2$
    1,4-Dichloro-2-methylbutane   A27.15

$C_5H_{10}N_2$
    3,4-Dimethylpyrazoline   A6.20
    2,5-Diazabicyclo[2.2.1]heptane   A17.14

$C_5H_{10}N_2O_2$
    3-Aminoproline   Y22.7
    Piperazic acid   A11.11
    Curcurbitine   A30.12

$C_5H_{10}N_2O_3$
    Glutamine   A9.1
    Isoglutamine   A9.8
    Methoxysuccinic acid, diamide   A2.12
    5-Hydroxypiperazic acid   A11.11

$C_5H_{10}O$
    3-Methoxybut-1-ene   A1.7
    2-Methylbutanal   A27.5
    2,3-Epoxy-2-methylbutane   A33.6
    Pent-3-en-2-ol   A12.11
    Pent-4-en-2-ol   A6.8
    2-Methyltetrahydrofuran   A6.16
    3-Methyltetrahydrofuran   A30.2

$C_5H_{10}O_2$
    3-Hydroxymethyltetrahydrofuran   A30.5
    2-Methylbutyric acid   A29.11
    4-Hydroxy-3-methylbutan-2-one   A28.16
    1,2-bis-(hydroxymethyl)-cyclopropane   A44.2

$C_5H_{10}O_2S$
    2-Mercaptoproprionic acid, S-methyl, methyl ester   A12.10

$C_5H_{10}O_3$
    2-Hydroxypentanoic acid   A6.5
    3-Hydroxypentanoic acid   A6.19
    4-Hydroxypentanoic acid   A6.2
    2-Hydroxy-2-methylbutyric acid   A33.8
    2-Hydroxy-3-methylbutyric acid   A24.11
    2-Hydroxybutyric acid, methyl ester   A1.22
    Lactic acid, ethyl ester   A1.9

$C_5H_{10}O_4$
    2,3-Dihydroxy-3-methylbutyric acid   A3.11
    2,3-Dihydroxy-2-methylbutyric acid   A33.5

$C_5H_{10}O_5$
    Lyxose   A11.17
    Arabinose   C1.8
    Xylose   C1.6
    Ribose   C1.9
    Threopentulose   C2.5
    Erythropentulose   C2.6

$C_5H_{11}Br$
    1-Bromo-2-methylbutane   A27.11

$C_5H_{11}BrO$
    1-Bromo-3-methoxybutane   A12.22

$C_5H_{11}Cl$
    1-Chloro-2-methylbutane   A27.11

$C_5H_{11}F$
    1-Fluoro-2-methylbutane   A27.11

$C_5H_{11}N$
    2-Methylpyrrolidine   A17.6

$C_5H_{11}NO$
    3-Hydroxypiperidine   A11.5
    2-Hydroxymethylpyrrolidine   A17.7

$C_5H_{11}NO_2$
    Valine   A4.19
    Norvaline   A10.15
    2-Amino-2-methylbutyric acid   A40.9
    4-Amino-2-methylbutyric acid   A34.5
    4-Amino-3-methylbutyric acid   A27.16

$C_5H_{11}NO_2S$
    Methionine   A8.17
    Penicillamine   A4.21

$C_5H_{11}NO_3$
    2-Amino-5-hydroxypentanoic acid   A9.6

$C_5H_{11}NO_3S$
    Methionine sulphoxides   Z8.14, Z8.18

$C_5H_{12}DN$
    [1-^{2}H]-1-amino-2,2-dimethylpropane   D1.12

$C_5H_{12}N_2$
    1,2-Diaminocyclopentane   A19.5

$C_5H_{12}N_2O_2$
    Ornithine   A11.15
    3,5-Diaminopentanoic acid   A9.10

$C_5H_{12}N_2O_2S$
    Methionine sulphoximines   Z8.15, Z8.17

$C_5H_{12}O$
    2-Methylbutan-1-ol   A27.10
    3-Methylbutan-2-ol   A58.13
    2-Methoxybutane   A1.16
    Pentan-2-ol   A6.7

$C_5H_{12}O_2$
    Pentane-1,2-diol   A6.6
    Pentane-1,4-diol   A6.3
    Pentane-2,4-diol   A6.9
    2-Methylbutane-1,2-diol   A33.7
    2-Methylbutane-1,4-diol   A27.20
    2-Methylbutane-2,3-diol   A12.9

$C_5H_{12}O_3$
    Pentane-1,2,3-triol   A11.7
    2-Methylbutane-1,2,3-triol   A33.1
    3-Methylbutane-1-2-3-triol   A3.14
    2-Methylbutane-1,2,4-triol   A33.13

$C_5H_{12}O_4$
    Pentane-1,2,3,5-tetraol   Y4.12

$C_5H_{12}O_5$
    Xylitol   C1.6
    Ribitol   C1.9
    Lyxitol   C1.7, C1.8

$C_5H_{13}N$
    2-Amino-3-methylbutane   A4.13

$C_5H_{13}NO$
    Valinol,   A4.18
    2-Amino-2-methylbutan-1-ol   A27.18
    3-Amino-2-methylbutan-2-ol   A4.14
    4-Amino-2-methylbutan-1-ol   A34.10

C_6

$C_6H_6O_3S$
2-Hydroxy-2-(2-thienyl)acetic acid — A21.16
2-Hydroxy-2-(3-thienyl) acetic acid — A21.13

$C_6H_6O_4$
3-Methylenecyclopropane-1,2-dicarboxylic acid — A37.18

$C_6H_6O_6$
Isocitric lactone — A30.6
Alloisocitric lactone — A30.3

$C_6H_7ClO_2$
2-Chlorocyclohexa-1,3-dien-5,6-diol — A16.3

$C_6H_8O_2$
Sorbinol — Y14.9
Cyclohexa-1,3-dien-5,6-diol — A16.8

$C_6H_8O_3$
3-Oxocyclopentanecarboxylic acid — A36.1
2,3-Dimethylsuccinic acid anhydride — A31.8

$C_6H_8O_4$
2-Hydroxy-2-methylglutaric acid lactone — A11.8
3,4-Dihydroxycyclohexane-1,2dione — A16.22
Cyclobutane-1,2-dicarboxylic acid — A37.5

$C_6H_8O_4S$
Thiophane-2,5-dicarboxylic acid — A18.3

$C_6H_8O_4S_2$
Dithiolane-3,5-dicarboxylic acid — A18.6

$C_6H_8O_4Se$
Selenophane-2,5-dicarboxylic acid — A18.4

$C_6H_8O_6$
Acetoxysuccinic acid — A1.25

$C_6H_8O_7$
Isocitric acid — A30.1

$C_6H_8O_8$
Hydroxycitric acid — A30.9
Allo-hydroxycitric acid — A30.10

$C_6H_9NO_2$
4-Methyleneproline — A17.4
2-(Methylenecyclopropyl)-glycine — A27.1
Baikiain — A10.9

$C_6H_9NO_3$
4-Oxopipecolic acid — A10.14

$C_6H_9N_3O_2$
Histidine — A20.16

$C_6H_9N_3O_2S$
2-Thiolhistidine — A20.20

$C_6H_9O_2$
3,4-Methyleneprolines — A30.14, A30.15

C_6H_{10}
2,3-Dimethylmethylenecyclopropane — A37.19

$C_6H_{10}Br_2$
1,2-Dibromocyclohexane — A18.15

$C_6H_{10}N_2O_2$
2,6-Diaminohex-4-ynoic acid — A10.1

$C_6H_{10}N_4O_2$
Viomycidine — Y22.3

$C_6H_{10}O$
3-Methylcyclopentanone — A36.6
Hex-1-yn-3-ol — A6.15
Hexa-3,4-dien-1-ol — X1.7
3-Methylpent-1-yn-3-ol — A33.4
4-Methylpent-1-yn-3-ol — A8.3

$C_6H_{10}O_2$
2-Methylpent-4-enoic acid — A34.7
3-Acetoxybut-1-ene — A1.6
Cyclohex-1-en-3,4-diol — A16.18
2-Hydroxycyclohexanone — A16.9
2,6-dioxabicyclo[3.3.0]-octane — X11.5
2,3-Dimethylcyclopropanecarboxylic acid — A44.5
5-Hydroxyhexanoic acid lactone — Y14.10

$C_6H_{10}O_3$
2-Methyl-4-oxopentanoic acid — A28.3
3-Methyl-4-oxopentanoic acid — A34.14
3,4-Dihydroxy-4-methylpentanoic acid γ-lactone — Y1.6
3,5-Dihydroxy-3-methylpentanoic acid lactone — A33.11
Glyceraldehyde, isopropylidene deriv. — A1.1
4,4-Dimethyl-3-hydroxy-γ-butyrolactone — A8.5
4,5-Dihydroxy-4-methylbutyric acid 1,5-lactone — A33.14
2,3-Epoxy-2-methylbutyric acid, methyl ester — A33.10

$C_6H_{10}O_4$
Isomannide — X11.4
2,3-Dimethylsuccinic acid — A31.8
2,3,4-trihydroxycyclo-hexanones — A16.20, A16.21
2-Ethylsuccinic acid — A28.9
3-Carbomethoxybutyric acid — A34.8
2-Methylglutaric acid — A32.12
Methyl acetolactate — A33.2

$C_6H_{10}O_4S$
Thiodilactic acid — A1.10
2-Thioethylsuccinic acid — A28.13

$C_6H_{10}O_4S_2$
2,5-Dimercaptoadipic acid — A18.7

$C_6H_{10}O_5$
2-Deoxygluconolactone — A26.8
Dilactic acid — A1.5
2,3,4,5-Tetrahydroxycyclohexanone — A16.10
Methyl hydrogen 3-hydroxyglutarate — A6.18
3-Hydroxyadipic acid — A25.8

$C_6H_{10}O_6$
3,4-Dihydroxyadipic acid — A3.3
Chitaric acid — Y20.7
Dimethyl tartrate — A2.6
2,3,4,5,6-Pentahydroxycyclohexanone — A16.6
Dimethoxysuccinic acid — A3.8

$C_6H_{10}O_7$
2,3,5-Trihydroxyadipic acid — A16.12

$C_6H_{10}S_2$
Dithiabicyclo[3.3.0]-octane — X11.6

$C_6H_{11}BrO_2$
2-Bromo-3-methylpentanoic acid — A27.7
2-Bromo-3,5-dimethylbutyric acid — A24.9

$C_6H_{11}ClO_2$
2-Chloro-3-methylpentanoic acid — A27.7
2-Chloro-3,3-dimethylbutyric acid — A24.9

Formulae Index

$C_6H_{11}DO$
    [5-^{2}H]-hexan-2-one   D1.19
$C_6H_{11}DO_2$
    [1-^{2}H]-Butyl acetate   D1.1
$C_6H_{11}F_3O$
    1,1,1-Trifluoro-3,3-dimethylbutan-2-o   A12.5
$C_6H_{11}NO$
    5-Methyl-2-piperidone   A26.16
$C_6H_{11}NOS_2$
    Sulphorophan   Z8.8
$C_6H_{11}NO_2$
    3-Methylprolines   A27.3, A27.4
    Pipecolic acid   A10.7
    Nipecotinic acid   A30.13
    2-Aminohex-4-enoic acid   A10.12
    Hygrinic acid   A11.14
$C_6H_{11}NO_2S$
    4-Thiomethylpyrrolidine-2-carboxylic acid   A17.10
    2,2-Dimethylthiazolidine-4-carboxylic acid   Y29.3
$C_6H_{11}NO_3$
    4-Hydroxypipecolic acids   A10.10, A10.11
    4-Hydroxymethylprolines   A17.3, A17.5
    4-Hydroxyproline, N-methyl   A17.11
$C_6H_{11}NO_3S$
    Cycloalliin   Z7.16
$C_6H_{11}NO_4$
    2-Aminoadipic acid   A25.13
$C_6H_{11}O_6$
    Cellobiose   C4.9
C_6H_{12}
    2-Ethyl-1-methylcyclopropanes   A44.8, A44.11
    3-Methylpent-1-ene   A27.19
$C_6H_{12}N_2$
    2,6-Diazabicyclo-[3.3.0]-octane   X11.9
$C_6H_{12}N_2O_4S$
    Lanthionine   A4.12
$C_6H_{12}N_2O_4S_2$
    Cystine   A4.16
$C_6H_{12}N_4O_2$
    3-Guanidinoproline   Y22.4
$C_6H_{12}O$
    4-Methoxypent-1-ene   A6.8
    3-Methylpentanal   A29.9
    3-Methylpentan-2-one   A29.12
    Hex-1-en-3-ol   A6.11
    Hex-5-en-2-ol   A58.12
    2,5-Dimethyltetrahydrofuran   A18.14
    2,3-Dimethylbutanal   A29.4
$C_6H_{12}O_2$
    Cyclohexane-1,2-diol   A16.13
    2-Methylpentanoic acid   A34.4
    3-Methylpentanoic acid   A61.21
    2,3-Dimethylbutyric acid   A61.1
$C_6H_{12}O_3$
    2-Hydroxy-2-methylbutyric acid, methyl ester   A33.8
    2-Hydroxyhexanoic acid   A1.24
    3-Hydroxyhexanoic acid   A6.13
    3-Hydroxy-2-methylpentanoic acid   A34.3
    3-Hydroxy-4-methylpentanoic acid   A8.2
    O-Methyl lactic acid, ethyl ester   A1.8
    1,2-isopropylideneglycerol   A14.7
    2-Hydroxyethyl-3-hydroxytetrahydropyran   Y15.3
    4-Methoxypentanoic acid   A12.18
    2,2-Dimethyl-3-hydroxybutyric acid   A12.14
    Cyclohexane-1,2,3-triol   A16.14
$C_6H_{12}O_4$
    Cyclohexane-1,2,3,4-tetraols   A16.15, A16.16, A16.19
    Cyclohexane-1,2,3,5-tetraols   A26.1, A26.6
    2,3-Dihydroxy-2-methylpentanoic acid   A34.16
    2,3-Dihydroxy-3-methylbutyric acid methyl ester   A3.11
    2,5-Dihydroxy-3-methylpentanoic acid   T31.5
$C_6H_{12}O_5$
    Cyclohexane-1,2,3,4,5-pentaols   A16.4, A16.5, A16.7, A16.11
$C_6H_{12}O_6$
    Glucose   A26.9
    Ribohexulose   C2.4
    Arabinohexulose   C2.3
    Xylohexulose   C2.2
    Lyxohexulose   C2.1
    Talose   C1.2
$C_6H_{13}NO_5$
    2-Amino-2-deoxyglucopyranose   C4.1
    1-Aminobutan-2-ol, oxalate   A8.10
    2-Aminobutan-1-ol, oxalate   A8.19
$C_6H_{14}N_2$
    1,2-Diaminocyclohexane   A19.1
$C_6H_{14}N_2O_2$
    β-Lysine   A11.16
    Lysine   A10.2
$C_6H_{14}N_2O_3$
    γ-Hydroxylysine   A10.3
$C_6H_{14}N_4O_2$
    Arginine   A11.10
$C_6H_{14}N_4O_3$
    3-Hydroxyadipic acid dihydrazide   A25.8
$C_6H_{14}O$
    2-Methylpentan-1-ol   A60.13
    2-Methylpentan-3-ol   A8.6
    3-Methylpentan-1-ol   A27.14
    4-Methylpentan-2-ol   A58.21
    2-Methoxypentane   A6.7
    2,3-Dimethylbutan-1-ol   A29.3
    Hexan-2-ol   A18.18
    Hexan-3-ol   A6.12
    Pinacolyl alcohol   A12.4
$C_6H_{14}O_2$
    2-Methylpentane-2,3,-diol   A8.7
    4-Methylpentane-1,3-diol   A8.1

Hexane-1,5-diol	A18.17	Exo-dehydronorborneol	**A46.3**
Hexane-3,4-diol	X11.8	Norcamphor	**A36.11**
2,3-Dimethylbutane-1,4-diol	A31.4	$C_7H_{10}O_2$	
$C_6H_{14}O_3$		Cyclohex-3-ene-1-carboxylic acid	A26.11
2-Methylpentane-1,2,3-triol	A34.15	4-Methylcyclohexane-1,2-dione	T58.4
2-Methylpentane-1,2,5-triol	A11.4	Cyclopent-2-eneacetic acid	A36.9
$C_6H_{14}O_4$		$C_7H_{10}O_3$	
2,3-Dimethoxybutane-1,4-diol	A3.9	3-Oxocyclopentaneacetic acid	A36.10
Hexane-1,2,5,6-tetraol	A11.2	2-Methyl-3-oxocyclopentanecarboxylic acid	A36.6
$C_6H_{14}O_6$			
Mannitol	A11.1	$C_7H_{10}O_4$	
Gulitol	C1.1	Caronic acid	A35.7
Talitol	C1.2, C1.4	Terebic acid	A35.1
Iditol	C1.5	4,5-Dihydroxycyclohexane-1-carboxylic acid	A3.2
Allitol	C1.10		
$C_6H_{15}N$		Cyclopentane-1,2-dicarboxylic acid	A37.2
Pinacolylamine	A24.8	2-Hydroxyglutaric acid lactone, ethyl ester	A11.13
$C_6H_{15}NO$			
Leucinol	A20.4	Isopilopic acid	A28.18
tert-Leucinol	A24.12	$C_7H_{10}O_5$	
$C_6H_{15}OP$		Shikimic acid	A26.7
Ethylmethylpropylphosphine oxide	Z6.5	2,5-Dioxo-5-methoxyhexanoic acid	T17.5
$C_6H_{16}N_2$		$C_7H_{10}O_6$	
1,2-Diamino-4-methylpentane	A20.9	Butane-1,2,4-tricarboxylic acid	A26.10
$C_6H_{16}N_2O_2$		Dehydroquinic acid	A26.3
2,3-Dimethoxybutane-1,4-diamine	A3.4	Butane-1,1,2-tricarboxylic acid	A28.14
C_7		$C_7H_{10}O_7$	
$C_7H_6O_4$		Homocitric acid	A26.5
Terreic acid	Y19.7	$C_7H_{11}Cl$	
C_7H_7DO		3-Chloro-5-methylcyclohexene	A38.13, A38.20
[1-^{2}H]-phenylmethanol	D1.7		
C_7H_7IOS		$C_7H_{11}ClO$	
Methyl-p-iodophenyl sulphoxide	Z7.9	2-Chloro-5-methylcyclohexanone	A38.7
C_7H_8ClNOS		$C_7H_{11}NO_2$	
N-Methyl-S-phenyl-sulphonimidoyl chloride	Z8.11	3-Ethylglutaric acid mononitrile	T58.7
		2-Ethyl-3-methylsuccinimide	A31.12
C_7H_8DN		$C_7H_{11}NO_3$	
[1-^{2}H]-phenylmethylamine	D1.8	Ecgoninic acid	A9.12
C_7H_8O		$C_7H_{11}NO_4$	
Dehydronorcamphor	A46.2	γ-Ethylideneglutamic acid	A10.13
C_7H_8OS		$C_7H_{11}NO_5$	
Methyl phenyl sulphoxide	Z8.5	N-Acetyglutamic acid	A9.7
$C_7H_8O_3$		$C_7H_{11}NO_6$	
2-Methylene-3-oxocyclopentanecarboxylic acid	A36.2	2-Aminobutane-1,2,4-tricarboxylic acid	A40.1
		$C_7H_{11}N_3O_2$	
$C_7H_8O_4$		3-Methylhistidine	A20.16
Epoxydone	Y19.5	C_7H_{12}	
Terremutin	Y19.6	3-Ethylcylopentene	A36.12
$C_7H_8O_6$		4-Methylcyclohexene	A26.13
Homocitric acid lactone	A26.5	$C_7H_{12}Cl_2O_4$	
C_7H_9NO		Methyl-4,6-dichloro-4,6-dideoxy-glucopyranoside	C4.3
1-(3-pyridyl)-ethanol	A25.16		
C_7H_9NOS		$C_7H_{12}O$	
N-Methylbenzenesulphinimide	Z8.6	Endo-norborneol	A46.5
$C_7H_9N_3O_2$		Exo-norborneol	A46.6
Spinacin	A20.22	2-Methylcyclohexanone	A38.12
$C_7H_{10}O$		3-Methylcyclohexanone	A38.4
5-Methylcyclohex-2-ene-1-one	A38.9	3-Methoxycyclohexene	A18.9

Formulae Index

    4-Hydroxymethylcyclohexene   A26.14
    5-Methylcyclohex-2-en-1-ols   A38.14, A38.15

$C_7H_{12}O_2$
    4-Hydroxy-4-methylhexanoic acid
      lactone   T3.9
    5-Hydroxy-3-methylhexanoic acid
      lactone   Y21.5, Y21.10
    2-Methylcyclopentanecarboxylic acid   A36.5
    2-Methylpent-4-enoic acid, methyl ester   A34.7

$C_7H_{12}O_3$
    3,4-Epoxy-4-methylpentanoic acid, methyl
      ester   A8.4
    3-Hydroxycyclohexanecarboxylic acid   A38.11

$C_7H_{12}O_4$
    3-Acetoxy-2-methylbutyric acid   A32.17
    2-Acetoxy-2-methylbutyric acid   A33.8
    2-Ethyl-2-methylsuccinic acid   A55.5
    2-Ethyl-3-methylsuccinic
      acids   A31.12, A31.20
    Lactic acid, *O*-acetyl, ethyl ester   A1.9
    2-Propylsuccinic acid   A28.9
    2-Ethylglutaric acid   A32.11
    2,4-Dimethylglutaric acid   A32.13
    2-Isopropylsuccinic acid   A28.5
    2,5-Dimethylglutaric acid   A18.5
    3-Methyladipic acid   A26.15
    2-Methyladipic acid   A38.3
    Methyl hydrogen 3-methylglutarate   A32.5

$C_7H_{12}O_4S$
    2-Thiopropylsuccinic acid   A28.13

$C_7H_{12}O_5$
    2-Methoxyadipic acid   A18.10
    3-Methoxyadipic acid   A25.8
    Methoxysuccinic acid, dimethyl ester   A2.12
    5-Deoxyquinic acid   A26.4

$C_7H_{12}O_6$
    Quinic acid   A26.2

$C_7H_{13}N$
    2-Aminonorbornane   A47.10

$C_7H_{13}NO$
    2-Methylcyclopentanecarboxylic acid,
      amide   A36.5
    1,5-Dimethyl-2-piperidone   A26.16
    5-Ethyl-5-methyl-2-pyrrolidone   A40.5

$C_7H_{13}NO_2$
    2-Aminocyclohexanecarboxylic
      acids   A37.11, A35.15

$C_7H_{13}NO_3$
    5-Hydroxymethyl-5-(2-hydroxyethyl)-2-
      pyrrolidone   A40.2
    4-Acetamido-3-methylbutyric acid   A27.16

$C_7H_{13}NOS_2$
    Alyssin   Z8.8

C_7H_{14}
    4-Methylhex-1-ene   A29.16
    4-Methylhex-2-ene   A29.10
    1,2-Dimethylcyclopentane   A36.4
    2,3-Dimethylpent-1-ene   A29.7

$C_7H_{14}Br_2$
    1,4-Dibromo-2-propylbutane   A29.14

$C_7H_{14}N_2O$
    2-Aminocyclohexanecarboxylic acid,
      amide   A37.11

$C_7H_{14}N_2O_3$
    Theanine   A9.2

$C_7H_{14}N_2O_4S_2$
    Djenkolic acid   A4.17

$C_7H_{14}O$
    2-Methylcyclohexanols   A38.17, A38.18
    3-Methylcyclohexanol   A38.10
    Hept-1-en-3-ol   A58.11
    4-Methylhexan-3-one   A29.12
    2-Ethyl-3-methylbutanal   A29.2

$C_7H_{14}O_2$
    2-Methylhexanoic acid   A61.8
    3-Methylhexanoic acid   A61.9
    4-Methylhexanoic acid   A61.10
    3,4-Dimethylpentanoic acid   A61.2
    3-Hydroxymethylcyclohexanol   A38.6
    2-Methoxy-4-methylpentan-3-one   A12.1
    2-Ethylpentanoic acid   A29.17

$C_7H_{14}O_3$
    3-Acetoxy-2-methylbutan-2-ol   A12.9
    2-Hydroxyheptanoic acid   A24.15
    2-Hydroxy-2,3,3-trimethylbutyric acid   X1.4
    1-Hydroxymethylcyclohexane-1,2-diol   Y19.11

$C_7H_{14}O_3S$
    2-Hydroxy-2-methyl-3-thiomethylbutyric
      acid, methyl ester   A33.9

$C_7H_{14}O_4$
    1-Hydroxymethylcyclohexane-
      1,2,3-triol   Y19.8
    Trachelanthic acid   K24.3
    Viridifloric acid   K24.5
    Oleandrose   Y25.1
    2,3-Isopropylidenethreitol   A13.12

$C_7H_{14}O_6$
    Pinitol   A16.2

$C_7H_{15}Br$
    1-Bromo-2-methylhexane   A62.8
    1-Bromo-3-methylhexane   A62.9
    1-Bromo-4-methylhexane   A62.10
    1-Bromo-5-methylhexane   A62.11
    1-Bromo-2,3-dimethylpentane   A62.2

$C_7H_{15}N$
    1,3-Dimethylpiperidine   A26.17
    2,6-Dimethylpiperidine   K19.6

$C_7H_{15}NO$
    2-Methylaminocyclohexanol   A3.15
    3-Methoxycyclohexylamine   A18.8

$C_7H_{15}NO_2$
    2-Aminoheptanoic acid   A24.14

C_7H_{16}
    3-Methylhexane   A29.13
    2,3-Dimethylpentane   A29.6

$C_7H_{16}NO_3$
    Carnitine   A2.16

Formulae Index

$C_7H_{16}N_2$
2-Aminomethylcyclohexy lamines ... A37.12, A37.14

$C_7H_{16}N_4O_2$
Blastidic acid ... A9.14

$C_7H_{16}N_4O_3$
γ-Hydroxyhomoarginine ... A10.4

$C_7H_{16}O$
2-Methylhexan-1-ol ... A60.5
3-Methylhexan-1-ol ... A60.6
4-Methylhexan-1-ol ... A60.7
2-Methylhexan-3-ol ... A58.16
5-Methylhexan-3-ol ... A58.17
Heptan-2-ol ... A22.12
Heptan-3-ol ... A58.10
2-Ethylpentan-1-ol ... A29.16
2-Ethyl-3-methylbutan-1-ol ... A29.1
2,3-Dimethylpentan-2-ol ... A29.8
3,4-Dimethylpentan-1-ol ... A59.13

$C_7H_{16}O_2$
2-Propylbutane-1,4-diol ... A29.15
2-Ethyl-2-methylbutane-1,4-diol ... A55.6
2-Ethyl-3-methylbutane-1,3-diols ... A31.16, A31.21

$C_7H_{16}O_3$
2-Ethyl-3-hydroxymethylbutane-1,4-diols ... A28.17

$C_7H_{17}N$
1-Amino-2-methylhexane ... A60.16

$C_7H_{18}NO_6P$
Glycerylphosphorylcholine ... A14.12

C_8

$C_8H_4N_2O_8S_2$
2,2′-Dinitro-3,3′-bithienyl-4,4′-dicarboxylic acid ... X5.12

C_8H_6DOT
[$^2H_1,^3H_1$]acetophenone ... D2.5

$C_8H_7BrO_2$
2-Bromo-2-phenylacetic acid ... A23.10

$C_8H_7ClO_2$
2-Chloro-2-phenylacetic acid ... A23.10

$C_8H_7F_3O$
1-Phenyl-2,2,2-trifluoroethanol ... A22.18

C_8H_8DOT
[$^2H_1,^3H_1$]-1-phenylethanol ... D2.8

C_8H_8O
Phenyloxiran ... A22.9

$C_8H_8O_3$
Mandelic acid ... A21.12

$C_8H_8O_4S$
2-(2-thienyl)-succinic acid ... A50.11

C_8H_8S
Phenylthiiran ... A22.8

C_8H_9Br
1-Bromo-1-phenylethane ... A22.3

C_8H_9Cl
1-Chloro-1-phenylethane ... A22.3

C_8H_9ClO
2-Chloro-1-phenylethanol ... A22.5

$C_8H_9NO_2$
Phenylglycine ... A23.14
2-Amino-2-phenylacetic acid ... A19.8

$C_8H_{10}ClNOS$
Methyl-p-tolyl sulphoximine, N-chloro- ... Z8.2

$C_8H_{10}O$
1-Phenylethanol ... A22.6
Bicyclo-[2.2.2]-oct-5-ene-2-one ... A46.14

$C_8H_{10}OS$
Methyl p-tolylsulphoxide ... Z7.11
Ethylphenylsulphoxide ... Z7.2

$C_8H_{10}O_2$
1-Phenylethane-1,2-diol ... A22.13
Norbornenecarboxylic acids ... A47.11

$C_8H_{10}O_3$
4,5-Dihydroxyocta-2,6-dienoic acid 1,5-lactone ... Y15.6
Terrein ... Y16.8

$C_8H_{10}O_4$
Anhydromonocrotalic acid ... K23.5
3-Methylenecyclopropane-1,2-dicarboxylic acid, dimethyl ester ... A37.18

$C_8H_{11}ClO_6$
Acetylchloromalic acid ... A2.4

$C_8H_{11}N$
1-Phenylethylamine ... A19.14

$C_8H_{11}NO$
2-Amino-1-phenylethanol ... A23.2
2-Amino-2-phenylethanol ... A19.10
1-(o-aminophenyl)-ethanol ... A22.6

$C_8H_{11}NO_2$
2-Amino-1-(3-hydroxyphenyl) ethanol ... K1.1

$C_8H_{11}NO_3$
Noradrenaline ... A22.14

$C_8H_{11}NOS$
N,S-Dimethyl-S-phenylsulphoximine ... Z8.10
Methyl-p-tolylsulphoximine ... Z8.2

$C_8H_{11}NO_4$
trans-Dihydrohaematinimide ... Y23.3

$C_8H_{11}OP$
tert-butylmethylphenylphosphine oxide ... Z4.14

$C_8H_{11}OPS$
S-Methyl-methylphenylphosphinothioate ... Z5.14

C_8H_{12}
Cycloocta-1,2-diene ... X2.2

$C_8H_{12}N_2$
1,2-Diamino-1-phenylethane ... A19.7

$C_8H_{12}N_2O_6$
Uridine ... C4.6

$C_8H_{12}N_2O_3S$
6-Aminopenicillanic acid ... Y29.2

$C_8H_{12}O$
2,2-Dimethylhexa-3,4-dienal ... X1.10
Bicyclo-[2.2.2]-oct-5-en-2-ol ... A46.15
Bicyclo-[3.2.1]-octan-2-one ... A46.11

Formulae Index

Bicyclo-[3.3.0]-octan-3-one	A37.3
3-Isopropenylcyclopentanone	T9.1
1-Methylnorbornan-2-one	T10.4
3-Methylcyclopentenyl methyl ketone	A36.7

$C_8H_{12}O_2$
Norbornane-2-carboxylic acids	A47.10

$C_8H_{12}O_3$
2-Methoxycarbonyl-3-methylcyclopentanone	T15.5
4-Methyl-3-oxocyclohexanecarboxylic acid	T17.10
2,3-Diethylsuccinic anhydride	A45.5

$C_8H_{12}O_4$
trans-norcaryophyllenic acid	T28.8
Dihydroanhydromonocrotalic acid lactone	K23.3
Cyclohexane-1,2-dicarboxylic acid	A37.6
3-Methylcyclopropane-1,2-dicarboxylic acid, dimethyl ester	A37.20
trans-umbellularic acid	A54.7
cis-umbellularic acid	A54.8
cis-trans-nepetic acid	T14.8
cis-cis-nepetic acid	T13.8
trans-cis-nepetic acid	T13.4
cis-Homocaronic acid	A35.8
trans-Homocaronic acid	A35.3

$C_8H_{12}O_6$
Dihydrohaematinic acid	T22.6
3-Carboxy-2,2-dimethylglutaric acid	T8.9
Methyl hydrogen 3-acetoxyglutarate	A6.14
Ethyl hydrogen 2-acetoxysuccinate	A1.23

$C_8H_{12}O_7$
Dimethyl acetyltartrate	A2.1

$C_8H_{13}Br$
4-Methylcyclohexylidenebromomethane	X3.10

$C_8H_{13}Cl$
1-Chloro-3,4,4-trimethylpenta-1,2-diene	X1.2

$C_8H_{13}D$
3-[^{2}H]-1-Isopropenylcyclopentane	D2.9

$C_8H_{13}N$
Isoheliotridene	K22.4

$C_8H_{13}NO$
Retronecanone	K22.9
Supinidine	K22.8
Tropan-2-one	K28.7
1-Acetyl-3-methyl-1,2,3,4-tetrahydropyridine	A26.18

$C_8H_{13}NO_2$
Trachelanthamidinic acid	K22.13
Isoretranecanolic acid	K22.10
Heliotridine	K22.8
Retronecine	K22.6

$C_8H_{13}NO_3$
Otonecine	K24.2

$C_8H_{13}NO_4$
Tropinic acid	A9.15

$C_8H_{13}N_3O_2S$
Ergothionine	A20.20

$C_8H_{13}N_5O_2$
Viocidic acid	Y22.2

C_8H_{14}
Cyclooctene	X2.9
Methylnorbornene	A47.11

$C_8H_{14}O$
Cyclooct-1-en-3-ol	X2.5
3-Methylcycloheptanone	A38.8
4-Methylcycloheptanone	A38.5
2,4-Dimethylcyclohexanones	A39.8, Y21.8
2,5-Dimethylcyclohexanone	A39.12
2,2,3-trimethylcyclopentanone	A36.17
3,4,4-trimethylpent-1-yn-3-ol	X1.3
3-Isopropenylcyclopentanols	T9.2, T9.3
Oct-3-yn-2-ol	A12.21
Hydroxymethylnorbonene	A47.11
2-Methylnorbornan-2-ol	A46.1
Bicyclo-[3.2.1]-octan-2-ol	A46.8, A46.12
Bicyclo-[2.2.2]-octan-2-ol	A46.13
3,4-dimethylcyclohexanone	A31.13

$C_8H_{14}O_2$
2-Hydroxymethyl-1-methylcyclohex-1-en-4-ol	T40.6
2,3-Diethyl-4-hydroxybutyric acid lactone	A45.12

$C_8H_{14}O_4$
2,3-Diethylsuccinic acid	A45.5
2-Butylsuccinic acid	A28.9
2-Isopropyl-2-methylsuccinic acid	A55.1
2-Isopropylglutaric acid	A28.6
3-Methylheptanedioic acid	Y27.4
2,5-Dimethyladipic acid	A32.7
3,4-Dimethyladipic acid	A31.14
2,2-Dimethyl-3-acetoxybutyric acid	A12.14
Ethyl hydrogen 3-methylglutarate	A32.5
2,3-Diacetoxybutane	A3.7

$C_8H_{14}O_6$
Dimethoxysuccinic acid, dimethyl ester	A3.8
Monocrotalic acid	K23.7

$C_8H_{14}O_2S_2$
α-Lipoic acid	A18.13

$C_8H_{14}O_4S$
2-Thiobutylsuccinic acid	A28.13
3-Mercaptooctanedioic acid	A18.12
Thiodiisobutyric acid	A34.12

$C_8H_{14}O_4S_2$
Dithiodiisobutyric acid	A34.13

$C_8H_{15}DO$
[1-^{2}H]-(cis-4-methylcyclohexyl)-methanol	D1.11

$C_8H_{15}N$
Heliotridane	K22.10
Pseudoheliotridane	K22.13

$C_8H_{15}NO$
Retronecanol	K22.7
Isoretronecanol	K22.10
Lindelofidine	K22.10
Trachelanthamidine	K22.13

Formulae Index

Isopelletierine	K19.9		Hygroline	K19.2
Hygrine	K19.1		Pseudoconhydrine	K19.4
Laburnine	K22.13		Pseudohygroline	K19.2
$C_8H_{15}NO_2$			Allosedridine	K19.7
Macronecine	K22.12		Conhydrine	K18.4
Platynecine	K22.7		2-Dimethylaminocyclohexanol	A3.15
Dihydroxyheliotridane	K22.14	$C_8H_{17}NO_2$		
2-Amino-2-cyclohexylacetic acid	A23.15		2-Aminooctanoic acid	A24.14
$C_8H_{15}NO_4$			N,γ-Dimethylalloisoleucine	A29.5
N-Carbethoxyethylalanine	A1.18	$C_8H_{17}NO_3$		
$C_8H_{15}NO_2S_2$			Desosamine	Y25.3
2-Aza-7-thiaspiro-[4.4]-nonane, methanesulphonyl derivative	X4.6	$C_8H_{17}NO_4S$		
			Ethyl-1-thioglucofuranoside	C4.2
$C_8H_{15}N_3O_2S$		C_8H_{18}		
2-Aminooctahydroquinazoline sulphates	A37.16, A37.8		3-Methylheptane	A59.11
			2,4-Dimethylhexane	A59.7
C_8H_{16}		$C_8H_{18}O$		
5-Methylhept-1-ene	A31.2		6-Methylheptan-2-ol	A12.8
$C_8H_{16}ClN$			Octan-2-ol	A12.16
2-Aminomethylnorbornane hydrochlorides	A46.9, A46.10		Octan-3-ol	A18.18
			Octan-4-ol	A58.24
$C_8H_{16}N_2$			2-Methylheptan-1-ol	A32.3
Decahydroquinazolines	A37.13, A37.17		2-Methylheptan-3-ol	A58.14
			2-Methylheptan-4-ol	A58.15
$C_8H_{16}N_2O_4$			3,3-Dimethyl-2-ethoxybutane	A12.4
Dimethoxysuccinic acid bis-methylamide	A3.8	$C_8H_{18}O_2$		
			2,3-Diethylbutane-2,3-diol	A45.4
$C_8H_{16}O$			2-Isopropyl-2-methylbutane-1,4-diol	A55.4
1-Methyl-1-(1-ethoxyethyl)-cyclopropane	A12.3	$C_8H_{19}N$		
1-Methoxy-3-methylcyclohexanone	A38.10		1-Amino-3-methylheptane	A60.15
1-Cyclohexylethanol	A22.2	$C_8H_{20}N_2O_2$		
Oct-2-en-4-ol	A58.25		1,4-(Dimethylamino) butane-2,3-diol	A2.11
2,3-Diethyltetrahydrofuran	A18.14	$C_8H_{20}O$		
$C_8H_{16}O_2$			3-Methylheptan-1-ol	A60.2
1,2-Bis-hydroxymethylcyclohexane	A37.10		4-Methylheptan-1-ol	A60.3
2-Methylheptanoic acid	A32.4		5-Methylheptan-1-ol	A60.4
Cyclooctane-1,2-diol	A3.16			
3-Methylheptanoic acid	A61.5		**C_9**	
4-Methylheptanoic acid	A61.6	C_9H_8O		
5-Methylheptanoic acid	A61.7		Marasin	X1.1
$C_8H_{16}O_3$		$C_9H_8O_2$		
2-Hydroxyoctanoic acid	A1.24		3-Methylphthalide	Y14.5
$C_8H_{16}O_4$		$C_9H_8O_3$		
Heliotrinic acid	K24.3		Dihydrocoumarilic acid	Y1.12
$C_8H_{16}O_5$			β-Phenylglycidic acid	A19.19
Lasiocarpic acid	K24.9	$C_9H_9BrO_2$		
$C_8H_{17}Br$			2-Bromo-2-phenylpropionic acid	A51.3
2-Bromooctane	A12.15	$C_9H_9Br_2NO_3$		
1-Bromo-2-methylheptane	A62.5		Aeroplysinin I	Y19.4
1-Bromo-3-methylheptane	A62.6	$C_9H_9ClO_2$		
1-Bromo-4-methylheptane	A62.7		2-Chloro-2-phenylpropionic acid	A51.3
$C_8H_{17}Cl$			3-Chloro-2-phenylpropionic acid	A41.9
2-Chlorooctane	A12.15	C_9H_9I		
$C_8H_{17}N$			1-Iodo-2-phenylcyclopropane	A44.6
Coniine	K18.1	$C_9H_9I_2NO_3$		
1-Cyclohexylethylamine	A19.15		3,5-Diiodotyrosine	A5.7
$C_8H_{17}NO$		$C_9H_9NO_2$		
Sedridine	K19.8		5-Phenyl-2-oxazolidone	A23.6

$C_9H_9NO_4$
    2-(p-Nitrophenyl)propionic acid   A41.10
    Cyclodopa   K17.9

$C_9H_{10}N_2O_5$
    3-Nitrotyrosine   A5.7

$C_9H_{10}N_2S$
    2-Mercapto-4-phenylimidazoline   Y30.6

$C_9H_{10}O$
    1-Phenyl-1,2-epoxypropanes   A21.6, A21.7
    Indan-1-ol   A25.5
    2-Phenylpropanal   A48.15

$C_9H_{10}O_2$
    2-Phenylpropionic acid   A41.10
    1-Phenylpropan-1-ol-2-one   A21.14

$C_9H_{10}O_3$
    2-Hydroxy-3-phenylpropionic acid   A3.5
    3-Hydroxy-2-phenylpropionic acid   A41.8
    3-Hydroxy-3-phenylpropionic acid   A23.1
    Atrolactic acid   A31.19
    2-Phenoxypropionic acid   A12.7

$C_9H_{10}O_4S$
    2-Thenylsuccinic acid   A28.15

$C_9H_{10}S$
    2-Methyl-1-phenylthiirans   A21.4, A21.5

$C_9H_{11}N$
    1-Aminoindane   A25.5
    2-Methyl-2-phenylaziridine   A40.16

$C_9H_{11}NO$
    2-Aminoindan-1-ols   A25.11, A25.12
    5-Hydroxy-5,6,7,8-tetrahydro-
        quinoline   A25.14
    2-Hydroxymethylindoline   A5.20

$C_9H_{11}NO_2$
    Phenylalanine   A5.12
    N-Phenylalanine   A13.3
    3-Hydroxy-2-phenylpriopionic acid,
        amide   A41.8
    2-Amino-2-phenylpropionic acid   A40.12
    3-Amino-3-phenylpropionic acid   A19.4
    2-(p-aminophenyl)-propionic acid   A41.10
    2-Methylamino-2-phenylacetic acid   A19.8

$C_9H_{11}NO_3$
    Tyrosine   A5.7
    Phenylserines   A19.17, A21.9
    Phenylisoserine   A19.11

$C_9H_{11}NO_4$
    DOPA   A5.7

$C_9H_{11}NO_6$
    Showdomycin   C4.5

$C_9H_{12}N_2$
    2-Methyl-1,2,3,4-tetrahydroquinoxaline   A8.8

$C_9H_{12}N_2O_3$
    3-Aminotyrosine   A5.7

$C_9H_{12}N_2O_6$
    Uridine   C4.6

$C_9H_{12}O$
    1-Phenylpropan-1-ol   A59.3
    2-Phenylpropan-1-ol   A62.26
    1-Phenylpropan-2-ol   A21.3
    1-Methoxy-1-phenylethane   A22.6

$C_9H_{12}OS$
    Ethyl-p-tolylsulphoxide   Z7.6

$C_9H_{12}O_2$
    1-Phenylpropane-1,2-diols   A21.1, A21.2
    Norbornenecarboxylic acids, methyl
        esters   A47.11
    Spiro-[4.4]-nonane-1,6-dione   X5.2

$C_9H_{12}O_4$
    Genipic acid   T15.9
    4-Hydroxy-1-methylcyclohexane-1,2-
        dicarboxylic acid 1,4-lactone   T40.3

$C_9H_{13}BrO_2$
    2-Bromo-2-(4-methylcyclohexyllidene)-
        acetic acid   X3.8

$C_9H_{13}ClO_5S$
    Ethylphenyloxosulphonium perchlorate   Z7.3

$C_9H_{13}DO_2$
    4-Methylcyclohexylideneacetic acid-α-d   X3.7

$C_9H_{13}N$
    Amphetamine   A5.10
    2-Amino-1-phenylpropane   A4.9

$C_9H_{13}NO$
    1-Amino-2-phenylpropan-2-ol   A51.10
    2-Amino-2-phenylpropan-1-ol   A40.11
    2-Amino-3-phenylpropan-1-ol   A5.13
    2-Amino-1-(p-hydroxyphenyl) propane   A5.3
    N-Methyl-N-ethylaniline oxide   Z3.1

$C_9H_{13}NO_2$
    3-(1, 4-cyclohexadienyl)-alanine   A5.9
    Tyrosinol   A5.2
    Synephrine   A22.14

$C_9H_{13}NO_3$
    Adrenaline   A22.14

$C_9H_{13}NO_5$
    4-Carbethoxy-2-pyrrolidone-4-acetic
        acid   X4.5

$C_9H_{13}N_3O_4$
    2'-Deoxycytidine   C4.8

$C_9H_{13}OP$
    Ethylmethylphenylphosphine oxide   Z6.2

C_9H_{14}
    Cyclonona-1,2-diene   X2.4

$C_9H_{14}Br_2$
    9,9-Dibromobicyclo-[6.1.0]-nonane   X2.8

$C_9H_{14}N_2$
    2-Hydrazino-1-phenylpropane   A5.4

$C_9H_{14}N_2OS$
    S-Dimethylamino-N-methyl-S-
        phenylsulphoximine   Z8.7

$C_9H_{14}N_4O_3$
    Carnosine   A20.21

$C_9H_{14}O$
    Camphenilone   T11.14
    Cryptone   T2.15
    Sabina ketone   T7.17
    Bicyclo-[4.3.0]-nonan-3-one   A37.7
    Bicyclo-[4.3.0]-nonan-8-one   A37.4

Formulae Index

$C_9H_{14}O_2$
- Boschnialactone — T15.4
- 3-Acetoxy-5-methylcyclohexene — A38.19
- 2-Methylocta-2,3-dienoic acid — X1.8
- 4-Methylcyclohexylideneacetic acid — X3.9
- 2-Formyl-3-methylcyclopentyl-acetaldehyde — T13.2
- 2-Acetoxynorbornane — A47.10
- Norbornane-2-carboxylic acids, methyl esters — A47.10
- Tetrahydroanhydrodeoxyaucubigenin — T15.12

$C_9H_{14}O_3$
- Tetrahydroanhydroaucubigenin — T15.12
- Nepetonic acids — T14.6, T14.8
- Pinononic acid — T8.14

$C_9H_{14}O_4$
- trans-Caryophyllenic acid — T28.10
- Boschnialinic acids — T15.6, T15.8
- Cyclopentane-1,2-diacetic acid — A36.15

$C_9H_{14}O_6$
- Camphoronic acid — T12.1

$C_9H_{15}DO_2$
- 2-[^{2}H]-2-(cis-4-methylcyclohexyl)-acetic acid — D1.9

$C_9H_{15}NO$
- Bicyclo-[4.3.0]-nonan-8-one, oxime — A37.4
- 1-Oxoquinolizidine — A15.8
- 2-Oxoquinolizidine — A15.9
- 3-Oxoquinolizidine — A15.10

$C_9H_{15}NO_2$
- 6-Methoxytropinone — K28.5

$C_9H_{15}NO_6$
- N-Carbethoxyethylaspartic acid — A4.6

C_9H_{16}
- Apofenchene — T10.10
- Bicyclo-[6.1.0]-nonane — X2.7
- 1-Methylene-2,2,3-trimethylcyclopentane — A36.18
- Pulegene — T7.15

$C_9H_{16}O$
- 1-(1-Cyclohexenyl)-1-methoxyethane — A22.1
- 2,3,6-Trimethylcyclohexanone — A39.7
- 3,4-Diethylcyclopentanone — A45.2
- 1,5,5-Trimethylbicyclohexan-6-ols — T12.7
- 3-Methoxycyclooctene — X2.5
- 3-Ethoxy-5-methylcyclohex-1-enes — A38.14, A38.15

$C_9H_{16}O_2$
- 3,4-Diethyl-5-hydroxypentanoic acid lactone — A45.3
- Spiro-[4.4]-nonane-1,6-diol — X5.1
- Camphonanic acid — T12.6

$C_9H_{16}O_3$
- 4,5-Dimethoxycycloheptanone — A3.13
- 4-Hydroxy-1,2,2-trimethylcyclopentane-1 carboxylic acid — T55.3
- Cinenic acid — T3.10
- 2-Isopropyl-5-oxohexanoic acid — A28.2
- 3-Methoxycyclohexanecarboxylic acid methyl ester — A38.16

$C_9H_{16}O_4$
- 3-Isopropyladipic acid — A28.1
- 2,2,3-Trimethyladipic acid — A36.16
- 3-Ethyl-3-methyladipic acid — A55.7
- 2,6-Dimethylheptanedioic acid — A32.8
- 3-Methyloctanedioic acid — A32.9
- 2,3-Diethylsuccinic acid monomethyl ester — A45.10
- 2,3-diethylglutaric acid — A45.11
- 2-Pentylsuccinic acid — A28.9
- 2-Isopropyl-2-methylglutaric acid — A55.2

$C_9H_{16}O_4S$
- 2-Thiopentylsuccinic acid — A28.13

$C_9H_{16}O_5$
- 3-Methoxyadipic acid, dimethyl ester — A25.8

$C_9H_{17}N$
- Pinidine — K19.12

$C_9H_{17}NO$
- N-Methylisopelletierine — K19.9

$C_9H_{17}NO_2$
- Hexahydrophenylalanine — A5.8
- 2-Aminocyclohexanecarboxylic acid ethyl ester — A37.11
- 2-Amino-2-cyclohexylpropionic acid — A40.15

$C_9H_{17}NO_3$
- 1-Acetamido-2-acetoxypentane — A6.10
- Hexahydrotyrosine — A5.1

$C_9H_{18}Br_2$
- 1,5-Dibromo-2,3-diethylpentane — A34.18

$C_9H_1N_2O$
- 1-Methyl-4-propylprolinamides — Y21.2, Y21.3

$C_9H_{18}O$
- 1-Cyclohexylethanol — A22.2
- 7-Methoxyoct-1-ene — A12.17
- 1-Cyclohexylpropan-1-ol — A58.6

$C_9H_{18}O_2$
- 2-Methyloctanoic acid — A61.14
- 3-Methyloctanoic acid — A61.15
- 4-Methyloctanoic acid — A61.16
- 5-Methyloctanoic acid — A61.17
- 6-Methyloctanoic acid — A27.17
- 2,6-Dimethylheptanoic acid — T1.7

$C_9H_{18}O_3$
- 3-Hydroxynonanoic acid — A7.6

$C_9H_{19}Br$
- 1-Bromo-3-methyloctane — A62.14
- 1-Bromo-4-methyloctane — A62.15
- 1-Bromo-5-methyloctane — A62.16
- 1-Bromo-6-methyloctane — A62.17

$C_9H_{19}N$
- 3,4-Diethylpiperidine — A34.19
- 1,2,2,3-Tetramethylcyclopentylamine — A36.19

$C_9H_{19}NO$
- 2-Ethylpentanoic acid dimethylamide — A45.6

$C_9H_{19}NO_5S$
- Methyl thiolincosaminide — Y21.7

C_9H_{20}
- 3-Methyloctane — A27.13

4-Methyloctane	A59.13
3-Ethyl-4-methylhexane	A34.17
2,5-Dimethylheptane	A59.8

$C_9H_{20}ClNO_2$
Muscarine	Y20.8

$C_9H_{20}O$
Nonan-2-ol	K19.5
Nonan-3-ol	A38.22
Nonan-4-ol	A58.23
2-Methyloctan-3-ol	A58.19
2-Methyloctan-4-ol	A58.20
3-Methyloctan-1-ol	A60.10
4-Methyloctan-1-ol	A60.11
5-Methyloctan-1-ol	A60.12
6-Methyloctan-1-ol	A27.13
2-Methoxy-6-methylheptane	A12.8

$C_9H_{21}N$
1-Amino-4-methyloctane	A60.18
2-(Dimethylamino)heptane	K19.3

C_{10}

$C_{10}H_6O$
Naphthalene-1,2-epoxide	A25.6

$C_{10}H_7BrO_2$
3-Bromo-4-phenylcrotonolactone	X3.4

$C_{10}H_7MnO_5$
Tricarbonylmanganese-2-methylcyclopentadienylcarboxylic acid	X9.17
Tricarbonylmanganese-3-methylcyclopentadienylcarboxylic acid	X9.14

$C_{10}H_8O_2$
4-Phenylbuta-2,3-dienoic acid	X3.3

$C_{10}H_8O_4$
3-Methylphthalide-3-carboxylic acid	A51.7

$C_{10}H_9NO_2$
2-Cyano-2-phenylpropionic acid	A55.12

$C_{10}H_{10}$
2-Phenylmethylenecyclopropane	A44.17

$C_{10}H_{10}Cl_2O_4$
2-*trans*-allyl-3,5-dichloro-1-hydroxy-4-oxocyclopentane-1-carboxylic acid, methyl ester	Y16.6

$C_{10}H_{10}O$
1,2-Epoxytetralin	A25.7
3-Methylindan-1-one	A49.3

$C_{10}H_{10}O_2$
3,4-Dihydro-4-methylcoumarin	A49.5
Indane-1-carboxylic acid	A25.4
1,2-Dihydro-1,2-dihydroxynaphthalene	A25.1
2-Phenylcyclopropanecarboxylic acids	A44.7, A44.9
1-(3-Hydroxy-2-tetrahydrofurylidene)-hexa-2,4-dione	Y15.12

$C_{10}H_{10}O_3$
Mellein	Y2.10

$C_{10}H_{10}O_4$
2-Phenylsuccinic acid	A50.12
Glutinic acid cyclopentadiene adduct	X3.2

$C_{10}H_{10}O_6$
Chorismic acid	A3.1

$C_{10}H_{11}DO$
[4-^{2}H]-4-phenylbutan-2-one	D1.5

$C_{10}H_{11}D_3O$
[1-^{2}H$_3$]-2-Benzyloxypropane	D1.16

$C_{10}H_{11}FeN$
2-Methylazaferrocene	X9.1

$C_{10}H_{11}NO$
Abikoviromycin	Y30.9
Boschniakine	T15.7

$C_{10}H_{11}NO_4$
3-Carboxy-6,7-dihydro-1,2,3,4-tetrahydroquinoline	A5.15

$C_{10}H_{12}$
1-Methylindane	A50.14
1-Methyl-2-phenylcyclopropane	A44.18
3-Phenylbut-1-ene	A49.12

$C_{10}H_{12}D_2O_2$
[1-^{2}H$_2$]-2-Benzyloxypropan-1-ol	D1.17

$C_{10}H_{12}N_2$
Anatabine	K18.9

$C_{10}H_{12}N_2O_3$
Kynurenine	A20.18

$C_{10}H_{12}N_2O_5$
Dinoseb	A49.15

$C_{10}H_{12}O$
1-Hydroxymethylindane	A50.15
1,2,3,4-Tetrahydro-2-naphthol	A25.3
1-Hydroxymethyl-2-phenylcyclopropane	A44.14
3-Phenylbutan-2-one	A41.1

$C_{10}H_{12}O_2$
2-Phenylbutyric acid	A25.15
3-Phenylbutyric acid	A49.7
1,2-Dihydroxytetralin	A25.2
3-Hydroxy-3-phenylbutan-2-one	A51.2

$C_{10}H_{12}O_3$
2-Methoxy-2-phenylacetic acid, methyl ester	A21.12
α-Methyltropic acid	A55.11
3-Methoxy-3-phenylpropionic acid	A21.15

$C_{10}H_{12}O_3$
2-Hydroxy-2-phenylbutyric acid	A51.14
3-Hydroxy-3-phenylbutyric acid	A51.11
2-(*p*-Methoxyphenyl)-propionic acid	Y5.7

$C_{10}H_{12}O_4$
Batatic acid	T17.8
Hygrophyllinecic acid dilactone	K23.2
Curvulol	T2.6

$C_{10}H_{12}OS$
Benzyl-(2,3-epithiopropyl)-ether	A14.14

$C_{10}H_{12}O_5$
Asperline	Y15.7

$C_{10}H_{13}As$
Allylmethylphenyl arsine	Z3.11

Formulae Index

$C_{10}H_{13}Br$
    1-Bromo-3-phenylbutane   A49.10
$C_{10}H_{13}Cl$
    1-Chloro-2-methyl-3-phenylpropane   A43.21
$C_{10}H_{13}D$
    [1-^{2}H]-1-Phenylbutane   D1.4
$C_{10}H_{13}N$
    2-Aminotetralin   A20.5
    Actinidine   T13.17
    1,2,3,4-Tetrahydronaphthylamine   A25.10
$C_{10}H_{13}NO$
    3-Hydroxymethyl-1,2,3,4-tetrahydro-isoquinoline   A5.16
    3-Amino-3-phenylbutan-2-one   A40.17
    2-Amino-1-tetralols   A20.1, A20.10
$C_{10}H_{13}NO_2$
    2-Amino-2-methyl-2-phenylpropionic acid   A40.3
    2-Amino-2-phenylbutyric acid   A40.6
$C_{10}H_{13}NO_4$
    α-Methyl-DOPA   A40.3
$C_{10}H_{13}N_5O_4$
    Formycin   C4.7
$C_{10}H_{13}P$
    Allylmethylphenylphosphine   Z5.10
$C_{10}H_{13}PS$
    Allylmethylphenylphosphine sulphide   Z5.6
$C_{10}H_{14}$
    2-Phenylbutane   A42.16
    Twistene   X10.5
$C_{10}H_{14}N_2$
    3-Amino-1-phenylpyrrolidine   A20.6
    Nicotine   K20.4
    Anabasine   K18.9
    Cyclohexane-1,2-diacetic acid dinitrile   A37.9
$C_{10}H_{14}N_2O_4$
    2-(3,4-Dihydroxybenzyl)-2-hydrazino-propionic acid   A40.7
$C_{10}H_{14}O$
    Carvone   T5.8
    Pinocarvone   T9.11
    3-Phenylbutan-2-ols   A49.14, A49.19
    2-(2-hydroxyphenyl)-butane   A49.15
    Menthofuran   T1.3
    1-Ethoxy-1-phenylethane   A22.6
    Chrysanthenone   T8.13
    Perillaldehyde   T8.12
    Verbenone   T8.11
    1-Phenylbutan-1-ol   A59.1
    2-Phenylbutan-1-ol   A41.7
    3-Phenylbutan-1-ol   A49.6
    2-Phenylbutan-2-ol   A51.16
    2-Methyl-1-phenylpropan-1-ol   A25.18
    2-Methyl-3-phenylpropan-1-ol   A43.21
    Carquejol   T16.9
    Myrtenal   T9.7
    Umbellulone   T7.2
    Filifolone   T10.1
    Bicyclic ketone   X10.2
    Twistan-4-one   X10.3
$C_{10}H_{14}O_2$
    Camphorquinone   T12.8
    3-Acetoxycyclooctyne   X2.6
    Nepetalactone   T14.9
    Isonepetalactone   T14.10
    Cinerolone   T17.4
    2-Ethoxy-2-phenylethanol   A22.10
    Piperitenone oxide   T4.3
    Neonepetalactone   T14.7
    2-Phenylbutane-1,2-diol   A51.15
    8-Methylperhydroindane-2,5-dione   A53.15
    3-Phenylbutane-1,3-diol   A51.12
$C_{10}H_{14}O_3$
    3-O-Benzylglycerol   A14.10
    Piperinic acid   T1.6
    5-Isopropyl-5-methyl-3-oxocyclopentene-carboxylic acid   T7.3
$C_{10}H_{14}O_4$
    trans-Chrysanthemic acid   A35.11
    2-Methoxymethyl-3-oxocyclopentene-1-acetic acid   Y16.10
$C_{10}H_{14}O_5$
    Hygrophyllinecic acid monolactone   K23.2
    Seneciphyllic acid   K23.8
$C_{10}H_{14}OS$
    Propyl-p-tolylsulphoxides   Z7.6
    Mesitylmethylsulphoxide   Z8.5
$C_{10}H_{15}As$
    Methylphenylpropyl arsine   Z3.6
$C_{10}H_{15}BrO$
    2-Bromo-1-decalones   A39.3, A39.4, A39.9, A39.11
    9-Bromo-1-decalones   A39.6, A39.10
    endo-3-Bromocamphor   T11.7
$C_{10}H_{15}Br_3O$
    cis-Carvone tribromide   T5.6
$C_{10}H_{15}BrO_4S$
    3-Bromocamphor-9-sulphonic acid   T11.8
$C_{10}H_{15}N$
    2-Amino-3-phenylbutanes   A49.16, A49.18
    1-Amino-2-phenylbutane   A41.5
    2-Methylamino-1-phenylpropane   A4.9
    2-Amino-2-phenylbutane   A40.13
$C_{10}H_{15}NO$
    Ephedrine   A21.10
    ψ-Ephedrine   A21.11
    1-Amino-2-phenylbutan-2-ol   A51.13
    2-Methylamino-1-phenylpropan-1-ol   A21.11
$C_{10}H_{15}NO_2$
    2-Amino-1-(3,4-dihyroxyphenyl)-propane   A4.9
$C_{10}H_{15}OP$
    Methylphenylpropylphosphine oxide   Z4.13
    Isopropylmethylphenylphosphine oxide   Z4.7
$C_{10}H_{15}P$
    Methylphenylpropylphosphine   Z5.4
$C_{10}H_{15}PS$
    Methylphenylpropylphosphine sulphide   Z5.5

Formulae Index

$C_{10}H_{15}PSe$
    Methylphenylpropylphosphine
        selenoxide   Z5.3

$C_{10}H_{16}$
Achillene	T1.11
Sylvestrene	T6.9
Twistane	X10.4
Isosylvestrene	T6.8
Sabinene	T7.13
m-Mentha-1(7),8-diene	T6.7
α-Fenchene	T10.12
α-Phellandrene	T2.12
β-Phellandrene	T2.13
Limonene	T2.7
1,8-Iridadiene	T3.1
α-Hymentherene	T1.5
β-Hymenthrerene	T1.8
α-Pinene	T8.4
β-Pinene	T8.5
Bicyclo-[7.1.0]-decane	X2.3
Camphene	T11.13
Car-2-ene	T6.1
Car-3-ene	A35.4
Cis-Car-4-ene	T6.13
trans-Car-4-ene	T6.14

$C_{10}H_{16}BrNO_2$
    2-Bromo-2-nitrofenchane   T10.2

$C_{10}H_{16}KNO_9S_2$
    Sinigrin   C4.10

$C_{10}H_{16}N_2$
    3,4-Diethyladipic acid dinitrile   A45.1

$C_{10}H_{16}N_2O_3S$
    Biotin   Y30.1

$C_{10}H_{16}N_5O_4$
    Formycin   C4.7

$C_{10}H_{16}O$
trans-Carveol	T5.4
trans-Pinocarveol	T9.6
Dihydrocarvone	T5.4
Thujone	T7.7
Thujan-2-one	T7.1
Isothujone	T7.5
trans-Caran-2-one	T6.6
trans-Car-4-ene-3-ol	T6.15
Sabinol	T7.9
Camphor	T11.11
Pinocamphone	T9.13
Isopinocamphone	T9.9
Epicamphor	T12.11
Myrtenol	T9.7
Pulegone	A38.1
2-Methyl-4-isopropenylcyclohexanone	T9.4
Fenchone	T10.6
Isofenchone	T10.11
Piperitone	T2.14
Carvotanacetone	T2.8
8-Methyl-perhydroindan-1-one	A53.3
8-Methyl-perhydroindan-2-one	A53.7
8-Methyl-perhydroindan-5-one	A53.12
1-Decalones	A39.2, A39.5
α-Cyclocitral	T54.5
Verbenols	T8.8, T8.10
Perillyl alcohol	T8.12
Matatabiether	T14.3

$C_{10}H_{16}O_2$
Pyrocin	A35.5
3-Acetoxycyclooctene	X2.5
α-Fencholenic acid	T10.5
Iridomyrmecine	T13.12
Isoiridomyrmecine	T13.15
2-Methylnona-2,3-dienoic acid	X1.9
Iridodial	T13.13
trans-Pulegenic acid	A38.2
Piperitone oxide	T4.2
cis-Chrysanthemic acid	A35.10
trans-Chrysanthemic acid	A35.6
2,6,6-Trimethyl-2-vinyltetrahydropyran-5-one	T3.11
1,5,5-Trimethylbicyclo-[2.1.1]-hexane-6-carboxylic acids	T12.4
Dihydronepetalactone	T14.5
Isodihydronepetalactone	T14.4

$C_{10}H_{16}O_3$
α-Nepetalic acid	T14.12
δ-Nepetalic acid	T14.14
Homoterpenyl methyl ketone	T2.2

$C_{10}H_{16}O_4$
Homothujadicarboxylic acid	T7.4
Dihydrosenecic acid lactone	K23.4
Tetrahydroseneciphyllic acid lactone	K23.6
α-Nepetalinic acid	T13.16
β-Nepetalinic acid	T14.13
γ-Nepetalinic acid	T14.11
δ-Nepetalinic acid	T13.14
cis-Camphoric acid	T12.3
2-Methoxymethylcyclopent-4-ene-3-ol-1-acetic acid	Y16.11
2-Carboxy-2-methylcyclohexaneacetic acids	A53.11, A53.14
Cyclohexane-1,2-diacetic acid	A37.9

$C_{10}H_{16}O_4S$
Camphor-8-sulphonic acid	T11.2
Camphor-10-sulphonic acid	T12.2

$C_{10}H_{16}O_5$
Senecic acid	K23.1
Intergerrinecic acid	K23.1

$C_{10}H_{16}O_6$
Hydrophyllenic acid	K23.2
Jaconecic acid	K22.1

$C_{10}H_{17}Cl$
    Bornyl chloride   T11.12

$C_{10}H_{17}NO$
Abikoviromycin derivative	Y30
2-Dimethylamino-3-phenylpropan-1-ol	A5.13

$C_{10}H_{17}NO_3$
    Ecgonine methyl ester   K28.6

Formulae Index

Pseudoecgonine methyl ester	K28.9
3-Acetoxy-6-hydroxytropane	K28.3

$C_{10}H_{18}$

Pinanes	T8.2, T8.6
Thujanes	T7.14, T7.19
Cis-Carane	T6.4
trans-Carane	T6.5
Fenchane	T10.9
Dimethylocta-2,7-diene	T1.2
1,3-Dimethyl-3-isopropylcyclopentene	T7.16
p-Menth-3-ene	T4.11
p-Menth-2-ene	T4.7
Bicyclo-[7.1.0]-decane	X2.3

$C_{10}H_{18}N_2O_3$

Desthiobiotin	Y30.2
Antibiotic LL-BH-872a	Y30.5

$C_{10}H_{18}O$

Dihydrocarveol	T5.11
Linalool	T3.8
Pinocampheol	T9.14
Isopinocampheol	T9.5
Neopinocampheol	T9.12
Neoisopinocampheol	T9.10
cis-Myrtanol	T9.8
p-Menth-1-en-9-ol	T2.3

$C_{10}H_{18}O$

α-Fenchyl alcohol	T10.3
β-Fenchyl alcohol	T10.8
β-Isofenchyl alcohol	T10.13
α-Isofenchyl alcohol	T10.14
cis-Carvotanacetol	T2.5
Lavandulol	T2.11
Cyclolavandulol	T2.10
α-Terpineol	T2.4
Menthone	T1.1
Isomenthone	T4.8
Carvomenthone	T5.9
Isocarvomenthone	T5.10
cis-Piperitol	T4.1
trans-Piperitol	T4.5
Epiborneol	T12.15
Epi-isoborneol	T12.15
Pinan-2-ols	T8.1, T8.7
Myrtanol	T8.3
Citronellal	T1.4
Cyclogeraniol	T54.5
Thujyl alcohol	T7.11
Neothujyl alcohol	T7.12
Isothujyl alcohol	T7.8
Neoisothujyl alcohol	T7.6
1-Hydroxydecalin	A39.1
2-tert-Butylcyclohexanone	A39.18
2,5-Dimethyl-5-ethylcyclohexanone	A55.9
Neodihydrocarveol	T5.2
Borneol	T11.10
Isoborneol	T11.6
Carquejanone	T16.13
Plinols	T3.2, T3.3, T3.4, T3.7
Sabina hydrate	T7.18
4-Isopropyl-4-methyl-2-methylene-cyclopentanol	T7.10

$C_{10}H_{18}O_2$

Neomatabiol	T14.1
Isomatabiol	T14.1
Fencholic acid	T10.7
1,2,2,3-Tetramethylcyclopentanecarboxylic acid	A36.20
6-Oxo-3-isopropylheptanal	T48.1

$C_{10}H_{18}O_3$

Cinenic acid methyl ester	T3.10
trans-trans-puleganolic acid	T15.1
3-Ethyl-3-methyl-6-oxoheptanoic acid	A55.10
2-Isopropyl-2-methyl-5-oxohexanoic acid	A55.3

$C_{10}H_{18}O_4$

3,4-Diethyladipic acid	A45.1
3,6-Dimethyloctane-1,8-dioic acid	A32.6
2-Hexylsuccinic acid	A28.9

$C_{10}H_{18}O_6$

Trichodesmic acid	K23.9

$C_{10}H_{19}Cl$

Menthyl chloride	T4.13

$C_{10}H_{19}NO$

Lupinine	K21.1

$C_{10}H_{19}NO_3$

2,3-Diethylsuccinic acid monodimethyl-amide	A45.9

$C_{10}H_{19}NOS_2$

Hirsutin	Z8.8

$C_{10}H_{20}$

cis-m-menthane	T6.10
trans-m-menthane	T6.11
2-cyclohexylbutane	A59.4

$C_{10}H_{20}N_2$

1-(Aminomethyl)-quinolizidine	K21.2
2′,3-Dipiperidyl	K18.6

$C_{10}H_{20}O$

1-Cyclohexylbutan-1-ol	A58.1
m-Menthan-1-ol	T6.12
Menthol	A26.12
Neomenthol	T4.4
Isomenthol	T4.6
Neoisomenthol	T4.12
Carvomenthol	T5.13
Isocarvomenthol	T5.7
Neocarvomenthol	T5.1
Neoisocarvomenthol	T5.14
5-Ethyl-6-methylheptan-2-one	T48.3
2-tert-Butylcyclohexanols	A39.17, A39.14

$C_{10}H_{20}O_2$

2-Methylnonanoic acid	A61.11
2-Ethyloctanoic acid	A61.3
4-Methylnonanoic acid	A61.12
5-Methylnonanoic acid	A61.13
6-Oxo-3-isopropylheptanal	T48.2

Formulae Index

$C_{10}H_{20}O_3$
    2-Hydroxydecanoic acid   A1.24
    2-Hydroxy-2-ethyl-6-methylheptanoic acid   T3.6

$C_{10}H_{20}O_5$
    Nogalose   C4.12

$C_{10}H_{21}Br$
    1-Bromo-4-methylnonane   A62.12
    1-Bromo-5-methylnonane   A62.13

$C_{10}H_{21}N$
    Menthylamine   T4.10
    Isomenthylamine   T4.9
    Neomenthylamine   T4.14
    Neoisomenthylamine   T4.15

$C_{10}H_{22}$
    3-Methylnonane   A27.12
    4-Methylnonane   A59.12

$C_{10}H_{22}O$
    4-Methylnonan-1-ol   A60.8
    5-Methylnonan-1-ol   A60.9
    2-Methylnonan-5-ol   A58.18
    Decan-3-ol   A58.7
    Decan-4-ol   A58.8
    Decan-5-ol   A58.9
    Tetrahydrolinalool   T3.5

$C_{10}H_{23}N$
    1-Amino-5-methylnonane   A60.17

C_{11}

$C_{11}H_8CrO_5$
    Tricarbonylchromium *m*-toluic acid   X9.20
    Tricarbonylchromium-*o*-toluic acid   X9.21

$C_{11}H_9BrO_2$
    3-Bromo-4-methyl-4-phenylcrotonolactone   X3.4

$C_{11}H_9ClO_5$
    4-Chloro-3-methyl-7-methoxyphthalide-3-carboxylic acid   A51.4

$C_{11}H_{10}O$
    Benznorbornen-2-one   A46.4

$C_{11}H_{10}OS$
    Naphthylmethylsulphoxides   Z8.5

$C_{11}H_{10}O_2$
    4-Phenylpenta-2,3-dienoic acid   X3.3

$C_{11}H_{10}O_2S$
    2-(2-thianaphthenyl)-propionic acid   A48.3
    2-(3-thianaphthenyl)-propionic acid   A48.2

$C_{11}H_{10}O_3$
    2-Phenylglutaric acid anhydride   A50.10
    4-Oxotetralin-1-carboxylic acid   A50.13

$C_{11}H_{10}O_4$
    Adamantane-4,8-dione-2-carboxylic acid   A45.14

$C_{11}H_{10}O_5$
    3-Methyl-7-methoxyphthalide-3-carboxylic acid   A51.4

$C_{11}H_{11}NO_2$
    2-(3-Indolyl)-propionic acid   A48.4

$C_{11}H_{11}NO_3$
    5-Methyl-2-phenyl-2-oxazoline-4-carboxylic acids   A24.3, A24.4

$C_{11}H_{11}NO_3$
    N-Benzoylaspartic acid   A4.6

$C_{11}H_{12}$
    Phenylspiro-[2.2]-pentane   A44.16

$C_{11}H_{12}Cl_2N_2O_5$
    Chloramphenicol   A21.8

$C_{11}H_{12}N_2O_2$
    Tryptophan   A20.12

$C_{11}H_{12}N_2O_3$
    2-Hydroxytryptophan   A20.8
    5-Hydroxytryptophan   A20.13

$C_{11}H_{12}N_2S$
    Tetramisole   Y30.7

$C_{11}H_{12}O$
    3-Methyl-1-tetralone   A43.23
    2-Methyl-1-tetralone   A43.14
    4-Methyl-1-tetralone   A49.13
    1-Acetylindane   A50.16
    3-Ethylindan-1-one   A42.4

$C_{11}H_{12}O_2$
    Tetralin-1-carboxylic acid   A25.9
    1-Acetoxyindane   A25.5
    2-Phenylpent-4-enoic acid   A50.8

$C_{11}H_{12}O_3$
    5-Methylmellein   Y2.2

$C_{11}H_{12}O_4$
    Benzylsuccinic acid   A28.19
    3-*p*-Carboxyphenylbutyric acid   A49.7
    2-Phenylglutaric acid   A50.10

$C_{11}H_{12}O_5$
    p-Methoxyphenylsuccinic acid   Y6.8

$C_{11}H_{12}O_7$
    Piscidic acid   A30.11

$C_{11}H_{12}O_8$
    Fukiic acid   A30.11

$C_{11}H_{13}BrO_2$
    5-Bromo-2-phenylpentanoic acid   A50.9

$C_{11}H_{13}N$
    2-Methyl-2-phenylbutyronitrile   A56.7

$C_{11}H_{13}NO$
    2-(3-Indolyl))-)propan-1-ol   A48.1
    1-Acetamidoindane   A25.5

$C_{11}H_{13}NO_4$
    Evoninic acid   A31.9
    2-Hydroxy-2-(2-acetylaminophenyl)-propionic acid   A51.6

$C_{11}H_{14}$
    1-Methyltetralin   A49.17
    2-Methyltetralin   A43.18

$C_{11}H_{14}N_2O$
    Cytisine   K21.4

$C_{11}H_{14}O$
    2-Methyl-2-phenylbutanal   A56.1
    10-Methyl-$\Delta^{3,6}$-2-hexalone   A53.9
    2-Benzoylbutane   A29.12
    2-Methyl-1-tetralol   A43.17

Formulae Index

$C_{11}H_{14}O_2$
2-Methyl-2-phenylbutyric acid	A56.7
2-Methyl-4-phenylbutyric acid	A32.15
3-Methyl-4-phenylbutyric acid	A43.22
3-Methyl-2-phenylbutyric acid	A25.17
3-*p*-Tolylbutyric acid	A49.7
2-Phenylpentanoic acid	A41.11
3-Phenylpentanoic acid	A42.10
4-Phenylpentanoic acid	A49.9
Pyrethrolone	T17.4
Isopyrethrolone	T17.2
2-Hydroxy-2-phenylpentan-3-one	A51.5

$C_{11}H_{14}O_3$
2-Hydroxy-3-methyl-2-phenylbutyric acid	A51.1
Adamantan-4-one-2-carboxylic acids	A45.13, A45.15
2-(*p*-Methoxyphenyl)-butyric acid	Y4.7
3-Hydroxy-3-phenylpentanoic acid	A51.17
4-Methoxy-4-phenylbutyric acid	A21.18

$C_{11}H_{14}O_4$
Panepoxydon	Y19.3
Genipin	T13.3

$C_{11}H_{15}N$
1-Dimethylaminoindane	A25.5

$C_{11}H_{15}NO$
2-Methyl-2-phenylbutyramide	A56.7
1-Benzoyl-1-dimethylaminoethane	A4.10

$C_{11}H_{15}NO_2$
2-Dimethylamino-2-phenylpropionic acid	A40.10

$C_{11}H_{15}N_3O_6$
Erythropentulose *o*-nitrophenylhydrazone	C2.6

$C_{11}H_{15}N_5O_3$
Aristeromycin	Y21.4

$C_{11}H_{16}$
2-Phenylpentane	A50.5
2-Methyl-1-phenylbutane	A27.9
2-Methyl-3-phenylbutane	A41.6

$C_{11}H_{16}O$
2-Methyl-2-phenylbutan-1-ol	A56.3
1-Phenylpentan-1-ol	A59.2
2-Phenylpentan-1-ol	A50.4
3-Phenylpentan-1-ol	A62.24
4-Phenylpentan-1-ol	A62.25
3-Phenylpentan-2-ols	A48.9, A48.13
1-Phenylpentan-3-ol	A58.27
2-Phenylpentan-3-ols	A48.6, A48.14
9-Methyl-Δ^2-1-octalone	A53.4
9-Methyl-Δ^4-3-octalone	A54.2
10-Methyl-Δ^3-2-octalone	A54.1
10-Methyl-Δ^8-1-octalone	A54.4
10-Methyl-Δ^6-2-octalone	A53.13
2,2-Dimethyl-1-phenylpropan-1-ol	A25.20

$C_{11}H_{16}OS$
Butyl-*p*-tolylsulphoxides	Z7.6

$C_{11}H_{16}O_2$
9-Methyldecalin-1,2-dione	A53.1
3-Phenylpentane-1,3-diol	A51.18
2-Phenylpentane-1,5-diol	A50.6

$C_{11}H_{16}O_3$
3-3,5-Dihydroxy-2-methylphenyl)-butan-2-ol	Y13.9

$C_{11}H_{16}O_4$
Pyrethric acid	A35.11

$C_{11}H_{16}O_5$
4-Oxocyclopentane-1,2-dicarboxylic acid, diethyl ester	Y14.8

$C_{11}H_{17}NO$
Tecomanine	K31.9
2-Dimethylamino-1-phenylpropan-1-ol	A21.10

$C_{11}H_{17}NO_2$
2-Amino-1-(3,4-dimethoxyphenyl)-propane	A5.6

$C_{11}H_{17}NOS$
N-Methyl-*N*-(3-methylsulphinyl)-propylaniline	Z8.4

$C_{11}H_{17}OP$
tert-Butylmethylphenylphosphine oxide	Z4.3

$C_{11}H_{17}P$
tert-Butylmethylphenylphosphine	Z4.4

$C_{11}H_{18}$
Dictyopterene A	A44.4
Dictyopterene C'	Y15.11
Methyl isopulegene	A39.16

$C_{11}H_{18}N_2O_3S$
Cephalothin precursor	Y29.5

$C_{11}H_{18}O$
Homocamphor	T27.6
9-Methyl-1-decalone	A53.16
10-Methyl-1-decalones	A54.5, A54.6
10-Methyl-2-decalones	A53.10, A54.3

$C_{11}H_{18}O_2$
cis-Homochrysanthemic acid	A35.9
trans-Homochrysanthemic acid	A35.2
Tetrahydroactinidiolide	T54.2

$C_{11}H_{18}O_4$
1-Methylcyclohexane-1,2-diacetic acid	A53.6
2-Carboxy-2-methylcyclohexanepropionic acids	A53.2, A53.17

$C_{11}H_{20}$
1-Isopropylidene-2,4-dimethylcyclohexanone	A39.13

$C_{11}H_{20}O$
sec-Butyl cyclohexyl ketone	A29.12
2-*tert*-Butyl-5-methylcyclohexanone	A39.19

$C_{11}H_{20}O_2$
p-Menthane-3-carboxylic acid	T5.16
2-Cyclohexyl-2-methylbutyric acid	A56.8
3-Cyclohexylpentanoic acid	A42.11
2-Methyl-5-isopropylcyclopentanecarboxylic acid, methyl ester	T15.2
4,8-Dimethyl-4-hydroxynonanoic acid lactone	Y9.6

Formulae Index

$C_{11}H_{20}O_3$
   3,4-bis-(hydroxymethyl)-4-methyl-
      cyclohexanone   T40.2

$C_{11}H_{21}BrO_2$
   5-Bromo-3,4-diethylpentanoic acid, ethyl
      ester   A45.7

$C_{11}H_{21}N$
   Skytanthines   T14.15, T14.16, T14.17, T14.18

$C_{11}H_{21}NO$
   Alkaloid C   K31.10

$C_{11}H_{21}NOS_2$
   Arabin   Z8.8

$C_{11}H_{22}$
   4-Methyldec-1-ene   A27.8
   1-Cyclohexyl-2-methylbutane   A59.5

$C_{11}H_{22}O$
   1-Cyclohexylpentan-1-ol   A58.4
   1-Cyclohexylpentan-3-ol   A58.5
   4,8-Dimenthylnonanal   T46.1

$C_{11}H_{22}O_2$
   2-Methyldecanoic acid   A34.6
   5-Methyldecanoic acid   A61.4

$C_{11}H_{22}O_2S$
   Menthyl methanesulphinate   Z7.10

$C_{11}H_{23}Br$
   1-Bromo-5-methyldecane   A62.4

$C_{11}H_{23}N$
   2-Dimethylaminononane   K19.10

$C_{11}H_{23}NO_2$
   1-(2-Hydroxy-2-methylpropyl)-2-
      (1-hydroxy-1-methylethyl)pyrrolidine   K22.2

$C_{11}H_{23}O_2P$
   Menthyl methylphosphinate   Z6.4

$C_{11}H_{24}$
   3-Methyldecane   A59.9
   5-Methyldecane   A59.10

$C_{11}H_{24}O$
   Undecan-3-ol   A58.28

$C_{11}H_{25}N$
   1-Amino-6-methyldecane   A60.14

C_{12}

$C_{12}H_9MnO_4$
   Tricarbonylmanganese cyclopentadienyl
      derivs.   X9.13, X9.23

$C_{12}H_{10}O$
   7,8-Benzobicyclo[2.2.2]-oct-2-ene-6-one   A47.9

$C_{12}H_{10}O_7$
   3,4-Dihydro-6-hydroxy-3-methyl
      isocoumarin-7,8-dicarboxylic acid   Y2.5

$C_{12}H_{11}BrO_2$
   3-Bromo-4-ethyl-4-phenylcrotono-
      lactone   X3.4

$C_{12}H_{11}MnO_4$
   Tricarbonylmanganese cyclopentadienyl
      deriv.   X9.16

$C_{12}H_{12}$
   2-Methylenebenznorbornene   A46.7

$C_{12}H_{12}FeO_2$
   3-Methylferrocene-1-carboxylic acid   X9.10
   2-Methylferrocene-1-carboxylic acid   X9.8

$C_{12}H_{12}N_2O_3$
   N-Methyleudan   A55.13

$C_{12}H_{12}N_2O_4$
   1-p-Nitrobenzoyl-4-methyl-2-
      pyrrolidone   K35.8

$C_{12}H_{12}O$
   5,6-Benzobicyclo-[2.2.2]-octan-2-one   A47.8

$C_{12}H_{12}O_2$
   4-Phenylhexa-2,3-dienoic acid   X3.3

$C_{12}H_{13}NO_2$
   3-(N-Methylindolyl)-propionic acid   A48.4

$C_{12}H_{13}NO_3$
   Allylhippuric acid   A10.16
   Indolmycenic acid   Y27.8

$C_{12}H_{14}$
   3-Phenylcyclohex-1-ene   A52.11

$C_{12}H_{14}N_2$
   1,2,3,4-Tetrahydroharman   K16.5

$C_{12}H_{14}N_2O_2$
   Abrine   A20.12

$C_{12}H_{14}O$
   3-Ethyl-3-methylindan-1-one   A56.10
   2,2,3-Trimethylindan-1-one   A42.5
   2-Phenylcyclohexanone   A52.5
   1,2-Epoxy-1-phenylcyclohexane   A52.3
   1-Acetyltetralin   A50.17

$C_{12}H_{14}O_2$
   3-Butylphthalide   Y14.6
   2-Acetoxy-1,2,3,4-tetrahydro-
      naphthalene   A25.3

$C_{12}H_{14}O_4$
   Lactic acid, O-benzoyl, ethyl ester   A1.9
   2-Phenyladipic acid   A52.10
   Dihydrotubaic acid   Y1.7

$C_{12}H_{14}O_5$
   Hydroxydihydrotubaic acid   Y1.7

$C_{12}H_{15}ClO_3$
   3-(3,4-Dimethoxyphenyl)-2-methylpropionyl
      chloride   A31.22

$C_{12}H_{15}NO$
   2-Acetamidotetralin   A20.5
   1,3-Dimethyl-3-ethyloxindole   A56.18
   2-Oxo-4-ethyl-4-methyl-1,2,3,4-
      tetrahydroquinoline   A56.15

$C_{12}H_{15}NO_3$
   N-Benzoylvaline   A10.15
   Anhalonine   K2.1

$C_{12}H_{15}NO_2S_2$
   N-Tosyl-2-thia-5-azabicyclo-[2.2.1]-
      heptane   A17.13

$C_{12}H_{15}NO_4$
   2-Amino-2-benzylglutaric acid   A32.16

$C_{12}H_{16}$
   4-Phenylhex-1-ene   A42.14

$C_{12}H_{16}F_3N$
   Fenfluramine   A5.5

Formulae Index

$C_{12}H_{16}O$
    2-Phenylcyclohexanols   A52.6   A52.9
    3-Methyl-1-phenylpentan-2-one   A29.12
    3-Methyl-3-phenylpentan-2-one   A56.11
    4-Phenylhexan-2-one   A42.18

$C_{12}H_{16}O_2$
    3-Methyl-3-phenylpentanoic acid   A56.6
    1-Phenylcyclohexane-1,2-diols   A52.7, A52.8
    2,2-Dimethyl-3-phenylbutyric acid   A42.6
    3,3-Dimethyl-2-phenylbutyric acid   A25.19
    2-Phenylhexanoic acid   A50.1
    4-Phenylhexanoic acid   A42.15
    5-Phenylhexanoic acid   A63.2
    3-Methyl-5-phenylpentanoic acid   A61.20
    4-(p-Tolyl)-pentanoic acid   T16.7
    2-Acetoxy-2-phenylbutane   A51.16
    3-Methoxy-9-methyl-$\Delta^{2,6}$-1-hexalone   A53.8

$C_{12}H_{16}O_3$
    4-Methoxy-4-phenylbutyric acid, methyl ester   A21.18
    3,3-Dimethyl-2-hydroxy-2-phenylbutyric acid   A51.19

$C_{12}H_{16}O_4$
    3-(3,4-Dimethoxyphenyl)-2-methylpropionic acid   A31.22

$C_{12}H_{17}NO$
    2-Acetamido-2-phenylbutane   A40.13

$C_{12}H_{19}NO_2$
    Salsolidine   K2.3

$C_{12}H_{17}NO_3$
    Lophophorine   K2.1
    Calycotomine   K2.3

$C_{12}H_{18}$
    3-Phenylhexane   A42.19
    3-Methyl-1-phenylpentane   A59.14
    2,2-Dimethyl-3-phenylbutane   A41.3

$C_{12}H_{18}IN$
    1-Indolyltrimethylammonium iodide   A25.5

$C_{12}H_{18}N_2O_2$
    Pilocarpine   K30.2
    Isopilocarpine   K30.3

$C_{12}H_{18}O$
    4-Phenylhexan-1-ol   A62.22
    5-Phenylhexan-1-ol   A62.23
    6-Phenylhexan-3-ol   A58.26
    2-Methyl-3-phenylpentan-2-ol   A42.13
    3,3-Dimethyl-2-phenylbutan-1-ol   A41.4

$C_{12}H_{18}O_2$
    Cnidilide   Y14.1
    1,9-Dimethyldecalin-2,6-dione   T23.15
    Ekasantalic acid   T11.5
    Tricycloekasantalic acid   T11.4

$C_{12}H_{18}O_3$
    Jasmonic acid   A36.14

$C_{12}H_{19}N$
    1-Dimethylamino-3-phenylbutane   A49.11

$C_{12}H_{19}NO_2$
    Dioscorine   K29.6

$C_{12}H_{20}Cl_6PSSb$
    Ethylmercaptomethylphenylpropylphosphonium hexachloroantimonate   Z5.2

$C_{12}H_{20}O_2$
    9-Hydroxy-5,5-dimethyl-2-decalone   T42.2
    2-Hydroxy-2,6,6-trimethylcyclohexanepropionic acid lactone   T36.4
    β-Dihydrosedanolide   T14.2

$C_{12}H_{20}O_3$
    Butyl 2-carboxycyclohexyl ketones   Y14.3, Y14.4

$C_{12}H_{20}O_4$
    Drimic acid   T32.7

$C_{12}H_{22}N_2O_4S$
    Cephalothin precursor   Y29.6

$C_{12}H_{22}O$
    4-Cyclohexylhexan-2-one   A42.3

$C_{12}H_{22}O_{11}$
    Cellobiose   C4.9

$C_{12}H_{23}NO_3$
    2,3-Diethylsuccinic acid monodiethylamide   A45.5

$C_{12}H_{23}NOS_2$
    Canellinin   Z8.8

$C_{12}H_{24}$
    1-Cyclohexyl-3-methylpentane   A59.6
    2,2-Dimethyl-3-cyclohexylbutane   A41.2

$C_{12}H_{24}O$
    6-Cyclohexylhexan-3-ol   A58.3

$C_{12}H_{24}O_2$
    3-Methylundecanoic acid   A32.2

$C_{12}H_{24}O_3$
    3-Hydroxydodecanoic acid   A7.5

C_{13}

$C_{13}H_{10}N_2O_4$
    Thalidomide   A9.9

$C_{13}H_{11}MnO_4$
    Tricarbonylmanganese cyclopentadienyl derivs.   X9.18, X9.22

$C_{13}H_{12}Fe$
    1-Ethynyl-2-methylferrocene   X9.2

$C_{13}H_{12}OS$
    p-Biphenylylmethylsulphoxide   Z8.5
    Phenyl-p-tolylsulphoxide   Z7.6

$C_{13}H_{12}O_2$
    2-(1-naphthyl)-propionic acid   A48.5
    2-(2-Naphthyl)-propionic acid   A48.7

$C_{13}H_{12}O_2S$
    3-Phenyl-2-(2-thienyl)-propionis acid   A43.10
    2-Phenyl-3-(2-thienyl)-propionic acid   A50.3

$C_{13}H_{12}O_7$
    3,4-Dihydro-8-methoxy-3-methylisocoumarin-6,7-dicarboxylic acid   Y2.4

$C_{13}H_{13}BrO_2$
    3-Bromo-4-isopropyl-4-phenylcrotonolactone   X3.4

$C_{13}H_{13}ClO_5$
  2-Benzoyloxy-2-methylmalonic acid monocid chloride, ethyl ester — A33.15
$C_{13}H_{13}NO_2S$
  N-Methyl-S-phenyl-S-phenoxy-sulphoximine — Z8.16
$C_{13}H_{14}$
  2-Phenylnorborn-2-ene — A47.1
  2-Methylene-5,6-benzobicyclo-[3.3.2]-octane — A47.7
$C_{13}H_{14}FeO_2$
  1-1''-Dimethylferrocene-3-carboxylic acid — X9.4
$C_{13}H_{14}N_2O_8$
  2,2-Dimethyl-3-hydroxybutyric acid 3,5,-dinitrobenzoate — A12.14
$C_{13}H_{14}NOP$
  N-Phenyl methylphenylphosphinic amide — Z6.7
$C_{13}H_{14}N_2O_4S_2$
  Gliotoxin — Y24.7
$C_{13}H_{14}O$
  3-Phenylnorbornan-2-ones — A47.2, A47.5
$C_{13}H_{14}O_2$
  5-Methyl-4-phenylhexa-2,3-dienoic acid — X3.3
  Tremetone — Y1.8
$C_{13}H_{14}O_3$
  Hydroxytremetone — Y1.8
  Toxol — Y1.5
$C_{13}H_{14}OS$
  3-Phenyl-2-(2-thienyl)-propan-1-ol — A43.15
$C_{13}H_{14}O_5$
  Citrinin — Y13.7
$C_{13}H_{14}S$
  3-Phenyl-2-(2-thienyl)-propane — A43.15
$C_{13}H_{15}NO$
  3,4,11,11a-Tetrahydro-1H-benzo-[b]-quinolizin-2(6H)-one — A5.19
  3,4,11,11a-Tetrahydro-2H-benzo-[b]quinolizin-1(6H)-one — A5.17
$C_{13}H_{15}NO_2$
  Securinine — K18.2
$C_{13}H_{15}NO_3$
  N-Benzoylpipecolic acid — A10.7
$C_{13}H_{16}N_2O$
  1,2,3,4-Tetrahydroharmine — K16.5
$C_{13}H_{16}O$
  3-tert-Butylindan-1-one — A49.4
  2,2-Dimethyl-3-ethylindan-1-one — A42.8
  3-Phenylnorbornan-2-ol — A47.6
$C_{13}H_{16}O_2$
  2-Phenylnorbornane-1,-2-diol — A47.3
  3-(2-ethylbenzoyl)-tetrahydrofuran — A30.4
$C_{13}H_{16}O_9$
  Julimycin deriv. — Y18.5
$C_{13}H_{17}BrO_7$
  6β-Bromo-7β,7aβ-epoxy-4α,6α-dihydroxy-7α-methoxycarbonyl-3β,4β-dimethyl-3aβ,7aβ-octahydrobenzo[c]-furan-1-one — Y18.7

$C_{13}H_{17}N$
  2-Amino-3-phenylnorbornane — A47.4
  1,3,4,6,11,11a-Hexahydro-[2H]-benzoquinolizine — A5.18
  1,2-Benzoquinolizidine — K18.3
$C_{13}H_{17}NO$
  2-Oxo-1,4-dimethyl-1,2,3,4-tetrahydroquinoline — A56.15
$C_{13}H_{18}O$
  2-Methyl-1-phenylhexan-3-one — A43.20
  4-Methyl-4-phenylhexan-3-one — A56.13
  3,3-Dimethyl-4-phenylpentan-2-one — A42.1
$C_{13}H_{18}O_2$
  2-Phenylheptanoic acid — A50.2
  5-Phenylheptanoic acid — A63.1
  4-Methyl-6-phenylhexanoic acid — A61.19
$C_{13}H_{18}O_2$
  4,4-Dimethyl-3-phenylpentanoic acid — A49.8
  2,2-Dimethyl-3-phenylpentanoic acid — A42.9
  2-Benzylhexanoic acid — A43.5
  2-(Buta-1,3-dienyl)-3-hydroxy-4-(penta-1,3-dienyl)-tetrahydrofuran — Y15.9
$C_{13}H_{18}OP$
  Cyclohexylmethylphenylphosphine oxide — Z4.2
$C_{13}H_{19}N$
  Cyclohexylphenylmethylamine — A23.20
$C_{13}H_{19}NO_3$
  O-Methylanhalonidine — K2.2
$C_{13}H_{19}N_3O_5S_2$
  Sparsomycin — Y20.5
$C_{13}H_{20}$
  2-Methyl-1-phenylhexane — A43.19
$C_{13}H_{20}N_2O_6$
  Actinobolin — Y30.4
$C_{13}H_{20}N_2OS_2$
  Iberin thiourea deriv. — Z8.12
$C_{13}H_{20}O$
  α-Ionone — T54.8
  γ-Ionone — T54.6
  5-Phenylheptan-1-ol — A62.21
  8,8,10-Trimethyl-$\Delta^{1(9)}$-2-octalone — T26.18
$C_{13}H_{20}O_2$
  2-(3,5-Dimethoxy-2-methylphenyl)-butane — Y13.10
$C_{13}H_{20}O_3$
  3-(3,5-dimethoxy-2-methylphenyl)-butan-2-ol — Y13.18
$C_{13}H_{21}NO$
  Luciduline — K26.3
$C_{13}H_{21}NO_2$
  Anodendrine — K22.11
  Alloanodendrine — K22.11
$C_{13}H_{22}Cl_6OPSb$
  Ethoxy-tert-butylmethylphenylphosphonium hexachloroantimonate — Z4.9
$C_{13}H_{22}O$
  5,8,10-Trimethyl-2-decalone — T27.3

Solanone T58.2
Tetramethylperhydroindanone T26.16

$C_{13}H_{22}O_3$
   Methyl dihydrojasmonate A36.13

$C_{13}H_{23}NO$
   Valeroidine K28.4

$C_{13}H_{24}N_2O$
   Anaferine K19.11

$C_{13}H_{24}O$
   5-Cyclohexylheptan-3-one A42.3

$C_{13}H_{26}O$
   8-Methyl-5-isopropylnonan-2-one T58.3

$C_{13}H_{26}O_2$
   2-Methyldodecanoic acid A34.2

$C_{13}H_{27}NO$
   N-Methyl-N-(2,2-dimethylpropyl)-4-methylcyclohexylamines Z3.3, Z3.4

$C_{13}H_{27}O_2P$
   Menthyl ethylmethylphosphinate Z6.3

C_{14}

$C_{14}H_{12}O$
   1,2-Diphenyl-1,2-epoxyethane A23.12

$C_{14}H_{12}O_2$
   Benzoin A23.9
   9,10-Dihydro-9,10-dihydroxyphenanthrene A24.6

$C_{14}H_{12}S$
   1,2-Diphenylthiiran A23.5

$C_{14}H_{13}ClO$
   2-Chloro-1,2-diphenylethanol A23.17

$C_{14}H_{13}N$
   2,3-Diphenylaziridine A23.19

$C_{14}H_{13}NO$
   Desylamine A23.18

$C_{14}H_{13}NO_4$
   3-Benzamido-3-(2-furyl)-propionic acid K20.1

$C_{14}H_{14}$
   2,2'-Dimethylbiphenyl X5.9

$C_{14}H_{14}FeO$
   1,2-(α-oxotetramethylene)-ferrocene X9.7

$C_{14}H_{14}O$
   1,2-Diphenylethanol A23.16

$C_{14}H_{14}OS$
   Benzyl-p-tolylsulphoxide D2.11
   Tolyl p-tolylsulphoxides Z7.6

$C_{14}H_{14}O_2$
   Ichthyothereol Y15.2

$C_{14}H_{14}O_2S$
   Benzyl p-tolyl-[$^{16}O^{18}O$]-sulphone D2.12

$C_{14}H_{14}O_3$
   Kawain Y15.10

$C_{14}H_{14}O_4$
   Lomatin Y1.3
   Decursinol Y1.9
   6,6'-Dimethyl-2,2'-diphenic acid X5.12
   Dihydrooroselol Y1.10
   Marmesin Y1.11

$C_{14}H_{14}O_5$
   Khellactones Y1.1, Y1.2

$C_{14}H_{14}O_6$
   Decevinic acid K35.4

$C_{14}H_{15}As$
   Benzylmethylphenylarsine Z3.8

$C_{14}H_{15}BrO_2$
   3-Bromo-4-tert-butyl-4-phenylcrotonolactone X3.4

$C_{14}H_{15}N$
   1,2-Diphenylethylamine A20.15

$C_{14}H_{15}NO$
   2-Amino-1,2-diphenylethanol A20.14, A20.19

$C_{14}H_{15}NO_4$
   5-Methylcyclohex-2-en-1-ol, p-nitrobenzoate A38.15

$C_{14}H_{15}N_3O_2$
   Tryptophan deriv. K30.7

$C_{14}H_{15}N_3O_2$
   Indolmycin Y27.5

$C_{14}H_{15}OP$
   Benzylmethylphenylphosphine oxide Z5.7, Z6.8

$C_{14}H_{15}O_2P$
   Methyl p-methoxyphenylphosphine oxide Z4.16

$C_{14}H_{15}P$
   Benzylmethylphenylphosphine Z5.11, Z6.9

$C_{14}H_{16}Fe$
   2-Methyl-[3]-ferrocenophane X9.9
   3-Methyl-[3]-ferrocenophane X9.5

$C_{14}H_{16}FeO$
   α-endo-Hydroxytetramethyleneferrocene X9.12

$C_{14}H_{16}N_2$
   1,2-Diamino-1,2-diphenylethane A19.3 also A23.4
   2,2'-Diamino-6,6'-dimethylbiphenyl X5.8

$C_{14}H_{16}N_2O_6$
   Indicaxanthin K17.10

$C_{14}H_{16}O$
   2-Benzylidene-5-methylcyclohexanone X4.1

$C_{14}H_{16}O_2$
   5,5-Dimethyl-4-phenylhexa-2,3-dienoic acid X3.3

$C_{14}H_{16}O_3$
   Fraxinellone Y19.14
   Desmethyldesmotroposantonin T22.10

$C_{14}H_{16}O_4$
   1,2-Diacetoxytetralin A25.2

$C_{14}H_{18}$
   1-tert-Butyl-3-methylindene A49.2
   3-tert-Butyl-1-methylindene A49.1
   1-Benzylidene-4-methylcyclohexane X4.2

$C_{14}H_{18}Fe$
   1,2-Tetramethylene-3-methylferrocene X9.3

$C_{14}H_{18}N_2O_3$
   Physovenine K30.9

$C_{14}H_{18}O$
   Chamaecynone T20.7

$C_{14}H_{18}O_5$
   4,6-O-Benzylidene-3-deoxy-α-D-ribohexopyranoside Y30.3

Formulae Index

$C_{14}H_{19}N$
   9-Phenyl-9-azabicyclo-[6.1.0]-nonane   X2.10
$C_{14}H_{19}NO$
   2-Dimethylamino-2-phenylcyclo-
      hexanone   A52.1
$C_{14}H_{19}NO_4$
   Anisomycin   Y27.3
$C_{14}H_{19}N_2O_7P$
   1α-Ribofuranosyl-5,6-dimethylbenzimidazole
      phosphate   Y24.3
$C_{14}H_{20}N_2O_4S$
   Penicillin F   Y29.4
$C_{14}H_{20}O$
   3,3-Dimethyl-4-phenylhexan-2-one   A42.12
   Chamaecynenol   T20.11
$C_{14}H_{20}O_2$
   5-Methyl-7-phenylheptanoic acid   A61.18
$C_{14}H_{20}O_3$
   13-nor-3-dehydroisoiresin   T32.5
$C_{14}H_{21}NO$
   2-Dimethylamino-1-phenylcyclo-
      hexanol   A52.4
   2-Dimethylamino-2-phenylcyclo-
      hexanol   A52.2
   Sedamine   K18.8
   Allosedamine   K18.7
   1,3-Dimethyl-5-ethoxy-3-ethyl-2,3-
      dihydroindole   A55.8
$C_{14}H_{22}N_2$
   2-Piperidinoethylamine   A13.2
$C_{14}H_{22}N_2O_2$
   Lamprolobine   K21.3
$C_{14}H_{22}O$
   Norcedranone   T29.4
   Apoaromadendrone   T26.3
   α-Apoaromadendrone   T26.5
   cis-α-Irone   T54.7
   trans-α Irone   T54.4
   β-Irone   T54.1
   cis-γ-Irone   T54.3
   Norseychellanone   T27.1
   Khusimol deriv.   T27.12
   3,4-dimethyl-4-phenyl-
      hexan-3-ols   A56.12, A56.16
$C_{14}H_{22}O_2$
   Rishitin   T22.14
   2,5,5,9-Tetramethyldecalin-1,6-dione   T42.12
$C_{14}H_{22}O_3$
   Indanone deriv. from kolavic acid   T37.11
$C_{14}H_{24}N_2$
   Angustifoline   K21.7
$C_{14}H_{24}O$
   Hexahydrochamaecynone   T20.3
$C_{14}H_{25}N_3O_9$
   Kasugamycin   C4.11
$C_{14}H_{26}O_3$
   Lesquerolic acid   A7.16
$C_{14}H_{28}O$
   2,6,10-Trimethylundecanal   T56.6
$C_{14}H_{28}O_2$
   8-Hydroxytetradecanoic acid   A7.15
$C_{14}H_{30}O$
   Tetradecan-3-ol   A13.11

C_{15}

$C_{15}H_{10}N_2O_5$
   4',1''-Dinitrodibenz-1,3-cycloheptadiene-
      6-one   X6.1
$C_{15}H_{11}I_4NO_4$
   Thyroxine   A5.14
$C_{15}H_{12}$
   1,3-Diphenylallene   X1.11
$C_{15}H_{12}O_3$
   7-Hydroxyflavan-4-one   Y6.3
$C_{15}H_{12}O_4$
   Liquiritigenin   Y6.2
$C_{15}H_{12}O_5$
   Naringenin   Y6.6
$C_{15}H_{12}O_6$
   Fustin   Y4.3
   Hesperetin   Y6.5
   Taxifolin   Y4.8
$C_{15}H_{14}$
   1,2-Diphenylcyclopropane   A44.10
$C_{15}H_{14}O$
   1-Benzoyl-1-phenylethane   A48.11
$C_{15}H_{14}O_2$
   Flavan-7-ol   Y6.4
   Methylbenzoin   A51.9
   2,3-Diphenylpropionic acid   A43.11
$C_{15}H_{14}O_5$
   Fisetinidol   Y4.6
$C_{15}H_{14}O_6$
   Catechin   Y3.1
   Epicatechin   Y3.3
   Mollisacacidin   Y4.4
$C_{15}H_{14}O_7$
   Canescins   Y6.9
$C_{15}H_{15}NO_2$
   2-Anilino-2-phenylpropionic acid   A40.18
   3-Amino-2,3-diphenylpropionic
      acids   A43.6, A43.12
   1,2,3,4-Tetrahydro-4,6-dihydroxy-1-
      phenylisoquinolines   K1.2, K1.5
$C_{15}H_{15}NO_4$
   Thyronine   A5.11
$C_{15}H_{16}$
   1,2-Diphenylpropane   A48.12
$C_{15}H_{16}FeO$
   5-Methyl-1,2-(α-oxotetramethylene)-
      ferrocene   X9.6
$C_{15}H_{16}N_2O_5S$
   Penicillin X   Y29.4
$C_{15}H_{16}O$
   1,2-Diphenylpropan-1-ols   A48.10, A48.16
   2,3-Diphenylpropan-1-ol   A43.8
$C_{15}H_{16}O_2$
   1,2-Diphenylpropane-1,2-diol   A51.8

Formulae Index

$C_{15}H_{16}O_3$
    Linderalactone   T20.12
    Isolinderalactone   T20.13

$C_{15}H_{16}O_5$
    Pentalenolactone   T30.14
    "Quinone A"   Y10.5
    Visamminol   Y1.15

$C_{15}H_{16}O_6$
    Elephantol   T24.4
    Plumericin   T16.3
    Picrotoxinin   T31.9
    Isoplumericin   T16.3

$C_{15}H_{16}O_8$
    Leucodrin   Y6.7

$C_{15}H_{17}BrO_3$
    2-Bromo-β-desmotroposantonin   T21.14

$C_{15}H_{17}BrO_6$
    α-Bromoisotutin   T31.7

$C_{15}H_{17}N$
    1-Amino-1,2-diphenylpropanes   A43.1, A43.9

$C_{15}H_{17}NO$
    3-Amino-2,3-diphenylpropan-1-ol   A43.2

$C_{15}H_{17}NO_3$
    Platydesmine   K31.4

$C_{15}H_{17}NO_5$
    Methyl-4,6-O-benzylidene-3-cyano-
      3-deoxyaltropyranoside   C4.13

$C_{15}H_{17}NO_2S_2$
    Methyl-p-tolyl-N-tosysulphilimine   Z8.1

$C_{15}H_{17}NO_3S_2$
    N-tosyl-methyl-p-tolylsulphoximine   Z8.2

$C_{15}H_{18}FeO$
    α-endo-hydroxytetramethylene-5-methyl-
      ferrocene   X9.11

$C_{15}H_{18}N_2$
    1,2,3,4,6,7,12,12b-octahydroindolo-
      [2,3-a]-quinolizine   K16.4

$C_{15}H_{18}O$
    Lindestrene   T21.16
    Lindenene   T21.9

$C_{15}H_{18}O_2$
    Cacalol   Y17.4
    Ligularenolide   T23.3
    Furanoligularenone   T23.4
    Warburgiadione   T23.18
    Linderene   T21.9
    Cataponol   Y17.1

$C_{15}H_{18}O_3$
    α-Santonin   T21.15
    β-Santonin   T22.11
    Achillin   T22.13
    Ambrosin   T24.5
    Perezinone   T1.10

$C_{15}H_{18}O_4$
    Marasmic acid   T30.13
    Artemisin   T21.7
    Helenalin   T24.7
    Parthenin   T24.8

$C_{15}H_{18}O_5$
    Pulvilloric acid   Y13.4

$C_{15}H_{18}O_6$
    Tutin   T31.8

$C_{15}H_{19}Br$
    Laurenisol   T12.13

$C_{15}H_{19}BrO$
    Aplysin   T12.16
    Laurinterol   T12.12

$C_{15}H_{19}BrO_2$
    Aplysinol   T12.16

$C_{15}H_{20}$
    Laurene   T12.13

$C_{15}H_{20}N_2O$
    Anagyrine   K21.5
    Thermopsine   K21.11

$C_{15}H_{20}N_2O_2$
    Baptifoline   K21.5

$C_{15}H_{20}O$
    α-Cuparenone   T12.9
    Nardostachone   T23.9
    Isofuranogermacrene   T20.14
    ar-Turmerone   T16.6

$C_{15}H_{20}O_2$
    Bilobanone   T17.9
    Costunolide   T22.1
    Cuparenic acid   T12.10

$C_{15}H_{20}O_3$
    Asperilin   T21.5
    Balchanolide   T21.8
    α-Pipitzol   T1.13
    β-Pipitzol   T1.12
    Douglanine   T22.3
    Perezone   T1.9

$C_{15}H_{20}O_4$
    Helicobasidin   T12.5
    ψ-Santonin   T21.11
    Geigerin   T21.2
    6-(3-formylbutyl)-2-methoxy-3-methylbenzoic
      acid, methyl ester   Y11.4
    Isophotosantonic lactone   T22.12
    Vulgarin   T22.2
    Illudin S   T30.9
    Laccijalaric acid   T29.1
    Hirsutic acid   T31.12

$C_{15}H_{20}O_5$
    Solsitalin   T24.3

$C_{15}H_{20}O_6$
    Shellolic acid   T29.7
    2-Hydroxymethylnorbornane
      p-toluenesulphonate   A47.10

$C_{15}H_{21}Br_2ClO_2$
    Pacifenol   T30.15

$C_{15}H_{21}N$
    Patchoulipyridine   T27.4
    Epiguaipyridine   T25.4

$C_{15}H_{21}NO_2$
    1,5-Dimethyl-3-benzamidohex-
      5-en-2-ol   A20.7

Formulae Index

$C_{15}H_{21}N_3O_3$
   Geneserine   K30.8
$C_{15}H_{21}N_3O_7S$
   Cephalosporin C   Y29.8
$C_{15}H_{22}$
   Cuparene   T12.10
   α-Curcumene   T16.5
   γ-Curcumene   T16.8
$C_{15}H_{22}N_2O$
   Monspessaulanine   K21.14
$C_{15}H_{22}O$
   Eremophilone   T23.6
   Nootkatone   T23.13
   Isonootkatone   T23.13
   Ishwarone   T23.16
   Isoishwarone   T23.12
   α-Santalal   T11.3
   α-Cyperone   T20.1
   β-Cyperone   T20.1
   Aristolone   T26.15
   Cyclocolorenone   T22.15
   Occidol   T21.17
   Cuparenol   T12.10
   β-Vetivone   T29.11
   Rotundone   T25.1
   Cyperotundone   T27.10
   epi-α-Cyperone   T19.3
   Chiloscyphone   T18.12
   Mustakone   T18.1
$C_{15}H_{22}OS$
   2-Octyl-p-tolysulphoxides   Z7.8, Z7.2
$C_{15}H_{22}O_2$
   Helminthosporal   T30.8
   ψ-Longifolic acid   T28.1
   Petasalbine   T23.5
   Zizanoic acid   T27.8
   Rotunols   T20.15
   Furopelargone A   T14.2
   Hinokiic acid   T26.13
   Confertifolin   T32.6
   Drimenin   T32.2
$C_{15}H_{22}O_3$
   Ngaione   T17.6
   Epingaione   T17.7
$C_{15}H_{22}O_4$
   Verrucarol   T31.2
   Fukinolidiol   T23.1
   Iresin   T32.1
$C_{15}H_{22}O_6$
   Pseudoanisatin   T30.16
$C_{15}H_{22}O_9$
   Aucubin   T15.11
$C_{15}H_{23}BrO$
   Cyperene deriv.   T27.9
$C_{15}H_{23}ClN_2O$
   11β-Chloromatrine   K24.4
$C_{15}H_{23}N$
   Porantherine   K29.7

$C_{15}H_{23}NO$
   Deoxynupharidine   K24.11
$C_{15}H_{23}NO_2$
   1,5-Dimethyl-3-benzamidohexan-2-ol   A20.2
   Nupharidine   K24.11
$C_{15}H_{23}NO_4$
   Actidione   Y21.9
$C_{15}H_{24}$
   Seychellene   T27.2
   Alloaromadendrone   T26.8
   Longifolene   T28.2
   Longicyclene   T28.1
   Sativene   T28.5
   Cyclosativene   T28.6
   Caryophyllene   T28.13
   Neoclovene   T28.14
   Pseudoclovene A   T28.12
   Bicycloelemene   T19.1
   β-Selinene   T19.10
   α-Selinene   T19.10
   δ-Selinene   T19.8
   Bicyclogermacrene   T19.2
   α-Ferulene   T26.11
   Aristolene   T26.11
   Thujopsene   T26.13
   Himachalenes   T12.14
   β-Gorgonene   T31.10
   Cyperene   T27.10
   β-Gurjunene   T26.6
   γ-Gurjunene   T25.7
   iso-α-Gurjunene   T25.9
$C_{15}H_{24}$
   β-Cadinene   T18.9
   γ-Cadinene   T18.10
   Copaene   T18.1
   ε-Muurolene   T18.3
   Germacrenes   T18.13
   Cyclocopacamphene   T28.7
   ε-Bulgarene   T18.15
   α-Bourbonene   T18.14
   β-Bourbonene   T18.14
   Aromadendrene   T26.2
   α-Santalene   T11.3
   β-Santalene   T11.9
   α-Cedrene   T29.2
   β-Bisabolene   T2.6
   Zingiberene   T2.9
   Sesquiphellandrene   T2.9
   α-Chamigrene   T26.14
   Sesquicarene   T6.2
$C_{15}H_{24}BrO_3P$
   Rearranged cyperene bromophosphonate deriv.   T27.11
$C_{15}H_{24}N_2O$
   Matrine   K24.6
   Leontine   K24.7
   Lupanine   K21.6
   Isolupanine   K21.10
   Allomatrine   K24.7

Formulae Index

$C_{15}H_{24}O$		
α-Biotol	T29.5	
Khusinol	T27.8	
Copacamphor	T11.1	
Calacone	T16.4	
α-Betulenol	T28.9	
β-Betulenol	T28.11	
Humulenol I	T30.3	
Humelenol II	T30.6	
Humulene epoxide II	T30.5	
Humulene epoxide I	T30.2	
Epishyobunone	T22.7	
Preisocalamendiol	T22.9	
Occidentalol	T22.4	
Oxocalarane	T26.7	
Oxothujopsane	T26.12	
Shyobunone	T22.8	
α-Agarofuran	T19.7	
β-Agarofuran	T19.7	
Fukinone	T23.7	
Lanceol	T2.6	
Sirenin	T6.3	
α-Santalol	T11.3	
Vetiselinenol	T19.5	

$C_{15}H_{24}O_2$
- Cyperolone — T19.12
- Occidenol — T20.8
- Acorone — T29.9
- Santanolide C — T22.5
- Tetrahydroalantolactone — T21.6
- Humulene dioxide — T30.1
- 2-Cyclohexyl-2-phenylpropionic acid — A56.17
- Hydroxydihydroeremophilone — T23.11
- Daucone — T25.14

$C_{15}H_{24}O_2S$
- 2-Octyl-*p*-tolylsulphone — A12.19

$C_{15}H_{24}O_3$
- Todomatuic acid — T2.1
- 1-(3,5-dimethoxyphenyl)-heptan-2-ol — Y13.5
- Illudol — T30.11
- Cryptomerone — T16.14

$C_{15}H_{24}O_4$
- Shiromodiol acetate — T20.17
- Pulchellin A — T24.9

$C_{15}H_{24}S$
- 2-Octyl-*p*-tolylsulphide — A12.20

$C_{15}H_{25}NO_2$
- Nupharamine — K24.8

$C_{15}H_{26}$
- Seychellene — T27.2
- Tetrahydro-*epi*-α-cyperone — T20.6
- Calarane — T26.10
- Isopatchoulane — T27.7

$C_{15}H_{26}Br_2$
- Bulgarene dihydrobromide — T18.16
- Cadinene dihydrobromide — T18.4

$C_{15}H_{26}Cl_2$
- Cadinene dihydrochloride — T18.4

$C_{15}H_{26}N_2$
- Sparteine — K21.8
- Isosparteines — K21.12, K21.13

$C_{15}H_{26}N_2O$
- Retamine — K21.8
- Leontiformine — K21.9

$C_{15}H_{26}O$
- Bulnesol — T25.11
- Cubeb camphor — T18.2
- Widdrol — T26.17
- Viridiflorol — T26.9
- Cedrol — T29.3
- Ledol — T26.4
- Cubenol — T18.11
- epi-Cubenol — T18.8
- Torreyol — T18.5
- Carotol — T25.13
- Guaiol — T25.5
- Liguloxide — T25.3
- Guaioxide — T25.6
- 10-*epi*-Liguloxide — T25.10
- Hinesol — T29.10
- Allohimachalol — T12.17
- Octahydrodehydroxylinderene — T21.13
- Maaliol — T19.4
- Intermedeol — T19.13
- Neointermedeol — T19.14
- α-Eudesmol — T19.9
- β-Eudesmol — T19.9
- γ-Eudesmol — T19.9
- Valeranone — T19.6
- Hedycarol — T20.9
- Elemol — T20.5
- 5,10-Dimethyl-3-isopropyl-2-decalone — T23.8
- Juniperol — T28.3
- Patchouli alcohol — T27.5
- Acorenol — T29.6
- Globulol — T26.1
- Nerolidol — T3.12

$C_{15}H_{26}O_2$
- Culmorin — T28.4
- Tetrahydrosaussurea lactone — T20.4
- Daucol — T25.14
- α-Kessyl alcohol — T25.12
- Oplopanone — T18.6
- 2,5,5,9-Tetramethyldecalin-1-carboxylic acid — T42.4

$C_{15}H_{26}O_3$
- 8α-Hydroxy-4,5-11α(H)-eudesman-13-oic acid — T21.10
- α-Kessyl glycol — T25.12

$C_{15}H_{26}O_5$
- Laserol — T25.15

$C_{15}H_{27}NO_2$
- 2-(Cyclohexylamino)-2-cyclohexyl propionic acid — A40.19

$C_{15}H_{28}$
- Nootkatane — T23.10

Formulae Index

Selinane	T19.11	$C_{16}H_{16}O_2$		
Cadinane	T18.7	3,3-Diphenyl-2-methylpropionic acid		A35.15
$C_{15}H_{28}N_2$		2,3-Diphenylpropionic acid, methyl		
Desoxymatrinol	K24.10	ester		A43.11
$C_{15}H_{28}O$		$C_{16}H_{16}O_4$		
2,2-Dimethyl-5-cyclohexylheptan-3-one	A42.7	Angolensin		Y5.6
Dihydroguaiols	T25.8, T25.10	Eleutherin		Y2.3
Eleman-8-one	T20.10	Isoeleutherin		Y2.3
Fukinan-8-ol	T23.2	$C_{16}H_{16}O_6$		
$C_{15}H_{30}O$		Bostrychin		Y17.2
3,7,11-Trimethyldodecanal	T56.8	$C_{16}H_{17}NO_3$		
C_{16}		Caranine		K6.3
$C_{16}H_{12}O_5$		Crinine		K6.10
Maackiain	Y3.8	$C_{16}H_{17}NO_4$		
$C_{16}H_{13}BrO_2$		Isolunine		K31.6
1-Bromo-2,2-diphenylcyclopropanecarboxylic		$C_{16}H_{18}$		
acid	A35.14	2,3-Diphenylbutane		A43.4
$C_{16}H_{13}NO_3$		$C_{16}H_{18}N_2$		
Bridged biphenyl	X7.3	Lysergine		K17.2
$C_{16}H_{14}O_2$		Agroclavine		K17.1
10,12-Dihydro-4H-5,11-dioxadibenzo-		$C_{16}H_{18}N_2O$		
[ef,kl]-heptane	X10.7	Setoclavine		K17.5
2,3-Diphenylcyclopropanecarboxylic		Elymoclavine		K17.1
acid	A44.15	Lysergol		K17.2
2,2-Diphenylcyclopropanecarboxylic		$C_{16}H_{18}N_2O_2$		
acid	A35.13	Penniclavine		K17.5
$C_{16}H_{14}O_3$		$C_{16}H_{18}N_2O_4S$		
4-Methoxydalbergione	Y4.10	Penicillin G		Y29.4
$C_6H_{14}O_4$		$C_{16}H_{18}N_2O_5S$		
2,3-Diphenylsuccinic acid	A43.7	Penicillin V		Y29.4
6-Benzyloxycoumaran-2-carboxylic		$C_{16}H_{18}O$		
acid	Y1.12	4,4-Diphenylbutan-2-ol		A14.3
$C_{16}H_{14}O_5$		$C_{16}H_{18}O_2$		
Oxypeucedanin	T57.7	2,3-Diphenylbutane-1,4-diol		A43.3
Sakuranetin	Y6.6	$C_{16}H_{18}OS$		
Phyllodulcin	Y6.1	Mesityl-p-tolylsulphoxide		Z7.6
$C_{16}H_{14}O_6$		$C_{16}H_{19}NO_2$		
Peltogynol B	Y4.5	Elaiocarpine		K29.3
Peltogynol	Y4.1	Isoeliocarpine		K29.3
$C_{16}H_{14}OS$		$C_{16}H_{19}NO_3$		
10,12-Dihydro-4H-5,11-oxathiadibenzo		α-Erythroidine		K7.6
[ef,kl]-heptane	X10.10	$C_{16}H_{19}NO_4$		
$C_{16}H_{14}S_2$		Dihydrolycorine		K6.5
10,12-Dihydro-4H-5,11-dithiadibenzo-		Balfourodine		K31.1
[ef,kl]-heptane	X10.11	$C_{16}H_{19}N_3O_3$		
$C_{16}H_{16}$		Febrifugine		K20.2
2,2-Diphenyl-1-methylcyclopropane	A35.12	$C_{16}H_{19}N_3O_6$		
9,10-Dihydro-4,5-dimethylphenanthrene	X5.5	Mitomycin A		Y27.1
$C_{16}H_{16}Br_2$		$C_{16}H_{19}N_2O_2P$		
2,2'-bis(bromomethyl)-6,6'-dimethyl-		Methylphenylpropylphosphine		
biphenyl	X5.6	p-nitrophenylimine		Z5.1
$C_{16}H_{16}N_2O_2$		$C_{16}H_{20}IP$		
Lysergic acid	K17.6	Benzylethylmethylphenylphosphonium		
Isolysergic acid	K17.6	idoide		Z6.1
$C_{16}H_{16}N_2O_6S_2$		$C_{16}H_{20}N_2$		
Cephalothin	Y29.7	Festuclavine		K17.3
$C_{16}H_{16}O$		Costaclavine		K17.3
1-(p-hydroxyphenyl)-tetralin	A52.12	Pyroclavine		K17.3

Formulae Index

$C_{16}H_{20}N_2O$
    Chanoclavines — K17.4

$C_{16}H_{21}NO_2$
    Isoelaeocarpiline — K29.2

$C_{16}H_{21}NO_3$
    Dihydro-β-erythroidine — K7.5
    Annotinine — K25.6

$C_{16}H_{22}N_2$
    Lycodine — K25.3

$C_{16}H_{22}O_9$
    Sweroside — T13.6

$C_{16}H_{23}IOS$
    Menthyl-p-iodobenzenesulphinate — Z7.7

$C_{16}H_{23}NO$
    Fawcettidine — K25.7

$C_{16}H_{23}NO_2$
    Serratidine — K25.1

$C_{16}H_{23}NO_3$
    Annotinine deriv. — K25.9
    β-Tetrahydro-β-erythroidine — K7.3

$C_{16}H_{24}$
    2-Cyclohexyl-2-phenylbutane — A56.9

$C_{16}H_{24}O_2$
    Menthyl benzenesulphinate — Z8.9

$C_{16}H_{24}O_3$
    Brefeldin A — Y14.7

$C_{16}H_{24}O_5$
    Ovalicine — T31.3

$C_{16}H_{24}O_{11}$
    Monotropein — T15.10

$C_{16}H_{25}BrO_5$
    Rearranged ovalicine deriv. — T31.4

$C_{16}H_{25}NO$
    Lycopodine — K25.2
    Lycopecurine — K26.1

$C_{16}H_{25}NO_2$
    Fawcettimine — K25.10
    Annofoline — K25.5
    Dendrobine — K31.12

$C_{16}H_{25}NO_3$
    Serratinine — K25.8
    Annopodine — K26.2

$C_{16}H_{25}O_2P$
    Menthyl phenylphosphinates — Z4.11, Z4.12

$C_{16}H_{26}N_2$
    Flabellidine — K25.3

$C_{16}H_{26}N_2O_4S$
    Penicillin K — Y29.4

$C_{16}H_{26}O$
    3-Cyclohexyl-3-phenylbutan-2-one — A56.14

$C_{16}H_{26}O_2$
    Davanone — T3.13
    Norambreinolide — T34.1

$C_{16}H_{26}O_3$
    Juvabione — T2.1

$C_{16}H_{28}O$
    Drimenol — T32.3

$C_{16}H_{30}O$
    Muscone — A32.14

C_{17}

$C_{17}H_{12}O_6$
    Aflatoxin B_1 — Y13.2

$C_{17}H_{12}O_7$
    Aflatoxin G_1 — Y13.1

$C_{17}H_{14}O_5$
    Pterocarpin — Y3.8
    Pisatin — Y3.12

$C_{17}H_{15}BrSi$
    Methylnaphthylphenylbromosilane — Z1.4

$C_{17}H_{15}ClGe$
    Methylnaphthylphenylchlorogermane — Z2.4

$C_{17}H_{15}ClSi$
    Methylnaphthylphenylchlorosilane — Z1.11

$C_{17}H_{15}FSi$
    Methylnaphthylphenylfluorosilane — Z1.12

$C_{17}H_{15}OP$
    Methylnaphthylphenylphosphine oxide — Z4.16

$C_{17}H_{16}Ge$
    Methylnaphthylphenylgermane — Z2.5

$C_{17}H_{16}O$
    Dimethyldibenz-1,3-cycloheptadien-6-one — X5.7

$C_{17}H_{16}O_2$
    Hinokiresinol — Y9.1
    1-Methyl-2,2-diphenylcyclopropanecarboxylic acid — A54.9
    [2,2]-paracyclophanecarboxylic acid — X8.2

$C_{17}H_{16}O_4$
    4,4′-dimethoxydalbergione — Y4.9
    Homopterocarpin — Y3.11

$C_{17}H_{16}O_5$
    Quinone from dihydrohomoterocarpin — Y3.9

$C_{17}H_{16}O_6$
    Mucroquinone — Y3.6

$C_{17}H_{16}OSi$
    Methylnaphthylphenylsilanol — Z1.16

$C_{17}H_{16}Si$
    Methylnaphthylphenylsilane — Z1.8

$C_{17}H_{17}ClO_6$
    Griseofulvin — Y17.5

$C_{17}H_{17}NO_2$
    Bridged biphenyl — X7.1

$C_{17}H_{17}NO_5$
    Hippeastrine — K6.2

$C_{17}H_{18}$
    2-Methyl-[2.2]-paracyclophane — X8.3
    12-Methyl-[2.2]-metaparacyclophane — X8.4

$C_{17}H_{18}BrNO$
    2-(p-bromophenyl)-3,4-dimethyl-5-phenyloxazolidine — A21.17

$C_{17}H_{18}N_2$
    Tröger's base — X11.7

$C_{17}H_{18}O$
    1-Benzoyl-2-phenylbutane — A42.17

$C_{17}H_{18}O_4$
    Agatharesinol — Y9.2
    Dihydrohomopterocarpin — Y3.10
    Sugiresinol — Y9.3

$C_{17}H_{18}O_5$
    Melanoxin   Y5.2
    1-(3-hydroxy-4-methoxyphenyl)-2-
        (4-methoxy-2,5-benzoquinonyl)
        propane   Y5.1
    Sequirin B   Y9.3
    Sequirin C   Y9.2

$C_{17}H_{19}NO$
    2-Benzamido-2-phenylbutane   A40.13
    1,2,3,4-Tetrahydro-6-methoxy-2-methyl-1-
        phenylisoquinoline   K1.3
    2-Benzamido-3-phenylbutanes   A49.16, A49.18

$C_{17}H_{19}NO_3$
    Buphanisine   K6.10
    Morphine   K4.5
    Coclaurine   K3.1

$C_{17}H_{19}NO_4$
    Crinamine   K6.11
    Haemanthamine   K6.11
    Isohaemanthamine   K6.12

$C_{17}H_{19}NO_5$
    Clivonine   K6.1
    Haemanthidine   K6.7
    6-Hydroxycrinamine   K6.7

$C_{17}H_{19}N_3$
    Brevicolline   K20.3

$C_{17}H_{20}AsClO_4$
    Allylbenzylmethylphenylarsonium
        perchlorate   Z3.10

$C_{17}H_{20}BrP$
    Allybenzylmethylphenylphosphonium
        bromide   Z5.9

$C_{17}H_{20}N_2O$
    Normacusine A   K11.2

$C_{17}H_{20}N_2O_2$
    Sarpagine   K11.2

$C_{17}H_{20}O$
    1,2-Diphenyl-2-methylbutan-1-ols   A56.2, A56.4

$C_{17}H_{20}O_5$
    Zeylanine   T20.16

$C_{17}H_{21}BrO_5$
    Bromogeigerin acetate   T21.1

$C_{17}H_{21}NO_2$
    Linearisine   K3.14

$C_{17}H_{21}NO_3$
    Galanthamine   K6.9
    Pluviine   K6.3

$C_{17}H_{21}NO_4$
    Cocaine   K28.6
    Hyoscine   K28.1
    Cytochalasin degradation product   Y27.7

$C_{17}H_{22}BrP$
    Benzylmethylphenylpropylphosphonium
        bromide   Z3.5

$C_{17}H_{22}O_2$
    Deoxypodocarpic acid   T33.4

$C_{17}H_{22}O_3$
    Podocarpic acid   T33.9

$C_{17}H_{22}O_5$
    Isotenulin   T24.6
    Xanthumin   T24.10
    Gaillardin   T24.2
    Xanthinin   T24.11
    Pyrethrosin   T21.12

$C_{17}H_{22}O_6$
    Isophotoartemisin acetate   T21.3

$C_{17}H_{23}BrO_3$
    Laurencin   Y15.4

$C_{17}H_{23}NO_3$
    Littorine   K28.8

$C_{17}H_{23}NO_3$
    Hyoscyamine   K28.2
    Mesembrine   K20.5
    Lycoramine   K6.9

$C_{17}H_{23}NO_4$
    Annotinine deriv.   K25.4

$C_{17}H_{23}NO_5$
    Balfourolone   K31.2

$C_{17}H_{24}O$
    Cyclohexyl-(1-methyl-1-phenylpropyl)
        ketone   A56.5
    Falcarinol   A7.4

$C_{17}H_{24}O_{10}$
    Verbenalin   T13.9
    Secologanin   T13.10

$C_{17}H_{25}NO_4$
    Stemonine   K29.5

$C_{17}H_{26}N_2O$
    Phenampromid   A13.5

$C_{17}H_{26}O_5$
    Portentol   Y15.5

$C_{17}H_{26}O_{10}$
    Loganin   T13.5

$C_{17}H_{26}O_2S$
    Menthyl p-toluenesulphinate   Z7.5

$C_{17}H_{27}NO_2$
    Stenine   K29.4

$C_{17}H_{27}NO_3$
    Nobilinone   K31.8

$C_{17}H_{27}O_2P$
    Menthyl methylphenyl-
        phosphinates   Z4.1, Z4.8

$C_{17}H_{27}O_2PS$
    Menthyl S-methylphenylphosphonothioate   Z4.10

$C_{17}H_{28}O_2$
    Ambreinolide   T34.8

$C_{17}H_{29}NO_4$
    Dendrobine deriv.   K31.11

$C_{17}H_{30}O_3$
    Densipolic acid   A7.13

$C_{17}H_{32}O_4$
    3-Methylhexadecane-1,16-dioic acid   A32.10

$C_{17}H_{34}O_2$
    14-Methylhexadecanoic acid   A31.1
    4,8,12-Trimethyltridecanoic acid
        methyl ester   T56.7

$C_{17}H_{36}O$
    Heptadecan-3-ol   A7.3

$C_{17}H_{36}O_2$
    Heptadecane-1,3-diol   A7.2

Formulae Index

C_{18}

$C_{18}H_{12}O_6$
   Sterigmatocystin — Y13.3

$C_{18}H_{13}NO_3$
   3-Phthalimido-1-tetralone — A20.11

$C_{18}H_{14}O_2$
   4,5,6,10,11,12-hexahydrodibenzo-[ef,kl]-heptalene-5,11-dione — X10.6

$C_{18}H_{15}NO_2$
   2-Phthalimidotetralin — A20.5

$C_{18}H_{16}Br_2Ge$
   Dibromomethylmethylnaphthylphenyl-germane — Z2.1

$C_{18}H_{16}Br_2Si$
   Dimromomethylmethylnaphthylphenyl-silane — Z1.7

$C_{18}H_{16}GeO_2$
   Methylnaphthylphenylcarboxygermane — Z2.9

$C_{18}H_{16}N_2O_8$
   Betanidin — K17.11

$C_{18}H_{16}O_9$
   Formylolivinic acid — Y16.2

$C_{18}H_{17}BBrO_4$
   N-(p-bromophenyl)-α-D-ribopyranosylamine-2,4-phenylboronate — C4.4

$C_{18}H_{17}BrGe$
   Bromomethylmethylnaphthylphenyl-germane — Z2.2

$C_{18}H_{17}BrSi$
   Bromomethylmethylnaphthylphenyl-silane — Z1.9

$C_{18}H_{18}Ge$
   Ethylnaphthylphenylgermane — Z2.5

$C_{18}H_{18}GeO$
   Methoxymethylnaphthylphenylgermane — Z2.8

$C_{18}H_{18}O_5$
   Agrimonolide — Y2.11

$C_{18}H_{18}O_7$
   Senepoxide — Y19.13

$C_{18}H_{18}O_8$
   Crotepoxide — Y19.2

$C_{18}H_{18}OSi$
   Methoxymethylnaphthylphenylsilane — Z1.15

$C_{18}H_{19}IO_8$
   Crotepoxide iodohydrin — Y19.1

$C_{18}H_{19}NO_2$
   4,8-Dimethoxy-5,7-12,12a-tetrahydroisoindolo-[2,1-b]-isoquinoline — X4.9

$C_{18}H_{19}NO_3$
   Erythraline — X7.1
   Stepharine — K3.12
   Morphothebaine — K4.7
   Oripavine — K4.6

$C_{18}H_{20}BrNO_2$
   4,4'-Dimethoxy-1,1',3,3'-tetrahydrospiro[isoindole-2,2'-isoindolium]-bromide — X4.7

$C_{18}H_{20}ClN_3O_5S_2$
   Sporidesmin — Y24.4

$C_{18}H_{20}N_2O$
   Eburnamonine — K16.2
   2,6-Diphenyl-1-methyl-4-piperidone oxime — X4.3
   Diazacycloheptane deriv. — X4.4

$C_{18}H_{20}N_4O_2$
   3,4-Dihydroxycyclohexane-1,2-dione phenylosazone — A16.22

$C_{18}H_{20}O$
   2-Benzoyl-2-methyl-3-phenylbutane — A42.2

$C_{18}H_{20}O_3$
   Gibberic acid — T35.16

$C_{18}H_{20}O_4$
   Guaiaretic acid — A31.17
   2,4-Bis-(p-methoxyphenyl)-butyric acid — Y9.4

$C_{18}H_{20}O_6$
   Duartin — Y3.5
   Byssochlamic acid — T58.6
   Glaucanic acid — T58.5

$C_{18}H_{20}O_7$
   Glauconic acid — T58.5

$C_{18}H_{20}O_8$
   Inumakilactone — T38.10

$C_{18}H_{21}NO_3$
   6,7-Dimethoxy-4-(p-methoxyphenyl)-1,2,3,4-tetrahydroisoquinoline — K1.7
   Codeine — K4.5
   Budbocodine — K3.9

$C_{18}H_{21}NO_4$
   Homolycorine — K6.2

$C_{18}H_{21}NO_5$
   Tazettine — K6.6
   Criwelline — K6.6
   6-Hydroxybuphanidrine — K6.13

$C_{18}H_{22}N_2$
   Condyfoline — K12.11
   Tubifoline — K12.7

$C_{18}H_{22}O_2$
   Hexoestrol — A45.8

$C_{18}H_{22}O_3$
   1-(2,4-Dimethoxyphenyl)-2-(p-methoxyphenyl)-propane — Y5.5

$C_{18}H_{22}O_5$
   Zearalenone — Y14.11

$C_{18}H_{22}O_{11}$
   Asperuloside — T13.1

$C_{18}H_{23}N$
   3-Dimethylamino-1,1-diphenyl-2-methylpropane — A31.6

$C_{18}H_{23}NO$
   3-Dimethylamino-1,1-diphenyl-2-methylpropan-1-ol — A31.6

$C_{18}H_{23}NO_4$
   Mesembrine precursor — K20.6

$C_{18}H_{23}NO_5$
   Narcissidine — K6.4

$C_{18}H_{23}NO_7$
   Retusamine — K24.1

Formulae Index

$C_{18}H_{23}N_2O$
    Hunteracine (cation)   K14.8

$C_{18}H_{24}BrNO_4S$
    N-(3-endo-bromo-2-oxo-9-norbornane-sulphonyl)-methyl-p-toluene-sulphoximide   Z8.3

$C_{18}H_{24}N_2$
    Tubifoline derivative   K12.10
    16α-Strychindole   K12.1

$C_{18}H_{24}N_2O_3$
    Julocrotine   T29.9

$C_{18}H_{24}N_4O_7$
    Butyl 2-carboxycyclohexyl ketone, DNP derivative   Y14.4

$C_{18}H_{24}O_2$
    Nimbiol   T33.13

$C_{18}H_{24}O_4$
    Fusamarin   Y2.7

$C_{18}H_{25}NO_2$
    15,16-Dimethoxy-cis-5,6-erythrinan   K7.2

$C_{18}H_{25}NO_6$
    Jacobine   K22.3

$C_{18}H_{26}O_4$
    Trisporic acid C   Y19.10

$C_{18}H_{28}O$
    Perhydrotriphenylene-2-one   X10.12

$C_{18}H_{28}O_6$
    Acetyldihydroovalicine   T31.6

$C_{18}H_{30}$
    Perhydrotriphenylene   X10.8

$C_{18}H_{30}O_3$
    C_{18} juvenile hormone   T58.8

$C_{18}H_{30}O_5$
    Dimorphecolic acid   A7.8

$C_{18}H_{32}O_2$
    Laballenic acid   X1.5
    Chalmoogric acid   A36.8

$C_{18}H_{32}O_3$
    Coriolic acid   A7.12

$C_{18}H_{34}N_2O_6S$
    Lincomycin   Y21.6

$C_{18}H_{34}O_3$
    Ricinoleic acid   A7.7

$C_{18}H_{34}O_5$
    9-Hydroxyoctadecanedioic acid   A7.9

$C_{18}H_{35}NO_2$
    Cassine   K20.7

$C_{18}H_{36}N_4O_{11}$
    Kanamycin A   C4.14

$C_{18}H_{36}O$
    6,10,14-Trimethylpentadecan-2-one   T56.3

$C_{18}H_{36}O_3$
    9-Hydroxyoctadecanoic acid   A7.1
    12-Hydroxyoctadecanoic acid   A7.10
    13-Hydroxyoctadecanoic acid   A.7.11

C_{19}

$C_{19}H_{13}F$
    1-Fluoro-12-methylbenzo-[c]-phenanthrene   X7.5

$C_{19}H_{15}GeLi$
    Methylnaphthylphenylgermyllithium   Z2.10

$C_{19}H_{18}GeO_2$
    Ethylnaphthylphenylcarboxygermane   Z2.9

$C_{19}H_{18}O_6$
    Atrovenetin   Y12.1
    Tri-O-methylpeltogynone   Y4.2

$C_{19}H_{18}O_9$
    Formylchromomycinoic acid   Y16.1
    Diacetylcanescins   Y6.9

$C_{19}H_{18}O_2Si$
    Methylnaphthylphenylacetoxysilane   Z1.17

$C_{19}H_{19}NO_4$
    Bulbocapnine   K4.9
    Amurensine   K1.4

$C_{19}H_{20}Ge$
    Isopropylnaphthylphenylgermane   Z2.6
    Ethylmethylnaphthylphenylgermane   Z2.7

$C_{19}H_{20}N_2O$
    Meloscine   K31.5

$C_{19}H_{20}NO_4P$
    Ethoxynaphthylphenylphosphonium nitrate   Z4.17

$C_{19}H_{20}O_2$
    Hinokiresinol dimethyl ether   Y9.1

$C_{19}H_{20}O_4$
    Sugiresinone dimethyl ether   Y9.5

$C_{19}H_{20}O_5$
    Nutallin   Y1.3

$C_{19}H_{20}O_6$
    Coleone B   T33.16

$C_{19}H_{20}Si$
    Ethylmethylnaphthylphenylsilane   Z1.10

$C_{19}H_{21}ClO_5$
    2-Chloro-5,7,3′,4′-tetramethoxyisoflavan   Y3.4

$C_{19}H_{21}NO_2$
    Nuciferine   K3.13

$C_{19}H_{21}NO_3$
    Pronuciferine   K3.12
    Thebaine   K4.6

$C_{19}H_{21}NO_4$
    Orientalinone   K3.15
    Salutaridine   K4.4

$C_{19}H_{21}NO_5$
    Macronine   K6.8

$C_{19}H_{22}ClNO_2$
    Aposclerotioramine   Y11.6

$C_{19}H_{22}N_2O$
    Cinchotoxine   K8.8
    Cinchonine   K8.7
    epi-Cinchonine   K8.11
    Cinchonidine   K8.4
    epi-Cinchonidine   K8.1

$C_{19}H_{22}N_2O_2$
    Cupreine   K8.4
    Caracurine VII   K12.5

$C_{19}H_{22}N_2O$
    16-Yohimbone   K9.8

$C_{19}H_{22}N_2O_4$
    2-Methylbutane-1,4-diol,bisphenyl-
        urethane   A27.20

$C_{19}H_{22}O_3$
    Gibberic acid   T35.16

$C_{19}H_{22}O_4$
    Odoratin   T53.7
    Sugiresinol dimethyl ether   Y9.3

$C_{19}H_{22}O_5$
    5,7,3′,4′-Tetramethoxyisoflavan   Y3.7

$C_{19}H_{22}O_6$
    Gibberellic acid   T35.13
    Antheridiogen-An   T35.10
    Catechin tetramethyl ether   Y3.2

$C_{19}H_{22}O_7$
    Nagilactone C   T38.13

$C_{19}H_{22}O_8$
    Seneol   Y19.12

$C_{19}H_{23}BrO_4$
    Methyl-6α-bromo-12-methoxy-7-oxo-
        podocarpate   T33.12

$C_{19}H_{23}NO_3$
    Schellhammericine   K7.7
    Schellhammeridine   K7.8
    Armepavine   K3.8

$C_{19}H_{23}NO_4$
    Salutaridinol I   K4.4
    Sinomenine   K4.3
    Schellhammerine   K7.7
    Orientaline   K3.10

$C_{19}H_{23}N_3O_2$
    Ergonovine   K17.7

$C_{19}H_{23}N_3O_4S$
    Hetacillin   Y29.1

$C_{19}H_{23}O_7$
    Nagilactone deriv.   T38.12

$C_{19}H_{24}ClNO_6$
    Acutumine   K5.4

$C_{19}H_{24}N_2$
    Aspidofractinine   K14.3
    Ibogamine   K15.6
    Cleavamine   K15.2
    1,2-Dehydroaspidospermidine   K13.12

$C_{19}H_{24}N_2O$
    Cinchonamine   K8.12
    Antirhine   K8.10
    Adenocarpine   K18.5

$C_{19}H_{24}N_2O_2$
    Quinamine   K8.9

$C_{19}H_{24}N_2O_3$
    Aspidodispermine   K13.10

$C_{19}H_{25}NO$
    Histrionicotoxin   K26.9

$C_{19}H_{26}N_2$
    Dihydrocorynantheane   K2.11
    Aspidospermidine   K13.8
    Quebrachamine   K13.5
    Indoloquinolizidine deriv.   K16.3

$C_{19}H_{26}N_2O$
    Rhazidine   K13.9

$C_{19}H_{26}O_6$
    3-*O*-Benzyl-1,2-*O*-cyclohexylidene-
        α-glucofuranose   X1.6

$C_{19}H_{26}O_{12}$
    Daphylloside   T15.13

$C_{19}H_{27}NO$
    Cyclazocine   Y20.1

$C_{19}H_{27}NO_3$
    Protoemetine   K2.10

$C_{19}H_{28}O$
    A-norstach-15-ene-2-one   T35.2

$C_{19}H_{28}O_2$
    Epoxynorcafestone   T36.7
    Testosterone   T32.13

$C_{19}H_{28}O_4$
    Columbin deriv.   T38.6

$C_{19}H_{28}O_5$
    Tricyclic ketone from steviol   T35.15

$C_{19}H_{29}NO_3$
    Dihydroprotoemetine   K2.10

$C_{19}H_{29}NO_4$
    Dendrine   K31.12

$C_{19}H_{30}O$
    Ketone from cafestol   T36.13
    17-Norkauran-16-one   T36.1
    Phyllocladene norketone   T36.6
    5α-Androstan-17-one   T27.13

$C_{19}H_{30}O_2$
    Perhydrotriphenylene-2-carboxylic acid   X10.9
    Colensenone   T34.3

$C_{19}H_{30}O_4$
    Columbin deriv.   T38.7

$C_{19}H_{31}O_2P$
    Menthyl phenylpropylphosphinate   Z4.15
    Menthyl isopropylphenylphosphinate   Z4.6

$C_{19}H_{32}$
    Hydrocarbon   K32.1
    5α-Androstane   T43.3

$C_{19}H_{34}$
    Fichtelite   T32.8

$C_{19}H_{34}O$
    Indanone deriv. from Calciferol   T48.1

$C_{19}H_{36}N_2O_5$
    Lipoxamycin   Y22.5

$C_{19}H_{38}O_2$
    10-Methyloctadecanoic acid   A32.1

$C_{19}H_{38}O_3$
    9-Hydroxyoctadecanoic acid,
        methyl ester   A7.1

C_{20}

$C_{20}H_{14}$
    1,1′-Binaphthyl   X6.7

$C_{20}H_{14}O$
    2-Hydroxy-1,1′-binaphthyl   X6.3

$C_{20}H_{16}N_2O_4$
    Camptothecin   K29.9

Formulae Index

$C_{20}H_{16}N_4O_{13}$
   2-Hydroxymethyl-3-hydroxytetrahydropyran
      bis-3,5-dinitrobenzoate — Y15.3

$C_{20}H_{17}LiGe$
   Ethylnaphthylphenylgermyllithium — Z2.10

$C_{20}H_{18}ClNO_6$
   Ochratoxin A — Y2.9

$C_{20}H_{18}O$
   1,2,2-Triphenylethanol — A23.7

$C_{20}H_{18}O_5$
   Pillaronone — Y28.4

$C_{20}H_{19}N$
   1,2,2-Triphenylethylamine — A23.3

$C_{20}H_{19}NO_5$
   Chelidonine — K5.8

$C_{20}H_{19}NO_6$
   Ochratoxin B — Y2.9
   Rhoeagenine — K1.10

$C_{20}H_{20}N_2O$
   nor-C-fluorocurarine — K12.3

$C_{20}H_{20}O_6$
   Matairesinol — Y7.8

$C_{20}H_{20}O_7$
   Herqueinone — Y12.2

$C_{20}H_{21}NO_4$
   Bridged biphenyls from schell-
      hammeridine — X7.2, X7.4
   Kopsanone — K14.1

$C_{20}H_{22}N_2O_2$
   Akuammicine — K12.8
   Condylocarpine — K12.12

$C_{20}H_{22}N_2O_3$
   Perivine — K11.8

$C_{20}H_{22}O$
   1-Hydroxy-1,2,3,4-tetrahydrobenzo-
      [2.2]-paracyclophane — X8.1

$C_{20}H_{22}O_3$
   Nafenopin — A52.12

$C_{20}H_{22}O_4$
   9,10-Dihydro-3-hydroxy-4-methoxy-
      2,5,8-trimethylphenanthrene-
      1-acetic acid — T53.14

$C_{20}H_{22}O_6$
   Columbin — T38.2
   Isocolumbin — T38.1
   Pinoresinol — Y7.7

$C_{20}H_{22}O_7$
   Palmarin — T38.4
   Chasmanthin — T38.3
   Jateorin — T38.2
   Thujastandin — Y7.16

$C_{20}H_{22}O_9$
   3-Oxodihydroinumakilactone 15-
      acetate — T38.11
   Olivin — Y16.4
   Cervicarcin — Y18.6

$C_{20}H_{23}BrO_4$
   Fucoxanthin deriv. — T55.2

$C_{20}H_{23}NO_4$
   Cularine — K5.6

$C_{20}H_{23}NO_5$
   Capaurimine — K3.3

$C_{20}H_{24}ClNO_2$
   Tetrahydroisoquinoline deriv. — K2.4

$C_{20}H_{24}INO_4$
   Cryptaustoline iodide — K5.7

$C_{20}H_{24}N_2O$
   Anhydrovobasindiol — K11.10

$C_{20}H_{24}N_2O_2$
   Quinine — K8.4
   Epi-quinine — K8.1
   Quinidine — K8.7
   Gardnerine — K11.3
   Epi-quinidine — K8.11
   Tubotaiwin — K12.12
   Quinotoxine — K8.8

$C_{20}H_{24}N_2O_4$
   Compactinervine — K13.6

$C_{20}H_{24}O_3$
   Jatrophone — T41.1

$C_{20}H_{24}O_6$
   Glaucanol — T53.8
   Lariciresinol — Y7.6

$C_{20}H_{24}O_7$
   Euparotin — T24.1
   Olivil — Y7.2
   Cycloolivil — Y7.1

$C_{20}H_{24}O_8$
   Capenicin — T31.11

$C_{20}H_{24}O_9$
   Plicatic acid — Y7.13
   Ginkgolide A — T40.11

$C_{20}H_{24}O_{10}$
   Ginkgolide B — T40.11
   Ginkgolide D — T40.11

$C_{20}H_{24}O_{11}$
   Ginkgolide C — T40.11

$C_{20}H_{25}ClN_2O_4$
   2,3-Diethyl-1,2,3,4-tetrahydro-12-
      methylindolo-(2,3-a)-quinolizinium
      perchlorate — K11.1

$C_{20}H_{25}NO_2$
   Spradine A — K32.7

$C_{20}H_{25}NO_3$
   Delnudine — K33.7
   α-Methyldihydrothebaine — K27.5

$C_{20}H_{25}NO_4$
   N-norlaudanosine — K3.5
   Erythristemine — K7.4
   1-(4′,5′-dimethoxy-2-hydroxybenzyl)-
      7-methoxy-2-methyl-1,2,3,4-tetra-
      hydroisoquinoline — K5.5

$C_{20}H_{26}Br_2O_3$
   Jatrophone dihydrobromide — T41.2

$C_{20}H_{26}N_2O$
   N-(2-Benzylmethylamino)-propyl-

propionanilide	A4.15	Sandaracopimaric acid	T32.18
Ibogaine	K15.6	Trachylobanic acid	T35.6
$C_{20}H_{26}N_2O_2$		$C_{20}H_{30}O_3$	
Ajmaline	K11.5	Steviol	T35.11
$C_{20}H_{26}O_4$		Isosteviol	T35.12
Picrosalvin	T33.15	Petasin	T23.14
Fujenal	T35.4	$C_{20}H_{30}O_4$	
$C_{20}H_{26}O_5$		Eunicin	T41.4
1-(3,4-Dimethoxyphenyl)-2-(2,4,6-		Cassaic acid	T39.9
trimethoxyphenyl)-propane	Y5.3	Kolavic acid	T37.10
		Crassin	T41.5
$C_{20}H_{26}O_6$		$C_{20}H_{30}O_5$	
Eupatoriopicrin	T21.4	Andrographolide	T36.16
Enmein	T33.2	$C_{20}H_{30}O_7$	
1-(3,4-dimethoxyphenyl)-3-(2,4,6-		Andrographolide deriv.	T36.15
trimethoxyphenyl) propan-2-ol	Y3.2	$C_{20}H_{31}NO$	
3-(3,4-dimethoxyphenyl)-2-(2,4,6-		Thelepogine	K31.7
trimethoxyphenyl) propan-1-ol	Y5.3	$C_{20}H_{32}$	
$C_{20}H_{26}O_7$		Cembrene	T41.3
Chaparrinone	T53.13	Kaurene	T33.7
$C_{20}H_{27}IN_2O_2$		Rosenonolactone deriv.	T34.15
Hunterburnine β-methiodide	K8.6	Dolabradiene	T34.13
$C_{20}H_{28}N_2$		Rimuene	T34.11
Vallesamidine	K13.1	Hibaene	T35.1
$C_{20}H_{28}O_2$		Isohibaine	T32.17
Callitrisic acid	T33.8	Phyllocladene	T36.5
$C_{20}H_{28}O_3$		Isophyllocladene	T36.5
Hardwickiic acid	T37.9	Atisirene	T32.12
Rosenonolactone	T34.12	$C_{20}H_{32}O$	
Cinerin I	T17.3	Monogynol	T35.9
Cafestol	T36.9	$C_{20}H_{32}O_3$	
Solidagenone	T37.4	Corymbol deriv.	T36.2
$C_{20}H_{28}O_4$		Isocupressic acid	T34.10
Marrubiin	T37.2	Beyerol	T35.5
Callicarpone	T33.11	$C_{20}H_{32}O_4$	
Salvin	T33.17	Portulal	T40.12
$C_{20}H_{28}O_5$		Taxa-4(16),11-diene-5α,9α,10β,13α-	
Gibberellin A_{14}	T35.17	tetraol	T40.8
Ingenol	T39.2	$C_{20}H_{34}$	
$C_{20}H_{28}O_6$		Kaurane	T35.7
Oridonin	T33.6	Beyerane	T35.8
Phorbol	T39.4	Stevane B	T35.7
Liatrin diol deriv.	T30.7	$C_{20}H_{34}O$	
$C_{20}H_{28}O_7$		Manool	T34.6
Chaparrin	T53.13	Plathyterpol	T39.10
$C_{20}H_{29}ClN_2O_2$		13-Epimanool	T34.7
Ochrosandwine (chloride)	K8.3	Manoyl oxide	T34.2
$C_{20}H_{30}$		$C_{20}H_{34}O_2$	
Abietatriene	T33.5	Torulosol	T34.9
$C_{20}H_{30}O$		Allodevadarool	T34.14
Ferruginol	T33.5	Eperuic acid	T36.10
Totarol	T32.4	$C_{20}H_{34}O_3$	
$C_{20}H_{30}O_2$		Corymbol	T36.3
Neoabietic acid	T32.11	Labdanolic acid deriv.	T36.12
Levopimaric acid	T32.11	$C_{20}H_{34}O_5$	
Palustric acid	T32.11	Prostaglandin E_1	Y16.11
Abietic acid	T32.11	$C_{20}H_{34}O_6$	
Pimaric acid	T32.14	Grayanotoxin III	T38.8
Isopimaric acid	T32.18		

$C_{20}H_{35}Cl_3$
    Manool deriv.   T34.5
$C_{20}H_{36}$
    Abietane   T33.1
$C_{20}H_{36}O_2$
    Sclareol   T34.4
$C_{20}H_{36}O_3$
    Labdanolic acid   T36.11
$C_{20}H_{36}O_5$
    Prostaglandin F1α   Y16.11
    Prostaglandin F 1β   Y16.11
$C_{20}H_{40}O$
    Phytol   T56.9

C_{21}

$C_{21}H_{12}Fe_2N_2$
    N,N'-Diferrocenylcarbodiimide   X3.5
$C_{21}H_{18}N_2$
    Isoamarine   A23.11
$C_{21}H_{21}NO_6$
    β-Hydrastine   K3.11
$C_{21}H_{22}N_2O_2$
    Strychnine   K12.6
$C_{21}H_{22}N_2O_3$
    Vomilenin   K11.4
$C_{21}H_{22}N_8O_8$
    2-Formyl-3-methylcyclopentylacetaldehyde bis-2,4-dinitrophenylhydrazone   T13.2
$C_{21}H_{23}ClO_5$
    Sclerotiorins   Y11.7
$C_{21}H_{23}NO_5$
    2,3-Methylenedioxy-1,9,10-trimethoxy-tetrahydroberberine   K3.7
$C_{21}H_{23}N_2O_3$
    Serpentine (cation)   K9.10
$C_{21}H_{24}$
    3,3,3',3'-Tetramethylbis-1,1'-spiroindane   X5.3
$C_{21}H_{24}N_2O_2$
    Catharanthine   K15.3
    Tabersonine   K13.2
$C_{21}H_{24}N_2O_3$
    Vobasine   K11.8
    Akuammidine   K11.7
    Perakin   K11.6
    Macusine A   K11.7
    Tetrahydroalstonine   K9.6
    Ajmalicine   K9.6
    Minovincine   K14.4
$C_{21}H_{24}N_2O_4$
    Talbotine   K16.6
    Mitraphylline   K9.3
    Voacarpine   K11.9
    Isomitraphylline   K9.3
$C_{21}H_{24}O_4$
    Nitenin   T57.1
$C_{21}H_{24}O_6$
    Tinophyllone   T37.4
$C_{21}H_{24}O_9$
    Chromomycinone   Y16.3
$C_{21}H_{24}O_{10}$
    Neriaphin   Y10.1
$C_{21}H_{24}OSi$
    tert-Butyloxymethylnaphthylphenyl-silane   Z1.15
$C_{21}H_{25}NO_4$
    Norcoralydine   K3.2
    Glaucine   K3.4
    Argemonine   K1.11
    Benzyltetrahydroisoquinoline deriv.   K3.6
$C_{21}H_{25}NO_5$
    Multifloramine   K2.8
    N-Formyldihydropavine methine   K1.12
    Androcymbine   K4.1
    Capaurine   K3.3
$C_{21}H_{26}N_2O_2$
    Retuline   K13.3
    Vincadifformine   K13.2
$C_{21}H_{26}N_2O_3$
    3-Epi-α-yohimbine   K9.7
    Yohimbine   K9.9
    Vincamine   K16.1
    Neblinine   K13.7
$C_{21}H_{26}N_2O_4$
    Aspidodasycarpine   K12.13
$C_{21}H_{26}N_2O_4S_2$
    2,7-Diazaspiro-[4.4]-nonane, ditosyl deriv.   X4.8
$C_{21}H_{26}O_5$
    Myricanol   Y15.1
$C_{21}H_{26}O_6$
    3,4-Dihydrotinophyllone   T37.8
$C_{21}H_{27}NO$
    Methadone   A18.11
    Isomethadone   A31.10
$C_{21}H_{27}NO_4$
    Laudanosine   K3.5
$C_{21}H_{27}NO_5$
    Kreysiginine   K5.2
    4-Demethylhasubanonine   K5.1
    Tetrahydroisoquinoline deriv.   K2.5
    Cryptostyline III   K1.6
$C_{21}H_{28}N_2O$
    N-α-acetyl-7-ethyl-5-desethylaspidospermi-dine   K15.4
$C_{21}H_{28}N_2O_2$
    Illudol phenylhydrazone deriv.   T30.12
$C_{21}H_{28}O_2$
    17-Acetyl-5α-etiojerva-12,14,16-trien-3-ol   K35.6
$C_{21}H_{28}O_2$
    trans-tetrahydrocannabinol   T5.15
$C_{21}H_{28}O_3$
    Pyrethrin I   T17.3
$C_{21}H_{28}O_4$
    Nitenin deriv.   T57.2

Formulae Index

$C_{21}H_{28}O_5$
   Cinerin II   T17.1
$C_{21}H_{28}O_{12}$
   Plumieride   T16.1
$C_{21}H_{29}NO$
   α-Methadol   A31.15
   β-Methadol   A31.11
$C_{21}H_{30}O_2$
   Cannabidiol   T5.11
$C_{21}H_{30}O_3$
   Jasmolin I   T17.3
   Furospongin I   T57.4
$C_{21}H_{31}ClO_5$
   Goutierolide   T37.5
$C_{21}H_{32}O$
   4,4-Dimethylandrost-5-ene-7-one   T26.19
$C_{21}H_{32}O_{15}$
   Melittoside   T15.11
$C_{21}H_{33}NO_2$
   Himbacine   K18.11
$C_{21}H_{33}NO_3$
   Atisine deriv.   K32.6
$C_{21}H_{34}O_5$
   Trisporic acid deriv.   Y19.9
$C_{21}H_{39}N_7O_{12}$
   Streptomycin   C4.15
$C_{21}H_{40}O_2$
   Lactone from tocopherol   T56.2

C_{22}

$C_{22}H_{12}S_2$
   Benzo-[d]-naphtho[1,2-d']-benzo-[1,2-b;4,3-b']-dithiophene   X7.8
$C_{22}H_{14}$
   [5]-helicene   X7.9
$C_{22}H_{14}O_4$
   1,1'-Binaphthyl-2,2'-dicarboxylic acid   X6.5
$C_{22}H_{18}$
   2,2'-Dimethyl-1,1'-binaphthyl   X6.10
$C_{22}H_{18}O_2$
   1,12-Dimethylbenzo-[c]-phenanthrene-5-acetic acid   X7.6
$C_{22}H_{18}O_6$
   2,2'-Dihydroxy-1,1'-binaphthyl-3,3'-dicarboxylic acid, dimethyl ester   X6.6
$C_{22}H_{20}N_2O_8S_2$
   Acetylaranotin   Y24.5
$C_{22}H_{21}NO$
   1-acetamido-1,2,2-triphenylethane   A23.3
$C_{22}H_{22}$
   1,1,2-Triphenylbutane   A41.13
$C_{22}H_{22}O$
   1,1,2-Triphenylbutan-1-ol   A41.14
$C_{22}H_{22}O_8$
   Podophyllotoxin   Y7.15
$C_{22}H_{22}O_{10}$
   Trifolirhizin   Y3.8
$C_{22}H_{23}ClN_2O_8$
   Aureomycin   Y28.2
$C_{22}H_{23}NO_7$
   α-Narcotine   K3.11
$C_{22}H_{24}Ge$
   (1-Methylpropyl)-triphenylgermane   Z1.2
$C_{22}H_{24}N_2O_4$
   Kopsine   K14.2
   Vomicine   K12.2
$C_{22}H_{24}N_2O_8$
   Achromycin   Y28.2
$C_{22}H_{24}N_2O_9$
   Oxytetracycline   Y28.2
$C_{22}H_{24}O_6$
   α-Conidendrin dimethyl ether   Y7.11
   α-Retrodendrin dimethyl ether   Y7.12
$C_{22}H_{24}Pb$
   (1-Methylpropyl)-triphenylplumbane   Z1.2
$C_{22}H_{24}Si$
   (1-Methylpropyl)-triphenylsilane   Z1.2
$C_{22}H_{24}Sn$
   (1-Methylpropyl)-triphenylstannane   Z1.2
$C_{22}H_{25}NO_6$
   Oreodine   K1.9
   Glaudine   K1.9
   Epiglaudine   K1.8
   Colchicine   K30.1
$C_{22}H_{26}N_2O_4$
   Corymine   K14.6
   Picraphylline   K9.4
$C_{22}H_{26}N_4$
   Calycanthine   K26.7
   Chimonanthine   K26.8
$C_{22}H_{26}O_7$
   Gmelinol   Y7.4
   Isogmelinol   Y7.3
   Neogmelinol   Y7.3
$C_{22}H_{26}O_8$
   Liatrin A   T30.4
$C_{22}H_{27}NO_2$
   Lobeline   K18.10
$C_{22}H_{27}N_7O$
   Cypridina luciferin   Y20.6
$C_{22}H_{28}N_2O_2$
   Spermostrychnine   K12.4
$C_{22}H_{28}N_2O_3$
   Haplocine   K13.11
   Corynantheidine   K9.5
   Corynantheine   K8.14
$C_{22}H_{28}N_2O_4$
   Rhyncophylline   K9.2
   Corynoxine   K9.1
   N-Methyldihydrotalbotine   K16.7
$C_{22}H_{28}O_4$
   Galbegin   Y8.2
   Galbulin   Y8.1
$C_{22}H_{28}O_5$
   Pyrethrin II   T17.1
$C_{22}H_{28}O_6$
   Lariciresinol dimethyl ether   Y7.16
   Isolariciresinol dimethyl ether   Y7.10

Formulae Index

$C_{22}H_{28}O_7$
    Eupacunin — T24.12
    Isodonal — T33.3
    Nepetaefolin — T37.1

$C_{22}H_{29}NO_4$
    Elemol p-nitrobenzoate — T20.5

$C_{22}H_{29}NO_5$
    Stemofoline — K29.10

$C_{22}H_{29}N_2O_4$
    Echitamine — K14.7

$C_{22}H_{30}N_2O_2$
    Aspidospermine — K13.8
    Vincaminoreine — K13.4

$C_{22}H_{30}O_3$
    Siccanin — T32.16

$C_{22}H_{30}O_4$
    Tauranin — T36.8
    Dihydroguaiaretic acid dimethyl ether — A31.18
    Cannabidiolic acid — T5.11

$C_{22}H_{30}O_5$
    Jasmolin II — T17.1

$C_{22}H_{30}O_6$
    Secoisolariciresinol — Y7.9

$C_{22}H_{30}O_7$
    2,3-di(3,4-dimethoxybenzyl)-butane-1,2,4-triol — Y7.5

$C_{22}H_{31}NO_4$
    Pyroheteratisine — K33.9

$C_{22}H_{32}O_7$
    Cascarillin — T39.11

$C_{22}H_{33}NO_2$
    Atisine — K32.4
    Garryfoline — K32.8
    Veatchine — K32.8

$C_{22}H_{33}NO_3$
    Atidine — K32.3
    Ajaconine — K32.5

$C_{22}H_{33}NO_4$
    Tuberostemonine — K29.1

$C_{22}H_{33}NO_5$
    Heteratisine — K33.8

$C_{22}H_{34}O_2$
    11-Methoxyferruginol methyl ether — T33.10

$C_{22}H_{34}O_3$
    Coleone B deriv. — T33.14
    17 -Acetoxy-4α-methyl-5α-androstan-3-one — T32.9

$C_{22}H_{35}O_2P$
    Menthyl cyclohexylphenylphosphinate — Z4.5

C_{23}

$C_{22}H_{36}O_2$
    Bisnor-5α-cholanic acid — T48.7

$C_{23}H_{16}O$
    Dinaphthocycloheptadienone — X6.2

$C_{23}H_{19}NO_4Si$
    Methylnaphthylphenyl-p-nitrobenzoyloxysilane — Z1.17

$C_{23}H_{19}OP$
    o-Methoxyphenylnaphthylphenylphosphine — Z4.21

$C_{23}H_{19}O_2P$
    o-Methoxyphenylnaphthylphenylphosphine oxide — Z4.20

$C_{23}H_{20}O$
    2,2-Diphenyl-3-ethylindan-1-one — A41.17

$C_{23}H_{20}O_4$
    1-Phenylpropane-1,2-diol, dibenzoate — A21.1

$C_{23}H_{20}O_2Si$
    Methylnaphthylphenylbenzoyloxysilane — Z1.17

$C_{23}H_{22}O_2$
    2,2,3-Triphenylpentanoic acid — A41.16

$C_{23}H_{22}O_6$
    Rotenone — Y1.4

$C_{23}H_{24}O$
    2,2,3-Triphenylpentan-1-ol — A41.15

$C_{23}H_{24}O_9$
    Olivin deriv. — Y16.7

$C_{23}H_{26}Br_2O_2$
    7,7'-Dibromo-6,6'-dihydroxy-3,3,3',3',5,5'-hexamethylbis-1,1'-spiroindane — X5.4

$C_{23}H_{26}N_2O_4$
    Brucine — K12.6

$C_{23}H_{26}N_2O_5$
    Picraline — K12.9
    Gardneramine — K14.9

$C_{23}H_{26}OSi$
    Cyclohexyloxymethylnaphthylphenylsilane — Z1.15

$C_{23}H_{26}O_7$
    Isodesoxypodophyllotoxin — Y7.14

$C_{23}H_{28}N_2O_6$
    Rauvoxinine — K10.1

$C_{23}H_{28}N_4$
    Calycanthidine — K26.8

$C_{23}H_{29}NO_3$
    Alopecurine — K26.1

$C_{23}H_{30}BrNO_6$
    Miyaconitine hydrobromide — K33.10

$C_{23}H_{31}NO_6$
    Delphinine deriv. — K33.6

$C_{23}H_{34}O_3$
    3β-Acetoxypregn-5-ene-20-one — K34.3

$C_{23}H_{34}O_4$
    Digitoxigenin — T49.3

$C_{23}H_{34}O_7$
    Glaucarubin deriv. — T53.12

$C_{23}H_{35}NO_3$
    Garryfoline deriv. — K32.9

$C_{23}H_{36}O_3$
    Dehydroroyleanone trimethyl ether — T33.10

$C_{23}H_{36}O_4$
    'Isopropylidene norketone' — T38.9

$C_{23}H_{36}O_7$
    Zearalenone deriv. — Y14.12

$C_{23}H_{37}NO_2$
    Methyl homosecodaphniphyllate — K26.5
    3β-Acetoxy-20α-aminopregn-5-ene — K34.2

Formulae Index

$C_{23}H_{37}NO_6$
   Lappaconine   K33.11

$C_{23}H_{38}O_3$
   Rearranged taxane deriv.   T40.9
   3β-Hydroxy-5α-norcholanic acid   T48.7

C_{24}

$C_{24}H_{20}OSi$
   Benzoylmethylnaphthylphenylsilane   Z1.13

$C_{24}H_{24}O$
   3,3,4-Triphenylhexan-2-one   T41.12

$C_{24}H_{24}O_8$
   Daunamycinone trimethyl ether   Y28.6

$C_{24}H_{26}N_2O_{13}$
   Betanin   K17.11

$C_{24}H_{26}N_2O_8S_2$
   Bisdethiodi(methylthio)-acetylaranotin   Y24.8

$C_{24}H_{28}O_8$
   Isopeulustrin   Y1.13

$C_{24}H_{30}N_4$
   Folicanthine   K26.8

$C_{24}H_{30}O_4$
   Farnesiferol A   T57.3
   Farnesiferol B   T57.5

$C_{24}H_{30}O_7$
   Athamanthin   Y1.14

$C_{24}H_{32}O_4$
   Scillarenin   T49.1

$C_{24}H_{32}O_5$
   Maleopimaric acid   T32.15

$C_{24}H_{32}O_6$
   Blancoic acid   Y12.3

$C_{24}H_{32}O_8$
   Neophorbol 13,20-diacetate   T39.3

$C_{24}H_{34}O_4$
   Bufalin   T49.1

$C_{24}H_{34}O_7$
   Clerodin   T37.3
   ε-Caesalpin   T40.10

$C_{24}H_{35}NO_4$
   Lucidusculine   K32.2

$C_{24}H_{35}NO_5$
   Batrachotoxinin A   K36.9

$C_{24}H_{36}N_2O_9S$
   Celesticetin   Y21.1

$C_{24}H_{38}O_4$
   3β-Acetoxybisnorallocholanic acid   K34.4

$C_{24}H_{39}NO_4$
   Cassaine   T39.9

$C_{24}H_{39}NO_6$
   Des(oxymethylene)lycoctonine   K33.3

$C_{24}H_{40}N_2$
   Conessine   K34.1

$C_{24}H_{42}N_2O_4$
   Azimine   K20.11

C_{25}

$C_{25}H_{20}Fe_2O_2$
   7,7'-Spirobis-[3]-ferrocenophane-6,6'-dione   X9.15

$C_{25}H_{22}SiO$
   Phenyl triphenylsilyl carbinol   Z1.1

$C_{25}H_{24}Fe_2O_2$
   7,7'-Spirobis[3]-ferrocenophane-6,6'-diol   X9.19

$C_{25}H_{24}SiO$
   Methylnapthylphenyl(1-hydroxy-1-phenylethyl)silane   Z1.6

$C_{25}H_{25}NO_5$
   Phototheinehydroquinone   K4.10

$C_{25}H_{28}O_5$
   Pyronimbic acid   T53.4

$C_{25}H_{29}NO_5$
   Lagerine   K27.6
   Decodine   K27.2

$C_{25}H_{31}N_3O_4$
   Lunarine   K29.8

$C_{25}H_{33}NO_7$
   Bundlin A   Y26.6

$C_{25}H_{35}NO_9$
   Ryanodine   T58.1

$C_{25}H_{36}O_4$
   Ophiobolin A   T41.6
   Cephalonic acid   T41.7

$C_{25}H_{36}O_6$
   Papuanic acid   Y12.4
   Isopapaunic acid   Y12.4

$C_{25}H_{36}O_{10}$
   Glaucarubin   T53.9

$C_{25}H_{37}NO_4$
   Piericidin A   Y30.8

$C_{25}H_{38}O_3$
   Ophiobolin C   T41.8

$C_{25}H_{38}O_7$
   Laserpitine   T25.15

$C_{25}H_{39}NO$
   Cyclobuxosuffrine   K36.4

$C_{25}H_{40}O$
   Ceroplastol I   T41.9

$C_{25}H_{41}NO_7$
   Browniine   K33.4
   Lycoctonine   K33.4

$C_{25}H_{42}N_2O$
   Cyclobuxine D   K36.3

$C_{25}H_{43}N_{13}O_{10}$
   Viomycin   Y22.6

$C_{25}H_{44}O$
   Geranyllinalool   T3.12

$C_{25}H_{45}NO_9$
   Pederin   Y20.4

C_{26}

$C_{26}H_{16}$
   [6]-Helicene   X7.7

$C_{26}H_{20}O_{13}$
   Tetraacetoxydothistromin   Y13.6

$C_{26}H_{24}O_8$
   Daphnetoxin   T39.6

Formulae Index

$C_{26}H_{24}O_{11}$
    Olivin deriv. — **Y16.5**

$C_{26}H_{24}OSi$
    Silyl ether — **Z1.3**

$C_{26}H_{26}NO_7$
    (2-Piperidinoethyl)amine dipicrate — **A13.2**

$C_{26}H_{27}NO_3$
    Stephavanine — **K5.3**

$C_{26}H_{30}O_7$
    Obacunone — **T52.9**

$C_{26}H_{30}O_8$
    Limonin — **T52.8**

$C_{26}H_{31}NO_5$
    Decinine — **K27.4**
    Decaline — **K27.10**
    Decamine — **K27.8**
    Vertaline — **K27.9**

$C_{26}H_{31}NO_6$
    Lythridine — **K27.1**

$C_{26}H_{31}O_2P$
    Menthyl naphthylphenylphosphinate — **Z4.19**

$C_{26}H_{32}N_2O_3S$
    Cinchonamine tosyl deriv. — **K8.13**

$C_{26}H_{32}O_7$
    Angolensic acid — **T52.2**

$C_{26}H_{32}O_9$
    Ichangin — **T52.4**

$C_{26}H_{33}NO_5$
    Dihydroverticillatine — **K27.3**

$C_{26}H_{34}O_9$
    Lumiphorbol triacetate — **T39.8**
    Isophorbol triacetate — **T39.5**

$C_{26}H_{35}NO_4$
    19-Propylthevinol — **K4.2**
    Lythranidine — **K20.10**

$C_{26}H_{36}O_8$
    7-Hydroxylathyrol-3,5,7-triacetate — **T39.1**

$C_{26}H_{36}O_{12}$
    Menthiafolin — **T13.7**
    Foliamenthin — **T13.11**

$C_{26}H_{38}O_5$
    Havanensin — **T53.1**

$C_{26}H_{38}O_{12}$
    Jasminin — **T15.3**

$C_{26}H_{44}N_2$
    Buxenine G — **K36.5**

$C_{26}H_{44}N_2O_2$
    Buxidienine F — **K36.1**
    Cyclobuxidine F — **K36.2**

$C_{26}H_{46}N_2O_3$
    Dihydromicrophylline F — **K36.2**

C_{27}

$C_{27}H_{26}BrP$
    Benzyl(diphenylmethyl)methylphenylphosphonium bromide — **Z6.6**

$C_{27}H_{29}NO_{10}$
    Daunomycin — **Y28.5**

$C_{27}H_{29}NO_{11}$
    Adriamycin — **Y28.5**

$C_{27}H_{32}N_4O_8$
    Pyridomycin — **Y26.7**

$C_{27}H_{32}O_6$
    Deoxyandirobin — **T52.1**

$C_{27}H_{32}O_7$
    Mexicanolide — **T52.7**

$C_{27}H_{34}N_2O_9$
    Vincoside — **K2.12**
    Isovincoside — **K2.12**

$C_{27}H_{34}O_8$
    Swietenolide — **T52.11**
    Veprisone — **T52.10**
    Methyl ivorensate — **T52.3**
    Mahoganin — **T53.2**

$C_{27}H_{34}O_9$
    Verrucarin A — **T31.1**

$C_{27}H_{35}NO_4$
    Lythramine — **K20.9**

$C_{27}H_{35}NO_8$
    Bundlin B — **Y26.6**

$C_{27}H_{35}NO_{12}$
    Ipecoside — **K2.9**

$C_{27}H_{36}O_9$
    Simarolide — **T53.10**

$C_{27}H_{38}O_5$
    3β-Acetoxyisobufalin methyl ester — **T49.2**

$C_{27}H_{39}NO_2$
    Veratramine — **K35.9**

$C_{27}H_{39}NO_3$
    Jervine — **K35.7**
    Piperidine deriv. from decinine — **K27.7**

$C_{27}H_{41}NO_3$
    Veratrobasine — **K35.7**

$C_{27}H_{42}O$
    Tachysterol III — **T.40.7**

$C_{27}H_{43}NO$
    Solanidine — **T49.4**

$C_{27}H_{43}NO_2$
    Veralkamine — **T34.5**
    Solasodine — **T49.7**

$C_{27}H_{43}NO_3$
    Verticinone — **K35.5**

$C_{27}H_{43}NO_8$
    Cevine — **K35.1**
    Germine — **K35.3**

$C_{27}H_{44}O$
    Cholest-4-ene-3-one — **T23.17**

$C_{27}H_{44}O_3$
    Smilagenin — **T49.9**
    Sarsapogenin — **T49.6**

$C_{27}H_{44}O_6$
    Ponasterone A — **T48.9**
    Ecdysone — **T48.9**

$C_{27}H_{44}O_7$
    Crustecdysone — **T48.9**

Formulae Index

$C_{27}H_{45}NO$			Viopurpurin	Y2.8
Solacongestidine	K34.6	$C_{29}H_{22}O_{11}$		
5,6-Dihydrosolanidine	T49.4		Duclauxin	Y18.8
$C_{27}H_{45}NO_2$		$C_{29}H_{35}NO_5$		
Tomatidine	T49.8		Cytochalasin A	Y27.6
$C_{27}H_{46}$		$C_{29}H_{36}O_{10}$		
Cholest-4-ene	T32.10		Taxol derivative	T40.5
Cholest-5-ene	T32.19	$C_{29}H_{37}NO_5$		
$C_{27}H_{46}Br_2$			Cytochalasin B	Y27.6
$2\alpha,3\beta$-Dibromo-5α-cholestane	A18.16	$C_{29}H_{37}N_3O_2$		
$C_{27}H_{46}N_2O_2$			Deoxytubulosine	K2.7
Cyclobuxoxazine	K36.6	$C_{29}H_{37}N_3O_3$		
Solanocapsine	K34.7		Tubulosine	K2.7
$C_{27}H_{46}O$		$C_{29}H_{38}O_4$		
5α-Cholestan-1-one	T53.11		Pristimerin	T45.11
5β-Cholestan-1-one	T38.5	$C_{29}H_{39}N_3O_2$		
5α-Cholestan-2-one	T36.14		Echinulin	K30.5
5β-Cholestan-3-one	T35.14	$C_{29}H_{40}N_2O_4$		
Cholesterol	T46.7		Emetine	K2.6
5α-Cholest-14-ene-3β-ol	T46.3	$C_{29}H_{40}O_4$		
$C_{27}H_{48}N_2$			Jingullic acid	T43.9
Cycloprotobuxine C	K36.7		Pristimerol	T45.12
$C_{27}H_{48}O$		$C_{29}H_{42}O_4$		
5α-Cholestan-7β-ol	T46.6		Eupteleogenin	T45.10
C_{28}		$C_{29}H_{44}O_5$		
$C_{28}H_{22}N_2$			Ceanothic acid	T43.7
1,2-Dimethyl-6,7-diphenyl[e,g][1,4]-diazocine	X5.10	$C_{29}H_{47}NO_5$		
			Solaphyllidine	K36.8
$C_{28}H_{27}ClO_7$			Stigmasterol	T48.6
Cervicarcin derivative	Y18.6	$C_{29}H_{48}O$		
$C_{28}H_{28}BrN_3O_3$			Lanost-7-en-3-one	T44.7
Quinine derivs.	K8.5, K8.2	$C_{29}H_{48}O_2$		
$C_{28}H_{28}O_{11}$			Ketohakonanol	T45.8
Pillaromycin A	Y28.3		Adipedatol	T45.5
$C_{28}H_{30}O_9$		$C_{29}H_{50}O$		
Physalin A	T53.5		4α-Ethyl-5α-cholestan-3-one	T36.17
$C_{28}H_{30}O_{10}$			Adiantol	T45.4
Physalin B	T53.6	$C_{29}H_{50}O_2$		
$C_{28}H_{34}O_7$			α-Tocopherol	T56.4
Gedunin	T52.5	**C_{30}**		
$C_{28}H_{37}NO_5$		$C_{30}H_{18}$		
Lythranine	K20.10		[7]-Helicene	X7.10
$C_{28}H_{38}O_6$		$C_{30}H_{18}O_4$		
Withaferin A	T48.10		2,2'-Dimethyl-1,1'-bisanthraquinonyl	X6.9
$C_{28}H_{38}O_8$		$C_{30}H_{20}$		
Huratoxin	T39.7		9,10-Dihydrodinaphtho-2,3'-3,4'; 2'',3''-5,6-phenanthrene	X6.4
$C_{28}H_{44}O$				
Suprasterol II	T48.5	$C_{30}H_{22}$		
Calciferol	T48.4		2,2'-Dimethyl-1,1'-bianthryl	X6.8
Ergosterol	T48.8	$C_{30}H_{22}Br_2O_{10}$		
$C_{28}H_{48}O$			Dibromodehydrotetrahydrorugulosin	Y18.2
3,3-Dimethyl-A-norcholestan-2-one	T35.3	$C_{30}H_{22}O_8$		
$C_{28}H_{50}N_2O_4$			Erythroaphin-fb	Y10.8
Carpaine	K20.8	$C_{30}H_{22}O_{10}$		
$C_{28}H_{50}O_4$			Rugulosin	Y18.1
Clavatol	T44.11	$C_{30}H_{22}O_{12}$		
C_{29}			Xanthomegnin	Y2.1
$C_{29}H_{20}O_{11}$			Luteoskyrin	Y18.1
Purpurogenone	Y17.3		Rubroskyrin	Y18.3

Formulae Index

$C_{30}H_{24}O_9$
   Chrysophin-*fb*   Y10.7
$C_{30}H_{26}GeO$
   (Diphenylhydroxymethyl)methyl-
   naphthylphenylgermane   Z2.11
$C_{30}H_{26}O_{10}$
   Xanthoaphin-*fb*   Y10.6
$C_{30}H_{26}O_{14}$
   Ergoflavin   Y17.6
$C_{30}H_{28}N_4O_6S_4$
   Chaetocin   Y24.6
$C_{30}H_{28}O_{12}$
   Rhododactynaphin-*jc*-1   Y10.3
   Xanthodactynaphin-*jc*-1   Y10.2
$C_{30}H_{36}O_9$
   Nimbin   T53.3
$C_{30}H_{42}N_4O_2$
   Homaline   K30.4
$C_{30}H_{44}O_3$
   Flindissone lactone   T51.2
$C_{30}H_{44}O_4$
   Lansic acid   T44.12
$C_{30}H_{46}O_3$
   Ebelin lactone   T51.13
$C_{30}H_{46}O_4$
   Glycyrrhetic acid   T42.10
   Melianone   T51.11
$C_{30}H_{46}O_5$
   Melaleucic acid   T43.6
$C_{30}H_{47}NO_3$
   Secodaphniphylline   K26.6
$C_{30}H_{48}O$
   Hopenone I   T45.1
   Baurenone   T42.13
$C_{30}H_{48}O_2$
   21-Oxoserrat-13-en-3β-ol   T44.4
   Serratenolone   T44.5
   Nyctanthic acid   T43.2
$C_{30}H_{48}O_3$
   Betulinic acid   T43.5
   Bourjotinolone   T46.2
   Elemadienolic acid   T46.5
   Oleanolic acid   T42.6
   Ursolic acid   T42.9
$C_{30}H_{48}O_4$
   Odoratone   T51.6
$C_{30}H_{48}O_5$
   Cimigenol   T51.12
$C_{30}H_{48}O_7$
   Platicodigenin   T45.9
$C_{30}H_{49}NO_{11}$
   Demycarosyl-leucomycin A_3 Y25.6
$C_{30}H_{50}$
   Fernene   T43.11
   Filicene   T44.2
   Alnusene   T42.14
   Adianine   T43.10
   α-Lupene   T43.5

$C_{30}H_{50}Br_2O_2$
   3β-Acetoxy-7α,11α-dibromolanostane-
   8α,9α-epoxide   T46.9
$C_{30}H_{50}O$
   Davallol   T44.3
   Lanost-7-en-3-one   T44.7
   Arborinol   T45.7
   Cycloartenol   T50.2
   Lanosterol   T44.10
   Tirucallol   T46.5
   Euphol   T46.4
   Baurenol   T42.13
   Friedelin   T42.11
   Isoursenol   T42.8
   α-Amyrin   T42.9
   β-Amyrin   T42.6
   Gorgosterol   T50.13
   Taraxerol   T43.1
   ψ-Taraxasterol   T43.4
   Multiflorenol   T42.3
   Shionone   T42.15
   Filican-3-one   T44.1
$C_{30}H_{50}O_2$
   Dipterocarpol   T51.4
   Betulin   T43.5
   α-Onocerin   T44.8
   β-Onocerin   T44.8
   γ-Onocerin   T44.6
$C_{30}H_{50}O_3$
   6-Oxoleucotylin   T45.6
$C_{30}H_{50}O_4$
   Alisol A   T51.14
$C_{30}H_{51}BrO_3$
   Betulafolienetriol deriv.   T51.10
$C_{30}H_{52}$
   Gammacerane   T44.9
   Oleanane   T42.7
   Hopane   T45.2
   Isohopane   T45.2
$C_{30}H_{52}O$
   3β-Hydroxyprotost-13(17)-ene   T50.9
   Friedelinol   T42.11
   Tetrahymanol   T44.9
   Ambrein   T42.1
$C_{30}H_{52}O_2$
   Canaric acid   T43.8
   Zeorin   T45.3
   Dammarenediol I   T51.7
   Dammarenediol II   T51.4
$C_{30}H_{52}O_3$
   Betulafolienetriol   T51.9
   Ocotillol   T51.5

C_{31}^+

$C_{31}H_{42}O_{12}$
   Clerodendrin A   T37.6
$C_{31}H_{44}O_4$
   Tirucallol deriv.   T46.8
$C_{31}H_{46}O_2$
   Phylloquinone   T56.5

Formulae Index

$C_{31}H_{47}NO_4$
    Daphmacrine   K26.4

$C_{31}H_{48}O_3$
    Cyclograndisolide   T50.12
    Ambonic acid   T50.3

$C_{31}H_{48}O_6$
    Fusidic acid   T50.10

$C_{31}H_{50}O_3$
    Eburicoic acid   T50.5

$C_{31}H_{50}O_4$
    Polyporenic acid A   T50.7

$C_{31}H_{52}O$
    Isotirucallenol   T51.1
    Cycloeucalenol   T50.1

$C_{31}H_{52}O_2$
    Taraxerol acetate   T43.1

$C_{32}H_{38}N_2O_8$
    Deserpidine   K9.11

$C_{32}H_{40}O_9$
    Swietenine   T52.12

$C_{32}H_{42}O_{10}$
    Khivorin   T52.6

$C_{32}H_{46}O_9$
    Cucurbitacin A   T50.6

$C_{32}H_{48}O_7$
    Isooleanolic acid dimethyl ester lactone   T42.5

$C_{32}H_{50}O_4$
    Echinodol   T50.8

$C_{33}H_{24}O_2$
    2,3,2′,3′-Bis(α-oxotetramethylene)-9,9′-spirobifluorene   X11.1

$C_{33}H_{35}N_5O_5$
    Ergotamine   K17.8

$C_{33}H_{40}N_2O_9$
    Reserpine   K9.11

$C_{33}H_{41}NO_{10}$
    Demethanolaconitine   K33.1

$C_{33}H_{45}NO_6$
    Veratramine deriv.   K35.10

$C_{33}H_{45}NO_9$
    Delphinine   K33.5

$C_{33}H_{46}N_4O_6$
    Stercobilin   Y23.4

$C_{33}H_{50}O_8$
    Cephalosporin P_1 T50.11

$C_{33}H_{56}O$
    Cycloneolitsin   T50.4

$C_{34}H_{30}OSi_2$
    Disilyl ether   Z1.14

$C_{34}H_{47}NO_{11}$
    Aconitine   K33.2

$C_{34}H_{54}O_8$
    Antibiotic X-537A   Y11.2

$C_{34}H_{70}NO_7P$
    Dimyristoylcephalin   A15.1

$C_{35}H_{42}O_9$
    Taxinine   T40.1

$C_{35}H_{61}NO_{12}$
    Oleandomycin   Y25.2

$C_{36}H_{24}P$
    Tris-2,2′-bisphenylenephosphorus(V)-ion   X10.1

$C_{36}H_{38}AsB$
    Allylmethylphenylpropylarsonium-tetraphenylborate   Z3.7

$C_{36}H_{38}O_{16}$
    Protoaphin-*fb*   Y10.4

$C_{36}H_{56}O_{12}$
    Fusicoccin A   T41.10

$C_{37}H_{32}O_2$
    [6,6]-vespirone   X11.2

$C_{37}H_{36}$
    [6,6]-vespirene   X11.3

$C_{37}H_{40}BP$
    Allylmethylphenylpropylphosphonium tetraphenylborate   Z5.13

$C_{37}H_{40}N_4O_7$
    Haplophytine   K15.5

$C_{37}H_{43}NO_{13}$
    Tolpomycinone   Y26.4

$C_{37}H_{67}NO_{13}$
    Erythromycin A   Y25.4

$C_{37}H_{72}O_5$
    1,2-Dipalmitoylglycerol   A15.4

$C_{38}H_{34}O_{14}$
    Julimycin BII   Y18.4

$C_{38}H_{44}O_8$
    Gambogic acid   Y9.7

$C_{38}H_{68}NO_9P$
    Dimyristoyllecithin   A15.5

$C_{38}H_{78}NO_7P$
    Dipalmitoylcephalin   A15.1

$C_{39}H_{36}O_2$
    [7,7]-Vespirone   X11.2

$C_{39}H_{40}$
    [7,7]-Vespirene   X11.3

$C_{39}H_{47}NO_{15}$
    Rifamycin Y   Y26.1

$C_{39}H_{49}NO_{14}$
    Rifamycin A   Y26.1

$C_{40}H_{56}$
    α-Carotene   T54.10
    δ-Carotene   T54.12
    ε-Carotene   T54.9

$C_{40}H_{56}O_2$
    Semi-α-carotene   T54.11

$C_{40}H_{56}O_2$
    Kryptocapsin   T55.4

$C_{40}H_{56}O_3$
    Capsanthin   T55.4

$C_{40}H_{65}BO_{14}$
    Boromycin hydrolysis product   Y.27.2

$C_{40}H_{68}O_{10}$
    Grisorixin   Y11.5

$C_{40}H_{68}O_{11}$
    Nigericin   Y11.1

Formulae Index

$C_{40}H_{78}O_8P$			Cobyric acid	Y24.1
Phosphatidylglycerol	A15.6	$C_{45}H_{74}BNO_{15}$		
$C_{41}H_{40}O_2$			Boromycin	Y27.2
[8,8]-Vespirone	X11.2	$C_{45}H_{78}O_{13}$		
$C_{41}H_{42}AsB$			Antibiotic X-206	Y11.3
Benzylmethylphenylpropylarsonium		$C_{46}H_{56}N_4O_{10}$		
tetraphenylborate	Z3.9		Leurocristine	K15.1
$C_{41}H_{42}BN$		$C_{46}H_{94}NO_9P$		
Benzylmethylphenylpropylammonium			Distearoyllecithin	A15.5
tetraphenylborate	Z3.2	$C_{47}H_{51}NO_{14}$		
$C_{41}H_{44}$			Taxol	T40.4
[8,8]-Vespirene	X11.3	$C_{47}H_{73}NO_{17}$		
$C_{41}H_{52}O_{17}$			Amphotericin B	Y26.3
Utilin	T52.12	$C_{55}H_{74}MgN_4O_6$		
$C_{41}H_{80}O_5$			Bacteriochlorophyll a	Y23.2
1,2-Distearoylglycerol	A15.4	$C_{55}H_{72}MgN_4O_5$		
$C_{42}H_{58}O_6$			Chlorophyll a	Y23.1
Fucoxanthin	T55.1	$C_{56}H_{40}O_{12}$		
$C_{42}H_{69}NO_{15}$			Hopeaphenol	Y15.8
Leucomycin A_3	Y25.5	$C_{63}H_{87}CoN_{13}O_{15}P$		
$C_{42}H_{82}NO_9P$			Vitamin B_{12} monocarboxylic acid	Y24.2
Dipalmitoleyllecithin	A15.5	$C_{63}H_{88}CoN_{14}O_{14}P$		
$C_{42}H_{86}NO_7P$			Vitamin B_{12}	Y24.2
Disteraroylcephalin	A15.1	$C_{81}H_{158}O_{17}P_2$		
$C_{45}H_{66}CoN_{10}O_8$			Cardiolipin	A15.2

Author Index

Aasen, A. J., **K22**
Abd El Rahim, A. M., **T42**
Abe, J., **Y17**, **Y25**
Abe, M., **K17**
Abel, H., **T39**
Abraham, A., **T48**
Abrahamsson S., **Y16**
Acheson, G. G., **T2**
Achiwa, K., **A40**
Ackerman, D., **A20**
Acklin, W., **K17**
Adam, G., **T49**, **K34**
Adams, K. A. S., **K25**
Adams, M. B., **T18**
Adams, R., **A34**, **K22**
Adeoye, S. A., **T52**
Adesogan, E. K., **T52**
Adinolfi, M., **T37**
Adler, E., **K3**
Adolf, W., **T39**
Afonso, A., **T32**
Ageta, H., **T43**, **T44**, **T45**
Agosta, W. C., **X3**
Ahmann, G., **K20**
Ahmed, R. R., **A4**, **A18**, **T47**, **K4**, **K25**
Aimi, N., **K14**, **K34**
Akeda, K. T., **T25**
Akhtar, M., **D2**
Akimoto, H., **X6**
Akisanya, A., **T52**
Akiyama, T., **T45**
Alam, S. N., **K25**
Albers-Schönberg, G., **T16**
Albonico, S. M., **K3**
Alderton, G., **A8**
Aldridge, D. C., **A36**, **K24**, **Y27**
Alleaume, M., **Y11**
Allen, F. H., **T11**, **T44**, **T50**, **K2**
Allen, L. E., **A56**
Allinger, J., **A51**, **A56**
Allison, A. J., **T34**
Almy, J., **A49**
Al-Shamma, A. A., **T37**
Altenkirk, B., **K26**
Alves, H. M., **Y3**
Amai, R. L. S., **K11**
Amenechi, P. I., **Y3**
Ames, T. R., **T43**
Amiard, G., **T48**
Amirthaligam, V., **T29**
Andersen, K. H., **Z8**
Anderson, A. B., **A16**
Anderson, B. F., **T34**
Anderson, D. G., **Z1**, **Z2**
Anderson, H. W., **X1**

Anderson, N. H., **T19**, **T29**
Ando, M., **T22**
Anet, F. A. L. **K12**, **K25**
Angyal, S. J., **A16**
Anisuzzaman, A. K. M., **A14**
Annen, K., **T48**
Anthonsen, T., **T37**
Antkowiak, W., **T32**
Antosz, F. J., **Y30**
Aono, K., **K6**
Aota, K., **T19**, **T20**, **T24**, **T27**
Aplin, R. T., **T53**
Appel, H. H., **T14**, **T32**
Applequist, D. E., **A18**, **A37**
Appleton, R. A., **T37**
Applewhite, T. H., **A7**
ApSimon, J. W., **T32**
Arai, G., **K15**
Arakawa, H., **A3**, **A25**, **T21**, **Y2**, **Y6**
Aratani, T., **A44**, **X2**
Arcamone, F., **Y28**
Archer, J. F., **A10**
Archer, S., **K28**
Arene, E. O., **T52**
Argondelis, A. D., **Y21**
Arigoni, D., **A33**, **A39**, **A53**, **T1**, **T2**, **T13**, **T19**, **T21**, **T32**, **T34**, **T42**, **T44**, **T45**, **T46**, **T51**, **T52**, **T57**, **K17**, **K36**, **D2**
Armarego, W. L. F., **A37**
Armstrong, J. J., **K24**, **Y27**
Armstrong, M. D., **A20**
Arndt, R. R., **K20**
Arnott, S., **T52**
Arojan, A. A., **K18**
Arold, H., **A27**, **A33**
Arthur, H. R., **T43**
Asada, Y., **T23**
Asai, M., **Y28**
Asano, H., **A27**, **A30**
Asao, T., **T20**, **T22**
Aschan, O., **T12**
Asher, J. D. M., **T22**
Ashida, T., **A32**, **K4**, **Z1**
Audisio, G., **X10**
Auer, E., **A38**
Auernheimer, A. H., **A13**
Austen, D. J., **Y19**
Autrey, R. L., **K9**
Avitabile, G., **Y26**
Axelrod, M., **Z7**, **Z8**
Ayer, W. A., **K25**, **K26**, **K32**
Aylett, B. J., **Z1**

Bach, R. D., **X2**

Bachelor, F. W., **A24**
Badger, G. M., **X6**
Badin, E. J., **A27**
Baer, E., **A11**, **A14**, **A15**
von Baeyer, A., **T6**
Bailey, A. S., **K12**
Baker, B. R., **K20**
Baker, C. D., **A7**
Baker, C. G., **A8**, **A24**
Baldwin, J. E., **T58**
Balenovic, K., **A19**, **A31**
Balkenhol, W. G., **T39**
Ballio, A., **Y2**
Balmain, A., **T40**
Balzer, W. D., **Z3**, **Z4**
Bakhaeva, G. P., **Y16**
Ban, I., **A47**
Ban, Y., **A36**, **K9**, **K10**
Banks, A. J., **Y10**
Banninster, R., **Y21**
Banthorpe, D. V., **T8**, **T9**
Barash, L., **A35**, **A54**
Barcellona, S., **Y2**
Barclay, G. A., **T51**
Barcza, S., **A28**, **K30**
Barker, A. C., **K1**
Barkley, L. B., **A53**
Barneis, Z. J., **K4**
Barnes, C. S., **T46**
Barnes, R. A., **A48**
Barnes, W. H., **K4**, **K21**
Barrett, G. C., **Y13**
Barrow, F., **A4**
Barrow, K. D., **T41**
Barry, J., **A43**, **A49**
Bartlett, L., **T55**, **K10**
Bartlett, M. F., **K11**
Barton, D. H. R., **A45**, **T20**, **T21**, **T28**, **T41**, **T42**, **T44**, **T46**, **T50**, **T51**, **T52**, **T58**, **K3**, **K4**, **K6**, **K7**, **Y14**, **Y16**
Bartsch, H., **T39**
Bartuska, V., **A16**
Bartz, Q. R., **A4**, **A9**
Bastani, B., **A2**
Bates, R. B., **T10**, **T14**, **T19**, **T25**
Battail-Robert, D., **A25**
Battersby, A. R., **A45**, **T13**, **K1**, **K2**, **K3**, **K4**, **K11**, **Y7**
Bauer, K., **X9**
Baughn, C., **A8**
Baum, M. E., **A46**
Baumann, G., **Y17**
Baumann, P., **T48**
Beak, P., **K11**
Beal, J. K., **T37**

Author Index

Beal, J. L., **K3**
Bear, C. A., **Y13**
Beard, C., **A38**
Beaton, J. M., **T42**, **T43**
Becher, D., **Y19**
Beckett, A. H., **A4**, **A5**, **A31**, **K9**
Beecham, A. F., **K10**, **K29**, **Y24**
Begley, M. J., **Y15**
Bekoe, D. A., **T52**
Bekurdts, J., **A45**
Bell, E. A., **A5**
Bell, M. R., **K28**
Bellardini, L., **T34**
Belleau, B., **X7**
Bencze, W., **Y1**
Bencze, W. L., **A52**
Benjamin, B. M., **A47**
Benoiton, L., **A11**
Benschop, H. P., **Z4**, **Z5**
Bentley, H. R., **K12**
Bentley, K. W., **K4**
Bentley, R. K., **Y15**
Berg, C. P., **A20**
van den Berg., G. R., **Z5**
Bergel, F., **A19**
Berger, J. G., **A46**
Bergman, R. G., **A44**
Bergmann, W., **T48**
Bergström, S., **T48**, **Y16**
Berlin, Yu. A., **Y16**, **Y27**
Bernasconi, R., **T51**
Bernauer, K., **K3**
Berner, E., **A31**
Bernotat-Wulf, H., **Y1**
Bernstein, J., **A20**
Berova, N. D., **A43**
Berson, J. A., **A46**, **A47**, **T10**, **T11**, **K27**, **X6**
Berthelot, M., **T11**
Berti, G., **A22**, **A23**, **A52**
Bertrand, J. A., **Y22**
Bestmann, H. J., **X1**
Bethell, M., **A17**
Bettolo, G. B. M., **Y3**
Bevan, C. W. L., **T52**, **K14**
Bevan, K., **A11**
Beyerman, H. C., **A10**, **K18**
Bhacca, N. S. A., **K5**, **K30**
Bhakuni, D. S., **K3**, **K4**
Bhat, H. B., **Y9**
Bhattacharrya, S. C., **T11**, **T19**, **T20**, **T21**, **T23**
Bianchi, E., **K34**
Bible, R. H., **T33**
Bick, I. R. C., **K3**
Bickart, P., **Z7**, **Z8**
Biemann, K., **K12**, **K13**, **K34**
Bienert, M., **K6**
Bijvoet, J. M., **A2**, **A27**
Binder, R. G., **A7**
Binks, R., **K2**
Birch, A. J., **T1**, **T2**, **T3**, **T10**, **T11**, **T51**, **Y3**, **Y6**, **Y7**, **Y8**
Bird, P. H., **K14**
Birkenmeyer, R. D., **Y21**
Birkinshaw, J. H., **Y6**
Birnbaum, G. I., **K33**
Birnbaum, K. B., **K33**
Birnbaum, S. M., **A11**
Bisarya, S. C., **T12**
Bjamer, K., **T36**, **T40**

Black, K. T., **Z1**
Blackburn, E. V., **A27**
Blackstock, W. P., **K2**
Blackwood, R. K., **Y28**
Blaha, K., **A22**, **K16**, **K18**, **K19**, **K20**
Blanchard, E. P., **T10**
Blank, F., **Y2**
Bloch, K., **A7**
Blossey, E. C., **K13**
Blount, J. F., **T2**, **K1**, **K2**, **Y11**
Blumann, A., **T5**
Boar, R. B., **T42**
Böckman, O. C., **T21**
Bode, K., **T11**
Boekelheide, V., **K7**
Boelens, H., **T9**
Boer, A. G., **T1**
Bogri, T., **K29**
Bohlmann, F., **K21**
Böhm, H., **X5**
Boime, A., **T9**
Bollinger, H., **Y20**
van Bommel, A. J., **A2**
Bond, F. T., **T50**
Bond, R. P. M., **T32**
Bonner, W. A., **Y1**
Bonsma, G. F., **K7**
Borkowsky, F., **K4**
van den Bosch, S., **A10**
Bose, A. K., **T4**, **T32**, **T34**
Bosnjak, J., **A18**
Bosshard, H., **T42**
Boter, H. L., **Z5**
Bottari, F., **A22**, **A23**
Bourn, P. M., **A37**, **A53**
Bowie, J. C., **Y10**
Bowman, R. M., **K31**
Boyce, C. B., **K27**
Boyd, D. R., **A10**, **A25**
Brack, A., **K17**
Bradfield, A. E., **T20**
Braekman, J. C., **K11**
Brandl, F., **T39**
Brandt, C. W., **T34**, **T36**
Brannon, D. R., **T19**
Bratek, M. D., **K21**
Brauchli, E., **A5**
von Braun, J., **A27**, **T1**, **T46**
Braun, F., **K16**
Brauns, F., **A2**
Braunschweiger, H., **A20**, **Y6**
Brechbuhler, S., **Y13**
Bredenberg, J. B., **T33**
Bredt, J., **T12**
Breen, G. J. W., **T46**
Bregant, N., **A31**
Brehm, L., **T37**
Brenner, A., **A48**
Breuer, S. W., **A45**, **K2**
Breuker, J. H., **A10**
Brewster, J. H., **A21**, **A25**, **A50**, **A51**, **X4**
Brewster, P., **A1**
Brickner, W., **T11**
Brienne, M. J., **A41**, **A42**
Brieskorn, C. H., **T33**
Briggs, L. H., **T13**, **T35**, **T49**
Bright, A., **T14**
Briner, R. C., **K27**
Bringi, N. V., **K8**, **K9**, **K11**

Britton, R. W., **T11**
Brockman, H., **A31**, **Y23**, **Y28**
Broeke, J. T., **A40**
Brook, A. G., **Z1**, **Z2**
Brookes, L. G., **A5**
Brooks, C. J. W., **T23**, **T46**, **T47**, **T49**
Brooks, J. S., **A12**, **Y12**
Brossi, A., **K1**, **K2**
Brossmer, R., **A11**
Brown, A., **X7**
Brown, G. B., **A4**
Brown, H. C., **A27**, **T8**, **T9**
Brown, K. S., **K13**, **K36**, **Y10**
Brown, R. T., **K2**, **K15**
Brown, S. H., **K13**
Brown, T. H., **K3**
Brown, W. A. C., **T53**, **Y17**
Browne, P. A., **X6**
Brownlee, R. G., **T8**
Bruderer, H., **T21**, **T46**, **T50**, **T51**, **T52**, **T58**, **K3**, **K4**, **K6**
Brufani, M., **T41**, **Y26**
Brugger, M., **A20**
Brummel, R. N., **X2**
Brunner, R., **K17**
Bruns, K., **T36**
Bryan, R. F., **T30**, **T41**
Buccini, J., **Y27**
Buchan, R., **A43**
Buchanan, C., **A2**
Buchart, O., **T39**
Buchecker, R., **T36**, **T54**
Buchi, G., **A8**, **T8**, **T13**, **T19**
Buchsacher, P., **K4**
Buck, K. T., **K3**
Buckel, W., **D2**
Buckingham, D. A., **A11**
Budzikiewicz, H., **K12**, **K13**, **K34**
Budhiraja, R. P., **T50**
Bukhari, S. T. K., **A3**
Bull, L. B., **K22**
Bu'Lock, J. D., **Y19**
Bunn, C. W., **Y29**
Bunnenberg, E., **X5**
Burgstaler, A. W., **T32**
Burian, F., **A55**
Burke, N. I., **Y1**
Burkhardt, F., **K2**
Burkhardt, U., **K35**
Burlinghame, A. L., **T24**
Burnell, R. H., **K14**
Burnett, A. R., **T13**, **A2**
Burrell, J. W. K., **T56**
Burrows, B. F., **A38**
Bush, M. A., **A19**, **X9**
Buta, J. G., **A25**, **A50**
Butler, J., **A19**
Butsugan, Y., **T13**
Buyle, R., **A17**
Bycroft, W., **K12**, **K13**, **K14**

Caglioti, L., **T45**, **T57**
Cahn, R. S., **A16**, **A26**, **C1**, **X4**, **X8**
Cain, B. F., **T13**
Cais, M., **T36**
Calam, D. H., **T52**
Calame, J. P., **K36**
Calas, B., **A56**
Cambie, R. C., **T35**
Camerman, A., **K15**

Author Index

Camerman, N., **K15**
Cameron, A. F., **T12, T47**
Cameron, D., **Y22**
Cameron, D. W., **Y10**
de la Camp, U., **Z7**
da la Camp, V., **A2**
Campbell, E. F., **K20**
Campbell, W. P., **T33**
Campello, J., **K13**
Campiglio, A., **A21, A22**
Cannon, J. R., **K28**
Cantrall, W. E., **T42**
Canzanelli, A., **A5**
Cardillo, G., **Y6, Y9**
Carlisle, C. H., **T42, T46**
Carlson, A., **Y21**
Carmack, M., **A2**
Carman, R. M., **T34, T43**
Carnmalm, B., **A31**
Carrazzoni, E. P., **K13**
Carée, F., **Z2**
Carroll, F. I., **Y21**
Carter, H. E., **A11**
Carter, W. L., **A44**
Carter, O. L., **K8, X9**
Casciato, C. A., **A30**
Caserio, F. F., **A1**
Caserio, M. C., **X2**
Cassady, J. M., **T24**
Cassinelli, G., **Y28**
Catalfomo, P., **T50**
Caughlan, C. N., **T22, T24**
Cava, M. P., **T36, K3, K13**
Cavill, G. W. K., **T13**
Ceder, O., **A26**
Cekovic, Z., **A18**
Celmer, W. D., **A32, Y25**
Centoni, L., **A27**
Cerar, D., **A19**
Cerny, M., **C1**
Cerrini, S., **T41**
Cervinka, O., **A11, K23**
Chain, E. B., **T41**
Chakraborty, D. P., **T53**
Chakravarti, K. K., **T23**
Chakravorty, P. N., **T48**
Chamberlin, J. W., **T32**
Chan, D., **A39**
Chan, R. P. K., **A5, A19, K18**
Chan, T. H., **A48, K19, Y27**
Chan, W. K., **K1**
Chan, W. R., **T36, T46, T51, T53**
Chandrasekaran, R., **C1**
Chang, C. T., **A38**
Chang, I., **Y27**
Chang, M. Y., **K7**
Chang Sin-Ren, A., **T33**
Chao, T. H., **A27**
Chapelle, A., **K14**
Chaplen, P., **Y6**
Charronat, R., **T9**
Chauvette, R. R., **Y29**
Cheema, Z. K., **A47**
Chen, F. C., **Y11**
Cheney, L. C., **A36**
Cheng, Y. S., **T21**
Cheng-Chiung, L., **A30**
Cheo, K. L., **A39**
Chetty, G. L., **T12, Y15**
Cheung, H-C., **T2**
Cheung, K. K., **T38, Y13, Y15, Z8**

Chevalier, R., **X7**
Chiba, R., **Y14**
Chickos, J., **Z4, Z5, Z6**
Chidester, C. G., **C1, T30**
Chikamatsu, H., **T21**
Chikamoto, T., **T17**
Chin, C., **Y15**
Chopra, C. S., **T43**
Chou, T. S., **T50**
Chow, S. W., **T11**
Chow, W. Z., **T29**
Chow, Y-L., **T32**
Christensen, B. W., **A8, A12, A33, Z8**
Christie, J. B., **A23**
Chu, S. C., **K27**
Chuprunova, O. A., **Y16**
Cimino, G., **T57**
Claeson, G., **A18**
Clardy, J. C., **K6**
Clark, D. R., **A25, A41**
Clark, G. R., **T33**
Clark, K. J., **T13**
Clark, S. D., **X1**
Clarke, F. H., **T29**
Clarke, H. T., **Y29**
Clarke, R. L., **T39**
Clark-Lewis, J. W., **A10, Y3, Y4, Y5**
Clements, J. H., **K3**
Closse, A., **Y19**
Closson, W., **T42**
Coates, R. M., **T27**
Cockburn, G. B., **T10**
Cocker, W., **T19, T21**
Coggon, P., **T20, T22, T40, K10**
Cohen, S. G., **A6**
Coke, J. L., **A13, K20**
Collins, C. J., **A23, A47**
Collins, D. J., **A43, T51**
Collins, J. F., **A8, K31**
Cole, A. R. H., **T43**
Colombo, C., **Y22**
Comer, F. W., **T31**
Comin, J., **K3, K4**
Cone, C., **A5**
Cone, N. J., **K15**
Conlay, C., **T41**
Connell, D. W., **T2**
Connolly, J. D., **T32, T34, T40, T51**
Conover, L. H., **Y28**
Cook, C. E., **K29**
Cookson, R. C., **T29, K21**
Cooper, A., **T32, T47, T50**
Cope, A. C., **A3, X2, X11**
Coppola, J., **K15**
Coppola, J. C., **T27**
Corbella, A., **T31**
Corey, E. J., **A5, A13, T11, T22, T30, T42, T52**
Cornforth, J. W., **T3, D1, D2**
Cornforth, R. H., **T3**
Corradi, A. B., **Y20**
Corral, R. A., **Y29**
Corriu, R. J. P., **Z1, Z2**
Corrodi, H., **K3, K30**
Corsano, S., **T50, T51**
Corse, J., **A26**
Coscia, C. J., **T13**
Cottis, S. G., **X9**

Coverdale, C. E., **K35**
Cowdrey, W. A., **A1, A4, A11, A12, A22, A51**
Cowley, D. E., **T43**
Cox, M. R., **A55, T30**
Cox, J. S. G., **T50**
Crabbé, P., **T53, T56, X1, X5**
Cradwick, P. D., **T24, T30**
Craig, J. C., **A5, A10, A19, A41, K4, K5**
Craig, L. C., **K17, K35**
Craik, J. C. A., **Y10**
Cram, D. J., **A21, A40, A46, A48, A49, A51, A56, D1, D2, X8, Z6, Z8**
Craven, B. M., **T31, K13, K14**
Crawford, R. J., **A6**
Croft, L. R., **Y22**
Crombie, L., **A8, A27, A35, T17, T39, Y1**
Cron, M. J., **A36**
Crooks, H. M., **A4**
Cross, A. D., **T47, T48**
Cross, B. E., **T35**
Crossley, J., **A35**
Crout, D. H. G., **A28, A33, K22, K23**
Crowfoot, D. C., **T46, Y29**
Crowley, H. C., **K23**
Cruse, W. B. T., **T37**
Csepreghy, G., **A41**
Culvenor, C. C. J., **K22, K23, K24**
Cutler, M. C., **Y15**

Dahm, D., **K6**
Dahn, H., **A40**
Dainis, I., **Y3, Y5**
d'Albuquerque, I. D., **Y3**
Daly, J. W., **A16, A25, T7, K26, K36**
Damodaran, N. P., **T12, T30**
Dangschat, G., **A26**
Danieli, N., **T49**
Danilova, A. V., **K22**
Das, K. C., **T55**
Dastoor, N. J., **K10, K15**
Dauben, W. G., **T21, T41, T46, T48**
Daum, S. J., **T39**
Davidson, B. E., **C1**
Davidson, R. S., **T40**
Davie, A. W., **T52**
Davies, J. S., **A11**
Davies, V. H., **T30**
Davies, D. D., **Z1**
Davis, R. E., **C1**
Dawson, T. M., **T48**
Day, J., **Z8**
Deane, C. C., **A11**
De Bruin, K. E., **Z4, Z5**
De Grazia, C. G., **A55, T34**
de Jongh, H. A. P., **K35**
de la Camp, U. **Z7**
de la Camp, V., **A2**
Delepine, M., **T9**
Della, E. W., **T5**
Delmont, D. W., **A22**
Delong, D. C., **Y24**
Delton, M. H., **X8**
de Maindreville, M. D., **K13**
Demanczyk, M., **K11, K13**

Author Index

de Mayo, P., **T18, T21, T28, T43, T53, X5**
Denne, W. A., **K29**
Denney, D. B., **A12, A38**
de Roos, J. B., **T22**
De Silva, K. T. D., **K2**
De Stefano, S., **T57**
Deulofeu, V., **K3**
Dev, S., **T12, T18, T25, T28, T29, T30, T32, T35, T37**
De Ville, T. E., **T55**
de Vries, J. X., **Y14**
de Vries, K. S., **A43**
de Waal, H. L., **K20**
Diamond, R. D., **Y6**
Diassi, P. A., **K9**
Dick, A. T., **K22**
Dickel, D. F., **T46**
Dickerson, D., **T14**
Dickey, F. H., **A13**
Dilger, W., **A20**
Dillon, J., **T58**
Dion, H. W., **A9**
Dirlam, J., **A46**
di Sanseverino, L. R., **T45, T49**
Diversi, P., **A28**
Dixon, J., **T48**
Djerassi, C., **A28, A29, A38, A39, A53, A54, A55, T1, T2, T7, T8, T9, T13, T23, T25, T32, T34, T35, T36, T42, T45, T50, T51, T56, K4, K12, K13, K32, K34, Y1, Y25, Y29, D2, X5**
Dobler, M., **T27**
Dobrynin, V. N., **A51, Y28**
Dobson, T. A., **K4**
Doering, W. von E. A., **A1, A12, A37, A44**
Dolder, F., **T35**
Doldouras, G. A., **A27**
Dolejs, L., **T2, T25**
Donnelly, B. J., **Y5**
Donnelly, D. M. X., **Y4, Y5**
Donohue, J., **Y28, Z4**
Dopke, W., **K6**
Dornberg, M. L., **A30**
Dornhege, E., **A20, A25**
Van Dorp, D. A., **Y16**
Dorr, H., **A2**
Doskotch, R. W., **T37**
Doty, M. S., **A44**
Douglas, B., **K3, K15, K27**
Doyne, T., **A4**
Draber, W., **T16**
Draffan, G. H., **T23**
Drake, A. F., **X1**
Dreiding, J., **T42**
Dreiding, A. S., **K17**
Drewer, R. J., **X6**
Drewes, S. E., **Y3, Y4**
Dreyer, D. L., **T52, T57**
Dreyfus, H., **T27**
Dry, L. J., **K22**
Duarte, A. P., **K13**
Duchamp, D. J., **C1, T30, Z8**
Duff, J. M., **Z1, Z2**
Duggan, J. J., **K11, K13**
Dugat, D., **A33**
Dukes, M., **T40**
Dullforce, T. A., **A19, T24, X9**
Dummer, G., **K18**

Duncan, W. G., **T8**
Dunitz, J. D., **T27, T48, Y27, Y28**
Dunnenberger, M., **T46**
Durham, L. J., **K13, K14**
Durst, O., **T34, T42**
Dutcher, J. D., **Y26**
Dvornik, D., **K32**
du Vigneaud, V., **A4, Y30**
Dyer, J. R., **Y22**

Eaborn, C., **Z2**
Eade, R. A., **T43, T51**
Eastman, R. H., **T1, T7**
Eastwood, F. W., **A10**
Eberle, M., **A33**
Eble, T. E., **Y21**
Ebnother, A., **Y2**
Eckhardt, G., **A45**
Edwards, A. G., **A36, K20**
Edwards, J. D., **T7**
Edwards, O. E., **T32, K32**
Edwards, T. P., **K2**
Eenshuistra, J., **K18**
Eggerer, H., **D2**
Eggers, S. H., **Y3**
Ehl, K., **Y23**
Ehrenstein, M., **A11**
Ehret, C., **T25**
Ehrig, V., **A2**
Ehrlich, F., **A27**
Einhorn, A., **K28**
Eisenbeiss, J., **Y1**
Eisenbraun, E. J., **A28, A36, A38, A39, A55, T13, T14, T15 T25, T50, T56, Y25**
Ekong, D. E. U., **T52**
Elderfield, R. C., **T49**
Eliel, E. L., **A22, T5**
Elhafez, F. A. A., **A48, A49**
Ellesrad, G. A., **A55, T34, Y11, Y15**
Elliott, D. F., **A24**
Elliott, M., **T17**
Elliott, T., **A38**
Elliott, T. H., **A39**
Ellis, J., **T43**
El-Olemy, M. M., **K19**
Els, H., **Y25**
Emerson, M. T., **T22, T24**
Emoto, S., **Y30**
Endo, K., **T26**
Endo, M., **T35**
Engel, D. W., **T39**
Engel, L. L., **T34**
English, J., **A31**
Enslin, P. R., **T50**
Enzell, C., **T12, T26, T34, Y9**
Erdtman, H., **T12, T32**
Erhardt, K., **A5, A17**
Eschenmoser, A., **A20, A39, T5, T42**
Escobar, M., **T1**
Estlin, J. A., **K6**
Eugster, C. H., **A2, T33, T36, T54, Y20**
Evans, R. H., **Y11, Y15**
Evans, R. J. D., **X1**
Evans, W. C., **T28**
Eveleens, W., **K18**
Ewing, D. F., **A19**
Eyre, H., **T40**
Eyton, W. B., **Y4**

Fabbri, C., **K17**
Falcone, M. S., **T29**
Fales, H. M., **K6**
Falk, H., **X8, X9**
Faltis, F., **K3**
Farber, L., **T4**
Fardig, O. B., **A36**
Farina, M., **X10**
Farney, R. F., **T27**
Farnham, W. B., **Z4, Z6**
Farrar, M. W., **A53**
Farrier, D. S., **K6, K20**
Fattorusso, E., **T57**
Faulkner, D. J., **A33, T58**
Fawcett, J. K., **T46**
Fawcett, J. S., **T46**
Fazakerley, H., **T45**
Feairheller, S. H., **T18**
Fechtig, O., **T12**
Fedeli, W., **T41, X46**
Feeney, J., **Y3**
Fehlhaber, H-W., **T51**
Fehr, T., **K17**
Feigl, D. M., **A12, A22**
Feldstein, J., **T5, T6**
Fenical, W., **T30**
Fereday, P. L., **A23**
Ferguson, G., **T12, T28, T36, T40, T47, T58, K29, K31, Y15**
Ferguson, G. W., **A4**
Fernando, Q., **Y13**
Fernholz, E., **T48**
Ferrari, C., **K3**
Ferrari, F., **Y3**
Ferrari, M., **T40**
Ferranini, P. L., **A22, A23**
Ferretti-Aloise, M-G., **T16**
Ferris, J. P., **K27**
Ficken, G. E., **Y23**
Fickett, W., **A13**
Filho, J. M. F., **K16**
Finch, N., **A52, K9, K10**
Fischer, E., **A4, A20**
Fischer, E. G., **T51**
Fischer, H., **Y23**
Fischer, H. O. L., **A11, A13, A14, A15, A16, A26**
Fisher, G. H., **A8**
Fleck, W. E., **Y1**
Fleischer, E., **A9**
Fleming, I., **T22, Y23**
Fles, D., **A34, K22**
Flores, S. E., **T36, K13**
Floyd, J. C., **Y22**
Flynn, E. H., **Y29**
Fodor, G., **A2, A21, A41, K19, K28**
Foley, P. J., **A36, A38**
Folkers, K., **A4, A22, Y24, Y30**
Foltz, C. M., **A21**
Fonken, G. S., **A8**
Ford, L., **T13**
Forrester, J. D., **A4, A11**
Fouquey, C., **T53**
Fourie, L., **Y2**
Fowden, L., **A27**
Fowler, L. R., **K33**
Fox, J. A., **Y21**
Franceschi, G., **Y28**
Francis, J. E., **K25**
Franck, R. W., **K35**

Author Index

Francois, P., **K2**
Frank, B., **Y17**
Frank, G. W., **X7**
Fratini, A. V., **Y20**
Fray, G. I., **T13**
Fredga, A., **A1, A2, A4, A18, A28, A31, A32, A50, A57**
Freeman, H. C., **A11**
Freeman, J. P., **A22**
Freudenberg, K., **A1, A2, A4, A26, A29, T2, Y7**
Frey, A. J., **A33, K11, K17**
Friedli, H., **A16**
Freyss, G., **A5**
Fridrichsons, J., **T44, K3, K4, K5, K18, K22, K31, Y24**
Fried, J. H., **T48**
Frohardt, R. P., **A4**
Frostl, W., **X9**
Frye, C. L., **Z1**
Fuchs, C. A., **T33**
Fuchs, H., **Z3, Z5**
Fuji, K., **T15, K20**
Fujimoto, T., **T44**
Fujimoto, Y., **A30**
Fujino, A., **T13**
Fujise, S., **T15**
Fujise, Y., **Y3**
Fujita, E., **T33, K20**
Fujita, S., **A41**
Fujita, T., **A50, T24, T30, T33**
Fujita, Y., **A10**
Fujiwara, T., **T45, K5**
Fukazama, Y., **K35**
Fuks, Z., **A19**
Fukumoto, K., **K1**
Fukushima, S., **T45**
Fukuyo, M., **T15**
Fulke, J. W. B., **T37**
Fullerton, D. S., **T50**
Fulmor, W., **Y19**
Furst, A., **A53, T19**
Furusaki, A., **T23, Y17, Y20, Y25**
Furuta, S., **A16, A26**
Fusari, S. A., **A4**

Gabarino, J. A., **A40**
Gabe, E. J., **T21, T29, K28, D2**
Gadola, M., **T1**
Gaffield, W., **A4, A7, A8, A24, A27, Y3**
Gagniare, D., **A25**
Gajewski, J. J., **A37**
Galasko, G. G., **T55**
Galbraith, M. N., **T38**
Galetto, W. G., **A4, A8, A24, A27**
Galik, V., **T32, T34**
Galt, R. H. B., **T35**
Gallagher, M. J., **Z3**
Gallagher, R. T., **Y13**
Gallina, C., **Y22**
Galt, S., **A36**
Games, M. L., **T39**
Gandhi, S. S., **T49**
Ganguli, G., **K20**
Ganis, P., **Y26**
Gaoni, Y., **T4, T5**
Garbers, C. F., **T50**
Gardello, L. A., **A36, A38, Y13**
Gardner, J. A. F., **Y7**

Garner, H. K., **A13**
Garratt, S., **A45, K2**
Garwood, D. C., **Z8**
Gascoigne, R. M., **T50**
Gassmann, I., **T39**
Gassman, P. G., **T12**
Gaston, L. K., **A40**
Gaudemer, A., **Y19**
Gaudemer, M. A., **T53**
Gautschi, F., **A53, K35**
Geise, H. J., **A18**
Geissman, T. A., **T22, T24**
Geismar, W., **A55**
Geller, L. E., **A29, A38**
Gemenden, C. W., **K11**
Gempeler, H., **Y4**
Van Der Gen, A., **T9**
Gentes, F. H., **A27**
George, T., **T25**
Gerald, M. C., **A20**
Gerlach, H., **D1, X3, X5**
Ghisalberti, E. L., **T40**
Ghislandi, V., **A21, A22, A23, A25**
Ghosh, C. K., **D1**
Giacomello, G., **Y26**
Giacopello, D., **T50**
Gibbon, G. A., **A17**
Gibbons, C. S., **K26**
Gibbs, J. A., **T53**
Gibson, K. H., **K2**
Giddings, W. P., **A46**
Giersch, W., **T3**
Gigg, J., **A15**
Gigg, R., **A15**
Giguere, J., **A6**
Gilardi, R. D., **K36, Y20**
Gilbert, B., **K13, K14**
Giles, D., **A36**
Gillam, A. E., **T20**
Gillard, R. D., **A19**
Gilman, R. E., **X8**
Giral, L., **A56, T1**
Girotra, N. N., **T32, K32**
Glass, M. A. W., **T8, X5, X10**
Glattfield, J. W. E., **A3**
Glauert, R. H., **Y28**
Gloor, U., **T56**
Glotter, E., **T48**
Glusker, J. P., **A30**
Godfredson, W. O., **T31**
Godfrey, J. C., **K7**
Godin, P. J., **A8, T17, Y1**
Godinho, L. D. S., **A45, T58**
Goering, H. L., **A12, A34, A38**
Goffinet, B., **T48**
Gokel, G., **X8**
Gold, A. M., **A53, T19**
Goldberg, S. I., **A26, Z3**
Goldman, I. M., **T25, T27**
Goldschmidt, S., **A5**
Gollnich, K., **T6**
Gomez, F., **T41**
Gonzales, M. P., **T1, Y17**
Goodfellow, D., **T55**
Gopalakrishna, E. M., **T32, T47**
Gordon, J. T., **T44**
Gorman, A. A., **K14, K15**
Gorman, M., **K11, K15**
Gosche, R., **K26**
Gosteli, J., **Y29**

Goto, G., **Y32**
Goto, K., **K4, K5**
Goto, M., **T45**
Goto, T., **Y20**
Gottarelli, G., **A21, A22, A23, A24, A25**
Gottlieb, O. R., **Y3, Y4**
Gotz, M., **K29**
Gough, L. J., **T33**
Goutarel, R., **K8, K9, K30, K36**
Govindachari, T. R., **T23, T52**
Gowal, H., **X9**
Graham, E. M., **T36**
Grant, D. F., **T23, T29**
Grant, I. J., **K26**
Grant, P. K., **T34, T36, T43**
Gray, A. H., **K29**
Gray, D. O., **A27**
Gray, E., **A9**
De Grazia, C. G., **A55, T34**
Green, B., **A55, T34**
Green, C. D., **T25**
Green, F. C., **T7**
Green, M. M., **A12, D1, Z7, Z8**
Green, N. M., **Y30**
Greenbaum, M. A., **X6**
Greenfield, S., **T48**
Greenstein, J. P., **A11**
Gregory, B., **K2**
Grethe, G., **K1**
Greuter, F., **T26**
Grimshaw, J., **A49**
Griot, R., **K17**
Grinter, R., **X6**
Groen, M. B., **X7**
Gronowitz, S., **A1, A21, X5**
Grossert, J. S., **K12**
Grove, J. F., **T31, T35, Y17**
Gruber, W., **A41**
Grundon, M. F., **A8, A10, K31**
Gubler, B., **T27**
Guette, J. P., **A25, A50**
Guggisberg, A., **K14**
Guglielmetti, L., **T34**
Guha, P. C., **T11**
Gulewitsch, W., **A5**
Gumlich, W., **K12**
Gunay, G. E., **T18**
Gunstone, F. D., **A7**
Gurbaxini, S., **T2**
Gurevich, A. I., **A51, Y28, Y30**
Gustafson, P., **X5**
Guthrie, R. D., **A3, C1, D2**
Gutmann, H., **T34, T42**
Gutsche, C., **D2**
Gutsche, C. D., **A38**
Gutzwiller, J., **T31**
Gwiner, R., **A2**

Haas, C., **X11**
Habaguchi, K., **T44**
Hadine, C. I., **T22**
Haefliger, W., **T5**
Haginawa, J., **K11**
Hagishita, S., **A47, A51, X3, X5**
Hai-fu, F., **A30**
Hale, R. L., **T50**
Hall, S. R., **T43**
Haller, G., **X9**
Haller, H. L., **A1, A6, A12, A22**
Halls, C. M. M., **T51**
Hallsworth, A. S., **T2**

Hally, D., **T34**
Halpern, B., **A4**
Halpern, O., **A28, T16, Y25**
Halsall, T. G., **T19, T20, T37, T43, T45, T52**
Haltiwanger, R. C., **T41**
Ham, P. J., **T45**
Hamada, Y., **K6**
Hamamoto, T., **K11**
Hamberg, K., **Y16**
Hamilton, J. A., **K27, Y30**
Hammer, C. F., **T53**
Hamor, T. A., **T37, T58, K26**
Hamsher, J. J., **T5**
Hanahan, D. J., **A15**
Hanaoka, M., **K11, K16**
Hanic, F., **T18**
Hannaford, A. J., **A55**
Hannaway, C., **T47**
Hansel, R., **Y15**
Hansen, J. F., **K6**
Hansen, S. E., **A8, A8**
Hanson, A. W. **A18, K7**
Hanson, G. C., **A46**
Hanson, J. R., **T35, T36, T47**
Hansson, B., **A26**
Hanzmann, E., **T58**
ul-Hague, M., **T22, T24**
Harada, H., **K29**
Harada, I., **Y1**
Harada, K., **A19, A24**
Harada, N., **T30, Y18**
Harada, S., **Y26**
Harayama, T., **K25**
Hardcastle, G. A., **Y29**
Hardegger, E., **A9, A20, K3, K30, Y4, Y6**
Hardy, A. D. U., **T24**
Hardy, D. G., **K4**
Harington, C. R., **A5**
Harispe, M., **T8, T9**
Harle, E., **T39**
Harper, P., **T43**
Harper, S. H., **A27, A35**
Harris, A., **T34**
Harris, D. A., **K14**
Harris, D. R., **Y25**
Harris, J., **A7**
Harris, M. M., **X6**
Harris, S. A., **A4, Y30**
Harrison, H. R., **T52**
Harrison, I. T., **T41**
Harrison, J. W., **T40**
Harrison, S., **T34, T41**
Hart, N. K., **K10, K21**
Hart, P. A., **A39**
Hartshorn, M. P., **T10**
Hartwell, J. L., **A5, A31, Y7**
Harvey, R. G., **T50**
Haskell, T. H., **A4**
Hassall, C. H., **A11, Y4**
Hassanali-Walji, A., **Y22**
Hata, T., **Y25**
Hatam, N. A. R., **Y9**
Hattori, T., **T50**
Hattori, Y., **Z7**
Hauk, F., **A5**
Hauser, D., **Y14, Y24**
Hauser, F. M., **K6**
Havinga, E., **A10**
Hawks, R. C., **K10**

Hawks, R. L., **K20**
Hawley, D. M., **T28, Y27**
Haworth, R. D., **Y7**
Haworth, W. N., **A3**
Hayashi, M., **T23**
Hayashi, S., **T18**
Hayashi, T., **T17**
Hayashi, Y., **T15, T23, T38**
Hayatsu, R., **A29**
Hayes, W. K., **T21**
Haynes, L. J., **T36, K3**
Hearn, W. R., **A11**
Heath, H., **A20**
Hecker, E., **T39**
Hedlund, I., **A32**
Hefelfinger, D. T., **X7**
Heffler, M., **A48**
Hegarty, B. F., **T17**
Hegde, B. H., **T20**
Hellwinkel, D., **X10**
van der Helm, D., **A30, T31, T41**
Helmcamp, G. K., **D2**
Hemingway, J. C., **T24**
Hemingway, R. J., **T24, Y19**
Henbest, H. B., **T2**
van den Hende, J. H., **K4, Y16, Y27**
Henderson, M. S., **T37**
Hendrickson, J. B., **K26**
Hendry, J. A., **T4**
Henrick, C. A., **T5, T36**
Henry, J. A., **K12**
Henry, T. A., **T2**
Herald, D. L., **Y30**
Herbert, R. B., **K4**
Herbst, D., **T23**
Herlem-Gaulier, D., **K36**
Hermann, H., **Y24**
Herout, V., **T14, T16, T18, T19, T21, T22, T23, T25, T26, T28, T29**
Herr, R. R., **Y21**
Herrán, J., **T36**
Herrin, J., **Y11**
Hertler, W. R., **K34**
Herz, W., **T21, T24, T35**
Herzog, E., **T39**
Hesse, M., **K11, K13, K15, K16**
Heusler, K., **Y29**
Heusner, A., **Y29**
Hever, D. B., **Y23**
Heyl, D., **Y24**
Van Heynningen, E., **Y29**
Hibbin, B. C., **A23**
Hickel, D., **Y11**
Hickernell, G. K., **Y12**
Hickernell, G. L., **T39**
Hiestand, A., **T46**
Hietala, P. K., **A11**
Highet, P. F., **K6**
Highet, R. J., **K6**
Hill, R. E. E., **Z2**
Hill, R. K., **A28, A30, A36, A38, A48, A56, K18, K19, K20, K30, Y1, Y13, Y27, X4**
Hiniko, H., **T19, T20, T25, T27, T38, T53**
Hiniko, Y., **T25**
Hirai, K., **T32**
Hiramatsu, M., **Y25**
Hirata, T., **T51**

Hirata, Y., **T22, T39, K22, K26, Y20**
Hirose, Y., **T6, T18, T20, T23, T29, Y1, Y9**
Hirabayashi, M., **T15**
Hirsch, H., **K4**
Hitchens, M., **A16**
Hobbs, J. J., **A43**
Hocheder, F., **Y23**
Hochstein, F. A., **Y25**
Hodder, O. J. R., **T37, T53**
Hodges, R., **T36, T50, Y13, Y24**
Hodgkin, D. C., **T48, T50, Y24, Y29**
Hodgkin, J., **T5**
Hodson, H. F., **K11**
Hoeksma, H., **Y21**
Hofer, O., **X9**
Hofmann, A., **A33, T49, K11, K17**
Hofmann, K., **T42**
Hoffmann, P., **X8**
Hofheinz, W., **T8, T26**
Hoffsommer, R. D., **Y14**
Hoge, R., **C1**
Hohmann, W., **A26**
Hohne, E., **T49, K34**
Hoizey, M.-J., **K13**
Holker, J. S. E., **T50, Y13**
Hollands, R., **Y19**
Hollands, T. R., **T35, X5**
Holness, N. J., **T42**
Holst, J. P. C., **Z3**
Holub, M., **T18, T22, T24, T25, T28**
Holzapfel, C. W., **T50**
Homer, G. D., **Z7**
Honda, O., **T38**
Honda, T., **K3**
Honda, Y., **T15**
Honjo, M., **A21**
Honma, H., **T38**
Honmaru, S., **A24, A25**
Honwad, V. K., **A49, T16**
Hooper, I. R., **A36**
Hope, D. B., **A12**
Hope, H., **A2, T24, Z1, Z7**
Hoppe, W., **T38, T39, T48**
Hopps, H. B., **X5, X10**
Horak, M., **T28**
Horeau, A., **T8, K5**
Horibe, I., **T19, T29**
Horii, Z., **K18**
Horn, D. H. S., **A1, T38, K6**
Horner, L., **Z3, Z4, Z5**
Horrocks, W. de W., **X5**
Hortmann, A. G., **A22**
Hosanky, N., **K35**
Hoshino, T., **T42**
Hosking, J. R., **T34, T36**
Hoss, H. G., **A10**
Hossain, M. B., **T31, T41**
Houbiers, J. P. M., **X4**
Hough, E., **K7**
Hough, L., **C1**
Howe, R., **T10, T32**
Hrbek, J., **K5, K10**
Hruban, L., **K5, Y19**
Hsiung, V., **C1**
Hu, S., **T23**
Hub, L., **A11, K23**

Huber, C. S., **K1, K19, K23**
Huber, R., **T48**
Huber, W., **Y16**
Huckel, W., **A37, T10, T12**
Huckstep, L. L., **K15, Y24**
Hugel, G., **T35**
Hughes, E. D., **A1, A4, A11, A12, A22, A23, A51**
Hughes, G. K., **K5**
Hugo, J. M., **K12**
Hulbert, P. B., **Y3, Y4, Y7, X11**
Hundt, H. K. L., **Y6**
Hunt, D. J., **C1**
Hunt, G. E., **A9**
Hurst, J. J., **T8**
Hursthouse, M. B., **A55, C1**
Hutchins, R. O., **T5**
Hyeon, H. B., **T14**

Ibe, S., **T20**
Ichihara, A., **A30**
Ichikawa, S., **T51**
Igarashi, H., **T50**
Iguchi, K., **T44**
Iguchi, M., **T22**
Ihara, M., **K3**
Iitaka, Y., **C1, T24, T32, T33, T41, T45, T51, K14, K32, K35, Y18, Y26, X6**
Ikeda, M., **K18**
Ikekawa, N., **T22**
Ikekawa, T., **C1**
Ikuta, M., **T22**
Imado, S., **K18**
Imai, S., **T45**
Imaizumi, K., **T22**
Imaizumi, S., **A51**
Imamura, H., **T38**
Imamura, K., **A41**
Imanishi, M., **T45**
Impastato, F. J., **A35, A54**
Inayama, S., **T24**
Inch, T. D., **A11**
Ingold, C. K., **A1, A4, A11, A12, A22, A23, A26, A51, C1, X4, X8**
Ingrosso, G., **A29**
Inhoffen, H. H., **A45**
Inouye, H., **T13, T15, Y17**
Inouye, K., **T15**
Inouye, Y., **A35, A37, A44, T4, T17**
Insole, J. M., **X5**
Inubushi, Y., **T44, K6, K25, K31**
Irai, Y., **T45**
Irrevere, F., **A30**
Irie, H., **T17, K29**
Irie, T., **T12**
Irvine, D. S., **T46**
Isaacs, N. W., **K2**
Ishibashi, K., **T32, T41**
Ishii, H., **T25, K25**
Ishikawa, K., **A40**
Ishikawa, M., **T18, T21, K8**
Ishizaki, Y., **T23**
Isler, O., **T56**
Isobe, K., **T45**
Isoe, S., **T14**
Itai, A., **T41, K14**
Ito, M., **A19, T42**
Ito, S., **T16, T26, T38, Y3**

Itschner, V., **A10**
Ivanov, I. C., **K21**
Iverach, G. C., **K25, K26, K32**
Ives, D. A. J., **T46**
Iwasaki, H., **A19**
Iwasaki, S., **T50**
Iwata, K., **T43, T44, T45**
Iwata, T., **K6**

Jackman, L. M., **T56, Y4**
Jackson, B. G., **T33, Y29**
Jackson, W. R., **A10**
Jacobi, P., **T39**
Jacob, G., **T28**
Jacobs, W. A., **T49, K17, K35**
Jacobson, R. A., **K6**
Jacobus, J., **A12, Z7, Z8**
Jacot-Guillarmod, A., **T16**
Jaeger, D. A., **D2**
Jaeger, R. H., **T13**
Jaeggi, K. A., **T49**
Jager, H., **A40**
Jaggi, H., **K11**
Jaggi, W., **A20**
Jain, M. K., **T51**
James, A. N., **T13**
James, M. N. G., **T37**
James, R., **K4**
Janot, M. M., **K2, K8, K9, K12, K13**
Jansen, A. C. A., **A10**
Janssen, P. A. J., **Y30**
Janzso, G., **K28**
Jaques, J., **A41, A42**
Jarreau, F., **K30**
Jay, E. W. K., **K33**
Jay, L., **K33**
Jayco, M. E., **A15**
Jefferies, P. R., **T5, T35, T36**
Jeffrey, G. A., **K9, K15, K27**
Jeffreys, J. A. D., **K29**
Jeffs, P. W., **T32, K6, K20**
Jeger, O., **A26, A39, A53, T1, T2, T19, T21, T32, T34, T42, T44, T45, T46, T51, T52, T57, K4, K35, Y9**
Jellinek, F., **Y20**
Jemison, R. W., **Y5**
Jenkins, I. D., **Z3**
Jenkins, J. A., **K26**
Jenkins, P. N., **K7**
Jennings, J. P., **A28, K3**
Jensen, F. R., **Z1**
Jensen, H., **A19**
Jerina, D. M., **A16, A25**
Jimenez, F. G., **T36**
Joel, C. D., **D2**
Johns, R. B., **Y23**
Johns, S. R., **K7, K8, K10, K21, K28, K29, X7**
Johnson, A. W., **Y22**
Johnson, C. K., **A30, D2**
Johnson, C. R., **Z8**
Johnson, D. A., **A36, Y29**
Johnson, D. L., **A36**
Johnson, J. R., **Y29**
Johnson, F., **Y21**
Johnson, F. H., **Y20**
Johnson, J. L., **T48**
Johnson, L. F., **A16, A26, T35, T36**

Johnson, P. C., **A30, D2**
Johnson, R. R., **Y58**
Johnson, S. M., **T27, Y11, Y16**
Johnson, W. S., **K35**
Jommi, G., **T31**
Jonas, G., **T29**
Jones, D. I., **A3**
Jones, E. R. H., **T43, T45, Y15**
Jones, G., **K31**
Jones, R. G., **Y21**
Jones, R. V. H., **T20**
Jones, W. H., **A22, Y24**
Jones, W. M., **A44, X1, X2**
de Jongh, H. A. P., **K35**
Jonsson, E. U., **Z8**
Jonsson, H.-G., **A18**
Jordan, P. M., **D2**
Joseph, T. C., **T12**
Joseph, J. P., **K20**
Joseph-Nathan, P., **T1, Y17**
Joshi, B. S., **T52**
Joshi, K. R., **K28**
Jostes, F., **A27**
Joule, J. A., **K13, K19, Y1**
Juliano, B. R., **A48**
Junginger, H., **A55**
Just, G., **Y2**

Kabuki, K., **T15**
Kaczka, E. A., **Y24**
Kagan, F., **Y21**
Kagan, H. B., **A43, A49**
Kaisin, M., **K11**
Kakisawa, H., **T38, T44**
Kakudo, M., **A32**
Kaltenbronn, J. S., **A8, Y1**
Kalvoda, J., **A53, T19, K4**
Kalyanpur, M. G., **A26**
Kamal, A., **Y2**
Kamata, S., **K31**
Kametani, T., **K1, K3**
Kamikawa, T., **T15, T52**
Kamiya, K., **T45, T51, K5, Y21, Y26**
Kamiyawa, K., **Z7**
Kaneko, T., **A2, A27, A30, K24**
Kaneko, Y., **A17**
Kanematsu, K., **A20**
Kano, H., **C1**
Kapadia, V. H., **T17, T25**
Kapil, R. S., **T13**
Karady, S., **A40**
Karapetyan, M. G., **A51, Y28**
Kapadi, A. H., **T32, T35**
Karle, I. L., **A30, K6, K26, K36, Y20**
Karle, J. M., **A30, K6, Y20**
Karrer, P., **A4, A5, A10, A11, A17, A19, A20, A45, T15, K12, K20, Z8**
Kartha, G., **K4, Y9**
Kasturi, T. R., **T49**
Katagiri, M., **Y25**
Kataoka, H., **A26, K21**
Kataoka, K., **K35**
Katarao, E., **K31**
Katayama, C., **T23, T37**
Katayama, H., **T33**
Kates, M., **A15**
Katekar, G. F., **Y4**
Kato, K., **T22**

Author Index

Kato, M., **T15**
Kato, N., **T37**
Katsu, M. A., **K25**
Katsuda, Y., **T17**
Katsuhara, J., **T4**
Katsui, N., **T22**
Katsumura, R., **T14**
Katsumura, S., **T23**
Katsura, H., **A2, A27, A30**
Kavkova, K., **K16**
Kawaguchi, K., **T44**
Kawaguchi, Y., **T23**
Kawai, M., **T53**
Kawasaki, I., **K24**
Kawashima, K., **T36**
Kawazoe, Y., **K31**
Kawasu, K., **A50, T33, T39**
Keefer, L. K., **K8**
Kelley, C. J., **A2**
Kelly, J. R., **T17**
Kelly, R. B., **T46**
Kelly, W., **Y7**
Kelsey, J. E., **T21, T24**
Kemp, C. M., **X7**
Kennard, O., **T27, T45, T49, K2**
Kenner, G. W., **A17**
Kenyon, J., **A29, T10**
Keresztesy, J. C., **Y30**
Kerigan, A., **K15**
Kerling, K. E. T., **A10**
Kern, H. J., **T10**
Kessar, S. V., **T49**
Keziere, R. J., **T11**
Khalil, M. F., **K14**
Khan, W. A., **A10**
Kharasch, M. S., **A27**
Khastgir, H. N., **T42**
Khedouri, E., **A6**
Khoung-Huu, F., **K36**
Khoung-Huu, Q., **K36**
Khurana, R. G., **T29**
Kiang, A. K., **T44**
Kiang, A. N., **K11**
Kidd, D. A. A., **A4**
Kido, F., **T27, T28**
Kiechel, J. R., **Y27**
Kikuchi, T., **K5**
Kimland, B., **T3**
Kimoto, W. I., **A12, A34**
Kimura, H., **T17**
Kimura, T., **T29, Y25**
Kimura, Y., **Y30**
King, F. E., **A4, A10, T50**
King, R. W., **K6**
King, T. J., **A10, T39, T50, Y17**
Kinstle, T. H., **K4**
Kirby, G. W., **K3, K4, K6**
Kirk, G., **A4, A31**
Kirkpatrick, J. L., **K15, K27**
Kirmse, W., **A27, A33, A41, A44**
Kirson, I., **T48, T51**
Kis, Z., **Y19**
Kishi, T., **Y21, Y26**
Kishi Y. **Y20**
Kisis, B., **A52**
Kishida, Y., **A29, T24, T48**
Kishner, N., **T1**
Kiss, J., **A21**
Kistner, J. F., **T4**
Kitagawa, I., **T45**
Kitagawa, M., **K14**

Kitagawa, T., **A40, K6**
Kitahara, Y., **T34**
Kitahonaki, K., **A47**
Kjaer, A., **A8, A12, A33, Z8**
Klasek, A., **K23**
Klaui, H., **T8, T29**
Klein, E., **T1, T4, T8**
von Klemperer, M. E., **K22**
Kloppenburg, C. A., **T11**
Klötzer, W., **K1**
Kloubek, J., **A22, K19**
Klusacek, H., **X8**
Klyne, W., **A28, A37, A53, A57, T19, T43, T55, K3, K10, K12, K13, K14, K19, K22, K23, Y4, X11, Z8**
Knabe, J., **A40, A55**
Knight, D. C., **T42**
Knobloch, G., **Y23**
Knowles, G. D., **T13**
Knowles, W. S., **A55**
Knox, J. R., **T24**
Kobayashi, K., **K35**
Kobayashi, M., **T23, Z7**
Kobayashi, N., **Y18**
Kobayashi, T., **A37**
Kobel, H., **K17**
Kobelt, D., **A19**
Koblicova, Z., **K16, K20**
Koch, H. P., **A55**
Kochetkov, N. K., **A12, K22, K24**
Kocourek, J., **C1**
Koczor, I., **K28**
Kodaira, S., **K3**
Kodama, H., **T16**
Kodama, M., **T24, T26, T38**
Kodera, K., **K18**
Koekemoer, M. J., **K22**
Koelliker, U., **Y16**
Koga, K., **A11, A17**
Kögl, F., **T1**
Kolbe-Haigwitz, M., **T11**
Kollonitsch, J., **A10, A27**
Kolonits, P., **K2**
Kolosov, M. N., **A51, Y16, Y28, Y30**
Kompis, I., **K10, K13**
Konno, K., **A51**
Koop, H., **K18**
Kopp, D., **A45**
Kopecky, K. R., **A51**
Koreeda, M., **T48, T58**
Korman, O., **T18**
Kornfeld, E. C., **Y21**
Kornrumpf, B., **A27, A33**
Korobka, V. G., **Y30**
Korpiun, O., **Z4, Z5, Z6**
Korte, W. D., **Z1**
Korver, O., **A29**
Korytnyk, W., **Y4**
Kostic, R. B., **T49**
Kotake, M., **K24**
Kotera, K., **K6**
Kovar, J., **A22, K18, K19**
Kovats, E., **T3, T7**
Kowala, C., **K7**
Koyama, G., **C1, Y26**
Koyama, H., **C1, T22, T40, T45, K6, K24, K29, Y11, Y18**
Kozima, T., **T38**
Kreibich, G., **T39**

Krepinsky, J., **T18, T19, T26**
Krestinski, V., **T6**
Krow, G., **X1, X4**
Kubo, A., **K11**
Kubo, I., **T33**
Kubota, T., **T15, T17, T33, T54**
Kuck, A. M., **K3, K4**
Kudo, Y., **A51**
Kuhn, R., **A2, A11, Y14**
Kühn, W., **T1**
Kulakov, V. N., **A12, K24**
Kulkarni, K. S., **T19**
Kullnig, R. K., **T39**
Kum, K., **Y14**
Kuna, M., **A19, A32, A52**
Kuneida, T., **A5**
Kung, H. P., **T33**
Kunimoto, J., **K3**
Kunstmann, M. P., **Y15, Y30**
Kupchan, S. M., **T21, T24, T30, T41, K5, K35, K36, Y19**
Kurosawa, K., **Y3**
Kuo, C. H., **Y14**
Kuriyama, K., **A47, K6**
Kurtev, B. J., **A43**
Kusano, E., **K34**
Kusaka, T., **Y21**
Kusumoto, S., **K24**
Kuthan, J., **T32, T34**
Kutney, J. P., **T50, K5**
Kuwano, D., **T20**
Kuzovkov, A. D., **K22**
Kuzuhara, H., **Y30**

Lacroix, W., **A23**
Ladd, M. F. C., **T49**
Lahey, F. N., **T42**
Laird, W., **T42**
Lalonde, R. T., **A38**
Lam, F. K., **Z3**
La Manna, A., **A20, A21, A22**
Lamberton, J. A., **K7, K8, K10, K21, K28, K29, X7**
Landor, S. R., **A6, A8, A12, A13, X1**
Lane, J. F., **A27**
Langemann, A., **A51**
Langenbeck, W., **A20**
Lanneau, G. F., **Z1**
Lansford, E. M., **A11**
Lardicci, L., **A27, A29, A42**
Lardon, A., **T48**
Lardy, H. A., **A13**
Larsen, P. K., **A7**
Larson, D. L., **A4**
Latham, H. G., **T49**
Lauinger, C., **A5**
Laur, P., **A12, Z7**
Lavagnino, E. R. **Y29**
Lavie, D., **T48, T51**
Law, D. A., **K25**
Lawrie, W., **T46**
Lawson, A., **A20**
Lawson, W. B., **A19, A29**
van Lear, G. F., **Y19**
Leard, M., **Z1**
Lederer, E., **T34, T42**
Ledouble, G., **K9**
Lee, C. M., **K9, K10**
Lee, G. K., **K2**
Lee, J., **Y15**

Author Index

Lee, K. S., **Z7**
Leeding, M. V., **T42**
Leeman, H. G., **K17**
Legrand, M., **K9**
Leithe, W., **A19, A21, K18**
Leitich, J., **A32, Y26**
Leme, L. A. P., **K14**
Le Men, J., **K9, K12, K13**
Lemieux, R. U., **A1, A6, C1**
Lemmich, E., **Y1**
Lemmich, J., **A3, A7, T57, Y1**
Lentz, P. J., **T13**
Leonard, N. J., **K21**
Leonardsen, R., **A31**
Le Quesne, P. W., **T13**
Leroux, P. J., **A13**
Lessard, J., **T50**
Lestinger, R. L., **A27**
Letcher, R., **K7**
Letourneau, F., **K19**
Levene, P. A., **A1, A6, A8, A12, A14, A19, A22, A29, A32, A41, A42, A57**
Levisalles, J., **K2**
Levy, H. R., **D1**
Levy, J., **K9, K12, K13**
Lewin, N., **T29**
Lewis, A., **T12**
Lewis, G. E., **X6**
Lewis, R. A., **Z4, Z5, Z6**
Libiseller, R., **A31**
Lichtenstein, N., **A9**
Lichti, H., **T35**
Liebermann, C., **K28**
Lightner, D. A., **X7**
Lim, F. Y., **T44**
Limburg, W. W., **Z1**
Lin, L-H., **X2**
Linde, H., **T33**
Linek, A., **T18, T22, T25**
Van der Linde, L. M., **T9**
Ling, N. C., **T50, K13**
Linstead, R. P., **Y23**
Lipscomb, W. N., **T32, K15, K19, K35**
Littlewood, P. S., **T40, T48**
Liu, S., **T56**
Liu, S. J., **Y11**
Lively, D. H., **Y24**
Locke, D. M., **K32**
Loder, J. D., **X1**
Lods, L., **T35**
Loeffler, P. K., **A27**
Loew, P., **T13**
Loewenthal, H. J., **T22**
Loewus, F. A., **D1**
Loh, S. K., **K11**
Longmore, R. B., **A55, K30**
Losee, K. A., **A20**
Love, W. E., **A30**
Loven, J. M., **A1**
Lowe, G., **X1**
Lucas, H. J., **A13**
Lukes, R., **A22, K18, K19**
Lundin, R. E., **A7, A26**
Lüthy, J., **D2**
Lwowski, W., **A29, Y2**
Ly, M. G., **A40**
Lyle, G. G., **A23, K8, X4**
Lyle, R. E., **A11**
Lynen, F., **A19, A28, D2**

Lythgoe, B., **T40, T48**
Maat, L., **A10, K19**
Macbeth, A. K., **T4**
Macchia, B., **A22, A23, A52**
Macchia, F., **A52**
MacDonald, C. G., **A16**
MacDonald, D. L., **A16**
MacDonald, P. L., **Y7**
Mackay, I. R., **T47, Y15**
Mackay, M. F., **K4, K5, K31**
Mackellar, F. A., **Y20**
MacLean, D. B., **K25**
MacLean, H., **Y7**
MacMillan, J., **Y17**
MacSweeney, D. F., **T27**
Maebayashi, Y., **T19**
Maeda, K., **C1, Y26**
Maeda, M., **T21**
Maeda, T., **T14, K31**
Magalhaes, M. T., **Y4**
Magerlein, B. J., **Y21**
Mahajan, J. R., **T32**
Mahajan, R. K., **T49**
Mahes-Wari, M L., **T23**
Mahishi, N., **A20**
de Maindreville, M. D., **K13**
Majerski, Z., **A12**
Mallams, A. K., **T55**
Mallikarjunan, M., **A5, A8, A20**
Mallory, F. B., **T44**
Mamuzic, R. T., **A18**
Manchand, P. S., **T37, Y11**
Manchanda, A. H., **Y6**
Mandava, N., **K28**
Mandel, N., **Z4**
Mangia, A., **Y20**
Mangoni, I., **T37**
Mangoni, M., **T34**
Manning, R. E., **T13**
Manohar, H., **K14**
Manor, P. C., **X2**
Manske, R. H. F., **K5**
Manwaring, D. G., **A8, A24**
Marckwald, W., **A29**
Maric, S., **A45**
Marion, L., **K21**
Marker, R. E., **A8**
Markey, F. X., **T23**
Markey, S., **K13**
Marquarding, D., **X8**
Marquardt, A., **K28**
Marsh, M. M., **Y24**
Marsh, R. E., **C1**
Marsh, W. C., **K31**
Marshak, M. L., **K14**
Marshall, D., **A54, A55**
Marshall, J. A., **T29**
Martin, D. G., **T30**
Martin, R. H., **K11**
Martin, R. J. L., **D1**
Martin, W. F., **K12**
Marta, C., **Y22**
Marukas, J., **A15**
Marumo, F., **A19, T40, Y26**
Marumo, S., **Y18**
Marusic, R., **Y27**
Maruyama, M., **T15, T24**
Marx, J. N., **T22, T32**
Marxer, A., **T42**
Marzilli, L. G., **A11**
Masaki, N., **T15, T39, K26, K29**

Masamune, S., **A34, K22, K35**
Masamune, T., **A26, T12, T22, K35**
Maslen, F. N., **T35, T43, T44, Y29**
Mason, D. J., **Y21**
Mason, S. F., **A15, A23, K1, K26, Y15, X1, X6, X7, X10, X11**
Masterman, S., **A1, A11, A12, A22, A51**
Masui, Y., **A25**
Matell, M., **A12, A28**
Matheson, A.McL., **T13, T44, K3, K4, K5, K18, K22, K29, K31, Y24**
Matheson, N. K., **A16**
Matikkala, E. J., **Z7**
Matsueda, S., **T22**
Matsumoto, T., **A2, T52, Y20**
Matsumura, H., **K8**
Matsuo, A., **T18**
Matsutani, S., **K24**
Matsuura, T., **T19, T52, T53**
Matz, M. J., **T41**
Mauli, R., **T23**
Mauperin, P., **K13**
Maxwell, I. E., **A11**
Maxwell, J. L., **D2**
Mayer, H., **T56**
Mayo, D. W., **T25, T27**
de Mayo, P., **T18, T21, T28, T43, T53**
Mazengo, R. Z., **X6**
Mazur, U., **X2**
Mazur, Y., **T49**
McCabe, P. H., **T37**
McCapra, F., **T35, T36**
McCarty, J. E., **A49**
McCasland, G. E., **A16, A26, A31**
McCaully, R. J., **A5**
McChesney, J. D., **T33**
McClure, R. J., **T20, K5**
McConnell, J. F., **T13**
McCreadie, T., **T34**
McCrindle, R., **T34, T37, T52**
McDonald, C. E., **K32**
McEachan, C. E., **T29, T39**
McElvain, S. M., **A36, A38, T14, T15**
McEvoy, F. J., **K20**
McGahren, W. J., **Y30**
McGeachin, S. G., **Y25**
McGhie, J. F., **T42, T50**
McGrath, M. J. A., **T43**
McGrew, J. G., **D1**
McKilliop, T. F. W., **T28**
McLaughlin, G. M., **Y27**
McLeod, W. D., **T23, T27**
McMurry, J. E., **T28**
McMurry, T. B. H., **T19**
McNab, A. S., **T46**
McNiven, N. L., **T5**
McOmie, J. F. M., **Y13**
McPhail, A. T., **C1, T20, T21, T24, T29, T31, T35, T39, T40, T48, T52, K5, K8, K11, K20, K29, Y15, Y17, Y19, X9**
McQuillin, F. J., **T20**
Mea, D., **T8**

Author Index

Mechlinski, W., **Y26**
Mechoulam, R., **T4, T5**
Mechtler, H., **X3**
Meck, R., **Y21**
Medcalfe, T., **T40**
Meehan, G. V., **K28**
Meek, B., **K4**
Mehta, A. S., **A3, X2**
Mehta, P. P., **Y13**
Meinwald, J., **T12**
Meister, A., **A8, A24**
Meister, W., **D1, D2**
Melera, A., **A34, T42**
Melillo, J. T., **A12, Z7**
Mellor, J. M., **T35, T51**
Meluch, W. C., **A18, A32**
Melville, D., **T36, T38**
Melville, D. V., **Y30**
Menard, E., **T46**
Menicagli, R., **A42**
Menshikov, G. P., **K22**
Mercier, D., **T42**
Merchant, J. E., **K2**
Merlini, L., **Y6, T9**
Messing, A. W., **Z7**
Meyer, A. S., **T58**
Meyer, C. E., **A8**
Meyer, E., **Z3**
Meyer, G. M., **A29**
Meyer, K., **T49**
Mhasalkar, S. E., **Y7**
Mhaskar, V. V., **T29**
Miana, G. A., **A24**
Michael, K. W., **Z1**
Michelski, J., **Z5**
Miettinen, J. K., **A28, A57**
Mihailovic, M. L., **A18**
Mikhail, A. A., **A17**
Miklos, D., **Y27**
Milborrow, B. V. **T16**
Millar, P. G., **A49**
Miller, B. J., **A6, A8, A12, A13, X1**
Miller, E., **Y16**
Milligan, B., **Y8**
Mills, H. H., **K14, K29, Y25**
Mills, J. A., **A57, T2, K19**
Mills, J. S., **T51**
Mills, R. D., **Y19**
Milne, G., **Y13**
Minale, L., **T57, K17**
Minato, H., **A28, T18, T19, T20 T21, T25, K24, Z7**
Mincione, E., **T50**
Minkin, J. A., **A30**
Mislow, K., **A12, A18, A22, A32, A36, A46, A47, A48, T8, K4, D1, X5, X6, X10, Z4, Z5, Z6, T27, Z8**
Mirrington, R. N., **T35**
Mironov, A., **T25**
Mishra, A., **A6**
Mishra, R., **T37**
Mita, I., **A20**
Mitchard, D. A., **A35**
Mitchell, M. J., **K3**
Mitra, R. B., **A13**
Mitscher, L. A., **T36, Y16**
Mitsuhashi, H., **K35, Y14**
Mitsui, R., **K6**
Mitsui, S., **A40, A51, K4**

Mitsui, T., **A50, T33, T39**
Mittenzwei, H., **Y23**
Murai, A., **A26**
Miura, R., **A24, A25**
Miyamoto, M., **T51**
Miyamoto, S., **Y30**
Miyawaki, M., **T21**
Miyasaki, M., **T24**
Miyazaki, S., **A47**
Mizsak, S., **T30**
Mizuno, H., **A40, A51**
Mizuno, K., **Y21**
Mo, L., **Y6**
Moffett, R. H., **T28**
Mokry, J., **K13**
Moldowan, J. M., **D1**
Molin-Case, J. A., **A9, A21**
Mollov, N. M., **K21**
Momber, F., **A3**
Momose, T., **K18**
Monache, F. D., **Y3**
Moncrief, J. W., **C1, K15, Y24**
Mondelli, R., **Y28**
Moniot, J. L., **K1**
Montgomery, J. A., **A9**
Monti, L., **A52**
Moorcroft, D., **K30**
Moore, H. W., **Y24**
Moore, J. A., **T49**
Moore, M., **K35**
Moore, R. E., **A44, Y15**
Moore, W. R., **X1, X2**
Morales, J. J., **Y17**
Morell, S. A., **A13**
Moretti, I., **A21, A22, A23**
Morgan, J. W., **K19**
Mori, A., **Y3**
Mori, H., **T48**
Mori, Y., **K35**
Morita, K., **T45**
Moriyama, M., **K6**
Morin, R. B., **T39, Y29**
Morisaki, M., **T41**
Morozumi, S., **K3**
Morimoto, A., **T44**
Morris, A. J., **A20**
Morris, C. J., **A9, A29, A41, A42, A57, Z7**
Morrison, A., **T29**
Morrison, G. A., **A12, Y12**
Morrison, G. C., **K7**
Morsingh, F., **K10**
Moore, M., **K9**
Mortimer, P. I., **A10**
Morton, G. O., **Y19**
Morton, R. B., **A11**
Moscowitz, A., **T8, Y23**
Mose, W. P., **T55, K22, K23, Y30**
Mosettig, E., **T35, T49**
Motherwell, W. D. S., **T27, K2**
Mosher, H. S., **A12, A22, A25, A27, A41**
Moss, G. P., **T55**
Motl, O., **T29**
Mozingo, R., **A4, A8, Y30**
Mueller, R. A., **Y29**
Muhammad, S., **T51**
Muhi-Eldeen, Z., **A20**
Mulheirn, J., **Y13**
Mulholland, T. P. C., **Y17**

Mullenback, G., **Y14**
Müller, E., **K18**
Müller-Enoch, D., **A31**
Mumukata, K., **T37**
Munk, M. E., **Y30**
Muntz, R. L., **Z3**
Murai, A., **T22, K35**
Murai, F., **T15**
Murai, K., **Y25**
Murakami, Y., **A17**
Murata, T., **T45, T51**
Muroi, M., **Y21**
Murphy, C. F., **K7**
Murphy, R. C., **T58**
Murray, R. D. H., **T34, T37**
Murray, R. K., **Z4, Z6**
Myers, C. W., **K26**

Naef, H., **T57**
Naegeli, P., **Y29**
Näf, U., **T35**
Nagahama, S., **T26**
Nagai, M., **T3, T51**
Nagai, U., **Y14**
Nagarajan, K., **A45, T23, K12**
Nagarajan, R., **Y24**
Nagasampagi, B. A., **T18**
Naik, V. G., **T18, T25**
Nair, P. M., **Y9**
Nakadaira, Y., **T44**
Nakagawa, A., **Y25**
Nakagawa, M., **A51, X3**
Nakagawa, T., **T22**
Nakagawa, Y., **C1, K13**
Nakai, H., **T45, Y18**
Nakajima, M., **T15**
Nakajima, S., **Y13**
Nakaruma, M., **T25**
Nakamura, S., **T23**
Nakamura, Y., **T13**
Nakanishi, K., **T30, T35, T36, T38, T44, T48, T58, K1, Y18**
Nakanishi, T., **T45**
Nakanishi, Y., **A44, T44, X2**
Nakano, T., **A53, T13, K36, Y29**
Nakashima, R., **K30**
Nakatsuka, T., **T29**
Nakayama, Y., **T41**
Nakazaki, M., **A20, A24, A25, T21, Y1, Y6**
Nanaumi, K., **T22**
Nangle, B. J., **Y4**
Narayanan, C. E., **T19**
Narayanan, C. R., **T33, T53, K35**
Narayanan, P., **T38, T39**
Nardelli, M., **Y20**
Narita, S., **T16**
Nasipuri, D., **D1**
Nasini, G., **Y6**
Natori, S., **T12, T43, T44**
Natsume, N., **T33**
Naumann, K., **Z4, Z5**
Naves, Y. R., **T14, T16**
Naya, H., **Y25**
Naya, K., **T17, T23, Y25**
Nayak, U. R., **T28**
Neelakantan, L., **A21**
Neidle, S., **A26, C1, T21, Z8**
Neilsen, B. E., **A3:, A7, T57, Y1**
Neilson, D. G., **A19**
Nelson, D., **T14**
Nelson, D. M., **Y30**

Author Index

Nelson, N. R., **K4**
Nerdel, F., **A37**
Neuberger, **A17**
Neuss, N., **K15, Y24**
Nevitt, T. D., **A38**
Newkome, G. R., **A56, K30**
Newman, H., **T35**
Newman, P., **X6**
Ng, A. S., **Y2**
Nicholas, A. F., **T41**
Nicholson, J. A., **T58**
Niemeyer, J., **Y28**
Niggli, A. **Y1**
Nikolai, F., **A4**
Nisbet, M., **T53, X5**
Nisbet, M. A., **T21**
Nishida, T., **Y18**
Nishikawa, H., **T12, T36**
Nishikawa, M., **T45, T51, K5, Y21, Y26, Y28**
Nishio, K., **K25**
Nishioka, S., **A36**
Nishiya, H., **T16**
Nishiyama, A., **T22**
Nitta, I., **T23, T45, K5, Y25**
Nkunika, D. S., **K26**
Nockolds, C. K., **Y24**
Noda, T., **Y17, Y25**
Nomura, K., **K3, K13**
Nordman, C. E., **T21**
Norin, T., **T3, T7, T26**
Northolt, M. G., **T11**
Norton, D. A., **T32, T47**
Norton, K. B. **T50**
Novak, C., **T22**
Novak, E. R., **A6**
Novotny, L., **T23**
Nowacki, W., **K7**
Noyce, D. S., **A18, A38**
Noyori, R., **Y16**
Nozaki, H., **A41, A44, X2**
Nozoe, S., **T22, T30, T32, T41**
Nozoe, T., **T20, T25**
Nudelman, A., **Z6**
Nugent, M. J., **X8**
Nugteren, D. H., **Y16**
Nulu, J. R., **A5**
Nwaji, N. M., **T52**

Oberhansli, W. E., **T47, K31**
O'Brien, J., **K2**
O'Brien, R. E., **D1, X5**
Occolowitz, J. L., **K8, K12**
Ochiai, E., **K8, K24**
O'Connell, A. M., **T35**
Oda, K., **K29**
O'Donnell, E. A., **T49**
O'Donovan, G. M., **K24**
Ogawa, H., **T12**
Ogawa, Y., **Y26**
Ogihara, Y., **Y18**
Ogura, M., **T38**
Oh, Y. L., **T44**
Ohashi, M., **T30**
Oh-hashi, J., **A24**
Ohloff, G., **A36, T3, T4, T5, T7, T8, T16, T36, T54**
Ohno, K., **K29**
Ohno, M., **A35, A37, A54**
Ohnsorge, U., **Y17**
Ohrt, J. M., **A26**
Ohrui, H., **Y30**

Ohsawa, T., **T51**
Ohta, M., **K3**
Ohta, T., **T38, T53**
Ohta, Y., **T6, T18**
Ohtani, Y., **K11**
Oishi, Y., **Z7**
Oka, Y., **K8**
Okabe, K., **T40**
Okamoto, T., **K25, K31**
Okamoto, Y., **K5**
Okaya, Y., **K4, Z1**
Oken, A., **T7**
Okigawa, M., **T30**
Okruszek, A., **Z5**
Okuda, S., **A26, T41, T50, K21, K24, K25**
Okuda, T., **K35, Y17**
Okugun, J. I., **T52**
Olagbemi, E. O., **T52**
Olander, C. R., **K4**
Olin, S. M., **A1, C1**
Olivier, L., **K12, K13**
Ollis, W. D., **Y1, Y3, Y4, Y5**
Omura, S., **Y25**
O'Murchu, C., **A40**
Ona, H., **T38**
Onaka, T., **K31**
Onoprienko, V. V., **Y30**
Onore, M. J., **T10**
Oonk, T. J., **T11**
Openshaw, H. T., **K2**
Opferkuch, H. J., **T39**
Opliger, C. E., **A9**
Oppolzer, W., **A32, Y26, Y29**
Orazi, O. O., **Y29**
Orban, I., **Y9**
Orezzi, P., **Y28**
Organ, T. D., **T47**
Osaki, K., **T15, T53, K5, K29**
O'Shea, T., **T49**
Osiecki, J., **A39, A55, T42**
O'Sullivan, A. M., **Y5**
Otake, N., **A9, Y26**
Otani, N., **T17**
Ott, H., **A9, K17**
Ouannes, C., **A41, A42**
Ourisson, G., **T23, T25, T27, T28, T35, T51**
Ovakimian, G., **A19**
Overberger, C. G., **A38**
Overton, K. H., **T32, T34, T36, T38, T44, T52**
Overwein, H., **K21**
Owen, L. N. **A12, A14**
Owellen, R. J. **K13**
Oyeo, S., **K29**

Pacak, J., **C1**
Pachaburkar, R. V., **T53**
Pachler, K. G. R., **Y6**
Padilla, J., **T1, T27**
Page, A. C., **A11**
Page, C. B., **K29**
Page, H., **K21**
Paget, H., **T2**
Pagnoni, V. M., **T40**
Pailer, M., **A31**
Pais, M., **K30**
Paknikar, S. K., **T10, T19**
Pakrashi, S. C., **T45**
Palerimiti, F. M., **A36**
Palm, J. H., **T11**

Palm, O., **A28**
Palmer, K. H., **K29**
Palmer, K. J., **Z7**
Pandey, R. C., **T37**
Panetta, C. A., **Y29**
Papariello, G. J., **K11**
Paquette, L. A., **A22**
Park, R. J., **T17**
Parke, T. V., **Y21**
Parker, G. A., **Z1**
Parkes, A. S., **X2**
Parsons, P. G., **T13, K2**
Parthasarathy, P. C., **T23**
Parthasarathy, R., **A5, A26, C1**
Pascard-Billy, C., **K10**
Pascu, E., **A27**
Patchett, A. A., **A17, T46**
Patel, M. B., **K13, K14**
Patil, F., **T23**
Paton, A. C., **Y21**
Patterson, A. L., **A30**
Patterson, W. I., **A4**
Paukstellis, J. V., **T8, T26**
Paul, I. C., **T27, T37, T58, Y11, Y16, Z3**
Paulus, E. F., **A19**
Paulus, H., **A9**
Pawson, B. A., **T2**
Pecher, J., **K11**
Peck, G. Yu, **Y16**
Peddle, J. D., **Z2**
Pederson, J., **A18**
Pedersen, P. A., **Y1**
Peerdeman, A. F., **A1, K12, Y13, Z3**
Pelizzi, G., **Y20**
Pelizzoni, F., **T40**
Pelosi, E. T., **X4**
Pelletier, S. W., **K32, K33**
Pelter, A., **Y3, Y6, Y7**
Penco, S., **Y28**
Pepinsky, R., **A4, K4, Z1**
Pereira, W. E., **A41**
Pérezamador, M. C., **T36**
Perold, G. W., **Y6**
Peter, H., **A20**
Peters, H. M., **A12, A22**
Petersen, M. R., **A33, A58**
Petru, F., **T32, T34**
Petrzilka, T., **T5**
Petterson, K., **A41, A43, A48, A50**
Pettit, G. R., **T49**
Pettus, J. A., **A44, Y15**
Peyer, J., **T49**
Phillips, D. A. S., **A11**
Phillips, H., **A29**
Phillipson, J. D., **K9**
Piers, E., **T11, K15**
Piers, K., **K25**
Pignolet, L. H., **X5**
Piatelli, M., **K17**
Pijewska, L., **K4**
Pinar, M., **K16**
Pines, S. H., **A40**
Pinder, A. K., **T35**
Pinder, A. R., **T23, K29**
Pinhey, J. T., **T21**
Pino, P. **A27**
Pioch, R. P. **Y29**
Piotrovich, L. A., **Y16**

Author Index

Pirkle, W. H., **Z3**
Pitt, C. G., **A14, A35**
Pittman, M. P., **A29**
Plat, M., **K13**
Plattner, J. J., **A3, K6**
Plattner, P. A., **T8, T29**
Plieninger, H., **Y23**
Pointer, D. J., **T39**
Poisson, J., **K9, K10**
Poling, M., **T39**
Pollard, D. R., **T47**
Polonsky, J., **T53, Y19**
Popjak, G., **T3, D1**
Porath, J., **A55**
Porte, A. L., **T28**
Portmann, P., **A4, A17, A19, A20**
Portoghese, P. S., **A4, A13, A17, A31**
Pospisek, J., **K16, K20**
Posternak, T., **A3, A16**
Potier, P., **K2**
Pousset, J. L. **K9, K10**
Powell, A. D. J., **T50**
Powell, J. W., **T52**
Powers, T. W., **X7**
Pracejus, H., **A24**
Pratesi, P., **A20, A21, A22**
Pravda, Z., **A11**
Prelog, V., **A26, A32, A34, A49, A53, C1, T3, T46, K8, K9, K35, Y26, Y27, X4, X6, X8, X11**
Prendergast, J. P., **Y5**
Pretorius, Y. Y., **A1**
Priston, H. E. M., **T10**
Privett, J. E., **X4**
Proskow, S., **A31**
Prota, G., **K17**
Przybylska, M., **T58, K14, K21, K25, K33**
Puckett, R. T., **A52, T47, K36**

Quitt, P., **T35**
Qureshi, A. A., **Y2**
Qureshi, I. H., **K27**

Raa, J. T., **Y27**
Raaen, V. F., **A23**
Radlick, P., **T30**
Rae, I. D., **K15**
Raeymakers, A. H. M., **Y30**
Raffauf, R. F., **K3**
Raffelson, H., **A53**
Raghavan, V. K. V., **Y9**
Rahman, M. B., **A12**
Rall, G. J. H., **K20**
Ramachandran, G. N., **A21, C1, Y9**
Ramage, G. R., **T28**
Ramage, R., **T27, Y29**
Raman, S., **A10, A21**
Ramaseshan, S., **K14, Y24**
Rampal, A. L., **T49**
Ramsay, G. C., **Y3, Y5**
Ramsay, M. V. J., **Y5**
Ramuz, M., **K4**
Randall, S. S., **A5**
Ranganathan, S., **Y29**
Rapoport, H., **A3, T6**
Rapp, A., **Z5**
Rao, A. S., **A49, T16, T21**
Rao, B. S., **T20**

Rao, G. V., **A14**
Rao, M. V., **K25**
Rao, P. A. D. S., **A1**
Rao, S. T., **A5, A8, A20**
Rapala, R. T., **T50**
Raske, K., **A4**
Rassat, A., **T28**
Rautenstrauch, V., **A36, T54**
Rayner, D. R., **Z8**
Read, J., **T4, T5**
Records, R. **X5**
Redman, B. T., **Y3**
Redmond, J. W., **D2**
Reece, C. A., **T8**
Reeke, G. N., **K35**
Rees, A. H., **K14**
Rees, K. R., **K26**
Regan, J. P., **X1**
Reich-Rohrwig, P., **X8**
Reichstein, T., **T48, K3**
Reid, J., **Y26**
Reinäcker, R., **T8**
Reinhold, D. F., **A40**
Remanick, A., **A46, A47, T10, T11**
Renauld, J. A. S., **T41**
Renfrew, A. J., **T34**
Renner, U., **K11, K13, K15**
Renwick, J. D., **K22, K23**
Rerat, M. C., **T10**
Resche, T., **Y19**
Resnick, P. R., **T41**
Ressler, C., **A5, A9, A20**
Retey, J., **A10, A28, D2**
Reymond, D., **A16**
Reynolds-Warnhoff, P., **A46, A47, T10, T11**
Rhodes, C. A., **Y1**
Riano-Martin, M., **Y6**
Ribi, M., **T33, T54**
Rice, W. Y., **A13, K20**
Rickards, R. W., **A8, A34, Y2**
Richardson, A. C., **C1**
Richey, J. M., **K10**
Riebel, A., **T39**
Rimington, C., **A20**
Rimmer, B., **T21**
Rimmer, B. M., **T48**
Rindone, B., **T31**
Rinehart, K. L., **A16, Y16**
Riniker, B., **A53, T1, T2, T7, T19, T25, T46**
Riniker, R., **A53, T1, T2, T7**
Rios, T., **T41**
Ripperger, H., **A10, A27, A30, T49, K34**
Ritchie, E., **T46, T50, K5**
Robb, E. W., **K9**
Roberts, J. C., **Y17**
Roberts, J. S., **T28**
Roberts, P. J., **K2**
Robertson, A., **T50**
Robertson, A. V., **A17, Y3**
Robertson, J. M., **T12, T21, T28, T37, T38, T40, T47, T52, T58, K8, K11, K26, Y15, Y17**
Robins, D. J., **A28, A33, K23**
Robinson, B., **A55, K30**
Robinson, R., **T13, K12, Y29**
Rodewald, P. G., **Z1**
Rodig, O. R., **Y13**

Rodin, J. O., **T8**
Rodrigo, S., **T39**
Roeske, R. W., **Y29**
Roevens, L. F. C., **Y30**
Rogers, D., **A26, A55, C1, T11, T21, T24, T28, K6, K7, Y6, Z8**
Rogers-Low, B. W., **Y29**
Röhrl, M., **T38, T39**
Rojahn, W., **T1**
Rollett, J. S., **T45, T49**
Romanuk, M., **T14, T19**
Romeo, A., **Y22**
Romers, C., **A18**
Romo, J., **T1, Y17**
Rönsch, H. S., **T49**
de Roos, J. B., **T22**
Rose, I. A., **D2**
Rose, W. C., **A8**
Rosenberger, M., **K15**
Rosenstein, R. D., **K9**
Rosich, R. S., **T35**
Rossi, R., **A28, A29**
Rossmann, M. G., **T5, T13, T32**
Roth, H. D., **A37**
Rothen, A., **A29, A57**
Rothweiler, W., **Y27**
Roux, D. G., **Y3, Y4**
Rowe, J. W., **T31**
Roy, S. K., **A5, A10, A19, K4, K5, K18**
Rubin, B., **A20**
Rubin, L. J. **A13**
Rudinger, J., **A11**
Rüegg, R., **T56**
Rupe, H., **T11**
Ruppert, J., **Y23**
Russell, M., **A15**
Russell, R. C., **Y1**
Russell, S. W., **T55**
Rutkin, P., **X5, X6**
Rutledge, P. S., **T35**
Ruveda, E. A., **K2**
Ruzicka, L., **A35, A39, T11, T34, T42, T46, T51**
Ryan, C. W., **Y29**
Ryback, G., **T16**
Rydon, H. N., **T7**

Sachdev, H. S., **A5**
Sachs, M., **A37**
Safe, S., **Y15**
Safir, D., **Y24**
Sahli, M. S., **A26**
Sair, M. I., **T50**
Saito, N., **C1**
Saito, S., **K18**
Saito, T., **C1**
Saito, Y., **A19, T30, T40, Y26**
Sakabe, N., **T40, K32**
Sakai, S., **K11, K14**
Sakai, T., **T18**
Sakan, T., **T13, T14, T38**
Sakata, K., **T39**
Saksena, A. K., **T48**
Sakuma, R., **T38**
Salgar, S. S., **K2**
Sallay, I., **A21**
Salvadori, P., **Y6**
von Saltza, M. H., **Y26**
Salzman, N. P., **A11**
Samek, Z., **T18, T24**
Samori, B., **A24, A25**

Author Index

Thomas, U., **A26**
Thompson, D. J., **Y17**
Thompson, J. F., **A9**
Thompson, Q. E., **A53**
Thornton, I. M. S., **T52**
Tichy, M., **X10**
Tobita, S., **T13**
Toda, M., **K26**
Toda, T., **T20**
Todd, A., **A10**
Todd, D., **T33**
Tokoroyama, T., **T26, T52**
Tokuyama, T., **K26**
Tomiie, Y., **T45, K5, Y25**
Tomita, B., **T20, T29**
Tomita, K., **T45, K25**
Tomita, M., **K3, K5**
Tomiyama, K., **T22**
Tomko, J., **K34**
Tomoskosi, I., **A44, X1**
Toome, V., **K1**
Tori, K., **T23, K6**
Torimoto, N., **A25, Y2**
Torre, G., **A21, A22, A23**
Toshioka, N., **A20**
Toth, G., **T55**
Tozyo, T., **T25**
Trager, W. F., **K9, K10**
Trave, R., **T14**
Traverso, G., **Y7**
Traynham, J. G., **A27**
Treibs, W., **T1**
Trier, G., **A9**
Tristram, E. W., **A40**
Trivellone, E., **T57**
Trojanek, J., **K16, K20**
Trommel, J., **A27**
Trotter, J., **C1, T31, T46, T50, K5, K26, Y30**
Trueb, W., **A2**
Trueblood, K. N., **C1, T48, Y28, X7**
Tscharner, Ch., **T36, T54**
Tschesche, R., **T48**
Tsuda, K., **A26, A29, T22, T32, T41, K21, K24, K35**
Tsuda, Y., **T44, T45, K31**
Tsucuda, Y., **C1**
Tsuneda, K., **T48**
Tsuyuki, T., **T42**
Tulinsky, A., **K11, Y27**
Tsuzuki, Y., **A20**
Tullen, P., **T14**
Turnbull, J. P., **T51**
Turner, J. C., **K2**
Turner, R. B., **T39, K8**
Turner, W. B., **A36, K24, Y27**
Turner-Jones, A., **Y29**
Tursch, B., **T9, D2**

Uchimaru, F., **K31**
Uda, H., **T15, T18, T28**
Uda, K., **T27**
Ueda, S., **T15**
Uemura, D., **T39**
Ugi, I., **X8**
Uhda, G., **T7**
Uhde, G., **T36, T54**
Uhle, F. C., **T49**
Uhlrich, L., **Y1**
ul-Haque, M., **T22, T24**

Ullmann, F., **T11**
Ulrich, S. E., **A27**
Umezawa, H., **C1, Y26**
Untch, K. G., **A17**
Uramoto, M., **Y26**
Urbahn, C., **A40, A55**
Urry, D. W., **A9**
Urry, W. H., **Y14**
Ursprung, J. J., **T42**
Uskokovic, M., **K1**
Usubilliga, A., **K36**
Utkin, L. M., **K22**
Utzugi, N., **K24**
Uyeo, S., **T17, K6, K29**

Vaciago, A., **T41, Y26**
Valenta, Z., **K25**
Valiant, J., **A22, Y24**
Valverde-Lopez, S., **K26**
van Bommel, A. J., **A2**
van den Berg, G. R., **Z5**
van den Bosch, S., **A10**
van den Hende, J. H., **K4, Y16, Y27**
Van der Gen, A., **T9**
van der Helm, D., **A30, T31**
Van der Linde, L. M., **T9**
Van der Merwe, K. J., **Y2**
Van Dorp, D. A., **Y16**
Van Tamelen, E. E., **A11**
Vane, G. W., **K1, K26, Y15, X1, X11**
Van Heyningen, E., **Y29**
van Lear, G. E., **Y19**
Van Soest, T. E., **Y13**
van Tamelen, E. E., **A11**
Van Veen, A., **K19**
Vasina, I. V., **Y16**
Vassova, A,. **K34**
Velarde, E., **X1**
Velick, S. F., **A31**
Velluz, L., **T48**
Venkataraman, K., **Y9**
Venkatesan, K., **A5, A8, A20**
Venkateswarlu, A., **Y16**
Vennesland, B., **D1**
Verbit, L., **D1**
Vercesi, D., **A23, A25**
Verghese, J., **T6**
Verkade, P. E., **A43**
Vernengo, J., **K3, K4**
Verny, M., **A33**
Vessiere, R., **A33**
Vieles, P., **A1**
du Vigneaud, V., **A4, Y30**
De Ville, T. E., **T55**
Vincent, J. L., **K35**
Virtanen, A. I., **A11, Z7**
Visser, G. J., **X7**
Viswanathan, N., **T21**
Vitali, G., **A20**
Viterbo, R., **T32, T44, T46**
Vlad, P., **T3**
Vlahov, R., **T18**
Vogler, K., **A21**
Vonasek, F., **T18**
von Baeyer, A., **T6**
von Braun, J., **A27, T1, T46**
von Klemperer, M. E., **K22**
von Phillipsborn, W., **K15, K19**
von Saltza, M. H., **Y26**
von Schrilz, D. M., **Z8**

von Wittenau, M. S., **T19, Y28**
Vorbruggen, H., **T45, K32, Y29**
Voticky, Z., **K36**
de Vries, J. X., **Y14**
deVries, K. S., **A43**
Vrcoc, J., **T16, T26, T29**

de Waal, H. L., **K20**
Wada, M., **T23**
Wada, Y., **Y26, Y28**
Waddell, T. G., **T22, T24**
Waddington-Feather, S. M., **T40**
Wadia, M. S., **T25, T29**
Wagh, A. D., **T20**
Wagner, G., **T2, T11**
Wagner-Jauregg, T., **A1**
Wahl, G. H., **X10**
Waight, E. S., **T41**
Waiss, A. C., **Y3**
Waite, M. G., **K4, K36**
Wakabayashi, K., **A27, A30**
Wakabayashi, M., **K11**
Walborsky, H. M., **A14, A35, A37, A44, A46, A54, A56**
Walbrick, J. M., **A44, X2**
Walia, J. S., **A46, A47, T10, T11**
Wall, M. E., **T40, K29**
Wallach, O., **A38, T2, T10**
Wallis, A. F. A., **T10**
Wallis, E. S., **A27**
Walls, F., **T1**
Wallwork, I. R., **A55**
Wallwork, S. C., **T39, Y15**
Walshaw, K. B., **T19, T20**
Walther, A., **T49**
Walti, A., **A1, A14, A22**
Wälti, M., **A12**
Wani, M. C., **T40, K29**
Warawa, E. J., **A39**
Warnell, J. L., **A20**
Warnet, R. J., **K4**
Warnhoff, E. W., **T51, K6, X7**
Warren, F. L., **K12, K22, K23**
Warren, K. E. H., **K2**
Warwick, A. J., **A10**
Waser, E., **A5**
Waser, J., **C1**
Washecheck, P. H., **T31**
Watanabe, E., **T3**
Watanabe, H., **T24**
Watanabe, I., **T41**
Watanabe, M., **T44**
Watanabe, T., **A4, Y17, Y20, Y25**
Waters, J. A., **K26**
Waters, J. M., **Y13, Y24**
Waters, T. N., **T33, T34, Y13, Y24**
Waters, W. L., **X2**
Watson, C. J., **Y23**
Watson, D. G., **T27, T52**
Watson, M. B., **A23, A43, A48**
Weatherston, J., **Y4**
Webb, T., **Y22**
Weber, H., **D2**
Weber, H. P., **A11, A33, T27, T31, Y7, Y14, Y24**
Webster, M. S., **Y28**
Wedekind, E., **T12**
Weedon, B. C. L., **T55, T56**
Weiberg, O., **A20**

Author Index

Weidmann, R., **A25, A50**
Weigang, O. E., **X8**
Weinges, K., **Y3, Y7**
Weinheimer, A. J., **T31**
Weinschenker, M. N., **Y16**
Weinstein, B., **T36**
Weir, N. G., **T38**
Weisbach, J. A., **K3, K15, K37**
Weisblat, D. I., **A1, C1**
Weisenborn, F. L., **K9**
Weiss, J., **X5**
Weiss, R., **T27, T28**
Weiss, U., **T53, K27, Y10**
Weissmann, Ch., **A45, K12**
Wellman, K., **X5**
Wells, R. J., **A15, X11**
Welzel, P., **T48**
Wendler, N. L., **Y14**
Wenkert, E., **T32, T33, K8, K9, K11, K15**
Wenziger, G. R., **K7**
Wepster, B. M., **A43**
Werner, G., **K18**
Werner, N. D., **A18, A37**
Werstiuk, N. H., **T28**
Werth, R. G., **A47**
West, T. F., **T2**
Westcott, N. D., **T50**
Westfelt, L., **T11, T18**
Westley, J. W., **A41, Y11**
Westman, L., **A50, A52**
Whaley, H. A., **Y22**
Whalley, W. B., **A55, T34, T37, Y11, Y13, Y17**
Wheatley, P. J., **Y27**
Wheatley, W. B., **A36**
Wheeler, D. M. S., **K4**
White, D. E., **T35, T43**
White, D. M., **T19**
White, D. N. J., **T24, T47, K5, Y27**
White, E. H., **T22**
White, E. P., **K21**
White, J. D., **T37**
Whitehurst, J. S., **K1, X11**
Whitham, G. H., **T8, T43**
Whiting, D. A., **A8, T45, Y1, Y9, Y15**
Whiting, M. C., **X1**
Whitman, P. J., **A8**
Whittaker, D., **T8, T9**
Whittaker, N., **K2**
Wiberg, K. B., **A1**
Widdowson, D. A., **K7**
Widmer, R., **K20**
Wieland, H., **K12**
Wieland, T., **A20**
Wiesner, K., **T58, K25, K33**
Wiewiorowski, M., **K21**
Wilcox, M. E., **K17**
Wildman, W. C., **K6**
Wiley, P. F., **C1, Y20**
Wilkinson, D. I., **A28, Y25**
Wilkinson, R. G., **A8**

Williams, D. A., **A31, T40**
Williams, D. E., **A40**
Williams, D. J., **K6**
Williams, J. H., **K20**
Williams, J. R., **K28**
Williams, R. E., **T18, T28**
Williams, T., **Y11**
Willing, R. I., **K29**
Willis, C., **T36**
Willis, C. R., **K15**
Willis, T. C., **A47**
Willner, D., **A46, A47, T10, T11**
Willstätter, R., **K18, Y23**
Wilson, F. B., **K29**
Wilson, J. M., **K13**
Wilson, J. W., **A44, X1**
Wing, R. M., **T30**
Winitz, M., **A11**
Winker, H., **Z3, Z5**
Winn, A. V., **T7**
Winter, S., **A24**
Winterfeldt, E., **K21**
Winters, T. E., **T24**
Wintersteiner, O., **K29**
Wiss, O., **T46**
Witiak, D. T., **A20**
Witkop, B., **A10, A17, A21, A30, K8, K13, K26, K36**
von Wittenau, M. S., **T19, Y28**
Witters, R. W., **T24**
Witteveen, J. G., **T9**
Wohl, A., **A2**
Wolf, D. E., **A4, A22, Y24, Y30**
Wolf, H., **Y23**
Wolfe, J. R., **D1**
Wolff, G., **T27**
Wolff, I. A., **A7**
Wolff, L., **T10**
Wolfrom, M. L., **A1, C1**
Wolinsky, J., **A39, T5, T14**
Wollenberg, G., **K3**
Wong, C. F., **A38**
Wong, C. M., **Y27**
Woo, P. W. K., **A9**
Woods, M. C., **T15**
Woodward, R. B., **A26, A53, T1, T19, T46, K8, K35, Y25, Y28, Y29**
Wootton, M., **T43**
Worch, H. H., **T49**
Wren, H., **A23**
Wright, R. S., **T50**
Wrigley, T. I., **T2**
Wrixon, A. D., **A2**
Wunderlich, J. A., **T11, K7, K24**
Wüst, W., **A23, K18**
Wyler, H., **T46, K17**
Wylie, A., **T38**
Wynberg, H., **X4, X7**

Yabuuchi, H., **A27**
Yagi, H., **K1**
Yamada, S., **A5, A11, A40, A51, K20, X6**

Yamada, Y., **T30**
Yamaguchi, K., **T12**
Yamamoto H., **Y16**
Yamamura, S., **T22, K26**
Yamatodani, S., **K17**
Yamauchi, H., **T45**
Yamauchi, M., **K18**
Yamawaki, Y., **K18**
Yamazaki, S., **T40**
Yanai, H. S., **K19**
Yanai, M., **T38**
Yanagiya, M., **Y20**
Yankee, E. W., **A21**
Yanschulewitsch, J., **A37**
Yasui, B., **K25, K31**
Yasunari, Y., **T12**
Yatagai, M., **T26**
Yates, P., **K15**
Yee, T., **T46**
Yeowell, D. A., **K11**
Yokoyama, H. N., **T54**
Yokoyama, N., **K6**
Yonehara, H., **A9, Y26**
Yonemitsu, O., **A36, K9**
Yoshida, R., **A2**
Yoshida, T., **T13, T26**
Yoshihara, K., **T18**
Yoshihara, T., **A30**
Yoshikoshi, A., **T18, T27, T28**
Yoshimoto, M., **K24**
Yoshino, A., **K32**
Yosioka, I., **T29, T45**
Young, A. E., **A35, A54**
Young, D. W., **T34, T35, T36, Y19**
Young, R. W., **A1**
Young, W. G., **A1, A12**
Youngson, G. W., **A23, A43, A48**
Youssef, A. A., **A46**

Zaar, B., **T8**
Zacharias, D. E., **K9, K13, K15**
Zachau, H. G., **A29, A39**
Zalan, E., **A34, K8**
Zalkin, A., **A4, A11**
Zalkow, J. H., **Y1**
Zalkow, L. H., **T13, T19, T23, T32**
Zalkow, V. B., **T19**
Zealley, J., **Y17**
Zechmeister, K., **T38, T39**
Zeitlin, A., **T13**
Zeitschel, O., **T4**
Ziffer, H., **A16, T53**
Zigic-Mamuzic, L., **A18**
Zinnes, H., **K9**
Zon, G., **Z4, Z5**
Zussman, J., **A17**
Züst, A., **Y4**
Zweifel, G., **T8, T9**
Zweistra, A., **K19**
Zymalkowski, F., **A19, A29**

Subject Index

Abietane, **T33**
Abietatriene, **T33**
Abietic acid, **T32**
Abikoviromycin, **Y30**
Abrine, **A20**
Abscisin, **T16**
Acetolactic acid, methyl ester **A33**
Acetylaranotin, **Y21**
Acetylchloromalic acid, **A1**
Acetylenes, **A6–8, A10, A12–13, A33, Y15, X1–2, X9**
Achillene, **T1**
Achillin, **T22**
Achromycin, **Y28**
Aconitine, **K33**
Acorenol, **T29**
Acorone, **T29**
Actidione, **Y21**
Actinidine, **T13**
Actinobolin, **Y30**
Acutumine, **K5**
2-acylindole alkaloids **K11**
Adamantanes, **A45**
Adenocarpin, **K18**
Adianine, **T43**
Adiantol, **T45**
Adipedatol, **T45**
Adonose, **C2**
Adrenaline, **A22**
Adriamycin, **Y28**
Aeroplysinin I, **Y19**
Aflatoxins, **Y13**
Agarofurans, **T19**
Agatharesinol, **Y9**
Agrimonolide, **Y2**
Agroclavine, **K17**
Ajaconine, **K32**
Ajmalicine, **K9–10**
Ajmaline, **K11**
Akuammicine, **K12**
Akuammidine, **K11**
Akuammine, **K12**
Alangium alkaloids, **K2**
Alanine, **A1**
Alaninol, **A19**
Alantolactone, tetrahydro-, **T21**
Albolic acid, **T41**
Alditols, **A11, C1**
Aldohexoses, **C1**
Aldoheptoses, **C3**
Aldooctoses, **C3**
Aldopentoses, **C1**
Aldoses, **C1**
Aldotetroses, **C1**
Alisol A, **T51**
'Alkaloid C', **K31**

Alkylidenecycloalkanes, **X3–4**
Alkynes, **A6–8, A10, A12–13, A33, T20, Y15, X1–2, X9**
Allenes, **T55**
Allitol, **C1**
Alloanodendrine, **K22**
Alloaromadendrene, **T26**
Allodevardarool, **T34**
Allohimachalol, **T12**
Allohydroxycitric acid, **A30**
Alloisocitric lactone, **A30**
Alloisoleucine, **A27**
Allomatrine, **K24**
Allose, **C1**
Allosedamine, **K18**
Allosedridine, **K19**
Allothreonine, **A24**
Alloxanthin, **T55**
Alloyohimbane, **K10**
Allulose, **C2**
Alnusene, **T42**
Alolycopine, **K25**
Alopecurine, **K26**
Alstonine, tetrahydro-, **K9**
Altritol, **C1**
Altrose, **C1**
Amaryllidaceae alkaloids, **K6**
Ambonic acid, **T50**
Ambrein, **T42**
Ambreinolide, **T34**
Ambrosin, **T24**
6-aminopenicillanic acid, **Y29**
3-aminoproline, **Y22**
Amphetamine, **A5**
Amphotericins, **Y26**
Amurensine, **K1**
β-amyrane, **T42**
Amyrins, **T42**
Anabasine, **K18**
Anaferine, **K19**
Anagyrine, **K21**
Anatabine, **K18**
Androcymbine, **K4**
Andrographolide, **T36**
Angolesin, **Y5**
Angolensic acid, **T52**
Angustifoline, **K21**
Anhalonidine, O-methyl, **K2**
Anhalonine, **K1**
Anhydromonocrolactic acid, **K23**
Anhydrovobasindiol, **K11**
Anisomycin, **Y27**
Annofoline, **K25**
Annopodine, **K26**
Annotinine, **K25**
Anodendrine, **K22**
Antheridiogen, **T35**

Antibiotic X-537A, **Y11**
Antibiotic X-206, **Y11**
Antibiotic LL-BH-872a, **Y30**
Antirhine, **K8**
Aphid pigments, **Y10**
Aplysin, **T12**
Aplysinol, **T12**
Apoaromadendrene, **T26**
Apoaromadendrone, **T26**
Apofenchene, **T10**
Aporphine alkaloids, **K3–4**
Aposclerotioramine, **Y11**
Arabin, **Z8**
Arabinitol, **C1**
Arabinohexulose, **C2**
Arabinose, **C1**
Arborinol, **T45**
Archangelicin, **Y1**
Argemonine, **K1**
Arginine, **A11**
Aristeromycin, **Y21**
Aristolene, **T26**
Aristolone, **T26**
Armepavine, **K3**
Aromadendrene, **T26**
Arsenic compounds, **Z3**
Artemisin, **T21**
Asparagine, **A4**
Asparaginol, **A4**
Aspartic acid, **A4**
Asperilin, **T21**
Asperline, **Y15**
Asperuloside, **T13**
Aspidodasycarpine, **K12**
Aspidodispermine, **K13**
Aspidofractinine, **K14**
Aspidospermidine, **K13**
Aspidospermine, **K13**
Athamanthin, **Y1**
Atidine, **K32**
Atisine, **K32**
Atisirine, **T32**
Atractylon, **T21**
Atrolactic acid, **A31**
Atrovenetin, **Y12**
Aucubin, **T15**
Aureolic acid group antibiotics, **Y16**
Aureomycin, **Y28**
Azabicyclic systems, **A5, A15, A17, A20, A30, A37, T14, K1–3, K5, K8, K20–22, K28, K31, Y20, Y22, Y29–30, X2, X4, X11**
Azaferrocenes, **X9**
Azaphilones, **Y11**
Azaserine, **A4**

Subject Index

Azcarpine, **K20**
Azetidines, **A19**
Azides, **A1, A11**
Azimic acid, **K20**
Azimine, **K20**
Aziridines, **A13, A17, A23, A40**

Bacteriochlorophyll, **Y23**
Baikiain, **A10**
Balchanolide, **T21**
Balfourodine, **K31**
Balfourolone, **K30**
Baptifoline, **K21**
Batatic acid, **T17**
Batrachotoxinin A, **K36**
Baurenol, **T42**
Baurenone, **T42**
Benchrotrenes, **X9**
Benzoin, **A23**
Benzoquinolizines, **A5**
Benzoyl allylhippuric acid, **A10**
Benzylglutamic acid, **A32**
3-O-benzylglycerol, **A14**
Betanidin, **K17**
Betanin, **K17**
Betulafolienetriol, **T51**
Betulenols, **T28**
Betulinic acid, **T43**
Betulin, **T43**
Beyerane, **T35**
Beyerol, **T35**
Bianthryls, **X6–7**
Biaryls, **K2–4, K20, K27, Y2, Y10, Y15, Y17–18, X3, X6, X10**
Bicyclo[4.4.0]decanes, **A39, A53–54, T18–23, T26, T31–32, T34, T37–39, T42, T53–54, T57**
Bicyclo[5.3.0]decanes, **T22, T24–26, T40**
Bicyclo[7.1.0]decanes, **X2**
Bicycloelemene, **T19**
Bicyclogermacrene, **T19**
Bicyclo[2.2.1]heptanes, **A36, A46–7, T10–13, X13**
Bicyclo[3.1.1]heptanes, **T8–10**
Bicyclo[3.2.0]heptanes, **T10**
Bicyclo[4.1.0]heptanes, **A35, T5–6**
Bicyclo[2.1.1]hexanes, **T12**
Bicyclo[3.1.0]hexanes, **T7, T12**
Bicyclo[4.3.0]nonanes, **A37, A53, T18–19, T23, T26, T30–31, T37, T48**
Bicyclo[6.1.0]nonanes, **X2**
Bicyclo[2.2.0]octanes, **A46–47**
Bicyclo[3.2.1]octanes, **A46, T27, T30**
Bicyclo[3.3.0]octanes, **A37**
Bicyclo[10.3.0]pentadecanes, **T41**
Bicyclo[5.4.0]undecanes, **T12, T30**
Bicyclo[7.2.0]undecanes, **T28**
Bile pigments, **Y23**
Bilobanone, **T17**
Binaphthyls, **X6**
Biotin, **Y30**
α-biotol, **T29**
Biphenyls, **X3, X7**
β-bisabolene, **T2**
Bisepoxylignans, **Y7–8**
Blancoic acid, **Y12**

Blastidic acid, **A9**
Borneol, **T11**
Bornyl chloride, **T11**
Boromycin, **Y27**
Boschniakine, **T15**
Boschnialactone, **T15**
Boschnialinic acids, **T15**
Bostrychin, **Y17**
Bourbonenes, **T18**
Bourjotinolones, **T46**
Brefeldin A, **Y14**
Brevicolline, **K20**
3-bromocamphor, **T11**
3-bromocamphor-9-sulphonic acid, **T11**
2-bromo-β-desmotropasantonin, **T21**
Bromogeigerin acetate, **T21**
α-bromoisotutin, **T31**
β-bromolactic acid, **A1**
Browniine, **K33**
Brucine, **K12**
Bulbocapnine, **K4**
Bulbocodine, **K3**
Bulgarenes, **T18**
Bulnesol, **T25**
Bundlins, **Y26**
Buphanisine, **K6**
tert-butylglycine, **A24**
Butyrospermol, **T46**
Buxenine G, **K36**
Buxidienine F, **K36**
Buxus alkaloids, **K36**
Byssochlamic acid, **T58**

Cacalol, **Y17**
Cadinane, **T18**
Cadinenes, **T18**
δ-cadinol, **T18**
ε-caesalpin, **T40**
Cafestol, **T36**
Calacone, **T16**
Calarane, **T26**
Calciferol, **T48**
Caldariomycin, **Y16**
Callicarpone, **T33**
Callitrisic acid, **T33**
Calycanthidine, **K26**
Calycanthine, **K26**
Calycotomine, **K2**
Camphene, **T11**
Camphenilone, **T11**
Campholic acid, **A36**
Camphonanic acid, **T12**
Camphor, **T11**
cis-camphoric acid, **T12**
Camphoronic acid **T12**
Camphorquinone, **T12**
Camphorsulphonic acids, **T11–12**
Camptothecin, **K29**
Canallinin, **Z8**
Canadine, **K3**
Canaric acid, **T43**
Canescins, **Y6**
Cannabidiol, **T5**
Cannabidiolic acid, **T5**
Capaurimine, **K3**
Capaurine, **K3**
Capenicin, **T31**
Capsanthin, **T55**
Car-3-ene, **A35**

Car-4-ene, **T5**
trans-car-4-ene-3-ol, **T6**
Caracurine VII, **K12**
trans-caran-2-one, **T6**
Caranes, **T6**
Caranine, **K6**
Carbodiimides, **X3**
Carbomycin, **Y25**
Carbonates, **A14-15**
Carbonyls, **X9**
β-carboxybutyramide, **A34**
Cardenolides, **T49**
Cardiolipin, **A15**
Carnitine, **A2**
Carnosine, **A29**
Carnosol, **T33**
Caronic acid, **A35**
Carotenes, **T54**
Carotenoids, **T54-55**
Carotol, **T25**
Carpaine, **K20**
Carpamic acid, **K20**
Carquejanone, **T16**
Carquejol, **T16**
trans-carveol, **T5**
Carvomenthol, **T5**
Carvomenthone, **T5**
Carvone, **T5**
cis-carvotanacetol, **T1**
Carvotanacetone, **T1**
Caryophyllene, **T28**
trans-caryophyllenic acid, **T28**
Cascarillin, **T39**
Cassaic acid, **T39**
Cassaine, **T39**
Cassine, **K20**
Catalponol, **Y17**
Catechin, **Y3**
Catharanthine, **K15**
Ceanothic acid, **T43**
Cedrelanol, **T18**
α-cedrene, **T29**
Cedrol, **T29**
Celesticetin, **Y21**
Cellobiose, **C4**
Cembrene, **T41**
α-cephalins, **A15**
Cephalonic acid, **T41**
Cephalosporin C, **Y29**
Cephalosporin P$_1$, **T50**
Cephalothin, **Y29**
Ceroplasteric acid, **T41**
Ceroplastol I, **T41**
Cervicarcin, **Y18**
Cevine, **K35**
Chaetocin, **Y24**
Chalcomycin, **Y25**
Chalmoogric acid, **A36**
Chamaecynol, **T20**
Chamaecynone, **T20**
α-chamigrene, **T26**
Chanoclavines, **K17**
Chaparrinone, **T53**
Chasmanthin, **T38**
8-epi-chasmanthin, **T38**
Chelidonine, **K5**
Chiloscyphone, **T18**
Chimonanthine, **K26**
Chitaric acid, **Y20**
Chiral methyl group, **D2**
Chloramphenicol, **A21**

302

Subject Index

Chloromalic acid, **A2**
Chlorophylls, **Y23**
Chlortetracycline, **Y28**
Cholesterol, **T46**
Chorismic acid, **A3**
Chromium compounds, **X9**
Chromycinone, **Y16**
Chrysanthemic acids, **A35**
Chrysanthemumdicarboxylic acid, **A35**
Chrysanthenone, **T8**
Chrysoaphins, **Y10**
Cimigenol, **T51**
Cinchona alkaloids, **K8**
Cinchonamine, **K8**
Cinchonidine, **K8**
epi-cinchonidine, **K8**
Cinchonine, **K8**
epi-cinchonine, **K8**
Cinenic acid, **T3**
Cinerins, **T17**
Cinerolone, **T17**
Citramalic acid, **A33**
β-citraurin, **T55**
Citrinin, **Y13**
Citronellal, **T1**
Clavatol, **T44**
Cleavamine, **K15**
Clerodendrin A, **T37**
Clerodin, **T37**
Clivonine, **K6**
Cnidilide, **Y14**
Cocaine, **K28**
Co-carcinogens, **T39**
Coccinine, **K6**
Coclaurine, **K3**
Codeine, **K4**
Colchicine, **K30**
Colensenone, **T34**
Coleone B, **T33**
Columbianadinoxide, **Y1**
Columbin, **T38**
epi-columbins, **T37**
Compactinervine, **K13**
Condyfoline, **K12**
Condylocarpine, **K12**
Conessine, **K34**
Confertifolin, **T31**
Conhydrine, **K18**
Coniine, **K18**
Conoflorin, **K13**
Copacamphor, **T11**
Copaene, **T18**
Coriolic acid, **A7**
Coryamyrtin, **T31**
Corymbol, **T36**
Corymine, **K14**
Corynanthe alkaloids, **K8-10**
Corynantheidine, **K9**
Corynantheine, **K8**
Corynoxine, **K9**
Costaclavine, **K17**
Costunolide, **T22**
Coumarans, **Y1**
Coumarins, **A49**
Crassin, **T41**
Crinamine, **K6**
Crinamine, 6-hydroxy, **K6**
Crinine, **K6**
Criwelline, **K6**
Crotepoxide, **Y19**

Crotocin, **T31**
Crotylglycine, **A10**
Crustecdysone, **T48**
Cryptaustoline, O-methyl, methiodide, **K5**
Cryptomerone, **T16**
Cryptone, **T2**
Cryptostyline II, **K1**
Cryptoxanthins, **T55**
Cubeb camphor, **T18**
Cubenol, **T18**
epi-cubenol, **T18**
Cucurbitacin A, **T50**
Cucurbitine, **A30**
Cularine, **K5**
Culmorin, **T28**
Cuparene, **T12**
Cuparenic acid, **T12**
Cuparenol, **T12**
α-cuparenone, **T12**
Cupreine, **K8**
Curcumenes, **T16**
Curvulol, **Y2**
Cyanacobalamin, **Y24**
Cyclazocine, **Y20**
Cycloallenes, **X2**
Cyclalkenes, chiral, **X2**
Cycloalkynes, **X2**
Cycloalliin, **Z7**
Cycloartenol, **T50**
Cyclobutane derivatives, **A28, A37, T8, T28**
Cyclobuxidine F, **K36**
Cyclobuxine D, **K36**
Cyclobuxosuffrine, **K36**
Cyclobuxoxazine, **K36**
α-cyclocitral, **T54**
Cyclocolorenone, **T22**
Cyclocopacamphene, **T28**
Cyclodecane derivatives, **T18-22, T24, T30**
Cyclodopa, **K17**
Cycloeucalenol, **T50**
α-cyclogeraniol, **T54**
Cyclograndisolide, **T50**
Cycloheptane derivatives, **A3, A38, T24, Y15**
Cyclohexane derivatives, **A3, A16, A18-19, A22-23, A26, A37-42, A52-53, A55-56, A58-59, T1-2, T4-6, T8-9, T16-20, T22, T31, T40, T54, T57-58, Y13-14, Y17-19, Y21, D1, X3-4, Z3**
Cycloheximide, **Y21**
Cyclolavandulol, **T2**
Cyclolignans, **Y7-8**
Cycloneolitsin, **T50**
Cyclononane derivatives, **T58, X2**
Cycloöctane derivatives, **A3, X2**
Cycloölivil, **Y7**
Cyclopentadecane derivatives, **A32**
Cyclopentane derivatives, **A19, A31, A36-38, A45, T3, T7, T9-10, T12-15, T17, T55, Y14-15, Y21, D2**
Cyclopentanoid monoterpenes, **T13-17**

Cyclophanes, **X8**
Cyclopropane derivatives, **A12, A27, A30, A35, A37, A44, A54, T17**
Cycloprotobuxine C, **K36**
Cyclosativene, **T28**
Cyclosteroids, **T50-51**
Cyclotetradecane derivatives, **T41**
Cycloundecanes, **T30**
Cyperene, **T27**
Cyperolone, **T19**
epi-α-cyperone, **T19**
Cyperones, **T20**
Cyperotundone, **T27**
Cypridina luciferin, **Y20**
Cysteine, **A4**
Cystine, **A4**
Cytidine, 2'-deoxy, **C4**
Cytisine, **K21**
Cytochalasins, **Y27**

Dalbergione, 4-methoxy- and 4,4'-dimethoxy-, **Y4**
Dammarenediols, **T51**
Dammarenolic acid, **T51**
Daphmacrine, **K26**
Daphnetoxin, **T39**
Daphniphyllum alkaloids, **K26**
Daphylloside, **T15**
Daucol, **T25**
Daucone, **T25**
Daunomycin, **Y28**
Daunomycinone trimethyl ether, **Y28**
Daunosamine, **Y28**
Davallol, **T44**
Davanone, **T3**
Decaline, **K27**
Decalins, **A39, A53, A54, T18-23, T26-27, T31-32, T34, T37-39, T42, T53-54, T57**
Decamine, **K27**
Decevinic acid, **K35**
Decinine, **K27**
Decodine, **K27**
Decursinol, **Y1**
1,2-dehydroaspidospermidine, **K13**
Dehydronorcamphor, **A46**
Exo-dehydronorborneol, **A46**
5-dehydroquinic acid, **A26**
3,4-dehydroproline, **A17**
Dehydroroyleanone trimethyl ether, **T33**
Delnudine, **K33**
Delphinine, **K33**
4-demethylhasubanonine, **K5**
Dendrine, **K31**
Dendrobine, **K31**
Densipolic acid, **A7**
Deoxyandirobin, **T51**
2'-deoxycytidine, **C4**
Deoxypodocarpic acid, **T33**
5-deoxyquinic acid, **A26**
Deoxytubulosine, **K2**
Deserpidine, **K9**
Desmethyldesmotroposantonin, **T22**
Desosamine, **Y25**
Desoxyephedrine, **A4**
Desoxymatrinol, **K24**

Subject Index

Desoxynupharidine, **K24**
Desthiobiotin, **Y30**
Desylamine, **A23**
Deuterated compounds, **X3, D1-2**
Diaboline, **K12**
Diacetoxyscirpenol, **T31**
Diatoxanthin, **T55**
Dibenzopyrrocoline alkaloids, **K5**
Dictyopterene A, **A44**
Dictyopterene C′, **Y15**
Digitoxigenin, **T49**
Diglycerides, **A15**
Dihydroanhydromonocrotalic acid, **K23**
Dihydrocarveol, **T5**
Dihydrocarvone, **T5**
Dihydrocorynantheane, **K2**
Dihydrocoumarilic acid, **Y1**
Dihydroguaiols, **T25**
Dihydroguaiaretic acid dimethyl ether, **A31**
Dihydrohaematinic acid, **T22**
trans-dihydrohaematinimide, **Y23**
Dihydrohomopterocarpin, **Y3**
Dihydrojasmonic acid, methyl ester, **A36**
Dihydrokaurenes, **T35**
Dihydrolycorine, **K6**
3,4-dihydro-4-methylcoumarin, **A49**
Dihydromicrophylline F, **K36**
Dihydronepetalactone, **T14**
Dihydronitenin, **T57**
Dihydrooroselol, **Y1**
Dihydropavine, N-formyl, methine, **K1**
Dihydroprotoemetine, **K2**
β-dihydrosedanolide, **Y14**
Dihydrosenecic acid, **K23**
Dihydrotalbotine, N-methyl, **K16**
Dihydrotubaic acid, **Y1**
Dihydroumbellone, **T7**
Dihydroverticillatine, **K27**
Dihydroxyheliotridane, **K22**
Dilactic acid, **A1**
Dimeric flavonoids, **Y3**
4,4′-dimethoxydalbergione, **Y4**
N,γ-dimethylalloisoleucine, **A29**
Dimorphecolic acid, **A7**
Dimyristoylcephalin, **A15**
Dimyristoyllecithin, **A15**
Dinoseb, **A49**
Dioscorine, **K29**
Diosgenin, **T49**
Dipalmitoylcephalin, **A15**
Dipalmitoylglycerol, **A15**
Dipalmitoleyllecithin, **A15**
Dipalmitoyllecithin, **A15**
Diphosphaditylglycerol, **A15**
Dipiperidyl, **K18**
Dipterocarpol, **T51**
Disaccharides, **C4**
Distearoylcephalin, **A15**
Distearoylglycerol, **A15**
Distearoyllecithin, **A15**
Disulphides, **A2, A18, A34, Y24**
Diterpene alkaloids, **K31-33**
Diterpenes, **T32-41**
Dithiodiisobutyric acid, **A34**
Dithiolanes, **A18**

Dithiothreitol, **A2**
Djenkolic acid, **A4**
Dolabradiene, **T34**
DOPA, **A5**
Dothistromin, **Y13**
Douglanine, **T22**
Drimenin, **T31**
Drimenol, **T31**
Drimic acid, **T31**
Duartin, **Y3**
Duclauxin, **Y18**

Ebelin lactone, **T51**
Eburicoic acid, **T50**
Eburna alkaloids, **K16**
Eburnamonine, **K16**
Ecdysone, **T48**
Ecgonine, **K28**
Ecgoninic acid, **A9**
Echinodol, **T50**
Echinulin, **K30**
Echitamine, **K14**
Edultin, **Y1**
Ekasantalic acid, **T11**
Elaiocarpine, **K29**
Elemadienoic acid, **T46**
Eleman-8-one, **T20**
Elemol, **T20**
Elephantol, **T24**
Eleutherin, **Y2**
Elymoclavine, **K17**
Emetine, **K2**
Endo-norborneol, **A46**
Enmein, **T33**
Eperuic acid, **T36**
Ephedrine, **A21**
Ψ-ephedrine, **A21**
epi-α-cyperone, **T19**
Epialloyohimbane, **K10**
Epiborneol, **T12**
Epicamphor, **T12**
Epicatechin, **Y3**
Epiglaudine, **K1**
Epiguaipyridine, **T24**
Epiisoborneol, **T12**
13-epimanool, **T34**
Epimines, **A13, A17, A23, A40**
Epingaione, **T17**
Episulphides, **A14, A21-22**
Epishyobunone, **T22**
Epoxides, **A2, A8, A13-14, A19, A21-25, A33, A52, T4, T21, T30-31, T37, T39, T45-46, T48, T55, T57-58, K28, Y19-20, Y30**
Epoxydone, **Y19**
Epoxynorcafestone, **T36**
1,2-epoxytetralin, **A25**
Ergochromes, **Y17**
Ergoflavin, **Y17**
Ergonovine, **K17**
Ergosterol, **T48**
Ergot alkaloids, **K17**
Ergotamine, **K17**
Ergothionine, **A20**
Ergot pigments, **Y17**
Erythraline, **K7**
Erythrina alkaloids, **K7**
Erythristemine, **K7**
Erythritol, **C1**
Erythroaphins, **Y10**

α-erythroidine, **K7**
β-erythroidine, dihydro-, **K7**
β-erythroidine, β-tetrahydro-, **K7**
Erythromycins, **Y25**
Erythropentulose, **C1**
Erythrose, **C1**
Erythroxydiol Y, **T34**
Erythrulose, **C2**
Ethambutol, **A8**
γ-ethylideneglutamic acid, **A10**
5β-etianic acid, **T49**
Eudesmanolides, **T21**
Eudesmols, **T19**
Eunicin, **T41**
Eupacunin, **T24**
Euparotin, **T24**
Eupatoriopicrin, **T21**
Euphol, **T46**
Eupteleogenin, **T45**
Evoninic acid, **A31**
Exo-norborneol, **A46**

Factor V-1a, **Y24**
Falcarinol, **A7**
Farnesiferols, **T57**
Fawcettidine, **K25**
Fawcettimine, **K25**
Febrifugine, **K20**
Feist's acid, **A37**
Fenchane, **T10**
α-fenchene, **T10**
α-fencholenic acid, **T10**
Fencholic acid, **T10**
Fenchone, **T10**
Fenchyl alcohols, **T10**
Fenfluramine, **A5**
Fernene, **T44**
Ferrocenes, **X3, X8**
Ferrocenophanes, **X9**
Ferruginol, **T33**
α-ferulene, **T26**
Festuclavine, **K17**
Fichtelite, **T32**
Filican-3-one, **T44**
Filicene, **T44**
Filifolone, **T10**
Fisetinidol, **Y4**
Flabellidine, **K25**
Flavan-4-one, 7-hydroxy-, **Y6**
Flavan-7-ol, **Y6**
Flavanoids, **Y3-6**
Flindissone lactone, **T51**
Foliamenthin, **T13**
Folicanthine, **K26**
Formycin, **C4**
Formylchromomycinoic acid, **Y16**
N-formyldihydropavine methine, **K1**
Formylolivinic acid, **Y16**
Fraxinellone, **Y2**
Friedelin, **T42**
Friedelinol, **T42**
Fructose, **C2**
Fucoxanthin, **T55**
Fujenal, **T35**
Fukiic acid, **A30**
Fukinan-8-ol, **T23**
Fukinanolide, **T23**
Fukinolide, **T23**

Fukinolidol, **T23**
Fukinone, **T23**
Funtumine, **K3**
Furanoeremophilanes, **T23**
Furanofuranoid lignans, **Y7-8**
Furanoid lignans, **Y7-8**
Furanoligularenone, **T23**
Furanoterpenes, **T1, T17, T20-21, T23, T30, T36-38, T52-53, T57**
Furopelargones, **T14**
Fusamarin, **Y2**
Fusicoccin A, **T41**
Fusidic acid, **T50**
Fustin, **Y4**

Gaillardin, **T24**
Galactitol, **C1**
Galactose, **A16**
Galanthamine, **K6**
Galbegin, **Y8**
Galbulin, **Y8**
Gambogic acid, **Y9**
Gammacerane, **T44**
Garcinia acid, **A30**
Gardneramine, **K14**
Gardnerine, **K11**
Garrya alkaloids, **K32**
Garryfoline, **K32**
Gazaniaxanthin, **T55**
Gedunin, **T52**
Geigerin, **T21**
Gelsedine, **K14**
Gelsemicine, **K14**
Genepin, **T13**
Geneserine, **K30**
Genipic acid, **T15**
Geranyllinalool, **T3**
Germacranolides, **T21**
Germacrenes, **T18**
Germanium compds., **Z2**
Germine, **K35**
Gibberellic acid, **T35**
Gibberellins, **T35**
Gibberic acid, **T35**
Ginkgolides, **T40**
Glaucanic acid, **T58**
Glaucanol, **T53**
Glaucarubin, **T53**
Glaucine, **K3**
Glauconic acid, **T58**
Glaudine, **K1**
Gliotoxin, **Y24**
Globulol, **T26**
Glucitol, **C1**
Glucose, **A26**
Glutamic acid, **A9**
Glutamine, **A9**
Glutinic acid, **X3**
Glyceraldehyde, **A1**
Glyceric acid, **A1**
Glycerol, **C1**
Glycerol 1,2-carbonate, **A15**
Glycerol 3-phosphate, **A14**
Glycerophosphoric acid, **A14**
Glycerotetrulose, **C2**
Glycerylphosphorylcholine, **A14**
Glycollic acid, **D2**
Glycyrrhetic acid, **T42**
Gmelinol, **Y7**
Goitrin, **A8**

β-gorgonene, **T31**
Gorgosterol, **T50**
Grayanotoxins, **T38**
Griseofulvin, **Y17**
Grisorixin, **Y11**
Guaianolides, **T22**
γ-guaiene, **T25**
Guaiol, **T25**
Guaioxide, **T25**
Guaiaretic acid, **A31**
3-guanidinoproline, **Y22**
Gulitol, **C1**
Gulose, **C1**
γ-gurjunene, **T25**
β-gurjunene, **T26**
Gutieride, **T37**

Haemanthamine, **K6**
Haemanthidine, **K6**
Haematinimide, dihydro-, **Y23**
Haplocine, **K13**
Haplophytine, **K15**
Hardwickiic acid, **T37**
Hastanecine, **K22**
Havanensin, **T53**
Hecogenin, **T49**
Hedycarol, **T20**
Helenalin, **T24**
Helicenes, **X7**
Helicobasidin, **T12**
Heliotridane, **K22**
Heliotridine, **K22**
Heliotrinic acid, **K24**
Helminthosporal, **T30**
Helvolic acid, **T50**
Hemmingsamine, **K12**
Heptahelicenes, **X7**
Heptoses, **C3**
Herqueinone, **Y12**
Hesperetin, **Y6**
Hetacillin, **Y29**
Heteratisine, **K33**
Heterohelicenes, **X7**
Heteroyohimbane alkaloids, **K7**
Hexahelicene, **X7**
Hexahydrochamaecynone, **T20**
Hexahydrophenylalanine, **A5**
Hexahydrotyrosine, **A5**
Hexitols, **C1**
Hexoestrol, **A45**
Hexoses, **C1**
Hibaene, **T35**
Hibiscus acid, **A30**
Himachalenes, **T12**
Himbacine, **K18**
Hinesol, **T29**
Hinokiic acid, **T26**
Hinokiresinol, **Y9**
Hippeastrine, **K6**
Hirsutic acid, **T31**
Hirsutin, **Z8**
Histidine, **A20**
Histrionicotoxin, **K26**
Homaline, **K30**
Homocamphor, **T27**
Homocaronic acids, **A35**
Homochrysanthemic acids, **A35**
Homocitric acid, **A26**
Homoerythrina alkaloids, **K7**
Homolycorine, **K6**
Homopterocarpin, **Y3**

Homosecodaphniphyllic acid, methyl ester, **K26**
Homoterpenyl methyl ketone, **T1**
Homothujadicarboxylic acid, **T7**
Hopane, **T45**
Hopeaphenol, **Y15**
Hopenone I, **T45**
Humulene dioxide, **T30**
Humulene epoxides, **T30**
Humulenols, **T30**
Hunteracine, **K14**
Hunterburnine, **K8**
Huratoxin, **T39**
Hydrastines, **K3**
Hydratropic acid, **A41**
3β-hydroxyallocholanic acid, **T49**
Hydroxycitric acid, **A30**
6-hydroxycrinamine, **K6**
Hydroxydihydroeremophilone, **T23**
Hydroxydihydrotubaic acid, **Y1**
γ-hydroxyglutamic acid, **A11**
β-hydroxyglutamic acid, **A2**
γ-hydroxyhomoarginine, **A10**
7-hydroxylathyrol 3,5,7-triacetate, **T39**
Hydroxylysine, **A10**
4-hydroxymethylproline, **A17**
4-hydroxypipecolic acid, **A10**
5-hydroxypiperazic acid, **A11**
Hydroxyprolines, **A17**
3β-hydroxyprotost-13(17)-ene, **T50**
Hydroxytremetone, **Y1**
2-hydroxytryptophan, **A20**
5-hydroxytryptophan, **A20**
Hygrine, **K19**
Hygrinic acid, **A11**
Hygroline, **K19**
Hygrophyllinecic acid, **K23**
Hymentherenes, **T1**
Hyoscine, **K28**
Hyoscyamine, **K28**

Iberin, **Z8**
Ibogaine, **K15**
Ibuga alkaloids, **K15**
Ibogamine, **K15**
Ichangin, **T52**
Ichthyothereol, **Y15**
Iditol, **C1**
Idose, **C1**
Illudin S, **T30**
Illudol, **T30**
Iminazole derivatives, **A20, A23, Y30**
Indane derivatives, **A25, A41, A42, A49, A50, A53, A56**
Indicaxanthin, **K17**
Indole alkaloids, **K8-17, K30**
Indoles, **A20, A48, A55, A56, K8-17, K30, Y20, Y27**
Indolines, **A5**
Indolmycenic acid, **Y27**
Indolmycin, **Y27**
Ingenol, **T39**
Inositol, **A16**
Inosose, **A16**
Intermedol, **T19**
Inumakilactone, **T38**

Subject Index

Ionones, **T54**
Ipecacuanha alkaloids, **K2**
Ipecoside, **K2**
Ipomeamarone, **T17**
Iresin, **T32**
Iridadiene, **T3**
Iridodial, **T13**
Iridomyrmecin, **T13**
Iron compounds, **X3, X8-9**
Irones, **T54**
Ishwarone, **T23**
Isoamarine, **A23**
Isoasparagine, **A20**
Isoborneol, **T11**
Isocarvomenthol, **T5**
Isocarvomenthone, **T5**
Isochromanones, **Y1**
Isocitric acid, **A30**
Isocitric lactone, **A30**
Isocolumbin, **T38**
Isocupressic acid, **T34**
Isodihydronepetalactone, **T14**
Isodocarpin, **T33**
Isodonal, **T33**
Isoelaeocarpiline, **K29**
Isoelaecarpine, **K29**
Isoeleutherin, **Y2**
Isoflavanoids, **Y3-6**
Isofenchone, **T10**
Isofenchyl alcohols, **T10**
Isofuranogermacrene, **T20**
Isoglutamine, **A9**
Isogmelinol, **Y7**
Iso-α-gurjunene, **T25**
Isohaemanthamine, **K6**
Isoheliotridene, **K22**
Isohibaene, **T32**
Isohopane, **T45**
Isoilludin S, **T30**
Isoiridomyrmecin, **T13**
Isoishwarone, **T23**
Isooleanonic acid, dimethyl ether, **Y7**
Isoleucine, **A27**
Isolinderalactone, **T20**
Isolunine, **K31**
Isolupanine, **K21**
Isolysergic acid, **K17**
Isomenthol, **T4**
Isomenthone, **T4**
Isomenthylamine, **T4**
Isomethadone, **A31**
Isomitraphylline, **K9**
Isoneomatabiol, **T14**
Isonepetalactone, **T14**
Isonootkatone, **T23**
Isooleanonic acid, dimethyl ether, **T42**
Isopapuanic acid, **Y12**
Isopatchoulane, **T27**
Isopavine alkaloids, **K1**
Isopelletierine, **K19**
Isopeulustrin, **Y1**
Isophorbol triacetate, **T39**
Isophotoartemisin acetate, **T21**
Isophotosantonic lactone, **T22**
Isophyllocladene, **T36**
Isopilocarpine, **K30**
Isopilopic acid, **A28**
Isopimaric acid, **T32**

Isopinocampheol, **T9**
Isopinocamphone, **T9**
Isoplumericin, **T16**
Isoquinolines, **A5, K1-2, K5, Y11**
Isoretronecanol, **K22**
Isoretronecanolic acid, **K22**
Isosapogenins, **T49**
Isoserine, **A1**
Isosparteines, **K21**
Isostevane, **T35**
Isosteviol, **T35**
Isosylvestrene, **T6**
Isotenulin, **T24**
Isothujone, **T7**
Isothujyl alcohol, **T7**
Isotirucallol, **T51**
Isotopically labelled compounds, **X3, D1-2**
Isoursenol, **T45**
Isovaline, **A40**
Isovincoside, **K2**
Ivalbin, **T24**
Ivorensic acid, **T52**

Jacobine, **K22**
Jaconecic acid,, **K22**
Jalaric acid, **T29**
Jasminin, **T15**
Jasmolin, **T17**
Jasmonic acid, **A36**
Jasmonic acid, dihydro, methyl ester, **A36**
Jateorin, **T38**
Jatrophone, **T41**
Jervine, **K35**
Jingullic acid, **T43**
Julimycin-BII, **Y18**
Julocrotine, **Y29**
Juniperol, **T28**
Juvabione, **T2**
Juvenile hormone, **T58**

Kanamycin A, **C4**
Kasugamycin, **C4**
Kaurane, **T35**
Kaurene, **T33**
Kawain, **Y15**
α-kessyl alcohol, **T25**
α-kessyl glycol, **T25**
Ketohakonanol, **T45**
Ketohexoses, **C2**
Ketopentoses, **C2**
Ketoses, **C2**
Ketotetroses, **C2**
Khellactones, **Y1**
Khivorin, **T51**
Khusimol, **T27**
Kolavenic acid, **T37**
Kolavenol, **T37**
Kolavic acid, **T37**
Kopsanone, **K14**
Kopsine, **K14**
Kreysiginine, **K5**
Kryptocapsin, **T55**
Kuromatsuol, **T28**
Kynurenine, **A20**

Laballenic acid, **X1**
Labdanolic acid, **T36**
Labelled compounds, **X3, D1-2**

Laburnine, **K22**
Laccijalaric acid, **T29**
Lactic acid, **A1**
Lactic aldehyde, **A1**
Lagerine, **K27**
Laksholic acid, **T29**
epi-laksholic acid, **T29**
Lambicin, **T31**
Lamprolobine, **K21**
Lampterol, **T30**
Lanafolein, **Y25**
Lanceol, **T2**
Lankamycin, **Y25**
Lanosterol, **T44**
Lansic acid, **T44**
Lanthionine, **A4**
Lappaconine, **K33**
Lariciresinol, **Y7**
Laserol, **T25**
Laserpitine, **T25**
Lasiocarpic acid, **K24**
Lathyrol, 7-hydroxy, 3,5,7-triacetate, **T39**
Laudanosine, **K3**
Laurencin, **Y15**
Laurene, **T12**
Laurensol, **T12**
Laurinterol, **T12**
Lavandulol, **T2**
α-lecithins, **A15**
Ledol, **T26**
Leonotin, **T37**
Leontiformine, **K21**
Leontine, **K24**
Lesquerolic acid, **A7**
Leucine, **A20**
Leucinol, **A20**
tert-leucinol, **A24**
Leucodrin, **Y6**
Leucomycin A₃, **Y25**
Leurocristine, **K15**
Leurosidine, **K15**
Levopimaric acid, **T32**
Liatrin A, **T29**
Lignans, **Y7–8**
Ligularenolide, **T23**
Liguloxide, **T25**
Limonene, **T2**
Limonin, **T52**
Limonoids, **T52**
Linalool, **T3**
Lincomycin, **Y21**
Lindelofidine, **K22**
Lindenene, **T21**
Linderalactone, **T20**
Linderene, **T21**
Lindestrine, **T21**
Linearisine, **K3**
α-lipoic acid, **A18**
Lipoxamycin, **Y25**
Liquiritigenin, **Y6**
Littorine, **K28**
Lobeline, **K18**
Loganin, **T13**
Lomatin, **Y1**
Longiborneol, **T28**
Longicyclene, **T28**
Longifolene, **T28**
Ψ-longifolic acid, **T28**
Lophophorine, **K2**

Lowe's Rule, **X1, X4**
Luciculine, **K32**
Luciduline, **K26**
Lucidusculine, **K32**
Luciferin, **Y20**
Lumiphorbol triacetate, **T39**
Lunarine, **K29**
Lupanine, **K21**
α-lupene, **T43**
Lupinine, **K21**
Lutein, **T55**
Luteoskyrin, **Y18**
Lycoctonine, **K33**
Lycodine, **K25**
Lycopecurine, **K26**
Lycopodine, **K25**
Lycopodium alkaloids, **K25–26**
Lycoramine, **K6**
Lycorine, dihydro- **K6**
Lysergic acid, **K17**
Lysergine, **K17**
Lysergol, **K17**
Lysine, **A10**
β-lysine, **A11**
Lythraceae alkaloids, **K27**
Lythramine, **K20**
Lythranidine, **K20**
Lythranine, **K20**
Lythridine, **K27**
Lyxitol, **C1**
Lyxohexulose, **C2**
Lyxose, **C1**
Lyxulose, **C2**

Maackian, **Y3**
Maaliol, **T19**
Macrocarpol, **T28**
Macrolide anitibiotics, **Y25-26**
Macronecine, **K22**
Macronine, **K6**
Macusine A, **K11**
Magnamycin, **Y25**
Mahoganin, **T53**
Maleopimaric acid, **T32**
Malic acid, **A1**
Malimide, **A3**
Mandelic acid, **A21**
Manganese compounds, **X9**
Mannide, **X11**
Mannitol, **A11, C1**
Manool, **T34**
Manoyl oxide, **T34**
Manthine, **K6**
Marasmic acid, **T30**
Marasin, **X1**
Marmesin, **Y1**
Marrubiin, **T37**
Matabiether, **T14**
Matairesinol, **Y7**
Matrine, **K24**
Melaleucic acid, **T43**
Melanoxin, **Y5**
Melianone, **T51**
Melittoside, **T15**
Mellein, 5-methyl, **Y2**
Mellein, **Y2**
Mellitoxin, **T31**
Meloscine, **K31**
p-menthane derivatives, **T4–5**
Menthiafolin, **T13**

Menthofuran, **C1**
Menthol, **A26**
Menthone, **C1**
ortho-menthone, **T16**
Menthylamine, **T4**
Menthyl chloride, **T4**
Mercaptans, **A1, A4, A13, A14, A20, A32**
Mesembrine, **K20**
Mevalonic acid, **A33**
Mexicanolide, **T52**
Metallocenes, **X3, X8**
Metaparacyclophanes, **X8**
Metasaccharonic acid, **A16**
Methadols, **A31**
Methadone, **A18**
Methionine, **A8**
Methionine sulphoxide, **Z8**
Methionine sulphoximine, **Z8**
4-methoxydalbergione, **Y4**
11-methoxyferruginol methyl ether, **T33**
Methyl acetolactate, **A33**
O-methylanhalonidine, **K2**
α-methylaspartic acid, **A40**
3-methylaspartic acid, **A28**
Methylbenzoin, **A51**
O-methylcryptaustoline iodide, **K5**
Methyldesoxybenzoin, **A48**
Methyldihydrojasmonate, **A36**
Methyldihydrothebaine, **K27**
α-methyl DOPA, **A40**
α-methyldopamine, **A4**
N-methyleudan, **A55**
4-methyleneproline, **A17**
3,4-methyleneproline, **A30**
Methyl homosecodaphniphyllate, **K26**
Methyl isopulegene, **A39**
O-methyllactic acid, **A1**
β-methylmalic acid, **Y5**
Methylmalonyl coenzyme-A, **A28**
5-methylmellein, **Y2**
α-methylphenylglycine, **A40**
3-methylproline, **A27**
α-methylserine, **A40**
Methyl thiolincosaminide, **Y21**
Milliamines, **T39**
Minovincine, **K14**
Mitomycin A, **Y27**
Mitragynine, **K9**
Mitraphylline, **K9, K10**
Miyaconitine, **K33**
Mollisacacidin, **Y4**
Monocrotalic acid, **K23**
Monoepoxylignans, **Y7–8**
Monoglycerides, **A14**
Monogynol, **T35**
Monosaccharides, **C1–4, Y21, Y24–26, Y28, X1**
Monoterpenes, **T1–T16**
Monotropein, **T15**
Monspessaulanine, **K21**
Montanine, **K6**
Morellin, **Y9**
Morphine, **K4**
Morphine alkaloids, **K4–5**
Morphothebaine, **K4**
Mucroquinone, **Y3**

Multifloramine, **K2**
Multiflorenol, **T42**
Muscarine, **Y20**
Muscone, **A32**
Mustakone, **T18**
Muurolenes, **T18**
Mycobactin, **Y22**
Mycosamine, **Y26**
Myricanol, **Y15**
cis-myrtanol, **T9**
trans-myrtanol, **T8**
Myrtenal, **T9**
Myrtenol, **T9**

Nafenopin, **A52**
Nagilactone C, **T38**
Napthalene derivatives, **A20, A25, A39, A43, A48–A50, A52, A54**
Napthalene-1, 2-epoxide, **A25**
Narcissidine, **K6**
Narcotines, **K3**
Nardostachone, **T23**
Naringenin, **Y6**
Neblinine, **K13**
Necic acids, **K23-24**
Neoabietic acid, **T32**
Neocarvomenthol, **T5**
Neoclovene, **T28**
Neodihydrocarveol, **T5**
Neoflavanoids, **Y4**
Neogmelinol, **Y7**
Neointermedol, **T19**
Neoisocarvomenthol, **T5**
Neoisomenthol, **T4**
Neoisomenthylamine, **T4**
Neoisopinocampheol, **T9**
Neoisothujyl alcohol, **T7**
Neomatabiol, **T14**
Neomenthol, **T4**
Neomenthylamine, **T4**
Neonepetalactone, **T14**
Neophorbol 13,20-diacetate, **T39**
Neopinocampheols, **T9**
Neosapogenins, **T49**
Neothujyl alcohol, **T7**
Neotigogenin, **T49**
Neoxanthin, **T55**
Nepetaefolin, **T37**
Nepetalactone, **T14**
Nepetalic acids, **T14**
Nepetalinic acids, **T13, T14**
Nepetic acids, **T13, T14**
Nepetonic acids, **T14**
Neriaphin, **Y10**
Nerolidol, **T3**
Ngaione, **T17**
Nicotine, **K20**
Nigericin, **Y11**
Nimbiol, **T33**
Nimbin, **T53**
Nipecotinic acid, **A30**
Nitenin, **T57**
Nobilinone, **K31**
Nodosin, **T33**
Nogalose, **C4**
Nonadrides, **T58**
Nootkatane, **T23**
Nootkatone, **T23**
Noradrenaline, **A22**

Subject Index

Norambreinolide, **T34**
Norbornane derivatives, **A46, A47, T10**
Norbornan-2-one, **A36**
Norborneols, **A46**
Norcamphor, **A36**
trans-norcaryophyllenic acid, **T28**
Norcedranone, **T29**
Nor-C-fluorocurarine, **K12**
Norcoralydine, **K3**
N-norlaudanosine, **K3**
Norleucine, **A10**
Normacusine, **K11**
Norseychellanone, **T27**
Nortriterpenoids, **T45–48, T50–53**
Norvaline, **A10**
Novacine, **K12**
N-oxides, **Z3**
Nuciferine, **K3**
Nucleosides, **C4**
Nupharamine, **K24**
Nupharidine, **K24**
Nutallin, **Y1**
Nyctanthic acid, **T43**

Obacunone, **T51**
Obscurines, **K25**
Occidenol, **T20**
Occidentalol, **T22**
Occidol, **T21**
Ochrosandwine, **K8**
Ochratoxins, **Y2**
Ocotillol, **T51**
Octahydrodehydroxylinderene, **T21**
Octoses, **C3**
Odoratin, **T53**
Odoratone, **T51**
Oleanane, **T42**
Oleandomycin, **Y25**
Oleanolic acid, **T42**
Olivil, **Y7**
Olivin, **Y16**
Onocerins, **T44**
Ophiobolins, **T41**
Oplopanone, **T18**
Oreodine, **K1**
Organoarsenic compounds, **Z3**
Organogermanium compounds, **Z2**
Organophosphorus compounds, **A14, A15, Z3–Z6**
Organosilicon compounds, **Z1**
Oridonin, **T33**
Orientaline, **K3**
Orietalinone, **K3**
Oripavine, **K4**
Ornithine, **A11**
Oroselol, dihydro-, **Y1**
Ostruthol, **T57**
Otonecine, **K24**
Ovalicine, **T31**
Oxazolidine derivatives, **A8, A19, A21, A23, A24**
Oximes, **X4**
Oxindole alkaloids, **K9–10**
Oxindoles, **A56**
6-oxoleucotylin, **T45**
4-oxopipecolic acid, **A10**
4-oxoproline, **A17**
Oxypeucedanin, **T57**
Oxytetracycline, **Y28**

Pacifenol, **T30**
Palmarin, **T38**
Palustric acid, **T32**
Panaxadiol, **T51**
Pancuronium bromide, **T47**
Panepoxydon, **Y19**
Papuanic acid, **Y12**
Paraconic acid, **T15**
Paracyclophanes, **X8**
Paravallarine, **K34**
Parthenin, **T24**
Patchouli alcohol, **T27**
Patchoulipyridine, **T27**
Pavine alkaloids, **K1**
Pederin, **Y20**
Pelletierine, **K19**
Peltogynols, **Y4**
Peltogynone, trio-*O*-methyl, **Y4**
Penicillamine, **A4**
Penicillins, **Y29**
Penniclavine, **K17**
Pentacyclic tritepenes, **T42–45**
Pentahelicene, **X7**
Pentalenolactone, **T30**
Pentitols, **C1**
Pentoses, **C1**
Perakin, **K11**
Peraksine, **K11**
Perezinone, **T1**
Perezone, **T1**
trans-perhydroindan-2-one, **A37**
trans-perhydroindan-5-one, **A37**
Perhydroindanes, **A53**
Perhydrotriphenylene, **X10**
Perillaldehyde, **T8**
Perillyl alcohol, **T8**
Perivine, **K11**
Petasalbine, **T23**
Petasin, **T23**
Phellandrenes, **T2**
Phenampromid, **A13**
Phenanthrene derivatives, **A24**
Phenylalanine, **A5**
N-phenylalanine, **A13**
β-phenylglycidic acid, **A19**
Phenylglycine, **A23**
Phenylserine, **A19**
β-phenylserine, **A21**
Phenylisoserine, **A19**
Phenyltetrahydroisoquinoline alkaloids, **K1**
Phomin, **Y27**
Phosphatidic acids, **A15**
Phosphatidylglycerol, **A15**
Phosphonomycin, **Y20**
Phosphorus compounds, **A14, A15, Y20, Y24, X10, Z3–Z6**
Phothebainehydroquinone, **K4**
Phthalideisoquinoline alkaloids **K3**
Phthalides, **A51**
Phthalimides, **A9**
Phyllocladene, **T36**
Phyllodulcin, **Y6**
Phylloquinone, **T55**
Physalins, **T53**

Physovenine, **K30**
Phytol, **T56**
Phytostigmine, **K30**
Picraline, **K12**
Picraphylline, **K9**
Picrasin A, **T53**
Picromycin, **Y25**
Picrosalvin, **T33**
Picrotoxinin, **T31**
Piericidin A, **Y30**
Pillaromycin, **Y28**
Pillaronone, **Y28**
Pilocarpine, **K30**
Pimaricin, **Y25**
Pimaric acid, **T32**
Pinacolyl alcohol, **A12**
Pinacolylamine, **A24**
Pinenes, **T8**
Pinitol, **A16**
Pinidine, **K19**
Pinocampheol, **T9**
Pinocamphone, **T9**
trans-pinocarveol, **T9**
Pinocarvone, **T9**
Pinonic acid, **T8**
Pinoresinol, **Y7**
Pipecolic acid, **A10**
Piperazic acid, **A11**
Piperidines, **A5, A10, A11, A13, A26, A30, A34, K18–20, Y21, Y29, X4**
Piperinic acid, **T1**
Piperitenone oxide, **T4**
Piperitols, **T4**
Piperitone, **T2**
Piperitone oxide, **T4**
Pipitzols, **T1**
Pisatin, **Y3**
Piscidic acid, **A30**
Plathyterpol, **T39**
Platicodigenin, **T45**
Platydesmine, **K31**
Platynecine, **K22**
Plicatic acid, **Y7**
Plinols, **T3**
Plumericin, **T16**
Plumieride, **T16**
Pluviine, **K6**
Podocarpic acid, **T33**
Podolactones, **T38**
Polyether antibiotics, **Y11**
Polyporenic acid A, **T50**
Ponasterone, **T48**
Porantherine, **K29**
Porphyrins, **Y23**
Portentol, **Y15**
Portulal **T40**
Preisocalamendiol, **T22**
Pretoxin, **T31**
Pristimerin, **T45**
Pristimerol, **T45**
Proaporphine alkaloids, **K3**
Proline, **A11**
Prolinol, **A17**
Pronuciferine, **K3**
19-propylthevinol, **K4**
Prostaglandins, **Y16**
Protoaphins, **Y10**
Protoemetine, **K2**
Protostanes, **T50**

Subject Index

Pseudoanisatin, **T30**
Pseudoclovene A, **T28**
Pseudoconhydrine, **K19**
Pseudoecgonine, **K28**
Pseudoguaianolides, **T24**
Pseudoheliotridane, **K22**
Pseudohygroline, **K19**
Pseudoyohimbane, **K10**
Psicose, **C2**
Pterocarpans, **Y3**
Pterocarpin, **Y3**
Pulchellin, **T24**
Puleganolic acid, **T15**
Pulegene, **T7**
trans-pulegenic acid, **A38**
Pulegone, **A38**
Pulvilloric acid, **Y13**
Purine derivatives, **Y21**
Purpurogenone, **Y17**
Pyrazolines, **A6**
Pyrethric acid, **A35**
Pyrethrins, **T17**
Pyrethrolone, **T17**
Pyrethrosin, **T21**
Pyridines, **A25, Y26, Y30**
Pyridomycin, **Y26**
Pyrimidine derivatives, **A55, Y21**
Pyrocin, **A35**
Pyroclavine, **K17**
Pyroglutamic acid, **A9**
Pyroheteratisine, **K33**
Pyronimbic acid, **T53**
Pyrrocolines, **K5**
Pyrrolidines, **A2, A3, A9, A11, A17, A20, A27, A30, A31, A40, K19, K35, Y21–23, Y27**
Pyrrolizidines, **K22, K24**

Quassinoids, **T53**
Quebrachamine, **K13**
Quercitols, **A16**
Quinamine, **K8**
Quinazolines, **A37**
Quinic acid, **A26**
Quinidine, **K8**
epi quinidine, **K8**
Quinine, **K8**
epi-quinine, **K8**
Quinolines, **A5, A25, A56, Y11**
Quinolizines, **A5, A15**
'Quinone A', **Y10**
Quintoxine, **K8**
Quinoxalines, **A8**

Rauvoxinine, **K10**
Reserpine, **K9, K10**
Retamine, **K21**
Reticulatoxanthin, **T55**
Retronecanol, **K22**
Retronecanone, **K22**
Retronecine, **K22**
Retuline, **K13**
Retusamine, **K24**
Rhazidine, **K13**
Rhododactynaphins, **Y10**
Rhoeadine alkaloids, **K1**
Rhoeagine, **K1**
Rhyncophylline, **K9**
Ribitol, **C1**
Ribohexulose, **C2**

Ribose, **C1**
Ribulose, **C2**
Ricinoleic acid, **A7**
Rifamycins, **Y26**
Rimuene, **T34**
Rishitin, **T22**
Rosenonolactone, **T34**
Rosmarinecine, **K22**
Rotenoids, **Y1**
Rotenone, **Y1**
Rotundone, **T25**
Rotunols, **T20**
Rubixanthin, **T55**
Rubroskyrin, **Y18**
Rugulosin, **Y18**
Ryanodine, **T58**

Sabina hydrate, **T7**
Sabina ketone, **T7**
Sabinene, **T7**
Sabinol, **T7**
Sakuranetin, **Y6**
Salsolidine, **K2**
Salutaridinol I, **K4**
Salvin, **T33**
Sandaracopimaric acid, **T32**
α-santalal, **T11**
Santalenes, **T11**
α-santalol, **T11**
Santanolide C, **T22**
Santonins, **T21, T22**
Sapogenins, **T49**
Sarkomycin, **A36**
Sarpagine, **K11**
Sarsapogenin, **T49**
Sativene, **T28**
Sceletium alkaloid A4, **K20**
Schellhammericine, **K7**
Schellhammeridine, **K7**
Schellhammerine, **K7**
Scillarenin, **T49**
Sclareol, **T34**
Sclerotiorins, **Y11**
Scopolamine, **K28**
Secodaphniphylline, **K26**
Secoisolariciresinol, **Y7**
Secologanin, **T13**
Securinine, **K18**
Sedamine, **K18**
Sedridine, **K19**
Selenium compounds, **A18**
Selinane, **T19**
Selinenes, **T19**
Semi-α-carotene, **T54**
Senecic acid, **K23**
Seneciphyllic acid, **K23**
Seneol, **Y19**
Senepoxide, **Y19**
Sequirins, **Y9**
Serine, **A4**
Serpentine, **K9**
Serratenolone, **T44**
Serratidine, **K25**
Serratinidine, **K25**
Serratinine, **K25**
Serratine,, **K25**
Sesquicarene, **T6**
Sesquiphellandrene, **T2**
Sesquiterpenes, **T16-32**
Sesterterpenes, **T41**

Setoclavine, **K17**
Seychellene, **T27**
Shellolic acid, **T29**
Shikimic acid, **A26**
Shionone, **T42**
Shiromodiol, **T20**
Showdomycin, **C4**
Shyobunone, **T22**
Siccanin, **T32**
Silanes, **Z1**
Silicon compounds, **Z1**
Simarolide, **T53**
Sinigrin, **C4**
Sinomenine **K4**
Sirenin, **T6**
Sisaustricin, **A8**
Skytanthines, **T14**
Smilagenin, **T49**
Solacongestidine, **K34**
Solanidine, **T49**
Solanocapsine, **K34**
Solanone, **T58**
Solaphyllidine, **K36**
Solasodine, **K34, T49**
Solidagenone, **T37**
Solsitalin, **T24**
Songorine, **K32**
Sorbinol, **Y14**
Sorbose, **C2**
Sparsomycin, **Y20**
Sparteine, **K21**
Spermostrychnine, **K12**
Spinacin, **A20**
Spiradine A, **K32**
Spiramycin, **Y25**
Spiro compounds, **A44, T26, T29, K6, K9-10, X4, X9, X11**
Sporidesmin, **Y24**
Stachane, **T35**
Stachene, **T35**
Stemofoline, **K29**
Stemonine, **K29**
Stenine, **K29**
Stepharine, **K3**
Stephavanine, **K5**
Stercobilin, **Y23**
Sterigmatocystin, **Y13**
Steroidal alkaloids, **K34**
Steroids, **A18, A53, T20, T23, T26, T27, T32, T35, T36, T38, T40, T46-51, K34-6**
Steviol, **T35**
Stigmasterol, **T48**
Streptomycin, **C4**
Strictosidine, **K2**
Strychindole, **K12**
Strychnine, **K12**
Strychnos alkaloids, **K12**
Styrene oxide, **A22**
Sugiresinol, **Y9**
Sugiresinone dimethyl ether, **Y9**
Sulphides, **A1, A4, A8, A12, A17-18, A21, A23, A28, A33-- 34, Y24, X4, X10-11**
Sulphinates, **Z8**
Sulphinimides, **Z8**
Sulphones, **A12, A17, A22, D2**
Sulphonium salts, **Z7**
Sulphorophan, **Z8**
Sulphoxides, **Z7-8**

Subject Index

Sulphoximines, **Z8**
Supinidine, **K22**
Suprasterol III, **T48**
Sweroside, **T13**
Swietenine, **T52**
Swietenolide, **T52**
Sylvestrene, **T6**
Synephrine, **A22**

Tabersonine, **K13**
Tachysterol III, **T40**
Tagatose, **C2**
Talbotine, **K16**
Talitol, **C1**
Talose, **C1**
Taraxasterol, **T43**
Taraxerol, **T43**
Tartaric acid, **A2**
Tartrimide, **A2**
Tauranin, **T36**
Taxicins, **T40**
Taxifolin, **Y4**
Taxinine, **T40**
Taxol, **T40**
Tazettine, **K6**
Tecomanine, **K31**
Terebic acid, **A35**
α-terpineol, **T1**
Terramycin, **Y28**
Terreic acid, **Y19**
Terrein, **Y16**
Terremutin, **Y19**
tert-butylglycine, **A24**
tert-leucinol, **A24**
Testosterone, **T32**
Tetracyclines, **Y28**
cis-tetrahydroactinidiolide, **T54**
Tetrahydroalantolactone, **T21**
Tetrahydroalstonine, **K9**
Tetrahydroanhydrodeoxy-
  aucubigenin, **T15**
Tetrahydroanhydroaucubigenin, **T15**
Tetrahydro-*epi*-α-cyperone, **T20**
β-tetrahydro-β-erythroidine, **K7**
Tetrahydrofurans, **A6, A18, A30, T17, Y7-8, Y11, Y15, Y20**
Tetrahydroharman, **K16**
Tetrahydroharmine, **K16**
Tetrahydroisoquinoline alkaloids, **K1-2**
Tetrahydrolinalool, **T3**
Tetrahydroprotoberberine
  alkaloids, **K3**
Tetrahyroquinolines, **A5, A56**
Tetrahydroquinoxalines, **A8**
Tetrahydrosaussurea lactone, **T20**
Tetrahydroseneciphyllic acid, **K23**
Tetrahydrosolanone, **T58**
Tetrahymanol, **T44**
Tetraline derivatives, **A20, A25, A43, A49, A50, A52, Y8, Y17**
Tetramisole, **Y30**
Tetraterpenes, **T54-55**
Tetroses, **C1**
Theanine, **A9**
Thebaine, **K4**
Thelepogine, **K31**
Thermopsine, **K21**

Thiabicyclic systems, **A17**
Thianaphthenes, **A48**
Thietanes, **A22**
Thiirans, **A14, A21, A22, A23**
Thiodiisobutyric acid, **A34**
Thiodilactic acid, **A1**
Thiolactic acid, **A1**
2-thiolhistidine, **A20**
Thiols, **A1, A4, A13, A14, A20, A32**
Thiones, **A8**
Thiophanes, **A18**
Thiophenes, **A21, A28, A43, A50**
Threitol, **C1**
Threonine, **A8, A24**
Threopentulose, **C2**
Threose, **C1**
Threulose, **C2**
Thujanes, **T7**
Thujan-2-one, **T7**
Thujastandin, **Y7**
Thujone, **T7**
Thujopsene, **T26**
Thujyl alcohol, **T7**
Thyronine, **A5**
Thyroxine, **A5**
Tigogenin, **T49**
Tinophyllone, **T37**
Tirucallol, **T46, T50**
α-tocopherol, **T56**
Todomatuic acid, **T1**
Tolpomycinones, **Y26**
Tomatidine, **T49**
Tomatillidine, **K34**
Torilin, **T21**
Torreyol, **T18**
Torulosol, **T34**
Totarol, **T32**
Toxol, **Y1**
Trachelanthamidine, **K22**
Trachelanthamidinic acid, **K22**
Trachelanthic acid, **K24**
Trachylobanic acid, **T35**
Tremetone, **Y1**
Trichodesimic acid, **K23**
Trichothecin, **T31**
Tricyclic bridged systems, **T11, T18, T23, T27-29**
Tricycloekasantalic acid **T11**
Tricyclovetivenol, **T27**
Triflorizin, **Y3**
Triglycerides, **A14**
Tri-O-methylpeltogynone, **Y4**
Triphenylenes, **X10**
Trisporic acids, **Y19**
Triterpenes, **T42-53**
Tritiated compounds, **D2**
Tröger's Base, **X11**
Tropane alkaloids, **K28**
Tropic acid, **A41**
Tropinic acid, **A9**
Tropin-2-one, **K28**
Tryptophan, **A20**
Tubaic acid, dihydro-, **Y1**
Tuberculostearic acid, **A32**
Tuberostemonine, **K29**
Tubifoline, **K12**
Tubotaiwin, **K12**
Tubulosine, **K2**
Turbicoryn, **T36**

Turmerone, **T16**
Tutin, **T31**
Twistane, **X10**
Twistan-4-one, **X10**
Twistene, **X10**
Tylosin, **Y25**
Tyrosine, **A5**
Tyrosinol, **A5**

Umbellularic acids, **A54**
Umbellulone, **T7**
Uridine, **C4**
Urobilin, **Y23**
Ursolic acid, **T42**
Utilin, **T52**

Valeranone, **T19**
Valeroidine, **K28**
Valine, **A4**
Valinol, **A4**
Vallesamidine, **K13**
Veatchine, **K32**
Veprisone, **T52**
Veralkamine, **K34**
Veratramine, **K35**
Veratrobasine, **K35**
Verbenalin, **T13**
Verbenols, **T8**
Verbenone, **T8**
Verrucarin A, **T31**
Verrucarinic acid, **T31**
Verrucarol, **T31**
Vertaline, **K27**
Verticillatine, dihydro-, **K27**
Verticinone, **K35**
Vespirenes, **X11**
Vespirones, **X11**
Vetiselinenol, **T19**
β-vetivone, **T29**
Viburnitol, **A16**
Vincoblastine, **K15**
Vincadifformine, **K13**
Vincamine, **K16**
Vincaminoreine, **K13**
Vincoside, **K2**
Vincristine, **K15**
Viocidic acid, **Y22**
Violoxanthin, **T55**
Viomycidine, **Y22**
Viomycin, **Y22**
Viopurpurin, **Y2**
Viridifloric acid, **K24**
Viridiflorol, **T26**
Visamminol, **Y1**
Vitamin B12, **Y24**
Vitamin B12 monocarboxylic
  acid, **Y24**
Vitamin E, **T56**
Vitamin K₁, **T56**
Voacarpine, **K11**
Vobasine, **K11**
Vomicine, **K12**
Vomilenin, **K11**
Vulgarin, **T22**

Warburgiadione, **T23**
Widdrol, **T26**
Wieland-Gümlich aldehyde, **K12**
Withaferin A, **T48**

Subject Index

Xanthanolides, **T24**
Xanthenol, **T24**
Xanthinin, **T24**
Xanthoaphins, **Y10**
Xanthodactynaphins, **Y10**
Xanthomegnin, **Y2**
Xanthumin, **T24**
Xylitol, **C1**

Xylohexulose, **C2**
Xylose, **C1**
Xylulose, **C2**

Yohimbane, **K10**
Yohimbe alkaloids, **K9-10**
Yohimbine, **K9**
epi-α-yohimbine, **K9**

16-yohimbone, **K9**

Zearalenone, **Y14**
Zeaxanthin, **T55**
Zeylanine, **T20**
Zingiberene, **T2**
Zizanin A, **T41**
Zizanoic acid, **T27**